Ocean Engineering

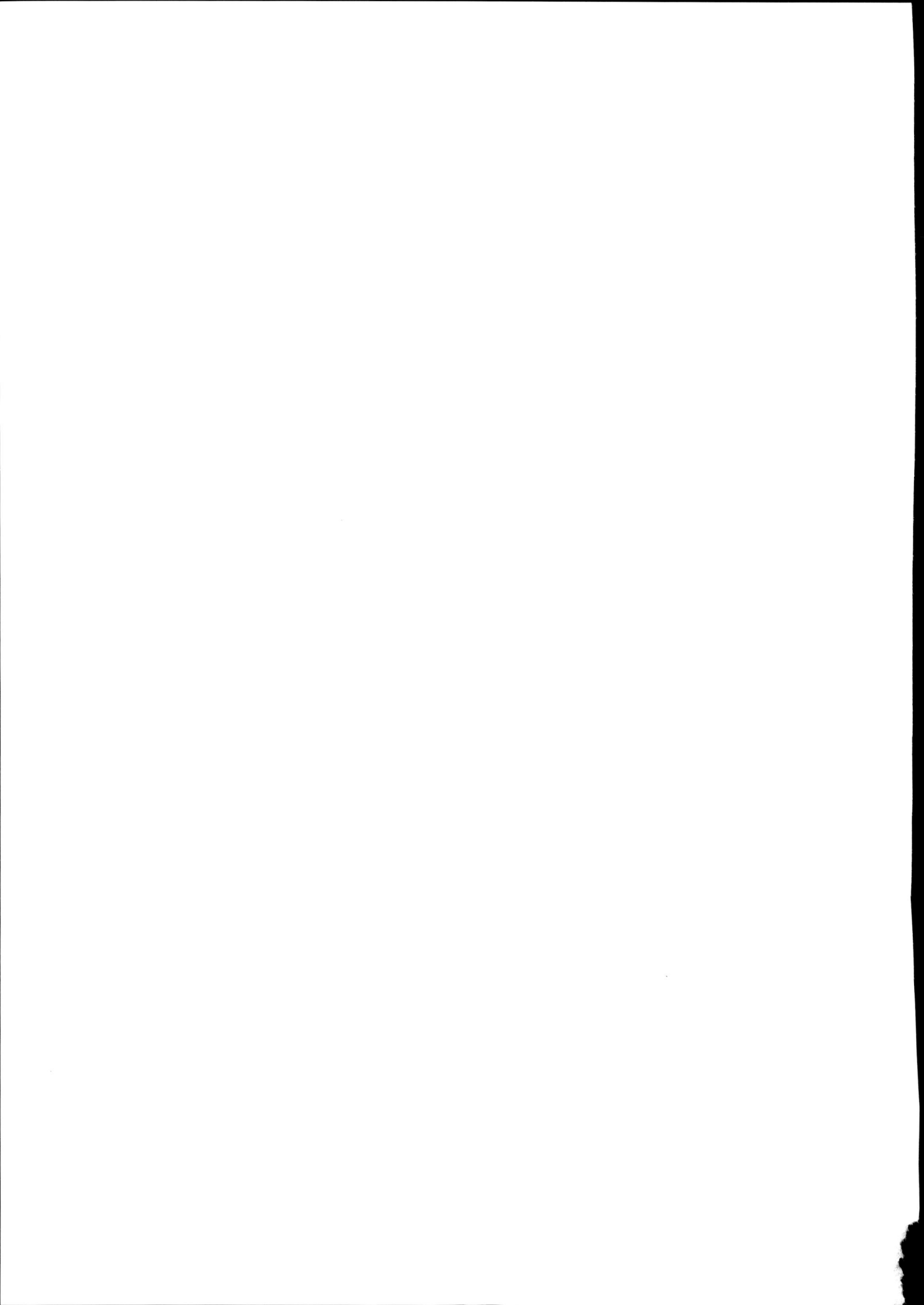

Ocean Engineering

Edited by **Theodore Roa**

SYRAWOOD
PUBLISHING HOUSE

New York

Published by Syrawood Publishing House,
750 Third Avenue, 9th Floor,
New York, NY 10017, USA
www.syrawoodpublishinghouse.com

Ocean Engineering
Edited by Theodore Roa

International Standard Book Number: 978-1-68286-161-5 (Hardback)

Printed in the United States of America.

Contents

Permissions

List of Contributors

Preface

Ocean science is a vast field encompassing all aspects of an ocean such as marine organisms, ecosystems, ocean currents, ocean physics, etc. This book focuses upon the interactions of the ocean with the atmosphere. It includes experimental as well as theoretical researches contributed by experts and scientists from across the globe. Some of the diverse topics presented in this book are tidal variability, air-sea gas exchange, etc. This book will serve as a resource guide for students of oceanography, climate sciences and allied disciplines. It will also prove beneficial to researchers and professionals involved in these fields.

This book is a comprehensive compilation of works of different researchers from varied parts of the world. It includes valuable experiences of the researchers with the sole objective of providing the readers (learners) with a proper knowledge of the concerned field. This book will be beneficial in evoking inspiration and enhancing the knowledge of the interested readers.

In the end, I would like to extend my heartiest thanks to the authors who worked with great determination on their chapters. I also appreciate the publisher's support in the course of the book. I would also like to deeply acknowledge my family who stood by me as a source of inspiration during the project.

Editor

Changes in ventilation of the Mediterranean Sea during the past 25 year

A. Schneider[1], T. Tanhua[1], W. Roether[2], and R. Steinfeldt[2]

[1]GEOMAR Helmholtz Center for Ocean Research Kiel, Kiel, Germany
[2]Institute of Environmental Physics, University of Bremen, Bremen, Germany

Correspondence to: T. Tanhua (ttanhua@geomar.de)

Abstract. Significant changes in the overturning circulation of the Mediterranean Sea has been observed during the last few decades, the most prominent phenomena being the Eastern Mediterranean Transient (EMT) in the early 1990s and the Western Mediterranean Transition (WMT) during the mid-2000s. During both of these events unusually large amounts of deep water were formed, and in the case of the EMT, the deep water formation area shifted from the Adriatic to the Aegean Sea. Here we synthesize a unique collection of transient tracer (CFC-12, SF_6 and tritium) data from nine cruises conducted between 1987 and 2011 and use these data to determine temporal variability of Mediterranean ventilation. We also discuss biases and technical problems with transient tracer-based ages arising from their different input histories over time; particularly in the case of time-dependent ventilation.

We observe a period of low ventilation in the deep eastern (Levantine) basin after it was ventilated by the EMT so that the age of the deep water is increasing with time. In the Ionian Sea, on the other hand, we see evidence of increased ventilation after year 2001, indicating the restarted deep water formation in the Adriatic Sea. This is also reflected in the increasing age of the Cretan Sea deep water and decreasing age of Adriatic Sea deep water since the end of the 1980s. In the western Mediterranean deep basin we see the massive input of recently ventilated waters during the WMT. This signal is not yet apparent in the Tyrrhenian Sea, where the ventilation seems to be fairly constant since the EMT. Also the western Alboran Sea does not show any temporal trends in ventilation.

1 Introduction

The circulation and ventilation in the Mediterranean Sea have been studied for several decades (e.g. Wüst, 1961; Malanotte-Rizzoli and Hecht, 1988; Roether and Schlitzer, 1991; Roether et al., 1998a; Lascaratos et al., 1999; Millot, 1999; Pinardi and Masetti, 2000; Theocharis et al., 2002). Figure 1 shows a map of the Mediterranean Sea with the main basins and geographical terms that are used in this report. The Mediterranean Sea is well ventilated in comparison to the world ocean, and is characterized by two deep overturning circulation cells and one shallow cell. Surface water from the Atlantic Ocean enters the Mediterranean basin through the Strait of Gibraltar, forming the upper arm of the shallow overturning circulation. This water flows towards the east and on its way salinity and density increase. In the Levantine basin in the eastern Mediterranean the density has reached such an extent that the water sinks to form the so-called Levantine Intermediate Water (LIW). The LIW flows back into the opposite direction in a few 100 m depth, to leave the Mediterranean Sea again through the Strait of Gibraltar. Exchange of the deep waters between the west and the east basin is inhibited by the shallow Sicily Channel ($< 500\,$m) and a rather independent deep water circulation takes place in the two basins. In the eastern basin, the Eastern Mediterranean Deep Water (EMDW) is regularly formed in the South Adriatic Sea, where high salinity waters get cooled in winter, reaching enhanced densities; after deep convection it flows over the Otranto sill into the Ionian basin to spread south and east (Wüst, 1961; Roether and Schlitzer, 1991; Schlitzer et al., 1991). In the western basin, a similar process takes place in the Gulf of Lyons forming the Western Mediterranean

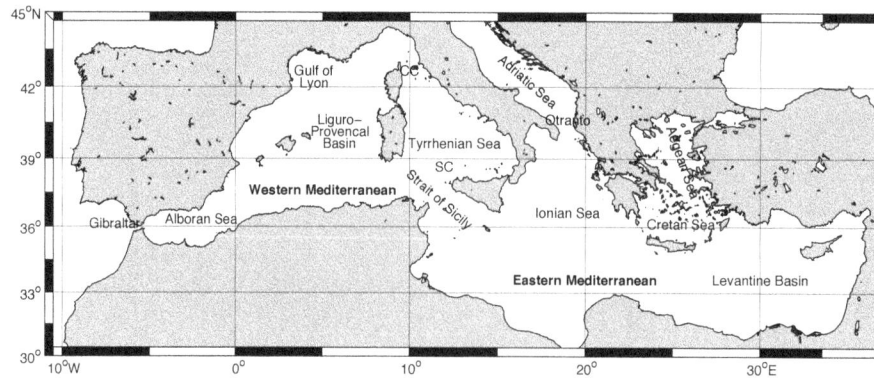

Fig. 1. Main basins and straits in the Mediterranean Sea mentioned in this article (CC = Corsica Channel, SC = Sardinia Channel). Schematic of main circulation patterns and water transformation regions can be found in the rich literature on the Mediterranean Sea (e.g. Wüst, 1961; Robinson et al., 2001; Bergamasco and Malanotte-Rizzoli, 2010).

Deep Water (WMDW). Here, the highly saline LIW acts as a preconditioner for deep water formation (MEDOC Group, 1970; Leaman and Schott, 1991).

In the following we further describe certain water masses and processes and explain changes that occurred in the "classical" circulation pattern. Although, as described above, the eastern basin is subject to relatively fast deep water formation, a slowly ventilated water body is found between 1200 m and 2600 m depth. It is replenished by upwelling of bottom water and mixing with overlying water (Roether and Schlitzer, 1991; Schlitzer et al., 1991). Up to the early 1990s, Aegean-derived water appeared to be limited to forming an intermediate water mass, the so-called Cretan Intermediate Water (CIW), which is found below the LIW at depths around 500–1200 m (Schlitzer et al., 1991). However, a cruise in 1995 noticed strong changes in the deep water formation and circulation compared to 1987 (Roether et al., 1996), now known as the Eastern Mediterranean Transient (EMT). Enhanced salinities of subducting surface water in the Aegean Sea had led to water flowing out of the Aegean Sea that was dense enough to sink down to the bottom of the main eastern basin. This new Aegean deep water was formed at distinctly larger rates (7 yr average $\sim$ 1 Sv (Roether et al., 1996) (Sverdrup; 1 Sv $= 10^6$ m^3 s^{-1}) and maximum rate $\sim$ 3 Sv in 1993 (Roether et al., 2007)) than the Adriatic source previously ($\sim$ 0.3 Sv (Schlitzer et al., 1991)). The EMT was triggered by the presence of unusually saline water in the Cretan Sea followed by unusually cold conditions (e.g. Lascaratos et al., 1999). The Aegean outflow led to an uplift of the EMDW and the overlying water masses. Furthermore, the massive input of near-surface water containing large amounts of dissolved organic carbon accelerated the oxygen consumption at depth considerably (Klein et al., 2003).

In the Tyrrhenian Sea, subsurface waters from the eastern and western Mediterranean meet. From the east, LIW and transitional Eastern Mediterranean Deep Water (tEMDW; mixing regime between LIW and EMDW) enter through the Sicily Channel, spreading along the northern Sicily slope (Sparnocchia et al., 1999). From the west, the WMDW enters at greater depth via the Sardinian Channel ($\sim$ 1900 m) (Astraldi et al., 1996). The tEMDW and the WMDW mix and form the Tyrrhenian Deep Water (TDW), whereas most of the shallow LIW leaves via the Sardinian and the Corsican Channels. The TDW circles cyclonically in the Tyrrhenian Sea and leaves via the Sardinian Channel against the inflowing WMDW (Millot, 1999). During the EMT, the inflow into the Tyrrhenian Sea through the Sicily Channel showed increased salinities with a maximum in 1992–1993, which goes along with enhanced input rates (Gasparini et al., 2005). The increased density of the Sicily Channel overflow consequently led to enhanced downward mixing into the deep Tyrrhenian basin and also changed the properties of the TDW (Gasparini et al., 2005).

As mentioned before, in the western basin the deep water formation is dependent on the salt content of the LIW layer as a cyclonic gyre in the Gulf of Lyon brings high-salinity intermediate water close to the surface. Preconditioned by this process, the stability of the water column is effectively reduced and during winter-cooling the dense water can sink to the bottom (Gascard, 1978). Accordingly, the salinity changes induced by the EMT influenced the deep water formation in the western basin. Recent observations showed a significant increase in salt and temperature in intermediate and deep waters of the western Mediterranean, which has been attributed to the propagation of the EMT signal from the eastern to western basin (Gasparini et al., 2005; Schröder et al., 2006). Due to extreme forcing in the winters 2004/2005 and 2005/2006 (high heat loss, little precipitation, persistent winds) the deep convection was remarkably strong, leading to enhanced salinity and temperature over almost the entire deep basin up to about 1600 m depth (Schröder et al., 2008). The result of these events is referred to as the Western Mediterranean Transition (WMT). Similar to the EMT, Schröder et al. (2008) also noted a WMT-induced uplift of the isopycnals in the western basin.

Ventilation age concepts

One way to estimate ventilation times is via water mass ages obtained with the help of transient tracers. The comparison of tracer data and the derived ages over time can provide information on temporal changes in ventilation. Water mass ages are defined as the time elapsed since a water parcel left the mixed layer. In the Mediterranean Sea the renewal timescales of different water bodies are relatively short. For example, the ventilation ages has been estimated to be 10–20 yr for the LIW (Stratford and Williams, 1997), 70–150 yr for the EMDW (Roether and Schlitzer 1991; Stratford et al., 1998) and 40 yr for the WMDW (Stratford et al., 1998), although based on different assumptions than the tracer framework used in this study, and therefor not directly comparable, see discussion below. The anthropogenic tracers CFC-12, SF_6 and tritium can provide information on decadal to inter-decadal timescales and thus they are suitable for the Mediterranean Sea. Their input histories are shown in Fig. 2. The concepts of the age calculations we use are shortly described in the following.

The numerous nuclear bomb tests in the 1950s and early 1960s introduced large amounts of tritium into the environment, most of which entered the ocean, primarily by vapor exchange between the low troposphere and the water surface (Weiss and Roether, 1980). Tritium has a half-life of 12.43 yr and decays into the stable isotope ^{3}He, which, combined with tritium, can yield water mass ages – the tritium-^{3}He age. We use the procedure of Roether et al. (1998b, 2013a) to separate the tritiugenic portion from a background of atmosphere-derived ^{3}He and a contribution released from the ocean floor (terrigenic ^{3}He). Together with the measured tritium concentration and its half-life, the time when the decay started (i.e. after leaving the mixed layer) can be determined. By now most of the nuclear-weapon produced tritium has decayed, limiting its future uses. The chemically inert tracers CFC-12 and SF_6 have (or had) monotonically increasing concentrations in the atmosphere (in case of CFC-12 concentrations peaked in 2002 and are now decreasing). The apparent CFC-12 (or SF_6) tracer age is defined as the elapsed time since the observed interior concentration was equal to the surface concentration, which is assumed to be in solubility equilibrium with the atmosphere. All tracer ages are affected (i.e. biased) by mixing (see below and Sonnerup, 2001).

Another way to calculate water mass ages is the transit time distribution (TTD) method (Hall and Plumb, 1994), which does account for mixing and thus represents a more realistic mean age. For steady transport, the TTD can be approximated by an inverse Gaussian function (IG) (Waugh et al., 2004) and with the help of two independent tracers the two variables (Δ = width and Γ = mean age) of the IG can be constrained. However, since we use the tracers separately, a constant ratio between Δ and Γ has to be assumed. An extension of the TTD method is the approximation by a 2IG TTD (the summation of two inverse Gaussian functions) (Waugh

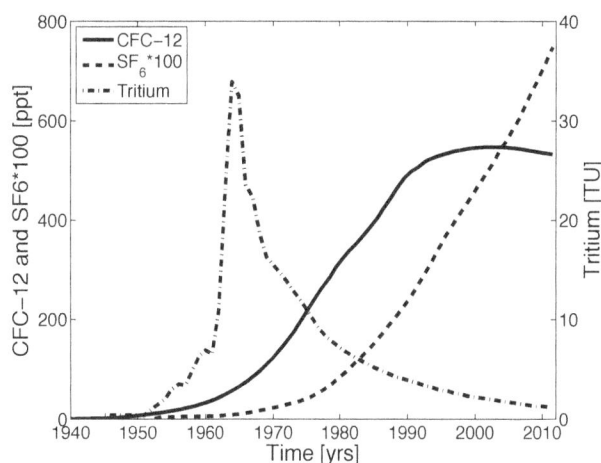

Fig. 2. Surface water input histories of CFC-12, SF_6 and tritium. For CFC-12 and SF_6 the atmospheric mixing ratios in ppt (Bullister, 2011) are assumed to be in 100 % equilibrium with the Mediterranean surface water. The tritium input history is based on observations and taken from R. Steinfeldt (unpublished data).

et al., 2003), which allows for mixing of two totally different water masses (e.g., a young and an old water mass which mix during deep water formation). In that case the percentages of both water masses can be constrained by an additional third tracer. For the Mediterranean Sea several scenarios of 1IG and 2IG TTDs have been tested for the year 2011 when the three tracers CFC-12, SF_6 and tritium were available (Stöven and Tanhua, 2013).

In this study we compare salinity, temperature, oxygen as well as apparent tracer ages and mean ages (from the 1IG TTD approach) over a period of 25 yr to detect changes in the circulation. We also address shortcomings of the different tracer age methods.

2 Data and methods

2.1 Tracer data

Table 1 lists the cruises from which tracer data were used for this work. On all cruises, the tracers CFC-12 and tritium were measured, which we use in this study. Additionally, for 2011 measurements of SF_6 are available, and these data are used as well.

Sampled stations of Meteor 31/1 (M31/1), Meteor 51/2 (M51/2) and Meteor 84/3 (M84/3) cover the entire Mediterranean Sea (although the focus of the first two was the eastern Mediterranean). Meteor 5/6 (M5/6), Aegaeo M4WF (Aegaeo98) and Meteor 44/4 (M44/4) are restricted to the eastern basin, whereas stations of Poseidon 234 (P234) are limited to the western Mediterranean Sea. During Urania MAI2 (Ura2) and Urania MAI7 (Ura7) only a few stations in the Adriatic

Sea were sampled. In Fig. 3 the locations of stations with tracer measurements of all cruises are shown.

CFC-12 and SF$_6$ samples were measured either directly on board or ashore from flame-sealed glass ampoules. The gas chromatographic instruments used for the analysis were similar to the ones described by Bullister and Weiss (1988), Bulsiewicz et al. (1998) and Vollmer and Weiss (2002). Precision for the CFC-12 measurements is $\pm 2\%$ or $0.02\,\mathrm{pmol\,kg^{-1}}$, whichever is greater. Precision for the SF$_6$ measurements is 1–2.5% (on board/cracker). For comparison with the tracers in the atmosphere, CFC-12 and SF$_6$ concentrations (in $\mathrm{pmol\,kg^{-1}}$ and $\mathrm{fmol\,kg^{-1}}$, respectively) were converted into equivalent mixing ratios and are given in ppt.

Tritium was measured using the ^{3}He ingrowth method and is reported in tritium units (TU, 1 TU represents a [^{3}H]/[H] ratio of 10^{-18}) (Clarke and Jenkins, 1976; Sültenfuß et al., 2009). Precision typically is $\pm 3\%$ or 0.02 TU, whichever is greater, but partly lower for the 1987 data set (Roether et al., 1999).

For details about the methods, precision and accuracy, as well as for descriptions of the salinity, temperature and oxygen measurements, the reader is referred to the references in Table 1.

2.2 Apparent tracer ages

The CFC-12 apparent tracer ages were determined by direct comparison of the tracer concentrations in water with the atmospheric time history of the gas (Bullister, 2011), assuming 100% saturation in surface water.

The tritium-^{3}He age is based on the radioactive decay of tritium into ^{3}He. Assuming that no tritiugenic ^{3}He (from tritium decay) is present in surface waters, the tritium-^{3}He age can be directly calculated, using the tritiugenic ^{3}He concentration in the water sample and the half-life of tritium. The tritiugenic ^{3}He in the water sample was determined by separation from the measured ^{3}He following the steps described by Roether et al. (2013a).

Any water sample represents a mixture of contributions that differ in their travel time from the formation region, and thus also in their tracer concentrations. The apparent tracer ages do not consider mixing, which explains the shift between tracer age and mean age, partly because the age information of water with zero tracer concentration (very old water) is lost. Furthermore, the tracer ages are biased depending on their specific non-linear atmospheric input functions. This means for example, the faster the growth rate (or decay rate), the younger is the tracer age (Waugh et al., 2003). The comparison of ages of the same tracer over time is biased by changes in growth rates. For CFC-12, for example, the atmospheric increase was rather exponential up to 1970 and then linear up to 1990. Thereafter the growth rate decreased and after a period of almost no change (2000–2005), the concentrations are now slowly falling. The input function of tritium

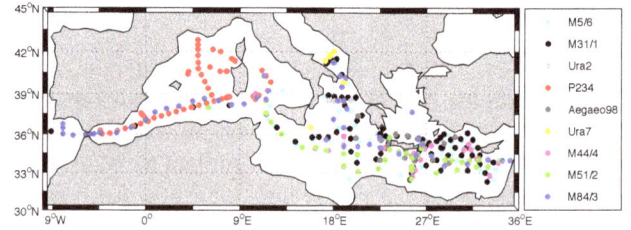

Fig. 3. The locations of all stations with tracer measurements of all cruises used in this work are shown: M5/6 in 1987, M31/1 in 1995, Ura2 in 1997, P234 in 1997, Aegaeo98 in 1998, Ura7 in 1999, M44/4 in 1999, M51/2 in 2001 and M84/3 in 2011.

shows a sharp peak in the 1960s (Fig. 2), which influences the tritium-^{3}He ages, although they directly depend on the non-changing decay rate. Subsequent to the peak, the tritium concentrations decreased, largely due to its radioactive decay. The input functions of tritium and CFC-12, thus, are of a distinctly different shape (Fig. 2).

2.3 The transit time distribution concept

Mean ages of CFC-12, SF$_6$ and tritium were determined using the transit time distribution (TTD) method (Hall and Plumb, 1994). The interior tracer concentration $c(t)$ is given by

$$c(t) = \int_0^\infty c_0(t - t') \cdot e^{-\lambda t'} \cdot G(t')\mathrm{d}t', \qquad (1)$$

where $c_0(t - t')$ is surface water tracer concentration, $e^{-\lambda t'}$ is the radioactive decay term in case of tritium, with the decay constant $\lambda = 0.05575$ and $G(t')$ is the transit time distribution. For steady transport an inverse Gaussian function is a good approximation for the TTD (Waugh et al., 2004):

$$G(t') = \sqrt{\frac{\Gamma^3}{4\pi \Delta^2 t'^3}} \cdot e^{\frac{-\Gamma(t' - \Gamma)^2}{4\Delta^2 t'}}, \qquad (2)$$

where Δ is the "width" of the TTD and Γ is the mean age. Even though it has been shown that the shape of the TTD in the Mediterranean was different from a IIG with Δ/Γ ratio of 1 in many places based on multiple-tracer analysis (Stöven and Tanhua, 2013), we chose a Δ/Γ ratio of 1, as proposed by Waugh et al. (2004) for ocean interior waters, since we do not have enough tracer data to calculate a possible variability of the shape of the TTD over time. This simplification will bias the results to some degree, and we discuss this bias below.

The atmospheric time history of SF$_6$ and CFC-12 were taken from Bullister (2011). The input function of tritium originated from Roether et al. (1992) and was extended in time by R. Steinfeld (unpublished data). It was newly adjusted by Stöven and Tanhua (2013) using tritium

Table 1. Summary of all cruises used in this work, showing dates, the sampling region and the respective references. EMed = eastern Mediterranean, WMed = western Mediterranean, NIonian = northern Ionian.

Cruise	Date	Basin/Sea	Reference
Meteor 5/6 (M5/6)	Aug/Sep 1987	EMed, Adriatic, Cretan, Tyrrhenian	Nellen et al. (1996) Schlitzer et al. (1991) (CFC12); Roether and Schlitzer (1991) (CFC12, Tri) Roether et al. (1999) (Tri)
Meteor 31/1 (M31/1)	Jan/Feb 1995	EMed, Adriatic, Cretan, (WMed)	Hemleben et al. (1996) Roether et al. (1996) (CFC12) Roether et al. (1998a) (CFC12, Tri)
Urania MAI2 (Ura2)	Aug/Sept 1997	Adriatic, NIonian	Manca et al. (2002) Roether et al. (1996, 2007) (CFC12)
Poseidon 234 (P234)	Oct/Nov 1997	WMed, Tyrrhenian	Rhein et al. (1999) (CFC12) Roether and Lupton (2011) (Tri)
Aegaeo M4WF (Aegaeo98)	Oct/Nov 1998	Cretan, Levantine	Theocharis et al. (2002) (CFC12)
Urania MAI7 (Ura7)	Feb 1999	Adriatic	Manca et al. (2002) Roether et al. (2007) (CFC12)
Meteor 44/4 (M44/4)	Apr/May 1999	EMed, Tyrrhenian	Pätzold et al. (2000) Theocharis et al. (2002) (CFC12) Roether and Lupton (2011) (Tri)
Meteor 51/2 (M51/2)	Oct/Nov 2001	EMed, Tyrrhenian, (WMed)	Hemleben et al. (2003) Roether et al. (2007) (CFC12) Schneider et al. (2010) (CFC12) Roether and Lupton (2011) (Tri)
Meteor 84/3 (M84/3)	Apr 2011	EMed, Adriatic, Tyrrhenian, WMed	Stöven (2011) (CFC12, SF_6) Stöven and Tanhua (2013) (CFC12, SF_6) Roether et al. (2013a) (Tri)

measurements in the Mediterranean Sea. This input function is almost identical to the one published by Roether et al. (2013a) derived from a slightly different approach. The tritium concentrations are generally 20 % higher in the eastern Mediterranean as compared to the western Mediterranean Sea due to increased continental influence for the former (over land masses, tritium concentrations in the lower troposphere have been distinctly higher; Weiss and Roether, 1980).

Generally, the mean ages present a more realistic estimate for the real water age than the apparent tracer ages but they are dependent on our assumption of the shape of the TTD. This can bias the results in both directions depending if mixing was over- or underestimated or if mixing increased or decreased over time. The direction is the same for all tracers but the magnitude can differ. SF_6 concentrations have increased rather steadily making it a good tracer for recent decades and probably also in the future. The usability of CFC-12 is limited to older waters because of the almost non-changing

atmospheric concentrations in the past 20 yr. A further uncertainty of the mean ages is the assumption of 100 % saturation in surface waters of CFC-12 and SF_6. Undersaturation is possible in areas where surface water quickly cools and subducts before equilibrium with the atmosphere is reached (e.g. in the deep water formation areas). The assumption of 100 % saturation would then lead to an overestimation of the water age. For instance Tanhua et al. (2008) suggest that the saturation of CFC-12 has increased over time as the atmospheric transient has decreased. Since we are analysing the temporal variability of ventilation ages of water masses, temporal variability of the surface saturation would lead to a bias in our analysis. We do not have sufficient information to realistically address this question for the Mediterranean but assume that the surface saturation has remained constant over the last decades, bearing in mind that a potential overestimation of ages (due to undersaturation) during intense convection periods could bias our analysis.

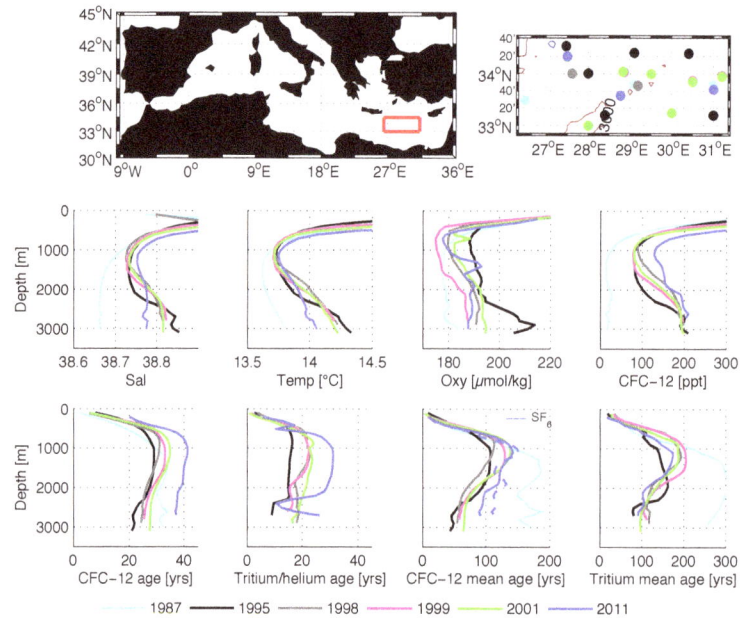

Fig. 4. The Levantine basin. In the upper left panel, a red box confines the respective area, where samples were taken on a map of the Mediterranean Sea. In the upper right panel, the exact locations of all sampled stations within this red box are shown. Each colour refers to one specific cruise/year specified in the legend. In the lower panels the same colours appear in the depth profiles of the measured parameters for each cruise. For the "mean-age panel", the mean-age calculated from SF_6 is shown as a blue dashed line; only available for 2011 data.

2.4 Combination and comparison of nine cruises

In the past, changes in the ventilation of the Mediterranean Sea have been documented (e.g. EMT, WMT, see above). Now, with the tracer data from nine cruises over the last 25 yr being available, a reasonable temporal and regional coverage has been obtained, which allows us to explore how those events are reflected in transient tracer data.

In the following we present and discuss tracer data and ages in nine areas in the Mediterranean Sea. As most of the campaigns focused on certain regions in the Mediterranean Sea, data from only three to seven cruises are available for the various areas. If more than one station during the same cruise falls into the boundaries of the chosen area, the mean profile is shown. The mean profile is calculated by first interpolating the individual profiles to standard depths using a piecewise cubic hermite interpolating method that do not allow too large vertical distances between data points, and then by taking the arithmetic mean of the interpolated profiles (e.g. Tanhua et al., 2010).The parameters are salinity, temperature, oxygen, CFC-12, CFC-12 age, tritium-^{3}He age, CFC-12 mean age, tritium mean age, and for the cruise of 2011 also the SF_6 mean age (mean ages means TTD-derived ages). A further instrument to identify changes in circulation is the comparison of CFC-12 and SF_6 ages determined in different years but at the same location. This approach (Tanhua et al., 2013) benefits from the almost identical atmospheric increase functions of both tracers with a time shift of 14 yr. In a steady-state situation, the tracer age of SF_6 should match

the tracer age of CFC-12 based on measurements taken 14 yr earlier in the same region. Since we only have SF_6 measurements for the cruise in 2011, the corresponding year for the CFC-12 measurements is 1997. For 6 regions the comparison of CFC-12 and SF_6 tracer age profiles could be performed. In two cases we used CFC-12 measurements from 1998, which results in a time gap of only 13 yr. The atmospheric increase of CFC-12 up to the year 1990 was almost linear (Fig. 2). Thus, the bias created by the changing rate of increase is expected to be small.

In this work we focus on relative changes in concentrations and ages over a time period of up to 25 yr. This is the reason why we do not quantify the uncertainties, but only discuss biases affecting the relative comparison and not the absolute numbers. For a detailed discussion of the uncertainties of the cruise in the year 2011, see Stöven and Tanhua (2013).

3 Results and interpretation

The Levantine basin results are shown in Fig. 4 and those of the area south of Crete (north of the east Mediterranean ridge) in Fig. 5. We find similar profiles in the two areas. Prominent in all parameters is the cruise of 1987, which represents the pre-EMT situation (Figs. 4 and 5). The later cruises consistently show higher temperatures, salinity and tracer concentrations and lower mean ages below 1000 m depth, due to the extraordinary amount of rather young, warm and saline waters from the Aegean Sea added by the EMT.

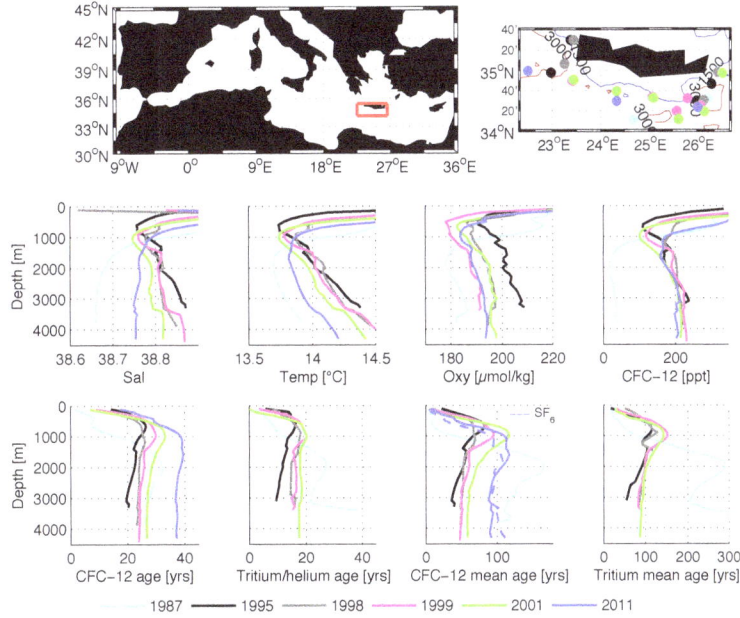

Fig. 5. Northern Levantine (south of Crete). For further explanation see caption of Fig. 4.

South of Crete, a gradual decrease in temperature and salinity is found after 1995. The oxygen profiles in the year 1995 also demonstrate the intrusion of new deep water with high concentrations, followed by a large drop in oxygen between 1995 and 1999, which has been attributed to enhanced oxygen consumption driven by large amounts of dissolved organic carbon (Klein et al., 2003).

In the CFC-12 profiles we also see a large difference between 1987 and 1995, whereas in the years after 1995 there is comparatively little change in CFC-12 concentration between 1000 m and 2500 m and no change at all below 2500 m (Figs. 4 and 5). As the atmospheric CFC-12 concentrations had been increasing continuously, until the year 1995, a continuing increase in deep waters is expected also after 1995. The constant concentrations at depth point to a period of stagnation following the EMT, confirming the analysis of Roether et al. (2013b). At the same time, some change in temperature and salinity is visible, pointing to the admixture of water with similar tracer concentrations. Both scenarios must have led to increased water ages from 1995 to 2011. If we look at the age profiles this is generally true (Figs. 4 and 5). Conspicuous is that the drop in the CFC-12 ages (and by trend also the tritium-^{3}He ages) between 1987 and 1995 is less than one would expect, presumably as a result of the age biases mentioned in Sect. 2.2, due to the non-linearity of the input function and the changes in the growth rate of CFC-12. We note that the TTD-based mean ages are much higher than the tracer ages and also closer to previous results (e.g., Roether and Schlitzer, 1991) that used a tracer constrained box model to estimate ventilation times. It thus appears that the tracer ages are rather a proxy of the real ages, but they

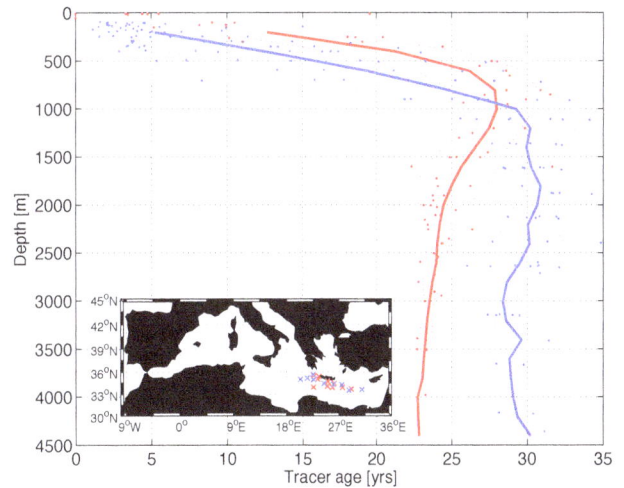

Fig. 6. The Levantine basin. CFC-12 and SF$_6$ ages (dots) at the stations shown on the map plus the averaged profiles (lines) for the area. CFC-12 ages are shown in red for the year 1998 and SF$_6$ ages in blue for the year 2011.

certainly have a diagnostic potential. The fact that the tritium mean age generally is higher than the CFC-12 mean age is possibly due to the strongly peaked tritium input, for which a Δ / Γ ratio different to the value actually used might be more adequate. In fact, Stöven and Tanhua (2013) found a Δ / Γ ratio of 0.5 for that region. A noteworthy feature is that the CFC-12 and SF$_6$ mean ages for 2011 agree rather well.

The uplift of water masses by the EMT can clearly be seen in all age profiles. The age maximum in Fig. 5 rises from 2000 m in 1987 to 800 m in 1995. At the same time the

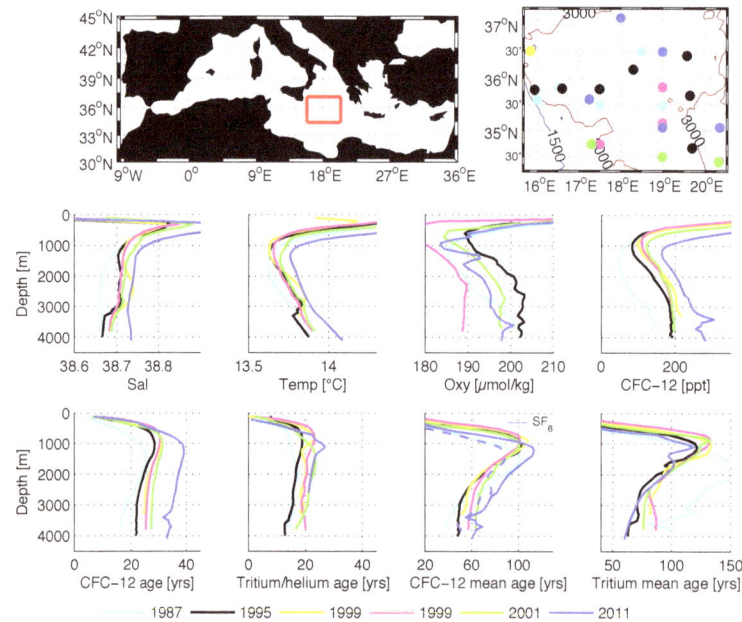

Fig. 7. The central Ionian. For further explanation see caption of Fig. 4.

CFC-12 mean age decreases from 150 yr to 80 yr, demonstrating the large amount of young water that has been transferred to depth during the EMT. In the course of time after 1995, the mean age and its centre depth increase again, perhaps moving slowly back to pre-EMT conditions.

Figure 6 displays the comparison of CFC-12 ages and of SF_6 ages 13 yr later for the Levantine basin. The 1998 profile represents the post-EMT situation with an uplifted age maximum at around 1000 m. In the years up to 2011 (the "stagnation" phase at depth), the water below 1000 m became older (5–7 yr). In the intermediate layer down to 1000 m, the shallower overturning circulation (LIW and CIW) reduces the age of the formerly uplifted old water, appearing as a vertical shift in the age gradient in the upper 1000 m of the water column.

The central Ionian is shown in Fig. 7. One notes that the parameter changes over time are less drastic than in the Levantine. The salinity and temperature profiles do not show any exceptional trends, while the enhanced oxygen consumption between 1995 and 1999 is even more pronounced than in the Levantine basin. The jump in the CFC-12 concentrations between 2001 and 2011 reflects the restarted deep-water supply from the Adriatic, after the mentioned "stagnation phase" (e.g. Manca et al., 2002). This is supported by the fact that a corresponding change is not yet apparent in the Levantine basin.

A further feature is noticeable in the profiles of the CFC-12 mean age in Fig. 7. In the year 1987 the young deep water is found at 3000–4000 m. In the years 1995 to 2001 young water occupies a wider depth range from 2000–4000 m because of the large volumes of deep water of Aegean origin added

by the EMT. This results in a more moderate vertical gradient in the mean age below 2000 m. In 2011, the mean-age minimum is again found closer to the bottom, approaching the pre-EMT situation. The difference between the CFC-12 and the SF_6 mean age in young waters can be explained by the different atmospheric histories. The almost constant concentrations of CFC-12 in the past 15 yr led to no great age difference when areas get ventilated. The SF_6 concentrations instead have increased continuously in the atmosphere and lead to younger mean ages in the same water mass. Another source of uncertainty in this analysis is the assumption of a Δ / Γ ratio of 1; Stöven and Tanhua (2013) have shown that this ratio was higher than 1 in the Adriatic overflow water during 2011. This would tend to bias our mean age estimates to be too low. However, the greatest difference between CFC-12 and SF_6 mean age in Fig. 7 is found between 1000 and 2000 m with water ages around 100 yr, whereas in the Levantine basin (Figs. 4 and 5) CFC-12 and SF_6 mean age fit well at these depths and ages. This might be caused by mixing of (two) different water masses (old and young) in the Ionian basin, that cannot be resolved with a 1IG TTD.

Profiles in the two regions of deep water formation in the eastern Mediterranean, i.e., the Adriatic Sea as the "classical" formation area and the Aegean Sea as the "additional" one during the EMT, are shown in Figs. 8 and 9. The Adriatic Pit data in Fig. 8 also demonstrates the "stagnation phase" that started there after 1987, as already noted previously (e.g., Roether et al., 2007). The deep CFC-12 concentrations hardly change between 1987 and 1999, while the CFC-12 age increases. By 2011 the "normal" ventilation has restarted and temperature and especially salinity and CFC-12 have

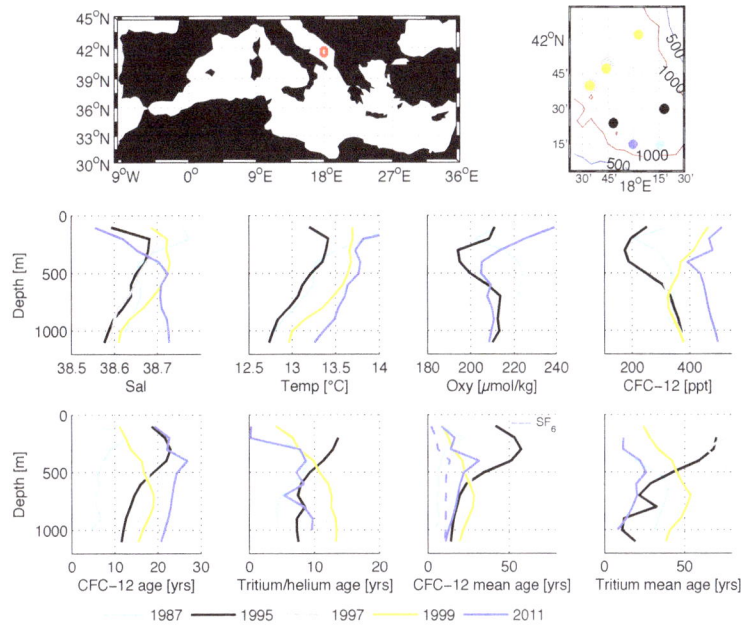

Fig. 8. The Adriatic Sea. For further explanation see caption of Fig. 4.

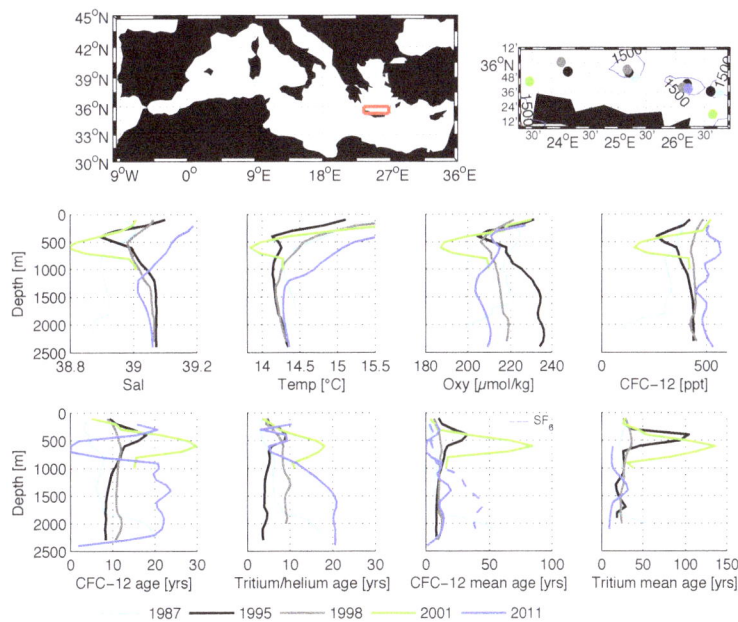

Fig. 9. The Cretan Sea. For further explanation see caption of Fig. 4.

increased and the mean age has decreased. Again, in the CFC-12 tracer age profiles a significant decrease in tracer age is not apparent because of the almost non-changing atmospheric concentrations since 1995. The tritium-^{3}He age instead displays a very similar trend to the CFC-12 and SF$_6$ mean age. This demonstrates the utility of the tracer tritium for relatively young waters.

The Cretan Sea is a small basin in the southern Aegean Sea. It is characterized by variable and fast changing

parameter values (Fig. 9). The water in the basin is generally very young (<30 yr) and high in oxygen, similar to the Adriatic Sea. The layer between 200 and 900 m appears to be especially variable. In the years 1995 and 2001 old water masses, low in salinity, temperature and oxygen are found at these depths, whereas in 2011 the opposite trend is visible. Below 1800 m the pre-EMT water mass characteristics can clearly be recognized in all parameters. This depth was ventilated after 1987 and a more homogenized deep water mass

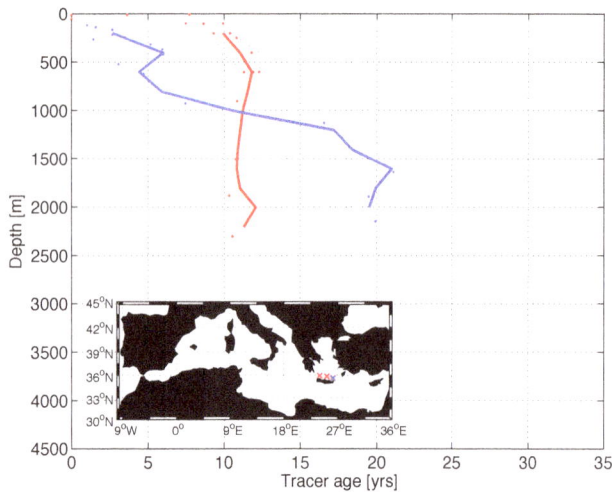

Fig. 10. The Cretan Sea. CFC-12 and SF$_6$ ages (dots) at the stations shown on the map plus the averaged profiles (lines) for the area. CFC-12 ages are shown in red for the year 1998 and SF$_6$ ages in blue for the year 2011.

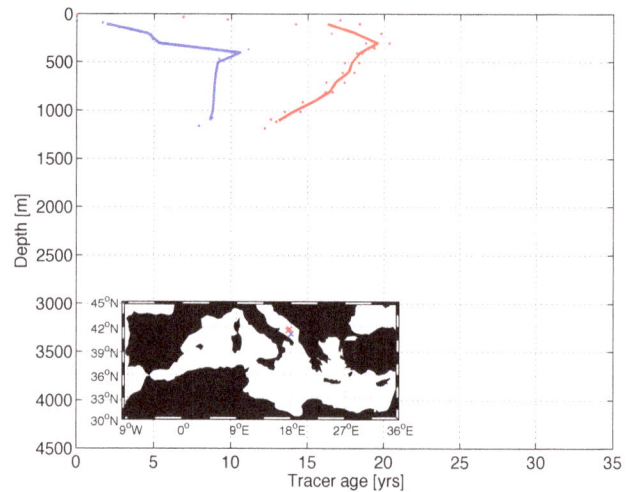

Fig. 11. The Adriatic Sea. CFC-12 and SF$_6$ ages (dots) at the stations shown on the map plus the averaged profiles (lines) for the area. CFC-12 ages are shown in red for the year 1997 and SF$_6$ ages in blue for the year 2011.

was created, spanning from around 1200 down to the bottom. Again, for these young waters, which have not undergone a lot of mixing, the tritium-^{3}He age and the SF$_6$ mean age show similar trends (increasing age at depth).

Figure 10 shows a well-mixed water column in the year 1998 for the Cretan Sea, with an age of around 11 yr. This is due to the intense deep water formation in the Aegean Sea in the early 90s, which ventilated the entire basin down to the bottom. By the year 2011 the restarted "normal" ventilation (formation of CIW, that reaches depth around 500–1200 m (Schlitzer et al., 1991)) has led to very young waters in the upper 1000 m. In contrast, the water below 1000 m apparently has not been ventilated since then and aged by 10 yr. The Adriatic Sea circulation (Fig. 11) also seems to have come back to its pre-EMT conditions in the year 2011. The age gradient below 500 m that was apparent in 1998 disappeared due to the restarted deep water formation and the tracer age decreased by ∼ 7 yr.

In the following we discuss three regions in the western Mediterranean Sea: the Alboran Sea (Fig. 12), the south Liguro-Provençal basin (Fig. 13) and the southeast Liguro-Provençal basin (Fig. 14). Concentrating first on the deep water (below 1500 m), we can observe the same tendency in all three figures. Hardly changing salinities, temperatures and CFC-12 concentrations in the years 1995, 1997 and 2001, followed by a large increase by the year 2011. This feature is to be ascribed to the WMT, in which, similar to the EMT, increased water density led to exceptionally intense deep water formation and thus ventilation of the deep western basin (Schröder et al., 2006, 2008). These changes of hydrographic properties can potentially have significant impact on water mass properties and mixing in the North Atlantic, although they are outside the scope of this work. In this event that

began about ten years after the EMT, the above-mentioned shortcomings of the tracer ages compared to the mean ages as well as the limits of the tracer CFC-12 itself become clear. The expected decrease in water age due to the recent significant ventilation of the deep water is neither seen in the CFC-12 ages, nor in the tritium-^{3}He ages and the CFC-12 mean ages. The tracer ages do not take mixing into account, which biases the results the more, the older a water mass is. In the Adriatic Sea (Fig. 8), where the tracer age at depth is only around 10 yr, the tritium-^{3}He age is similar to the SF$_6$ mean age, but here in the central western basin the water is older and tracer age and mean age do not match anymore. The CFC-12 ages are additionally biased because deep water ventilated between 2001 and 2011 was exposed to nearly constant or even decreasing atmospheric concentrations. This also influenced the CFC-12 mean age in 2011, although that bias is less strong: the CFC-12 mean age in Fig. 13 shows a continuous aging of the deep water mass from 1995 to 2001 while in 2011 the CFC-12 mean age decreased slightly. The CFC-12 concentration on the other hand, shows a marked increase between 2001 and 2011; a tell-tale of the WMT. The signal in the SF$_6$ mean age is more pronounced, in particular close to the bottom. The tritium mean age shows the same trend, but in general we think that the SF$_6$ mean age is more robust because the tritium input function is on the one hand so strongly peaked and on the other hand regional differences (e.g. increased tritium inputs in the deep water formation areas due to the vicinity to the continent) would probably change the tritium mean age substantially. However, Stöven and Tanhua (2013) found a 2IG distribution for the western Mediterranean Sea with 60 % old water and 40 % very young water, which would change our results of the mean ages of all tracers. Similar to the accelerated oxygen consumption in the

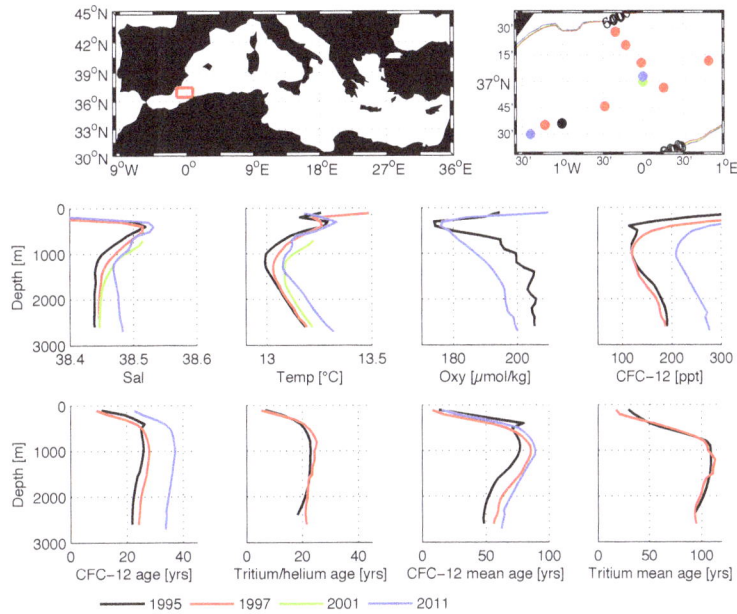

Fig. 12. The Alboran Sea. For further explanation see caption of Fig. 4.

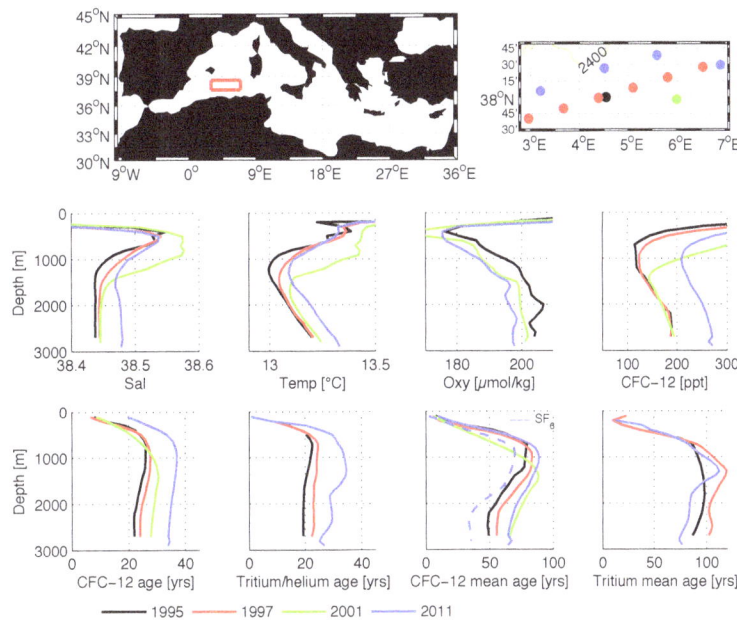

Fig. 13. The south Liguro-Provençal basin. For further explanation see caption of Fig. 4.

eastern basin after the EMT, we find low oxygen concentrations in the year 2011, which leads to the tentative conclusion that the WMT was also followed by increased consumption. However, an uplift of isolines is not obvious. Furthermore, the parameter profiles between 500 and 1500 m in the year 2001 (Fig. 13) suggest that the impact of the EMT (warmer and more saline water) had already reached the western basin in that year, influencing the entire intermediate layer.

The influence of the WMT in the Liguro-Provençal basin is also visible in Fig. 15, where the tracer age at depth decreased by about 5 yr in the period from 1997 to 2011. Furthermore, the entire water column shows a moderate age decrease, which in the intermediate waters could be an effect of the EMT. The western part of the Alboran Sea (Fig. 16) is the only basin where we find almost unchanged tracer age profiles, which suggest a steady circulation at least in the upper

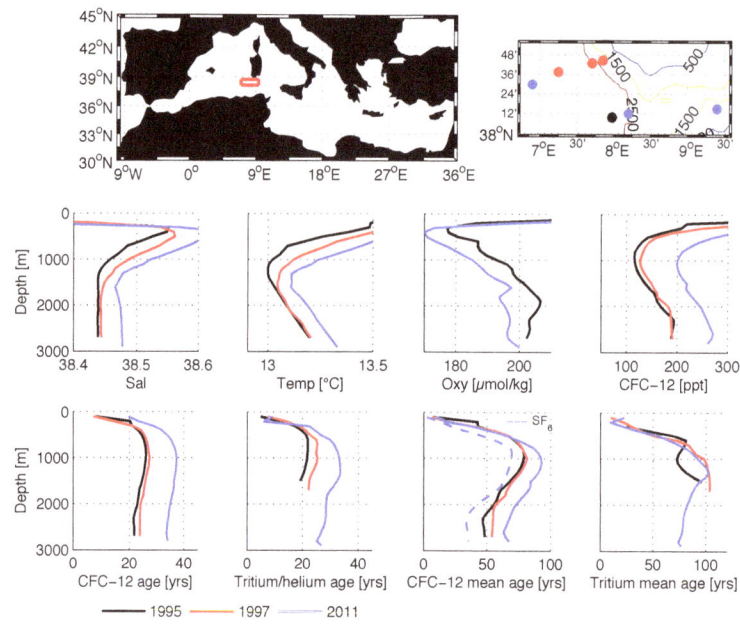

Fig. 14. The southeast Liguro-Provençal basin. For further explanation see caption of Fig. 4.

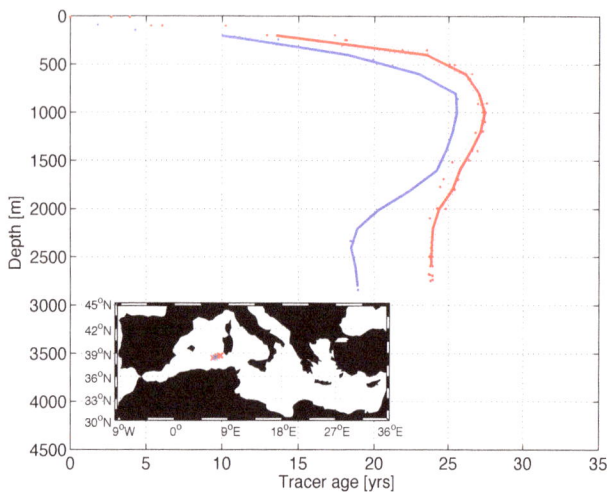

Fig. 15. The southeast Liguro-Provençal basin. CFC-12 and SF$_6$ ages (dots) at the stations shown on the map plus the averaged profiles (lines) for the area. CFC-12 ages are shown in red for the year 1997 and SF$_6$ ages in blue for the year 2011.

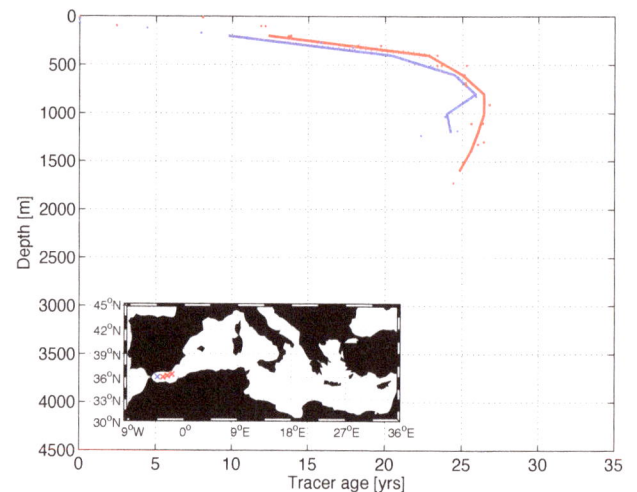

Fig. 16. The Alboran Sea. CFC-12 and SF$_6$ ages (dots) at the stations shown on the map plus the averaged profiles (lines) for the area. CFC-12 ages are shown in red for the year 1997 and SF$_6$ ages in blue for the year 2011.

800 m. Below that depth, the influence of the WMT might be visible by younger water in 2011.

Figure 17 shows the situation in the Tyrrhenian Sea, a basin where Western Mediterranean Deep Water mixes with the overflow waters of the Sicily Channel (LIW, CIW and tEMDW). There was EMT-induced enhanced mixing in the 1990s, but this involved rather old waters from the eastern basin (Roether and Lupton, 2011), so that insignificant changes in ventilation are apparent after 1987, keeping the deep water mean ages around 100 yr. The 1987 profiles,

being not yet affected by the EMT, slightly differ from the later ones. The signal of the WMT had not reached the basin even by 2011, when below 2500 m, an old water mass was still found in that same year. This can also be seen in Fig. 18, where water ages increase below 1800 m. Furthermore, we find a significant decrease in age of up to 20 yr between 500 and 1500 m. This layer is fed by the input of water over the Sicily Channel ($\sim$500 m) from the east. Since the EMT led to upwelling of old water reaching up into the LIW, the signal in the intermediate layer in the Tyrrhenian Sea in 1997 most

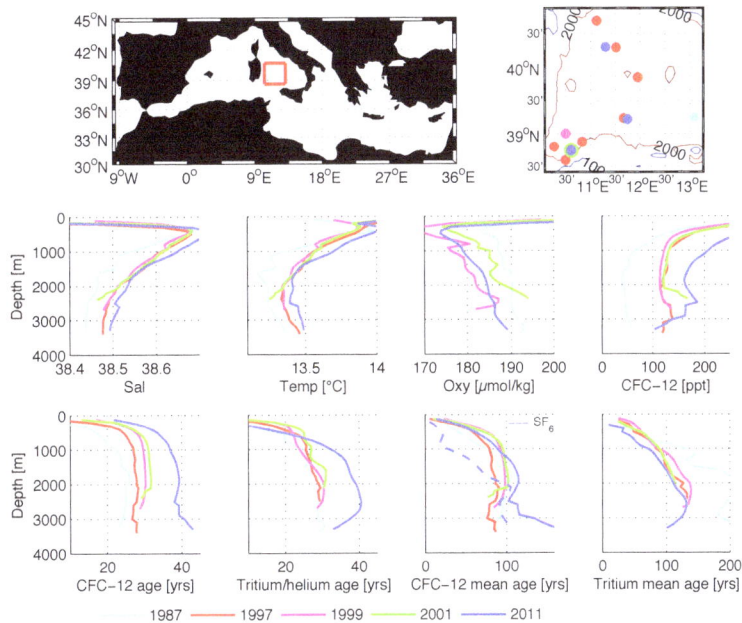

Fig. 17. The Tyrrhenian Sea. For further explanation see caption of Fig. 4.

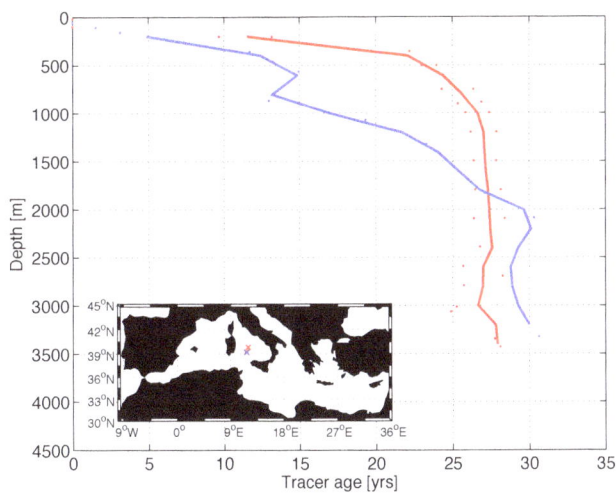

Fig. 18. The Tyrrhenian Sea. CFC-12 and SF$_6$ ages (dots) at the stations shown on the map plus the averaged profiles (lines) for the area. CFC-12 ages are shown in red for the year 1997 and SF$_6$ ages in blue for the year 2011.

probably is influenced by this older water. By 2011, the "normal" circulation/ventilation of the LIW was restored, which led to a decrease in age compared to 1997.

4   Conclusions

This study analyses a large data set of several transient tracers over a period of 25 yr in the Mediterranean Sea, from which we deduced an overview of ventilation and its temporal changes throughout the sea. The tracer data were used to obtain water mass ages, using different methods and assumptions, and regional as well as temporal comparisons were carried out. Furthermore, the different methods to estimate the ages were tested for applicability to the Mediterranean Sea.

The two dominating events that changed the ventilation, namely the Eastern Mediterranean Transient (EMT) in the eastern basin, and the Western Mediterranean Transition (WMT) in the western basin around 12–15 yr later can clearly be recognized in most parameters. The EMT led to an uplift of parameter isolines in consequence of its considerable addition of new deep water of Aegean origin, which was followed by a "stagnation phase" with respect to ventilation. Our analysis indicates a return toward pre-EMT conditions after 2001 with a restart of the classical Adriatic deep water source, a deepening of the isolines and recent ventilation of the Ionian deep waters. In the Levantine basin, in contrast, we see no signs of recent ventilation of the deep water since the EMT in the early 1990s. The signal of the EMT (e.g., high salinities and temperatures) propagated west, reaching the Tyrrhenian basin before 1997 and the intermediate depths of the western basin between 1997 and 2001, acting as a triggering factor of the WMT beginning in 2004/2005. Differently from the eastern basin, the data in the western basin do not show an uplift of isolines in consequence of the WMT. While by the year 2011 the signal of the WMT was visible in the entire western basin, it had yet not reached the Tyrrhenian Sea.

The main results concerning the age calculation methods used in this work are the following: apparent tracer ages, that is, the period between the sampling and the point back in time when the atmospheric tracer concentration was the same as the measured value, is a useful diagnostic but only a proxy of the real age. A better estimate of the real age is the mean age obtained using the TTD (transit time distribution) method. However, the ages obtained by both these methods vary between the different tracers. For the TTD-based ages this arises mainly from an inadequate choice of the strength of mixing and the fact that the Mediterranean Sea circulation is not in steady-state and for the tracer ages from its dependence on the specific input history and the fact that mixing is not considered. Particularly we want to point out the difficulties in estimating changes in age over time with apparent tracer ages. Ages can be biased in different directions and with different magnitude, when comparing tracer ages of the same tracer for different time periods. Time invariant tracer ages do not imply constant ventilation and vice versa. In the case of CFC-12, for example, the apparent tracer age underestimates the age prior to about 1970, when the atmospheric concentration increased almost exponentially, and overestimated it, since about the year 2000 when the atmospheric concentrations hardly changed or even decreased. The dependence on the input histories is particularly problematic if one deals with a transient situation such as that caused by the EMT or the WMT. The apparent tracer ages are most strongly affected, but an effect is also visible in the mean ages. We find that for recently ventilated waters SF_6-based ages are particularly realistic due to its smooth input history (Fig. 2). Tritium is problematic due to uncertainties of the tritium input function. That input varies also regionally depending on vicinity to land masses, and moreover, it strongly decreased in recent decades, so that the internal recirculation became dominant relative to new input. A special approach to identify temporal changes in ventilation was done by a comparison of apparent CFC-12 ages vs. SF_6 ages obtained 14 yr later, when the growth rate of the SF_6 input from the atmosphere was very similar to that of CFC-12 at the earlier time. Generally, the results of the other parameter comparisons were confirmed and also that the Mediterranean Sea is clearly not in a steady state. The only basin with almost no changes in ventilation ages was the Alboran Sea, which is close to the Strait of Gibraltar and mainly influenced by the Atlantic Ocean. However, the changes in salinity observed in the Alboran Sea can potentially have far reaching influence on North Atlantic water mass properties.

Basically, the TTD method assumes circulation/ventilation to be in a steady state, which makes evaluation of the unsteady ventilation in the Mediterranean a challenge. A more detailed analysis of the contemporary Mediterranean TTD has been made by Stöven and Tanhua (2013) for observations obtained from the Meteor cruise of 2011 using the tracers CFC-12, tritium and SF_6.

Acknowledgements. We sincerely acknowledge the work of all scientists and crews on the research cruises and helpers in the home laboratories who contributed to tracer sampling and measurements in the Mediterranean Sea since 1987. We acknowledge a grant from the German Science Foundation (DFG) TA3173/3 that supported A. Schneider during preparation of this manuscript, part of the publication costs and helium/tritium measurements.

The service charges for this open access publication have been covered by a Research Centre of the Helmholtz Association.

Edited by: M. Hoppema

References

Astraldi, M., Gasparini, G., Sparnocchia, S., Moretti, M., and Sansone, E.: The characteristics of the water masses and the water transport in the Sicily Strait at long time scales, in Dynamics of Mediterranean straits and channels, Bulletin de l'Institut Océanographique, edited by: Briand, F., 17, 95–115, 1996.

Bergamasco, A. and Malanotte-Rizzoli, P.: The circulation of the Mediterranean Sea: a historical review of experimental investigations, Adv. Oceanogr. Limnol., 1, 11–28, doi:10.1080/19475721.2010.491656, 2010.

Bullister, J. L.: Atmospheric CFC-11, CFC-12, CFC-113, CCl_4 and SF_6 Histories (1910–2011), Carbon Dioxide Information Analysis Center, available at: http://cdiac.ornl.gov/oceans/new_atmCFC.html (last access: March 2012), 2011.

Bullister, J. and Weiss, R.: Determination of CCL_3F and CCL_2F_2 in Seawater and Air, Deep Sea Res.-Pt. A, 35, 839–853, doi:10.1016/0198-0149(88)90033-7, 1988.

Bulsiewicz, K., Rose, H., Klatt, O., Putzka, A., and Roether, W.: A capillary-column chromatographic system for efficient chlorofluorocarbon measurements in ocean waters, J. Geophys. Res., 103, 15959–15970, 1998.

Clarke, W. and Jenkins, W.: Determination of Tritium by Mass-Spectrometric Measurement of ^{3}He, Int. J. Applied Ra., 27, 515–522, doi:10.1016/0020-708X(76)90082-X, 1976.

Gascard, J.-C.: Mediterranean deepwater formation, baroclinic instability and oceanic eddies, Oceanol. Acta, 1, 315–330, 1978.

Gasparini, G. P., Ortona, A., Budillon, G., Astraldi, M., and Sansone, E.: The effect of the Eastern Mediterranean Transient on the hydrographic characteristics in the Strait of Sicily and in the Tyrrhenian Sea, Deep Sea Res.-Pt. I, 52, 915–935, doi:10.1016/j.dsr.2005.01.001, 2005.

Hall, T. M. and Plumb, R. A.: Age as a diagnostic of stratospheric transport, J. Geophys. Res., 99, 1059–1070, 1994.

Hemleben, C., Roether, W., and Stoffers, P.: Östliches Mittelmeer, Rotes Meer, Arabisches Meer, Cruise No. 31, 30 December 1994–22 March 1995, METEOR-Berichte, Universität Hamburg, 1996.

Hemleben, C., Hoernle, K., Jørgensen, B. B., and Roether, W.: Ostatlantik, Mittelmeer, Schwarzes Meer, Cruise No. 51, 12 September–28 December 2001, METEOR-Berichte, Universität Hamburg., 2003.

Klein, B., Roether, W., Kress, N., Manca, B. B., D'Alcala, M. R., Souvermezoglou, E., Theocharis, A., Civitarese, G., and Luchetta, A.: Accelerated oxygen consumption in eastern Mediterranean deep waters following the recent changes in thermohaline circulation, J. Geophys. Res., 108, 8107, doi:10.1029/2002JC001454, 2003.

Lascaratos, A., Roether, W., Nittis, K., and Klein, B.: Recent changes in deep water formation and spreading in the eastern Mediterranean Sea: a review, Prog. Oceanogr., 44, 5–36, doi:10.1016/S0079-6611(99)00019-1, 1999.

Leaman, K. D. and Schott, F. A.: Hydrographic structure of the convection regime in the Gulf of Lions: Winter 1987, J. Phys. Oceanogr., 21, 575–598, doi:10.1175/1520-0485(1991)021<0575:HSOTCR>2.0.CO;2, 1991.

Malanotte-Rizzoli, P. and Hecht, A.: Large-scale properties of the Eastern Mediterranean?: a review, Oceanol. Acta, 11, 323–335, 1988.

Manca, B. B., Kovačević, V., Gačić, M., and Viezzoli, D.: Dense water formation in the Southern Adriatic Sea and spreading into the Ionian Sea in the period 1997–1999, J. Marine Syst., 34, 133–154, 2002.

MEDOC Group: Observation of formation of deep water in the Mediterranean Sea, 1969, Nature, 227, 1037–1040, 1970.

Millot, C.: Circulation in the Western Mediterranean Sea, J. Marine Syst., 20, 423–442, doi:10.1016/S0924-7963(98)00078-5, 1999.

Nellen, W., Bettac, W., Roether, W., Schnack, D., Thiel, H., Weikert, H., and Zeitschel, B.: MINDIK, Reise Nr. 5, 2 January–24 September 1987, Meteor-Berichte Universität Hamburg., 1996.

Pinardi, N. and Masetti, E.: Variability of the large scale general circulation of the Mediterranean Sea from observations and modelling: a review, Palaeogeogr. Palaeocl., 158, 153–173, doi:10.1016/S0031-0182(00)00048-1, 2000.

Pätzold, J., Halbach, P. E., Hempel, G., and Weikert, H.: Östliches Mittelmeer – Nördliches Rotes Meer, Cruise No. 44, 22 January–16 May 1999, METEOR-Berichte, Universität Hamburg, 2000.

Rhein, M., Send, U., Klein, B., and Krahmann, G.: Interbasin deep water exchange in the western Mediterranean, J. Geophys. Res., 104, 495–508, doi:10.1029/1999JC900162, 1999.

Roether, W. and Lupton, J. E.: Tracers confirm downward mixing of Tyrrhenian Sea upper waters associated with the Eastern Mediterranean Transient, Ocean Sci., 7, 91–99, doi:10.5194/os-7-91-2011, 2011.

Roether, W. and Schlitzer, R.: Eastern Mediterranean deep water renewal on the basis of chlorofluoromethane and tritium data, Dynam. Atmos. Oceans, 15, 333–354, doi:10.1016/0377-0265(91)90025-B, 1991.

Roether, W., Schlosser, P., Kuntz, R., and Weiss, W.: Transient-tracer studies of the thermohaline circulation of the Mediterranean, in Winds and Currents of the Mediterranean Basin, Proc. NATO workshop "Atmospheric and Oceanic Circulations in the Mediterranean Basin", 7–14 September 1983, 291–317, H. Charnock, Harvard University, Cambridge MA, 1992.

Roether, W., Manca, B. B., Klein, B., Bregant, D., Georgopoulos, D., Beitzel, V., Kovacevic, V., and Luchetta, A.: Recent changes in Eastern Mediterranean Deep Waters, Science, 271, 333–335, 1996.

Roether, W., Klein, B., Beitzel, V., and Manca, B. B.: Property distributions and transient-tracer ages in Levantine Intermediate Water in the Eastern Mediterranean, J. Marine Syst., 18, 71–87, doi:10.1016/S0924-7963(98)00006-2, 1998a.

Roether, W., Well, R., Putzka, A., and Rüth, C.: Component separation of oceanic helium, J. Geophys. Res., 103, 27931, doi:10.1029/98JC02234, 1998b.

Roether, W., Beitzel, V., Sültenfuß, J., and Putzka, A.: The Eastern Mediterranean tritium distribution in 1987, J. Marine Syst., 20, 49–61, doi:10.1016/S0924-7963(98)00070-0, 1999.

Roether, W., Klein, B., Manca, B., Theocharis, A., and Kioroglou, S.: Transient Eastern Mediterranean deep waters in response to the massive dense-water output of the Aegean Sea in the 1990s, Prog. Oceanogr., 74, 540–571, doi:10.1016/j.pocean.2007.03.001, 2007.

Roether, W., Jean-Baptiste, P., Fourré, E., and Sültenfuß, J.: The transient distributions of nuclear weapon-generated tritium and its decay product [3]He in the Mediterranean Sea, 1952–2011, and their oceanographic potential, Ocean Sci., 9, 837–854, doi:10.5194/os-9-837-2013, 2013a.

Roether, W., Klein, B., and Hainbucher, D.: The Eastern Mediterranean Transient: Evidence for similar events previously?, in AGU Monograph "The Mediterranean Sea: Temporal Variability and Spatial Patterns", edited by: Eusebi-Borzelli, G.-L., 2013b.

Schlitzer, R., Roether, W., Oster, H., Junghans, H.-G., Hausmann, M., Johannsen, H., and Michelato, A.: Chlorofluoromethane and oxygen in the Eastern Mediterranean, Deep Sea Res.-Pt. A, 38, 1531–1551, doi:10.1016/0198-0149(91)90088-W, 1991.

Schneider, A., Tanhua, T., Körtzinger, A., and Wallace, D. W. R.: High anthropogenic carbon content in the eastern Mediterranean, J. Geophys. Res., 115, C12050, doi:10.1029/2010JC006171, 2010.

Schröder, K., Gasparini, G. P., Tangherlini, M., and Astraldi, M.: Deep and intermediate water in the western Mediterranean under the influence of the Eastern Mediterranean Transient, Geophys. Res. Lett., 33, L21607, doi:10.1029/2006GL027121, 2006.

Schröder, K., Ribotti, A., Borghini, M., Sorgente, R., Perilli, A., and Gasparini, G. P.: An extensive western Mediterranean deep water renewal between 2004 and 2006, Geophys. Res. Lett., 35, L18605, doi:10.1029/2008GL035146, 2008.

Sonnerup, R. E.: On the relations among CFC derived water mass ages, Geophys. Res. Lett., 28, 1739–1742, doi:10.1029/2000GL012569, 2001.

Sparnocchia, S., Gasparini, G., Astraldi, M., Borghini, M., and Pistek, P.: Dynamics and mixing of the Eastern Mediterranean outflow in the Tyrrhenian basin, J. Marine Syst., 20, 301–317, doi:10.1016/S0924-7963(98)00088-8, 1999.

Stratford, K. and Williams, R. G.: A tracer study of the formation, dispersal and renewal of Levantine Intermediate Water, J. Geophys. Res.-Oceans, 102, 12539–12549, doi:10.1029/97JC00019, 1997.

Stratford, K., Williams, R. G., and Drakopoulos, P. G.: Estimating climatological age from a model-derived oxygen–age relationship in the Mediterranean, J. Marine Syst., 18, 215–226, doi:10.1016/S0924-7963(98)00013-X, 1998.

Stöven, T.: Ventilation processes of the Mediterranean Sea based on CFC-12 ad SF_6 measurements, Diploma, Leibniz-Institut für Meereswissenschaften der Mathematisch-naturwissenschaftlichen Fakultät, Christian-Albrecht-Universität zu Kiel, Kiel, 2011.

Stöven, T. and Tanhua, T.: Ventilation of the Mediterranean Sea constrained by multiple transient tracer measurements, Ocean

Sci. Discuss., 10, 1647–1705, doi:10.5194/osd-10-1647-2013, 2013.

Sültenfuß, J., Roether, W., and Rhein, M.: The Bremen mass spectrometric facility for the measurement of helium isotopes, neon, and tritium in water, Isot. Environ. Healt. S., 45, 83–95, doi:10.1080/10256010902871929, 2009.

Tanhua, T., Waugh, D. W. and Wallace, D. W. R.: Use of SF_6 to estimate anthropogenic CO_2 in the upper ocean, J. Geophys. Res., 113, C04037, doi:10.1029/2007JC004416, 2008.

Tanhua, T., van Heuven, S., Key, R. M., Velo, A., Olsen, A., and Schirnick, C.: Quality control procedures and methods of the CARINA database, Earth Syst. Sci. Data, 2, 35–49, doi:10.5194/essd-2-35-2010, 2010.

Tanhua, T., Waugh, D. W., and Bullister, J. L.: Estimating changes in ocean ventilation from early 1990s CFC-12 and late 2000s SF_6 measurements, Geophys. Res. Lett., 40, 927–932, doi:10.1002/grl.50251, 2013.

Theocharis, A., Klein, B., Nittis, K., and Roether, W.: Evolution and status of the Eastern Mediterranean Transient (1997–1999), J. Marine Syst., 33–34, 91–116, doi:10.1016/S0924-7963(02)00054-4, 2002.

Vollmer, M. K. and Weiss, R. F.: Simultaneous determination of sulfur hexafluoride and three chlorofluorocarbons in water and air, Mar. Chem., 78, 137–148, 2002.

Waugh, D. W., Hall, T. M., and Haine, T. W. N.: Relationships among tracer ages, J. Geophys. Res., 108, 3138, doi:10.1029/2002JC001325, 2003.

Waugh, D. W., Haine, T. W. N., and Hall, T. M.: Transport times and anthropogenic carbon in the subpolar North Atlantic Ocean, Deep Sea Res.-Pt. I, 51, 1475–1491, doi:10.1016/j.dsr.2004.06.011, 2004.

Weiss, W. and Roether, W.: The rates of tritium input to the world oceans, Earth Planet. Sc. Lett., 49, 435–446, doi:10.1016/0012-821X(80)90084-9, 1980.

Wüst, G.: On the vertical circulation of the Mediterranean Sea, J. Geophys. Res., 66, 3261–3271, doi:10.1029/JZ066i010p03261, 1961.

2

Tidal variability of the motion in the Strait of Otranto

L. Ursella, V. Kovačević, and M. Gačić

Istituto Nazionale di Oceanografia e di Geofisica Sperimentale (OGS), B.go Grotta Gigante 42/c, 34010 Sgonico (TS), Italy

Correspondence to: L. Ursella (lursella@ogs.trieste.it)

Abstract. Various current data, collected in the Strait of Otranto during the period 1994–2007, have been analysed with the aim of describing the characteristics of the tidal motions and their contribution to the total flow variance. The principal tidal constituents in the area were the semi-diurnal (M2) and the diurnal (K1), with the latter one predominant. The total flow was, in general, more energetic along the flanks than in the middle of the strait. Specifically, it was most energetic over the western shelf and in the upper layer along the eastern flank. In spite of the generally low velocities (a few cm s^{-1}) of the principal tidal constituents, the tidal variance has a pattern similar to that of the total flow variance, that is, it was large over the western shelf and low in the middle. The proportion of non-tidal (comprising the inertial and sub-inertial low-frequency bands) to tidal flow variances was quite variable in both time and space. The low-frequency motions dominated over the tidal and inertial ones in the eastern portion of the strait during the major part of the year, particularly in the upper and intermediate layers. In the deep, near-bottom layer the variance was evenly distributed between the low frequency, diurnal and semi-diurnal bands. An exception was observed near the western shelf break during the summer season when the contribution of the tidal signal to the total variance reached 77 %. This high contribution was mainly due to the intensification of the diurnal signal at that location at both upper and bottom current records (velocities of about $10 \, \text{cm s}^{-1}$). Local wind and sea level data were analysed and compared with the flow to find the possible origin of this diurnal intensification. Having excluded the sea-breeze impact on the intensification of the diurnal tidal signal, the most likely cause remains the generation of the topographically trapped internal waves and the diurnal resonance in the tidal response. These waves were sometimes generated by the barotropic tidal signal in the presence of summer stratification and the strong bottom slope. This phenomenon may stimulate diapycnal mixing during the stratified season and enhance ventilation of the near-bottom layers.

1 Introduction

The notion of tidal variability was historically better described from the measurements of the sea level than from measurements of the sea currents. The simplicity, low cost and easy maintenance of the tide gauges at the coast compared to current-meter moorings are the main reason for this historical discrepancy. Therefore, the study of the sea level tidal oscillations in the Adriatic Sea has a long and rich history (summarized by Cushman et al., 2001). The sea level tidal oscillations are of the mixed type, and are well reproduced, taking into account the seven major tidal constituents, namely, K1, O1, P1 (diurnal) and M2, S2, K2, N2 (semi-diurnal band).

Experimental studies dealing with the sea level were conducted by Polli (1961), Mosetti and Manca (1972), Buljan and Zore-Armanda (1976) and Orlić (2001). The tidal motion in the Adriatic is induced by the Mediterranean Sea tide, and co-oscillates with it. Polli (1961) drew charts of the cotidal lines and lines of equal phases for the surface elevation of the entire Adriatic Sea. He showed that the semi-diurnal tide has a sea-level amphidromic point located off Ancona (at about $14.5° \, \text{E}$; $43.5° \, \text{N}$), and that there is no amphidromic point for the diurnal tide. Theoretical studies (Taylor, 1921; Hendershott and Speranza, 1971) explained that semi-diurnal M2 tide (the principal one in the basin) propagates as a Kelvin wave entering from the Strait of Otranto, travelling to the north along the eastern coast, reflecting at the basin's end and descending the Adriatic along the western coast. As a result, an amphidromic point for the currents is positioned off the Gargano promontory (at around 16° E; 42° N), that is, south of the one for the sea level (Ursella and Gačić, 2001). The maxima in the current tide therefore correspond to zero amplitude in the sea level (Hendershott and Speranza, 1971). From numerical model studies, Malačič et al. (2000) explained the dynamics of semi-diurnal and diurnal tides in the northern Adriatic with a general theory of gravity and topographic waves. Thus, the M2 tide is well represented by a Kelvin wave that propagates along the basin, as gravity dominates, while K1 is well approximated by a continental shelf wave propagating across the basin as topographic effects dominate. Book et al. (2009b) showed that in the northern Adriatic the K1 tide displayed a behaviour of both a topographic Rossby wave and a Kelvin wave, with departure from Kelvin behaviour where bathymetry is steep. In addition to the above-mentioned papers there has been a considerable amount of other theoretical and observational research on tides; results and findings up to 2001 are summarized by Cushman-Roisin et al. (2001).

Current-meter measurements of sufficient duration to determine the tidal flow constituents have been something of a challenge all over the Adriatic Sea due to high costs, and instrument losses caused mainly by heavy fishery activities. Tidal currents derived from current-meter measurements as found in the literature refer predominantly to the northern Adriatic Sea. In particular, studies in the channels along the eastern coast (references in Cushman-Roisin et al., 2001) showed tidal flow ranges between 1 and 10 cm s^{-1}. Tidal characteristics of the surface current along the northern Italian coast were studied by Budillon et al. (2002), Kovačević et al. (2004) and Chavanne et al. (2007), while the detailed vertical pattern of the tidal flow throughout the water column was examined in the northernmost part of the basin by Book et al. (2009a) and in the central Adriatic by Martin et al. (2009). All these studies show that the values of the semi-major axes are a few cm s^{-1} (M2 was mostly between 5 and 7 cm s^{-1} and K1 between 2 and 4 cm s^{-1}).

However, a basin-wide survey in 1995 by a Vessel Mounted Acoustic Doppler Current Profiler (VM-ADCP) enabled a detailed description of the distribution of the M2 tidal current ellipses for a wider area of the Adriatic basin (Ursella and Gačić, 2001).

Poulain (2013) derived the characteristics of the surface tidal flow for the first time from the drifters that populated the Adriatic Sea between 1990 and 2006. The tidal ellipses of the M2, S2 and K1 constituents are given on a 0.25° × 0.25° grid over the entire Adriatic area. The author emphasized that despite all the shortcomings intrinsic to the measurements, the estimates of the surface tidal flow are generally in good qualitative agreement with the observations.

Recent numerical modelling studies on tides (Cushman-Roisin and Naimie, 2002; Janeković et al., 2003; Janeković and Kuzmić, 2005; Guarnieri et al., 2013) are based on 3-D, high-resolution models. In particular, Janeković and Kuzmić (2005) obtained satisfactory results,

as demonstrated from the comparison of numerical data with the observed sea levels from several tide gauges around the coast, and with some sporadic current-meter data from several locations in the northern Adriatic Sea. Guarnieri et al. (2013) established the importance of the tidal influence on the mean circulation, mixing, and consequently on the distribution of water masses in the basin. Particularly in the regions of freshwater influence (like south of the Po River delta) near-bottom transports of temperature and salinity in both along- and across-shore direction, are much better reproduced with tides included.

On the other hand, studies based on detailed observations of tidal currents across the Strait of Otranto have not been made before. From some sporadic and short-lasting measurements that were conducted in the eighties and nineties in that area (i.e., Michelato and Kovačević, 1991; Leder et al., 1992; Ferentinos and Castanos, 1988) it turned out that the flow has prominent diurnal and inertial fluctuations. From the current measurements in the adjacent southern Adriatic, Vilibić et al. (2010) found evidence of fortnightly oscillations of the flow over the shelf break in phase with Mf tide (period 13.66 days).

In numerical studies the strait has been considered either as an open boundary of the Adriatic Sea (i.e, Janeković and Kuzmić, 2005) or as a part of the wider area. A tidal model of the barotropic M2 for the whole Mediterranean was developed by Lozano and Candela (1995). While the results agreed well over most of the Adriatic with the findings of Ursella and Gačić (2001), a full comparison for the Strait of Otranto was not possible because of the lack of measurements.

The long-term current time-series collected in the Strait of Otranto intermittently from the 1990s until 2007 form a unique data set for studying the flow characteristics in the area, primarily as the exchange of water and biogeochemical properties between the Adriatic and Ionian Sea. In the past studies the emphasis was put on the low frequency (with periods of about a week), and seasonal flow and transport variability (Gačić et al., 1996; Vetrano et al., 1999; Kovačević et al., 1999; Ursella et al., 2011, 2012; Yari et al., 2012) or on the inter-annual variability related to the processes of the dense water formation (Manca et al., 2002). The interest in the transport rate and its seasonal and inter-annual variability did not include a thorough and exhaustive analysis of the flow in the diurnal and semi-diurnal tidal frequency bands. So, we addressed here these latter topics, and the phenomenon of baroclinic diurnal tide intensification in the deep layers over the western shelf edge.

A number of studies on baroclinic internal tide generation in the presence of abrupt topographic features are found in the literature (Beckenbach and Terril, 2008, and references therein). In particular, it has been found that interaction between barotropic tidal currents and topography generates a baroclinic internal tide; the diurnal one resembles the first baroclinic mode, which is topographically distorted, amplified and polarized in the near-bottom layer. Such effects were not observed in the semi-diurnal response.

However, in the study of the Adriatic tides some non-tidal phenomena should be taken into account because their typical frequencies are close to the tidal ones. In particular, in the Adriatic, there are two basin-wide resonant modes of sea-level oscillations, seiches, at the periods around 22 and 11 h with maximum for the amplitude in the sea level at the northern end of the Adriatic Sea (Cushman-Roisin et al., 2001). Seiches are an important response to atmospheric forcing in the Adriatic (Cerovečki et al., 1997; Leder and Orlić, 2004). The longest seiche period (21–22 h) is very close to the daily tidal constituents (such as OO1 and UPS1). As they are free modes of oscillation of the basin, they are theoretically present all throughout the year (Cushman-Roisin et al., 2001), but the probability that important events happen during summer at 22 h is really low. In fact, experimental evidence of strong events has been found almost exclusively during the late autumn/winter period of the year (from October to April) due to the presence of low-air-pressure/sirocco-wind events (Cerovečki et al., 1997; Comune di Venezia, 2013). Other important features present in the energy spectra are the inertial oscillations, whose period for the Adriatic is around 17–18 h (therefore, in between the two tidal bands) and the sea-breeze events whose period of variability is 24 h (Klaić et al., 2009). The energy of all these oscillations, if their frequencies are not properly resolved, can leak into the tidal range and create misleading interpretations of the phenomena.

This paper is structured as follows: data and methods are presented in Sect. 2, description of the spatial and seasonal variability of the tidal signal is given in Sect. 3, detailed considerations of the bottom layer characteristics are exposed in Sect. 4, intensification of the diurnal tide at the western shelf-break and slope is discussed in Sect. 5 and, finally, Sect. 6 contains concluding remarks.

2 Data and methods

2.1 Mooring arrangement and deployment

We used data collected from the mooring arrays in the Strait of Otranto (Fig. 1) during three long-term observations. The first set of observations from February 1994 to November 1995 is common to the Mediterranean Targeted Project (MTP) – OTRANTO and to the Otranto Gap EXperiment (OGEX), indicated as OTRANTO/OGEX projects (Fig. 2a). The second set of observations from March 1997 to July 1999 (Fig. 2b) comes from the MTP II-MATER (MAss Transfer and Ecosystem Response) project. The third one, from November 2006 to April 2007 (Fig. 2c), is obtained in the framework of the VECTOR (VulnErabilità delle Coste e degli ecosistemi marini italiani ai cambiamenti climaTici e loro ruolO nei cicli del caRbonio mediterraneo) project.

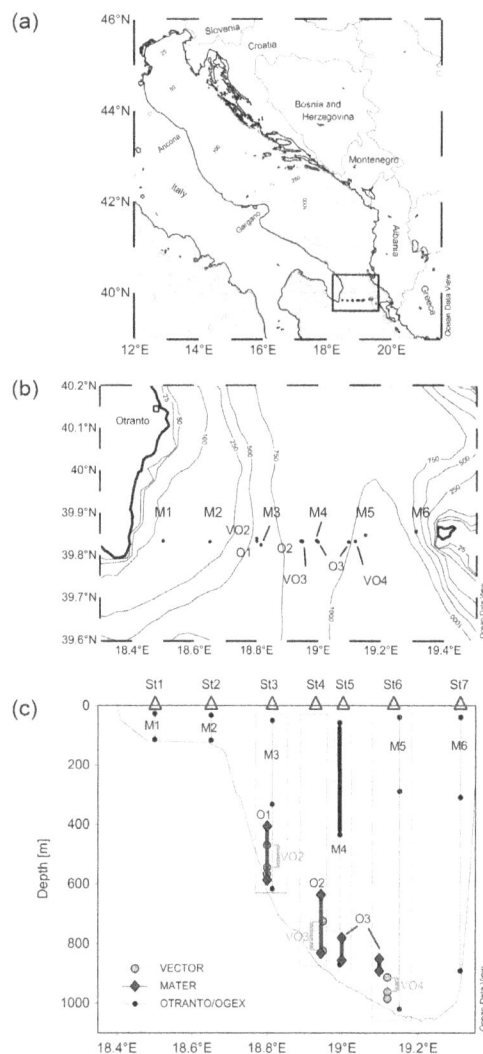

Fig. 1. (a) Study area in the Strait of Otranto at the southern end of the Adriatic Sea, depicted by a rectangle and expanded in (b). (b) Mooring locations with original station nomenclature. Sea level and wind are available at the coast in Otranto. (c) Vertical scheme of the mooring lines: thick lines indicate layers covered by ADCPs. Both the original station nomenclature and the one adopted in this paper (St1, ..., St7) are indicated. St3, St4, St5 and St6 enclose 2 or 3 moorings within dashed-line rectangles. Current measurements were conducted within the framework of different projects (see legend) during the time interval 1994–2007. Depth contours in (a) and (b) are in metres.

Some findings from the OTRANTO/OGEX projects, regarding the current data set used in this study, were reported by Kovačević et al. (1999). Briefly, currents were measured at six stations: M1, M2, M3, M4, M5 and M6 (Fig. 1). The stations were positioned along an east–west transect in the southernmost part of the Adriatic Sea at 39°50′ N. Typically, the currents were measured in three layers, the surface layer (28–42 m depth interval), the intermediate layer (300–330 m depth) and near-bottom layer, a few tens of meters above the seabed. There were some gaps in the data set due to instrument and battery malfunctioning. The moorings at locations M1 and M2 on the western side included two current meters, while moorings in the central and western parts at locations M3, M5 and M6 included three current meters (Fig. 1). Four types of self-recording current meters were used on the moorings. The NBA-DNC-2B with rotor and EG&G SMART acoustic current meters recorded a mean speed and instantaneous direction every 20 and 10 min, respectively. The Aanderaa RCM7 current meter recorded an hourly current vector averaged from 50 samples taken every 72 s. At M4 only, apart from the bottom current meter, an RDI ADCP operating at 75 KHz was deployed at about 430 m below the sea surface, looking upward to measure the currents in the water column in 8 m bins. The accuracy of the aforementioned instruments for measuring the speed and direction are $1 \, \text{cm} \, \text{s}^{-1}$ and 3–7.5 degrees, respectively. Because of the high vertical correlation of the ADCP current data, the time series from just the uppermost and the lowermost cells (near-surface layer and mid-depth) were considered representative of the upper and intermediate layers, respectively. All the moorings were recovered almost every 3 months and redeployed after maintenance. The available time series are indicated in Fig. 2a.

MATER moorings were equipped with RDI BB ADCPs that recorded the deep currents above the seabed, within about 160 m at stations O1 and O2 and about 50 m at O3, with a vertical resolution of 5 m and a sampling rate of 30 min. MATER moorings were deployed five times and each data series is about 6 months long (Fig. 2b).

VECTOR moorings were equipped with: (i) an Aanderaa current-meter, 17 m above the seabed, whose sampling rate was 30 min; (ii) a conductivity–temperature SBE37 CT probe 2 m above the current meter, with a sampling rate of 15 min; and (iii) an upward-looking RDI ADCP 10 m above the CT, with a sampling interval of 15 min. The ADCP measurements covered a layer of about 100 m at station V3 and of about 80 m at V2 and V4.

A detailed description of the mooring arrangement and schedule during the MATER and VECTOR projects can be found in Ursella et al. (2012) and references therein. The VECTOR stations V2, V3 and V4 correspond to O1, O2 and O3 in the MATER project, and stations M3 and M4 of the OTRANTO/OGEX projects roughly coincide with V2 and V3, respectively (Fig. 1). Henceforth, a unique nomenclature is introduced: St1 for station M1, St2 for station M2, St3 for stations M3, O1 and V2, St4 for O2 and V3, St5 for M4 and O3 for the first measurement phase, St6 for M5, O3 and V4 and, finally, St7 for M6.

2.2 Data processing

For all the analysis considered herein, each ADCP cell was considered as an independent time series and each measurement period separately analysed. ADCP data and current-meter series were treated in the same way, using the same procedure in data analysis. Quality control was carried out, removing spikes and bad data, while the missing data within the gaps lasting for less than 6 h were linearly interpolated.

Three periods were chosen for the OTRANTO/OGEX projects: PR1 (51 days), from 3 December, 1994 to 25 January, 1995; PR2 (97 days), from 20 May to 26 August, 1995; and PR3 (83 days), from 27 August to 18 November, 1995 (Fig. 2a). These three periods were considered roughly as winter, summer and autumn seasons. They were selected for the sake of the best possible time and space coverage along the Otranto section.

For the other two projects, MATER and VECTOR, only bottom-layer data were available. The periods for the analysis differ from the OTRANTO/OGEX ones, and they are simply the periods of each deployment (Fig. 2 b, c). It must be kept in mind that during the first phase of MATER (A), mooring O3 was positioned at station St5 while during the other measurement phases it was deployed at station St6 and was therefore deeper.

On the edited and interpolated data, rotary spectra were calculated for each instrument depth and period. The following different window lengths were applied: 512 points (21 days) for OTRANTO/OGEX hourly data and 1024 and 2048 points for raw, originally sampled MATER and VECTOR ADCP data, respectively. The overlapping period always corresponded to the window half-length. The length of the window in all cases permitted to resolve tidal signal from seiches and inertial oscillations. Harmonic analysis (Foreman, 1978) was applied in order to obtain tidal constituents. Harmonic analysis calculations were done using the Matlab programme t_tide (Pawlowicz et al., 2002), which automatically selects the tidal constituents resolvable with the data set used. Moreover, nodal corrections were taken into account in our analyses in order to properly consider the modulation and therefore intensification of tidal constituents. The method also enables calculation of the percentage of the total variance accounted for by the resolved tidal signal.

For the purposes of the analysis of Sect. 5 only, the periods of the OTRANTO/OGEX project were selected in a different way with respect to PR1, PR2 and PR3.

3 Spatial and seasonal variability of the tidal signal over the entire Otranto section

The data from the OTRANTO/OGEX projects were the only ones that covered the entire Otranto section and were therefore used to study the vertical and horizontal variability of the tidal motion.

As already indicated, the rotary spectral analysis and harmonic analysis of the OTRANTO/OGEX data set were ultimately limited to the three time intervals PR1, PR2 and PR3. The results address the most salient characteristics in the upper, intermediate and bottom layers.

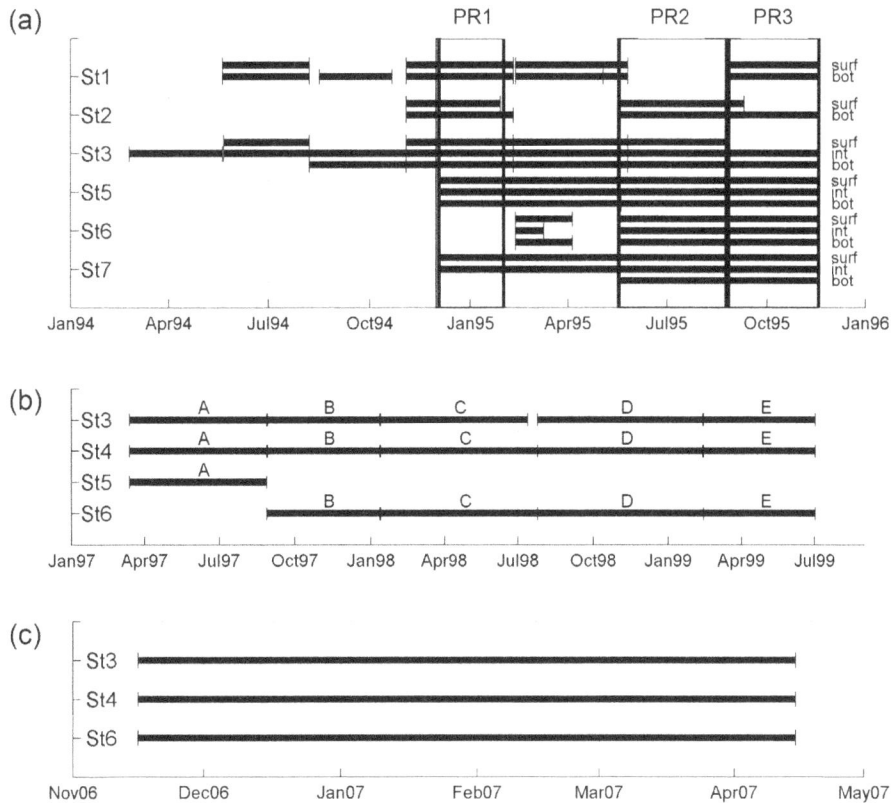

Fig. 2. Time diagram of the available current meter data within (**a**) the OTRANTO/OGEX projects. Shaded areas correspond to time intervals for which rotary spectral analysis was done (PR1, PR2, and PR3) (**b**) MATER project. Time intervals A, B, C, D and E are indicated. (**c**) VECTOR project.

3.1 Variance and spectral analysis

Figure 3 summarizes the variance in the three layers within each period at different moorings. The tidal signal is composed of all the constituents resolved by the harmonic analysis. In this way, owing to the finite length of the time series, tidal variance can be slightly overestimated (by about 3–4 %), due to the fact that some non-tidal phenomena are close in frequency to some tidal ones and part of their variance can enter in the tidal variance (low-pass signal versus MM/MFS or 22 h seiche versus UPS1/OO1). This must be kept in mind when interpreting the results. The percentage of the predicted variance to the total one is also depicted. The percentage was always below 40 %, apart from the stratified period PR2: at St2, it reached about 60 % at the top and 77 % at the bottom, and at St7-bottom the value was 47 %.

The upper-layer tidal analysis was based on current measurements at depths varying between 28 and 42 m below the sea surface. Moreover, the upper current meter at St2 during period PR2 was positioned at 56 m depth. The western flank of the strait (station St1) was characterized by the largest total variance (Fig. 3) and by the largest tidal variance as well. The central portion of the strait had total and tidal variances lower than at both flanks. The only comparison available for the nearby locations over the western shelf (St1 and St2) was for the period PR1, during which the total and tidal variances decreased rapidly in the offshore direction. A significant fluctuation in the variance in time and space is due to the varying energy levels of both the tidal and the non-tidal bands. The spectral characteristics of the upper layer show that the flow variance was distributed principally among the tidal oscillations (diurnal and semi-diurnal), inertial oscillations (0.053454 cph–18.7 h period) and a long-period motion (Fig. 4). The diurnal tidal band was more energetic than the semi-diurnal one. Moreover, at the shelf break (station St2) during period PR2 the diurnal one was the principal peak in the spectrum, with particularly high energy at negative frequencies (indicating a clockwise rotation of the tidal vector). Inertial oscillations, indicated by a dominant peak at a negative frequency, were more energetic than the diurnal one, except during the stratified (PR2) period at the shelf break (station St2). During PR1 and PR3 the low-frequency band was predominant and it was more energetic at the flanks, both eastern (station St7) and western (station St1), than in the middle of the strait (Fig. 4).

The measurements in the intermediate layer were carried out only in the deepest portion of the section (stations from St3 to St7). The total and tidal variances (Fig. 3) showed

(a)

(b)

(c)

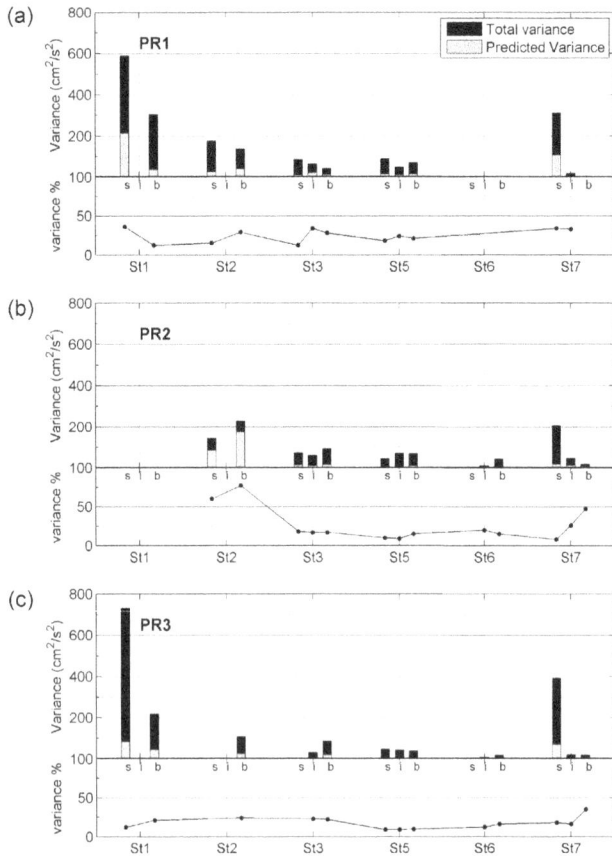

Fig. 3. Total variance (black bars), predicted variance (grey bars), and the contribution of predicted to total (%; continuous line) due to all resolved tidal constituents along the Otranto section, from OTRANTO/OGEX data for the three time intervals PR1 (**a**), PR2 (**b**) and PR3 (**c**). Symbols s, i, and b stand for surface, intermediate and bottom layers, respectively.

Fig. 4. Rotary spectral analysis in the upper layer along the Otranto section for the three time intervals PR1, PR2, and PR3 from the OTRANTO/OGEX project. The 95 % confidence level is indicated. f is the Coriolis parameter divided by 2π.

extremely low energetics at these depths. The spectral characteristics (not shown) indicate that low-frequency motion was the most energetic. Inertial oscillations were attenuated with respect to the upper layer (Fig. 4), and displayed values similar to those of the semi-diurnal and diurnal tides.

In the bottom layer, total variance (Fig. 3) was larger over the continental shelf (station St1), gradually diminishing in the offshore direction. The tidal contribution also diminished from the shelf toward the eastern flank. Tidal variance was exceptionally high over the shelf edge (station St2) during the stratified PR2 time interval. In general, the low-frequency motion (Fig. 5) was the most energetic; however, an exception to this was observed at station St2 during all the three periods, when the diurnal signal was higher than the low pass or even dominated. In addition, during PR2 the semi-diurnal band was as energetic as the low-frequency one. As for the upper layer, the sense of rotation was clockwise. In the central deep portion of the strait (stations St5 and St6), the low frequency dominated over the tidal and inertial ones.

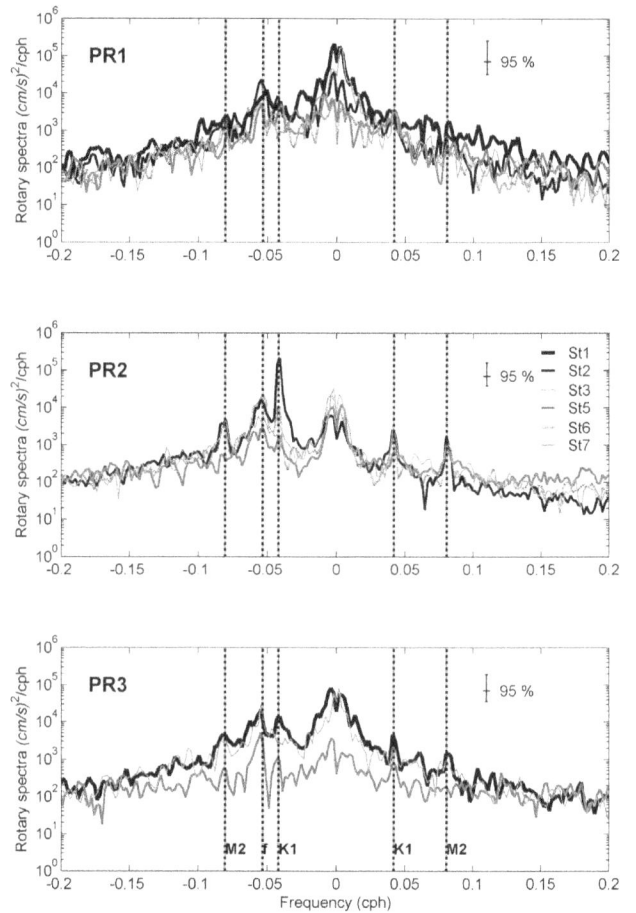

In conclusion, the total variance was maximum at the westernmost two stations and at the easternmost surface one. Predicted tidal variance behaved similarly to the total variance. A behaviour was different during the period PR2 at St2 (near the shelf break), when the tidal variance was large, and when its contribution to the total variance was maximum (77 %).

3.2 Harmonic analysis

The temporal and spatial characteristics of the tidal signal were studied by means of harmonic analysis of the OTRANTO/OGEX data sets. Three principal bands were distinguished: long-term, diurnal and semi-diurnal. Within each band it is not possible to distinguish among all the constituents, because of the limited duration of the three periods; therefore the energy of the unresolved constituents may leak into the variance of the nearest resolved ones. This in particular concerns the distinction between K1 and P1, and between K2 and S2.

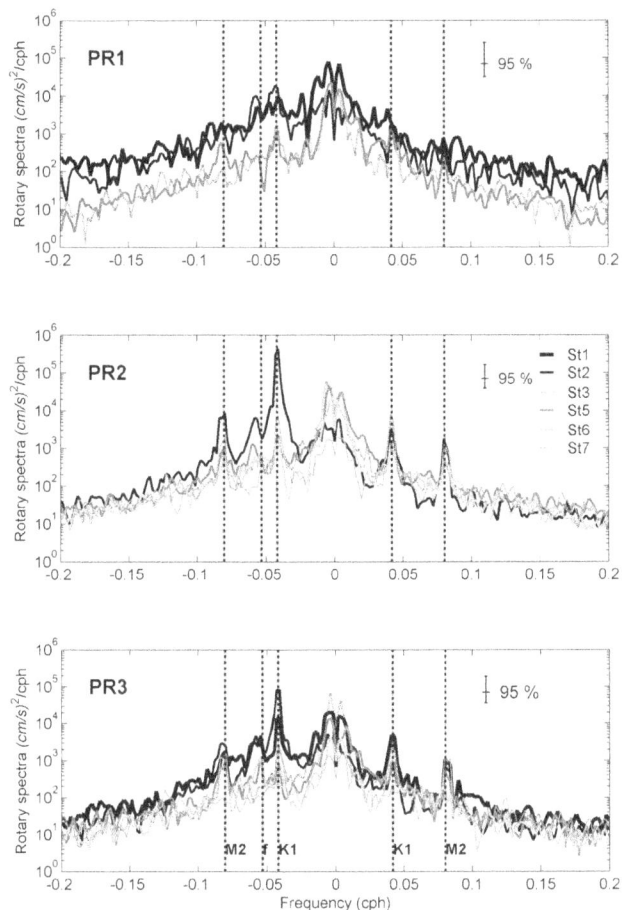

Fig. 5. Rotary spectral analysis in the near-bottom layer along the Otranto section for the three time intervals PR1, PR2, and PR3 from OTRANTO/OGEX project. The 95 % confidence level is indicated. f is the Coriolis parameter divided by 2π.

Tidal constituents MM and MFS, with frequencies of 0.0015 and 0.0028 cph, respectively (that is, with periods of 27.78 and 14.88 days), are grouped into the long-term tidal band. Their amplitudes were highly variable in time and space (not shown), probably due to the fact that with harmonic analysis it is not always possible to distinguish between the astronomical tidal forcing and the non-tidal low-frequency meteorological signal with periods of about 10 days or more. The largest amplitudes of this signal were encountered along the western and eastern flanks, while the smallest were in the central part, resembling the pattern of the entire non-tidal variance (Fig. 3).

Figure 6 summarizes the semi-major-axis amplitudes of some of the diurnal and semi-diurnal tidal ellipses. It shows the most relevant constituents for the Adriatic (Cushman-Roisin et al., 2001), except P1 and K2 that are not resolved. There are few other constituents with amplitudes in some cases comparable to that of the principal ones (i.e., OO1, UPS1, and L2). Concerning these last constituents, it must be kept in mind that their frequency is very close to the 22 h

seiche. The constituents that showed the largest value of the tidal ellipse semi-major axis, namely K1 (0.04178 cyc h^{-1}) and M2 (0.08051 cyc h^{-1}), were taken as representative of the corresponding band. Amplitude of the K1 constituent was always below 6 cm s^{-1} except at station St2, where it reached 10 and 15 cm s^{-1} in the upper and bottom records, respectively, during the stratified period PR2. Amplitudes of the M2 constituent were always below 3 cm s^{-1}. The K1 semi-major axes of tidal ellipses were almost twice those of M2 and were thus the main tidal signal in the flow. This kind of relationship between K1 and M2 is opposite to the one typical in the northern Adriatic Basin, where M2 prevails over K1, as observed also by Poulain (2013). The M2 tidal constituent had somewhat larger amplitudes along the flanks and smaller in the centre of the section. It did not show much change from PR1 to PR3. There was, in contrast and as already mentioned, an anomalous feature of the diurnal constituent K1: an amplification at the shelf break (station St2) just encountered in the OTRANTO/OGEX summer time series (PR2 period). Unfortunately, due to the lack of measurements on the shelf, no confirmation of this characteristic was encountered either during other periods of the OTRANTO/OGEX or during the MATER and VECTOR projects. The intensification at the shelf break is described and discussed in detail in Sect. 5.

4 Detailed look into the bottom layer in the deep portion of the strait

The data set from the MATER and VECTOR projects permits accurate description of the tides in the bottom layer. Rotary spectra (Fig. 7) show that inertial, diurnal and semi-diurnal bands were always present at all stations and during all periods. The energy of these peaks was lower at station St4 during the last MATER measurement phase (E). In particular, the diurnal constituent was always more energetic at location St3 than at other locations, where its energy could be almost as high as the energy of the low-frequency band. This characteristic may indicate an amplification of the diurnal tidal signal near the shelf break (see the position of St3 in Fig. 1). Moreover, no significant differences in the spectra were found between upper and lower cells at any current meter (not shown).

Total and tidal (predicted) variances are depicted in Fig. 8 as a function of depth for MATER and VECTOR data. In general the peak total variance was found at the central station St4 (see also Ursella et al., 2011), where it decreased toward the bottom. At the other two lateral moorings (St3 and St5/St6) it remained constant or increased while approaching the bottom. Tidal predicted variance was always less than 30 % of the total variance, and it was usually highest at St3. Its contribution to the total variance at this mooring was generally larger than 15 %, except for the period E when the non-tidal, and hence the total, variance was very high. This again shows that near the shelf break the tidal signal

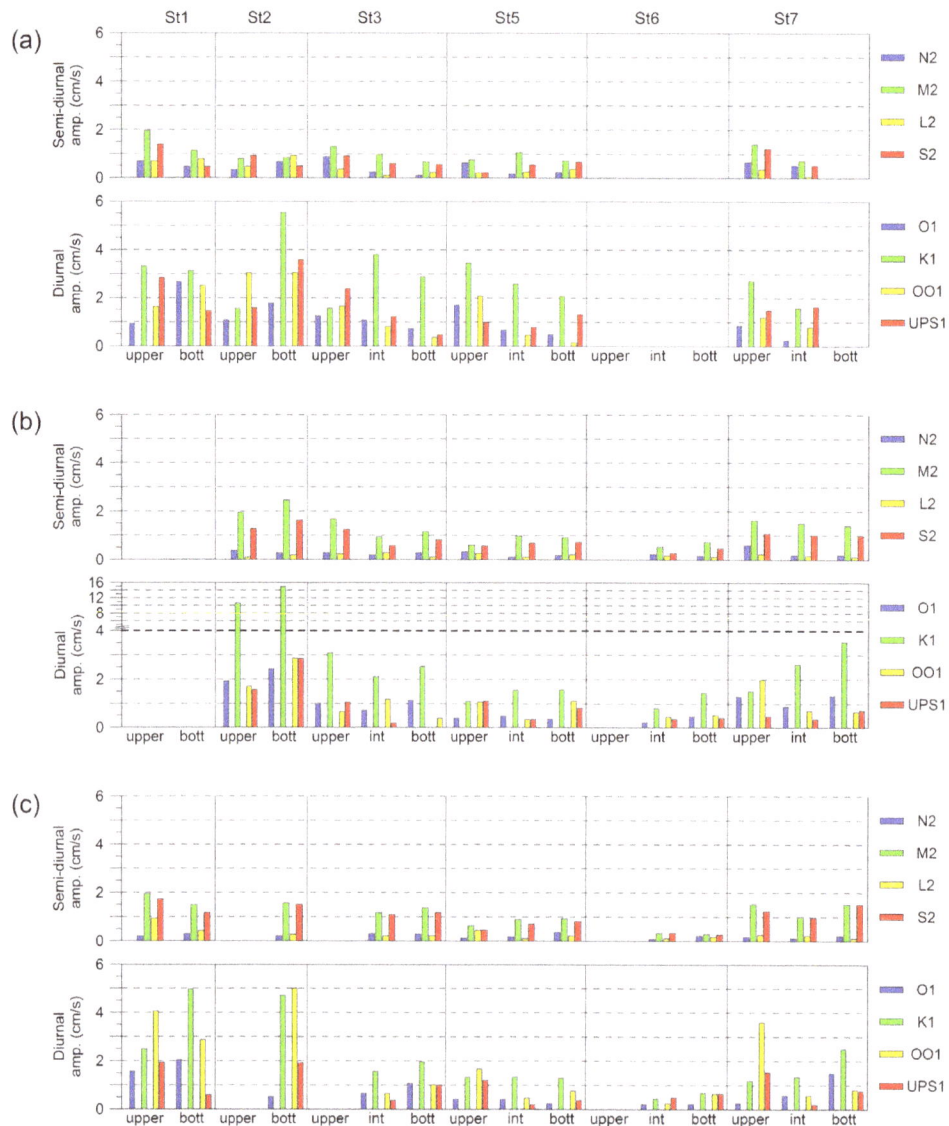

Fig. 6. Semi-major axes of the tidal ellipses for the diurnal and semi-diurnal tidal constituents, obtained from the harmonic analysis applied to the periods PR1 (**a**), PR2 (**b**) and PR3 (**c**) of the OTRANTO/OGEX projects. Upper, intermediate (int) and bottom (bott) layers are indicated. Please note the change of the amplitude scale for the diurnal plot in (**b**).

was more important than in the deep central part of the strait. The lowest tidal contribution to the total variance was found at St4, indicating a variability in the current field due principally to non-tidal phenomena. Finally, the lowest percentages at the three stations were found during period E, when the non-tidal variability was rather strong.

Tidal semi-major axes for the two main constituents, K1 and M2, are depicted in Fig. 9 together with the orientation of the tidal ellipses. The K1 amplitude was always larger than M2 and it assumed the greatest values at location St3, reaching 3 to 5 cm s^{-1} during all the periods except phase D when it was about 2 cm s^{-1}. The K1 amplitudes slightly increased with depth at location St3, decreased at St4 and were constant or increased at St5 and St6. In contrast, the M2 amplitudes

were almost constant over the entire measurement interval and equal to about 1 cm s^{-1}. Hence, the amplification of the tidal signal near the shelf break concerns only the diurnal K1 tidal constituent and not the semi-diurnal (M2) one. Semi-minor axes are not plotted in the figure, but they were always smaller than 0.7 cm s^{-1} and 0.3 cm s^{-1} for K1 and M2, respectively. Such a large ratio between the semi-major and the semi-minor axes means that the tidal flow was almost rectilinear. The orientation of the semi-major axis was parallel to the isobaths, and it was roughly 60° at St3 and 80° at the other two stations, anticlockwise from east (trigonometric system).

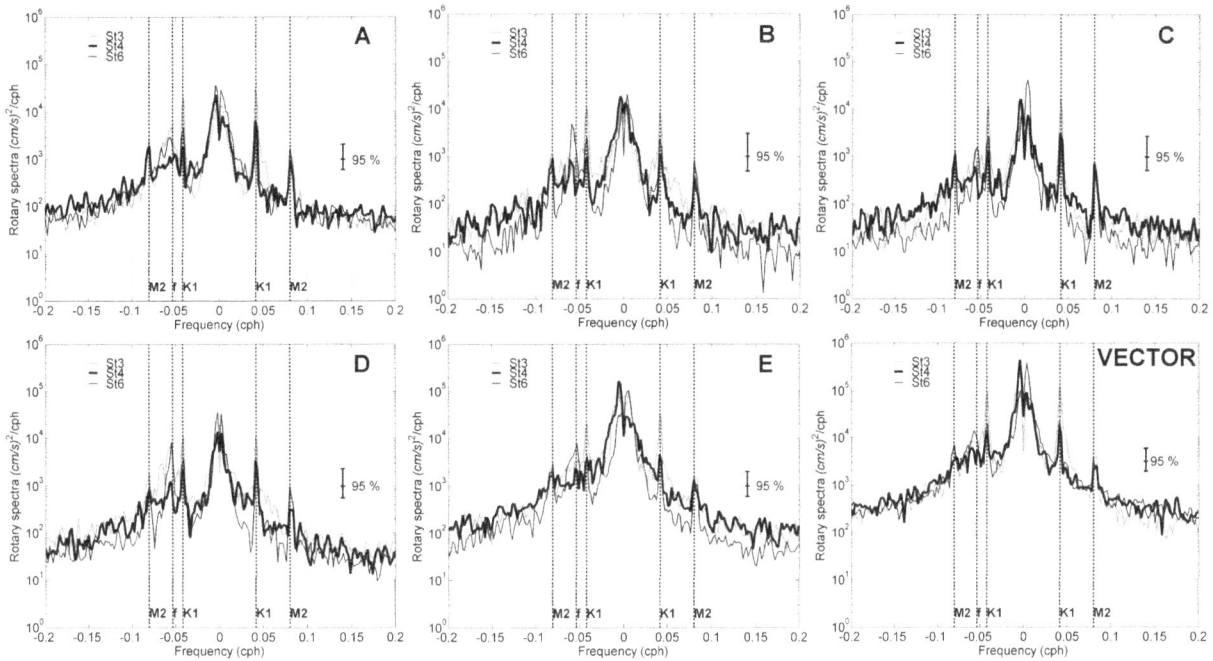

Fig. 7. Rotary spectra for the deepest ADCP cell during five MATER periods (A, B, C, D and E) and VECTOR. Inertial (f) and principal tidal frequencies (M2 and K1) are indicated by dashed lines. f is the Coriolis parameter divided by 2π.

5 Intensification of diurnal tide at the shelf-break and slope

As described in the previous sections, the diurnal K1 tide at the shelf break (station St2) had larger amplitude during the summer-like period PR2 (when the water column is stratified) compared to other sites. In addition, a weaker intensification just down-slope was observed at station St3 during the MATER and VECTOR periods. It seems that there was no similar manifestation at the inshore station St1 over the shelf. However, because of the measurement gaps it was not possible to compare the flow characteristics at the two locations, St1 and St2, for the same time period, but only for two periods similar as far as the density stratification is concerned, i.e. summers 1994 and 1995, as shown in Fig. 10. If we suppose that the circulation at location St1 during summer 1994 is representative of the stratified conditions in general, it is evident from Fig. 11 that during summer the diurnal signal predominated only at location St2 (shelf edge). The cross-shelf u component was of the same sign in the upper and in the bottom layers (Fig. 11a), even though the magnitude was different (larger at the bottom); the findings for the along-shelf v component were similar (Fig. 11b). The current components at location St1 were very different and occasionally of the opposite sign between the upper (depth 30 m) and the bottom layers (Fig. 11d, e). In addition, temperature measured at the current-meters at St2 has been considered (Fig. 11c). In the bottom, oscillations with relatively large amplitudes are evident in mid- and late July, and in

late August 1995; these oscillations have a diurnal period, as specified by means of the Continuous Wavelet Transform spectra (not shown). The Wavelet Transform performs a decomposition of the variability in time and scales, determining the time variability of the spectra at each scale. The same signal is not as evident in the upper series.

In the follow-up, we discuss the possible mechanisms responsible for the observed diurnal intensification near the shelf edge during summer.

5.1 Possible generation mechanisms

With the scope to examine if there is any time dependence on a long-term scale within our data, 30-day-long subsets shifted by 7 days were created and harmonic analysis was applied to each of them. In this way, we obtained a time series of the tidal ellipse parameters for the diurnal and semi-diurnal components. From this type of analysis, it emerged (not shown here) that for the MATER data set at St3 (the most complete in time coverage) and also for the bottom OTRANTO/OGEX data set at St3 a slight intensification of the diurnal signal was found twice a year: in June–July and in December–January. The amplitudes reached almost $6\,\mathrm{cm\,s^{-1}}$ for the MATER series and almost $5\,\mathrm{cm\,s^{-1}}$ for the Otranto one. This weak intensification (about $1\,\mathrm{cm\,s^{-1}}$) was not significant, and should occur because the P1 tidal constituent, which remained unresolved with a 30-day-long time series, beats with the K1 constituent at a 6-month period.

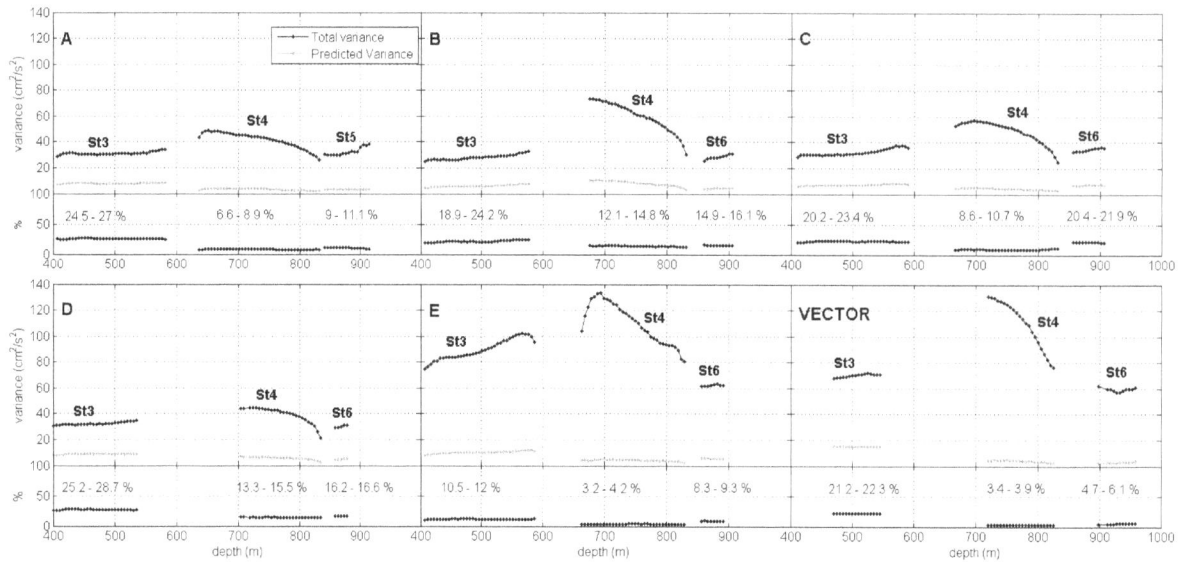

Fig. 8. Total variance (black dots), predicted variance (grey dots) and its contribution to the total one (%) as a function of depth. The predicted variance was calculated considering all the resolved tidal constituents (35) along the Otranto section, from MATER (periods A, B, C, D and E) and VECTOR data. Note that during period A, to avoid overlapping, the variances at St5 are plotted adding 60 m to the true depth.

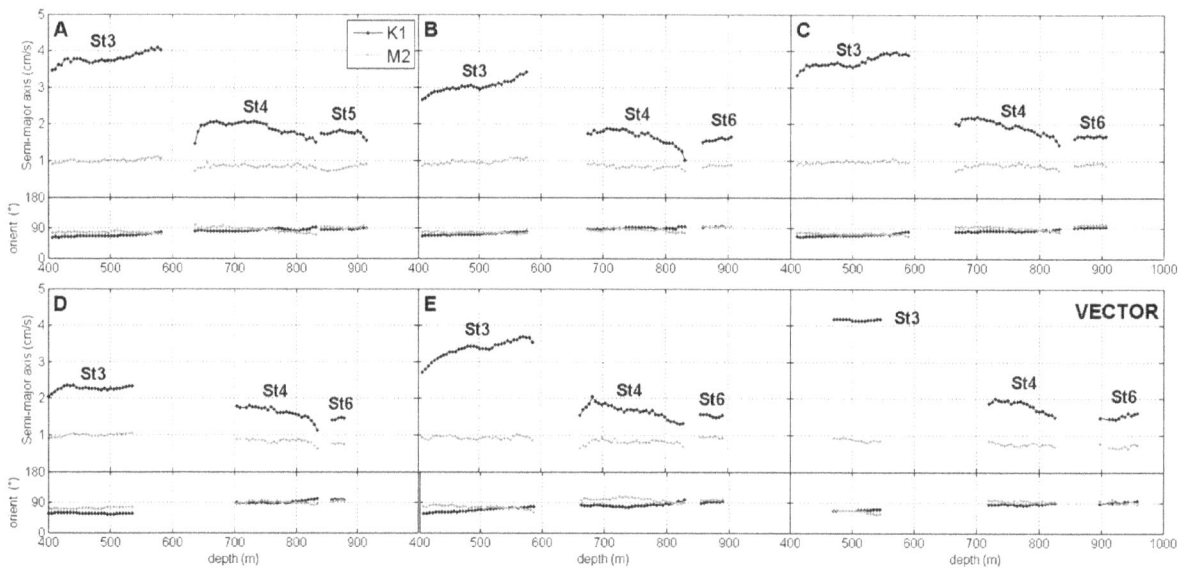

Fig. 9. Semi-major axes for K1 (black dots) and M2 (grey dots) with tidal ellipse orientation angle (in degrees anticlockwise from East) as a function of depth. The semi-axes were calculated with harmonic analysis for the bottom current meters along the Otranto section, from MATER (periods A, B, C, D and E) and VECTOR data. Note that during period A, to avoid overlapping, the semi-major axes at St5 are plotted adding 60 m to the true depth.

On the other hand, the amplitudes of the semi-major axis at the bottom of location St2 varied between $15\,cm\,s^{-1}$ and $20\,cm\,s^{-1}$ until the end of July 1995. From the beginning of August on, they started diminishing. The corresponding semi-minor axis amplitudes were also large, between $10\,cm\,s^{-1}$ and $20\,cm\,s^{-1}$, indicating a circular motion of the diurnal tide at this location. From the available data we saw that this intensification, however, appeared just once a year. Therefore, the different behaviour at the two stations (St2 and St3) should be of different origin or at least a different manifestation of the same phenomena. Thus, for better understanding the origin of the intensification, and in order to exclude possible interferences of unresolved tidal constituents, the mentioned calculations were repeated and

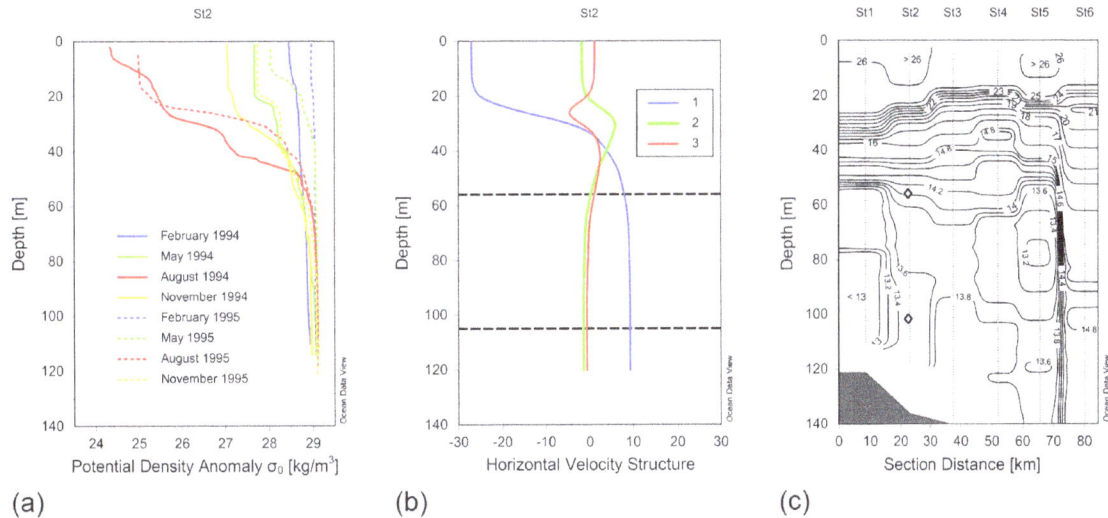

Fig. 10. (a) Potential density anomaly from the CTD surveys conducted in close proximity of St2 during OTRANTO/OGEX projects. **(b)** Vertical distribution of the first three horizontal velocity modes at station St2 calculated from the August 1995 CTD profile. Current-meter depths at St2 during the PR2 time period are indicated by horizontal dashed lines. **(c)** Vertical distribution of the potential temperature in the upper layer of the Otranto section on 27 August 1995.

further analyses were performed on the longest available time series for each current-meter. In particular, harmonic analysis performed at the bottom series at St2 from mid-May until mid-November 1995 (184 days long) permits a resolution between close frequencies of the diurnal tidal band (ALP1, 2Q1, Q1, O1, TAU1, BET1, NO1, P1, K1, PHI1, J1, SO1, OO1, UPS1). In this case, we still find an important intensification for K1 ($9.98\,\mathrm{cm\,s^{-1}}$). The intensification, however, is not as high as in the preceding May–August 1995 time interval ($14.98\,\mathrm{cm\,s^{-1}}$). This fact we explain with the inclusion of the autumn period into analysis, when no intensification was found ($4.71\,\mathrm{cm\,s^{-1}}$), and with the presence of the P1 constituent that was not resolved during PR2. As a conclusion, diurnal intensification cannot be attributed to the semi-annual beating provoked by the unresolved P1 tidal constituent with respect to the K1.

As far as seiches and their possible contribution to the intensification are concerned, we examined the conditions for their occurrence. As pointed out by several authors (Cerovečki et al., 1997; Leder and Orlić, 2004), influential seiche episodes occur mostly during the late autumn/winter period of the year due to the presence of low air pressure/sirocco wind events. Still, Cushman-Roisin et al. (2001), summarize that fundamental mode (peak at the 21–22 h period) might be active almost the whole year, but with amplitudes higher in winter than in summer. The spectral analysis of our velocity time series showed that there was no peak at the 21–22 h period at St2, and neither at St1. Moreover, the intensification in our data varies with depth reaching the maximum near the bottom, where seiches should be weaker due to bottom friction. Near the St2 position, according to Leder and Orlić (2004), seiches might reach up to

$20\,\mathrm{cm\,s^{-1}}$, triggered by a suddenly changing scirocco (SE) wind as during winter 1989. ECMWF wind time series for the year 1995 near location St2 indicated that no significant sirocco wind episodes occurred during that summer. Moreover, there was no evidence in the northern Adriatic (where the sea level seiche amplitude is maximum) of significant seiche events that could provoke an "acqua alta" phenomenon in Venice. Namely, for the entirety of 1995 there was only one such episode, observed on 31 December 1995 (Comune di Venezia, 2013). Furthermore, we analysed the sea level in Trieste, northern Adriatic, for the year 1995, and no significant seiche amplitudes were observed during the summer. Consequently, we claim that there were no significant seiche events during summer 1995, which could explain the bottom current intensification near the edge of the shelf break (St2).

Beckenbach and Terrill (2008) studied a similar diurnal phenomenon in the Southern California Bight, and they found that the structure of the diurnal internal tide resembled the vertical shape of the first normal baroclinic mode. The good vertical spatial resolution of their ADCP measurements enabled indication of such behaviour. In our case, the upper current meter at St2 was located at 56 m depth, therefore below the pycnocline, which during summer 1995 was located between 20 and 40 m depth (Fig. 10a). As shown in Fig. 10b, this depth is below the zero-crossing of the first baroclinic mode. Hence, the values for the diurnal amplitudes at the deeper current meter are expected to be slightly greater than at the upper one, and the two levels should be in phase, which is exactly what we observed.

The phenomena described in Beckenbach and Terrill (2008) took place in a situation with an abrupt bathymetric slope on the seaward side of the ridge. Their measurement

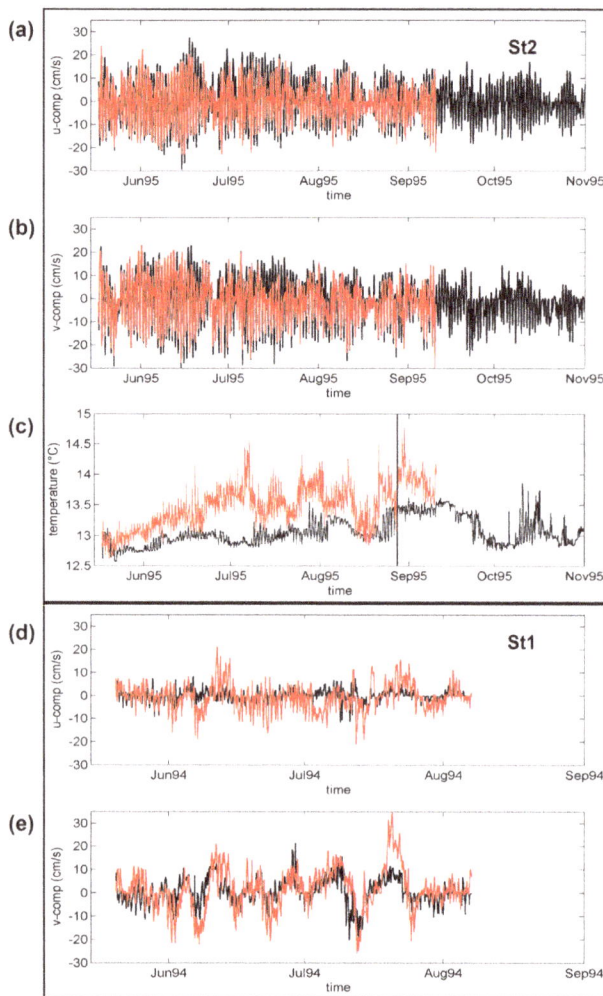

Fig. 11. u and v current components in the upper (red line) and bottom (black line) layers at station St2 (**a** and **b**) during summer 1995 and at location St1 (**d** and **e**) during summer 1994. The temperature time series at station St2 is plotted in (**c**); the time of the CTD cast at St2 referred to in Fig. 10 is indicated by a vertical black line.

site was poleward of the critical latitude for diurnal frequency internal waves, as was ours, and nevertheless currents were dominated by the diurnal signal. They explained this intensification in terms of a baroclinic internal tide generated through the interaction of a barotropic tide with topography. In these circumstances, subinertial internal waves can exist beyond the critical latitude and explain a consistent part of the variance. In particular, they observed an intensification of the across-shelf current component in accordance with the first baroclinic mode, with propagation of the signal from the shelf to the open sea. In their case, a possible effect of the sea breeze was excluded. They suggested that the origin of the regularity of the amplified diurnal signal could be found either in (a) the extension of the low limit in frequency of the internal wave spectrum when also including the horizontal component of the Coriolis parameter f, or in

(b) the presence of internal coastal-trapped waves. The first of these causes was excluded due to high stratification, that is, the Brunt-Väisälä frequency $N \gg \Omega$ (rotation frequency), while the second one was considered possible, whether of remote and/or local origin. The condition of strong stratification during summer 1995 (Fig. 10a), like in the case of Beckenbach and Terrill (2008), makes us exclude the first possible cause of diurnal intensification for our observations.

Another possible origin for the extension of the low-frequency limit could be related to a strong negative relative vorticity (Kunze, 1985); for example, a current shear of $50 \, \mathrm{cm \, s^{-1}}$ over a horizontal distance of 20 km at 40° latitude can shift the limit to 25 h. For what concerns the negative relative vorticity, it could occur on the Otranto Shelf due to either anti-cyclonic eddies propagating down the Italian coast or simply to the anti-cyclonic inshore side of a sheared slope current. This is an interesting alternative explanation but not very likely applicable at the study region, where horizontal current shears in both directions are not of such intensities (usually less than $20 \, \mathrm{cm \, s^{-1}}$ between St2 and St3 over a distance of 14 km, which extends the internal wave limit to about 20 h). On the other hand, anti-cyclonic eddies found in the deepest part of this transect (St3–St6) were estimated to have a peak azimuthal velocity around $12–15 \, \mathrm{cm \, s^{-1}}$ and a diameter of 24 km (Ursella et al., 2011), whose relative vorticity is thus not able to shift enough the longest period for internal waves (period equal to 19.7–19.9 h). This consideration led us to exclude the possibility of the low-frequency limit extension of the internal spectrum.

Orlić et al. (2011) and Mihanović et al. (2009) found that summertime stratification occasionally generates internal coastal waves that travel daily around an island in the southern Adriatic Sea, creating the conditions for resonant excitation of the diurnal frequency by sea breeze and/or diurnal tides. In particular, Orlić et al. (2011) forced the Princeton Ocean Model by real wind stress and found that an intensification of the diurnal signal, similar to that found in our data, is not possible without the topographic effect. In the Otranto area the topography is characterized by a gently increasing slope from the coast to the position of the station St2; seaward from that location the bottom slope abruptly increases, reaching its maximum value between St2 and St3 (Fig. 1).

To complete the picture of the intensification phenomena at the shelf break, temperature time series measured at current-meter locations during the OTRANTO/OGEX projects were considered. Their spectra have a weak diurnal peak that was evident almost only at station St2, predominantly near the bottom, during period PR2, and was practically absent in other periods and at the nearby stations (not shown).

In order to understand whether diurnal breeze can have some influence on the intensification, we analysed the local wind behaviour during the PR2 period. Neither onshore meteo station nor LAMI (Limited Area Model) wind data were available for the year 1995; hence, the ECMWF

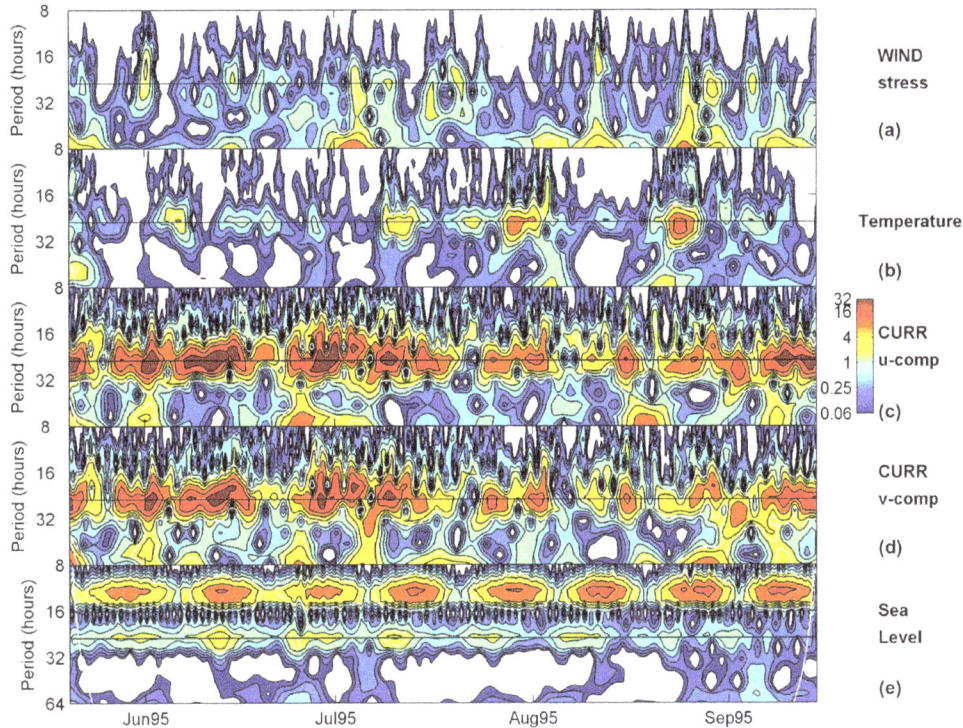

Fig. 12. Wavelet spectrum as a function of time and scale (period) for wind stress (**a**), temperature (**b**) and residual bottom current components (**c** and **d**) at St2, and sea level at Otranto (**e**) during summer 1995. Diurnal scale (24 h) is indicated by a horizontal black line.

(European Centre for Medium-range Weather Forecast) wind data were used. ECMWF data have a spatial resolution of 0.25 degrees in both latitude and longitude, and a temporal resolution of 6 h; data at the grid point nearest to the St2 location were extracted. As a test, the ECMWF wind data for year 2007 were compared with the 2007 on-shore station wind data (Otranto). Moreover, wavelet power spectra for both wind time series during 2007 were calculated and similar levels of energy were found at diurnal frequency. ECMWF winds thus proved useful in determining the importance of the diurnal signal when no other wind data are available, and for that reason, they were used for discussing the 1995 phenomena.

Interconnection among the wind, currents and sea temperature was examined by means of the Continuous Wavelet Transform. It performs a decomposition of the variability in timescales, thus enabling a comparison of their respective wavelet power spectra on a scale-by-scale basis (Torrence and Compo, 1998; Torrence and Webster, 1999). Performing a wavelet analysis of the bottom current and temperature data at St2, the wind stress and the sea level from the tide gauge located in Otranto (Rete Mareografica Nazionale, http://www.mareografico.it/) during summer 1995, interesting features emerged (Fig. 12). A strong diurnal signal with some modulation in time was observed in both current components, although it was more energetic for the u component (cross-shelf). In general there was a weaker

diurnal signal in the wind field (relatively more pronounced in the cross-shelf u component, not shown), occurring just occasionally. Temperature had a prominent but transient diurnal signal. In contrast, sea level had a strong signal at the semi-diurnal frequency (modulated at 15 days) rather than at the diurnal one; this is in accordance with the characteristics of the Adriatic tides (Hendershott and Speranza, 1971; Ursella and Gačić, 2001).

Multiple and partial coherences between wind stress, current and temperature on the diurnal scale were calculated as in Mihanović et al. (2009). That is, the diurnal scale was extracted from the Wavelet results. Multiple coherences were usually above the 95 % confidence level, often very close to 1, while partial coherences between current components and wind components showed few values above the confidence level (not shown). Finally, a wavelet spectrum of the currents and wind was calculated for the off-shore station St3 and only a weak diurnal signal was found during period PR2, while the coherence resulted in no correlation at all with wind data (not shown). This indicates that there was no wind influence on the diurnal scale of the current field.

We examined in more detail, but do not show here, the wind during summers 1994 and 1995: the sea–land breeze was present and it was more energetic during summer 1994 at St1 than during summer 1995. For this reason, any effect should be more evident in the 1994 current data. Moreover, its influence should be felt more strongly near the shore (St1) than further away (St2). However, the diurnal tide (K1 semi-major axis) at the near-shore station St1 in summer 1994 had values around 2.6–2.8 cm s^{-1}. These were much smaller than at the bottom of St2 in summer 1995 (15 cm s^{-1}) and similar to that found at the bottom of St3 in winter (4.2 cm s^{-1}). We argue therefore, that there was no evidence of the intensification of the diurnal signal during summer 1994 near the coast.

5.2 Evidences of coastal trapped internal wave

Having excluded the effect of wind in the intensification of the diurnal tidal signal, the most probable cause remains the generation of internal waves that are coastally trapped by topography and resonate at the diurnal frequency (e.g., Huthance, 1978; Thomson and Crawford, 1982). The mechanism for generating topographically trapped diurnal waves may arise from the interaction with the topography of the barotropic K1 wave that encounters the shelf edge. The reduced bottom depth at the shelf edge triggers the internal wave (as in the case of a stratified system) with frequency of the forcing K1 wave. As the diurnal frequency is below the inertial frequency for this latitude, the wave (hybrid between Kelvin and shelf wave), becomes trapped both in the vertical and in the horizontal, and "propagates" along the isobaths with the coast at its right side (Wunsch, 1975; Baines, 1986; Mihanović et al., 2006). The presence of these trapped internal diurnal waves can be proved by the investigation on the phase shift between the diurnal signal in the coastal sea level and the currents together with energy flux considerations. The harmonic analysis performed on the longest available current time series located far from the bottom boundary layer, showed that K1 tidal barotropic phases are between 326° and 350° for the stations St1–St4. The sea level phases across the strait were extracted from the co-tidal charts of Janeković et al. (2003): they range from 47° to 60° (54° at St2). The phase differences concerning the sea level and the currents are consequently between −273° and −300°. In particular, at station St2, the phase for the current is 326° and the phase difference with the sea level is −273°. Points with difference in phase close to −270° are important because they separate zones in the strait with opposite energy flux direction. Hence, we calculated the barotropic K1 tidal energy fluxes for the stations St1 to St4, as indicated in Book et al. (2009b), with the tidal ellipse values obtained from the above-mentioned harmonic analyses. The results are reported in Table 1. The sign of the Fv energy flux inverts between St2 and St3, but the most prominent characteristic is that at station St2 the energy flux becomes oriented cross-basin, indicating a cross-slope barotropic K1 tidal energy propagation. St2 seems therefore to be a critical point in the slope, a site of local generation of trapped internal diurnal waves. Wavelet spectra in Fig. 12 evidence that these internal waves were not always present, and they did not generate amplification on the on-shore side of the shelf break (for example at station St1 during summer 1994). Moreover,

stratification during summer 1995 was stronger than during summer 1994 (Fig. 10a). This indicates that the topographic slope as well as the density stratification is the necessary element for the resonance.

The diurnal spectra of u/v current component and temperature at St2, sea level and wind stress were extracted with wavelet analysis (Fig. 13). In very few events, the wind stress has a peak in correspondence of the ones for diurnal temperature. Moreover, both current components have a peak in agreement with the temperature peaks. Peaks in the diurnal u component are more pronounced (Fig. 13), indicating that the forcing of such waves should be in the same direction, i.e east–west propagation direction of the barotropic K1 tidal component. The peaks in the upper and bottom temperature diurnal spectra are often not coincident in time. This can be related to the temperature gradients: the top current meter is in a zone of vertical temperature gradient, while the bottom one is in a zone of horizontal temperature gradients, as shown from the CTD data, for the month of August 1995 (Fig. 10c). Also peaks in the diurnal coastal sea level are not always present during these events. We observe that during the time interval, diurnal temperature peaks in the upper layer (Fig. 13a) are decreasing, while those in the bottom layer (Fig. 13b) are increasing in amplitude. This, with an intensification of diurnal signal at bottom, should evidence generation by the topography of the wave trapped near the bottom along the bathymetry. Coherences were therefore calculated in order to better understand the possible cause of the diurnal peaks in the bottom temperature (Fig. 14). Here we will focus on the events characterized by the most prominent peaks in the temperature, leaving the study of the remaining events to a more detailed study foreseen in the future on the observed phenomena and their relationship with the coastal trapped waves. As already pointed out, only few events with significant partial coherence with wind are observed when considering current, temperature and wind stress, while coherence between current components and temperature is significant to a higher degree. This, together with coherence of current components with sea level (not shown) implies that the waves are locally generated in the area of the Strait of Otranto and southern Adriatic.

A 24 h-centred bandpass filter was applied to sea level and currents (Fig. 15a–b). The filtered data were zoomed for the event at the end of July 1995 and a very small phase shift is found between top and bottom current components, probably indicating a vertical component in the phase velocity of the trapped wave. A larger phase shift is found between sea level and v current component than between sea level and u current component. This implies that the across-shelf motion of the wave is limited, as it is for a trapped wave, along isobaths. Taking into account the general solution phase shift (of about 270°), we tried to evaluate if this phase shift is compatible with a wave travelling along the isobaths from Otranto station latitude, to station St2. The distance to travel is 31.5 km, in a time interval of 9.4–11.5 h (from the "phase shift difference"

Table 1. Barotropic K1 tidal energy fluxes (Fu-East and Fv-North) and parameters used for their calculation: phase (ph) and amplitude (amp) for the sea level, and phase/inclination (inc) and semi-major/semi-minor axis amplitudes of tidal ellipse for the currents.

	Period		Sea level		Current			
Station	initial date (dd/mm/yy)	duration (days)	ph (°)	amp (cm)	ph/inc (°)	maj/min (cm s^{-1})	Fu (W m^{-1})	Fv (W m^{-1})
St1 mean	03/11/94	147	60	2	347/42	2.25/−0.78	118.7	−15.5
St2 56 m	17/05/95	116	54	2	333/44	10.3/−8.6	937.8	−664.4
St3 300 m	24/02/94	455	50	2	350/66	2.88/−0.23	501.6	816.2
St5 435 m	03/12/94	350	47	2	333/78	1.45/0.02	86.5	348.9

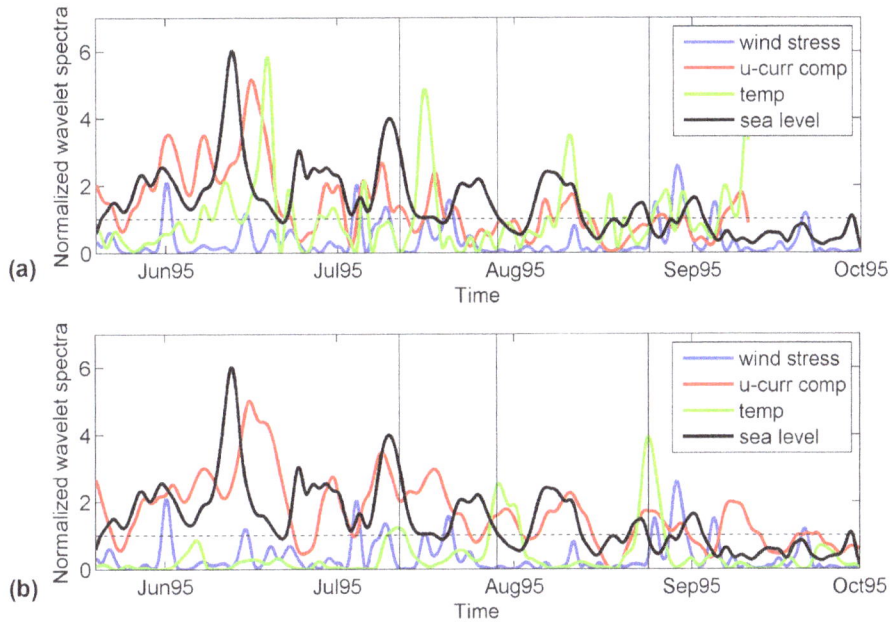

Fig. 13. Wavelet spectra of temperature, wind stress and u current component at station St2, and sea level at Otranto station, at the diurnal scale, for summer 1995. The spectra are normalized by the respective variance and significance levels. Black dashed line indicates 95 % level of confidence. (**a**) upper current-meter (56 m); (**b**) lower current-meter (105 m).

that varies in time during the length of the event): this gives a phase velocity of 0.8–0.9 m s^{-1}. If we consider, as a first approximation an internal Kelvin wave in a stratified system such as ours with density difference in the water column of 4 kg m^{-3} m^{-1}, mean density of 1029 kg m^{-3}, and $H = 30$ m (mixed layer), we obtain a velocity of 1.1 m s^{-1} that is compatible with the phase speed. Moreover, the first baroclinic mode calculated from the Brunt-Väisälä frequency profile has a velocity of 0.8 m s^{-1}. Finally, such time delays (11.5 h is almost half the diurnal period) can have the effect of superposing constructively (positive interference) with the wave generated at location St2, which could amplify the diurnal signal. Similar results are found for the event at the end of August 1995 (on 24th) for which a time interval of 9.8 h gives a phase velocity of 0.9 m s^{-1}.

At the end, we calculated the lagged cross-correlation between the sea level and the current at each of the two depths for the diurnal scale extracted by the wavelet analysis at station St2. Overall, 24 values for cross-correlation were calculated lagging sea level with respect to current by 1, 2, ..., 24 h. The two correlation functions show maxima shifted by one hour (see Fig. 15c for the u current component), with the one for the bottom preceding the upper-layer one. This indicates phase propagation from the bottom to the surface.

We looked in detail at the episode that occurred at the end of August 1995, when pronounced temperature and velocity diurnal oscillations were evident at St2 (Figs. 11, 12, 13). From Fig. 11 it emerges that the temperature oscillates between 12.9 and 13.5 °C. This range is compatible with the temperature differences observed in the bottom layer between St1 and St2 (Fig. 10c). The oscillating across-shelf (u) current, of about 15–20 cm s^{-1} in amplitude might be responsible for advecting back and forth over the bottom sensor the water of that temperature range.

As a final consideration, we try to derive what would be the consequences of such types of dynamic phenomena, relying

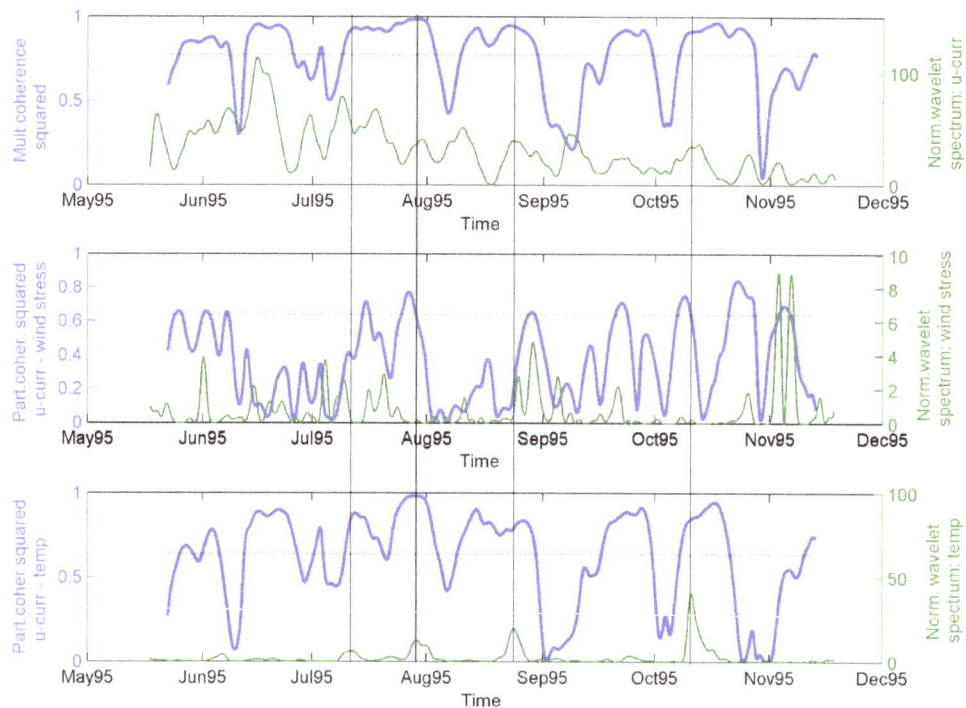

Fig. 14. Multiple and partial wavelet coherences squared (blue line) and Normalized Wavelet Spectra (green line) at St2 during summer 1995, at the diurnal scale. Multiple coherence is calculated with the u current component, the wind stress and the temperature. The 95 % confidence levels are indicated by the horizontal dashed lines.

on the analysis by Pereira et al. (2002). Their paper describes a modelling study of tidal effects at the shelf break of the Weddel Sea in the zone of critical latitude (75–80 °S, inertial period of about 12.4 h). The vicinity of the critical latitude is crucial for the resonance of the semi-diurnal tidal constituents (in their case M2–S2) and the tidal current and mixing enhancement. We are not close to the critical latitude for K1, that is 30°, and therefore the comparison between these two situations is not completely analogous, mainly because the mechanisms of the origin of the internal waves in our zone are different from those exposed in Pereira et al. (2002). They state that internal tides are expected to be generated at the shelf break because of the cross-slope barotropic velocities are strongest in that region (up to 30 cm s^{-1}, while we observe up to 15 cm s^{-1}). Besides the strength of the barotropic currents, the generation of internal tides also depends on the stratification and the steepness of the slope. These general ideas match very well to the conditions that we actually observed at the shelf break of the section in the Strait of Otranto. They estimated that, in general, the tidally induced mixing is an important phenomenon on the shelf and on the continental shelf break, and discerned its seasonal variability. In winter, the turbulent mixing involves a larger part of the water column, in summer it influences the bottom boundary layer and the pycnocline. Moreover, they report that there exist additional mechanisms, which increase friction, and, therefore mixing. Some of them are definitely present in our case,

such as diurnal continental shelf waves, internal tides, and waves trapped to the pycnocline. Although we are not able to quantify the vertical viscosity coefficients in our conditions, the effects on the mixing might be very similar. K1 tidal ellipses during summer at St2 show increase of both across-shelf and along-shelf components, especially in the bottom layer. This fact, as stated by Pereira et al. (2002), results in the more vigorous displacement of the stratified fluid up and down the slope region.

6 Conclusions

We analysed some important features of the tidal flow and its relationship with the total flow at the transect in the Strait of Otranto using all available current-meter records from the period 1994–2007. Along the strait flanks, the total variance was rather high, especially during the winter period, due to the generally strong meteorologically induced sub-tidal flow variability. Both tidal and low-frequency variances attained their minimum in the centre of the Strait of Otranto. However, in absolute terms the tidal flow, whose major contribution is represented by the diurnal (K1) constituent, reached its maximum at the channel flanks, in particular at the western continental shelf break. Moreover, the amplitude of the diurnal constituent showed annual variability over the shelf edge, reaching a maximum in the stratified season. The fact that the largest vertical amplitude of the diurnal constituent

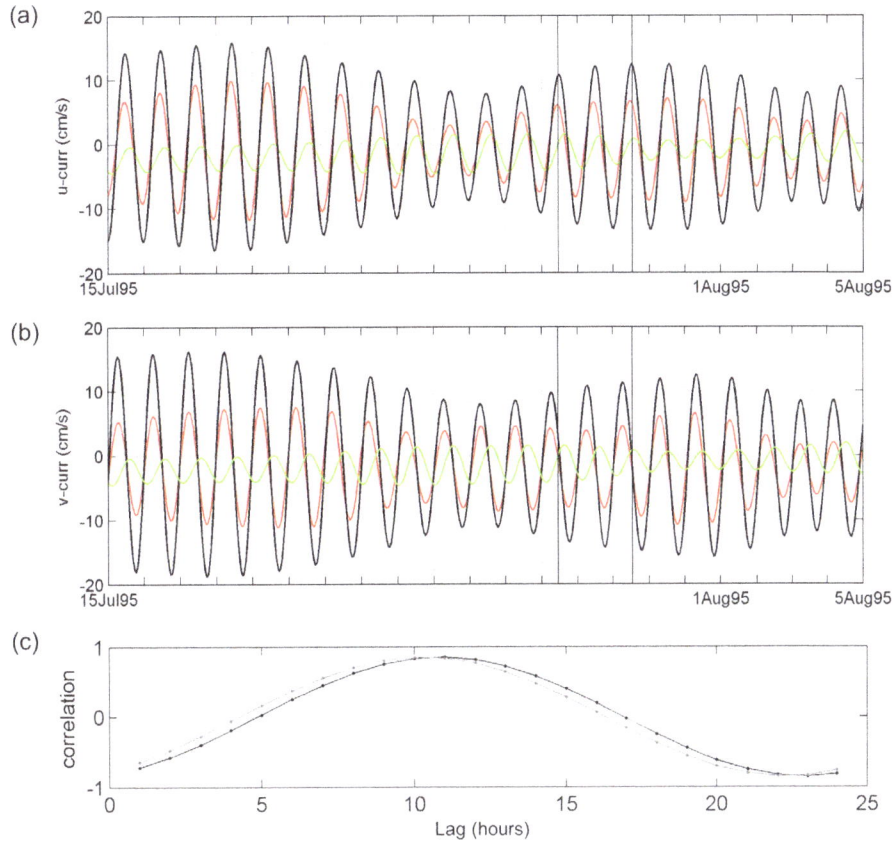

Fig. 15. 24 h-centered bandpass filtered data for u (**a**) and v (**b**) current components at the upper current-meter (red line) and bottom current-meter (black line); the 24 h-centered bandpass filtered sea level (in cm) is indicated in green. (**c**) lagged cross-correlation between the wavelet-extracted diurnal scales of the sea level and the u current component as a funtion of the lag interval; the upper layer is black and the bottom is grey.

appeared in the deepest layer and increased towards the bottom is explained in terms of the first baroclinic mode pattern. We excluded the possible sea-breeze impact on the intensification of the diurnal tidal signal and the most likely cause remains the generation of the topographically trapped internal waves resonant at diurnal period as a response to the tidal forcing and local interference. These waves (hybrid between Kelvin and shelf wave) are triggered by the reduced bottom depth at the shelf edge in condition of density stratification of the water column and have the frequency of the forcing K1 barotropic tidal signal. They become trapped both in the vertical and in the horizontal and propagate along the isobaths with the coast at their right side. These waves are only sometimes generated because they require the presence of density stratification in the vicinity of the topographic slope change. This phenomenon can stimulate both diapycnal mixing during the stratified season and sediment resuspension, and can enhance ventilation of the near-bottom layers.

Table A1. Abbreviations.

ADCP	Acoustic Doppler Current Profiler
BB	Broad Band
CT	Conductivity-Temperature
CTD	Conductivity-Temperature-Depth
ECMWF	European Centre for Medium-range Weather Forecast
MATER	MAss Transfer and Ecosystem Response
MTP	Mediterranean Targeted Project
NWS	Normalized Wavelet Spectra
OGEX	Otranto Gap EXperiment
RDI	RD Instruments
VECTOR	VulnErabilità delle Coste e degli ecosistemi marini italiani ai cambiamenti climaTici e loro ruolO nei cicli del caRbonio mediterraneo
VM-ADCP	Vessel Mounted Acoustic Doppler Current Profiler

Acknowledgements. We thank the anonymous reviewers very much for the careful reading of our manuscript, helping us to consider and improve various important aspects of this research. We express our thanks to M. Orlić for fruitful discussions on some topics. We greatly acknowledge the technical and scientific staffs of the OGS and of the former establishment SACLANTCEN of the Centre for Maritime Research and Experimentation (CMRE), Italy, as well as of the Hellenic Centre for Marine Research, Greece, who participated in the CTD, multi-beam and current meter data acquisition and pre-processing in the framework of the projects OTRANTO/OGEX, MATER and VECTOR. We are also grateful to the crew of the research vessels Urania, Aegeo, Alliance, ITS Magnaghi, OGS-Explora and Universitatis who helped in all the operations at sea. We are grateful to V. Cardin for extracting the ECMWF data. We thank I. M. Mosquera for offering his help regarding wavelet analysis and interpretation of results. We also acknowledge F. Raicich for providing 1995 sea level data at Trieste.

The central part of the bathymetry line used in Fig. 1c (~ 18.6–$19.25°$ E) is a product of the multi-beam survey with high spatial resolution conducted on board r/v OGS-Explora in April 2007 (VECTOR).

The work was partially supported by the Italian national project **R**icerca **it**aliana del **mare** (Ritmare) financed by the Ministry of Research and Education and by the EU project **P**olicy-oriented marine **E**nvironmental **R**esearch for the **S**outhern **Eu**ropean **S**eas (PERSEUS).

Edited by: N. Pinardi

References

Baines, P. G.: Internal tides, internal waves and near-inertial motions, in: Baroclinic Processes on Continental Shelves, edited by: Mooers, C. N. K., AGU, Washington, D.C., 19–31, 1986.

Beckenbach, E. and Terrill, E.: Internal tides over abrupt topography in the Southern California Bight: observations of diurnal waves poleward of the critical latitude, J. Geophys. Res., 113, C02001, doi:10.1029/2006JC003905, 2008.

Book, J. W., Martin, P. J., Janeković, I., Kuzmić, M., and Wimbush, M.: Vertical structure of bottom Ekman tidal flows: Observations, theory, and modeling from the northern Adriatic, J. Geophys. Res., 114, C01S06, doi:10.1029/2008JC004736, 2009a.

Book, J. W., Perkins, H., and Wimbush, M.: North Adriatic tides: observations, variational data assimilation modeling, and linear tide dynamics, Geofizika, 26, 115–143, 2009b.

Budillon, G., Grilli, F., Ortona, A., Russo, A., and Tramontin, M.: An assessment of surface dynamics observed offshore Ancona with HF radar, Marine Ecology PSZNI, 23, Supplement I, 21–37, 2002.

Buljan, M. and Zore-Armanda, M.: Oceanographical properties of the Adriatic Sea, Oceanogr. Mar. Biol., 14, 11–98, 1976.

Cerovečki, I., Orlić, M., and Hendershott, M. C.: Adriatic seiche decay and energy loss to the Mediterranean, Deep Sea Res.-Pt. I, 44, 2007–2029, 1997.

Chavanne, C., Janeković, I., Flament, P., Poulain, P.-M., Kuzmić, M., and Gurgel, K.-W.: Tidal currents in the northwestern Adriatic: High Frequency radio observations and numerical model predictions, J. Geophys. Res., 112, C03S21, doi:10.1029/2006JC003523, 2007.

Comune di Venezia: Previsioni delle altezze di marea per il bacino San Marco e delle velocità di corrente per il Canal Porto di Lido – Laguna di Venezia, Valori astronomici 2013, available at: http://93.62.201.235/maree/DOCUMENTI/Previsioni_delle_altezze_di_marea_astronomica_2013.pdf, p. 12, 2013.

Cushman-Roisin, B. and Naimie, C. E.: A 3D finite-element model of the Adriatic tides, J. Mar. Syst., 37, 279–297, 2002.

Cushman-Roisin, B., Malačič, V., and Gačić, M.: Tide, Seiches and Low-frequency oscillations, in: Physical Oceanography of the Adriatic Sea: Past, Present and Future, edited by: Cushman-Roisin, B., Gačić, M., Poulain, P.-M., and Artegiani, A., Kluwer Academic Publishers, Dordrecht/Boston/London, 217–240, 2001.

Ferentinos, G. and Kastanos, N.: Water circulation patterns in the Otranto Strait, eastern Mediterranean, Cont. Shelf Res., 8, 1025–1041, 1988.

Foreman, M. G. G.: Manual for Tidal Currents Analysis and Prediction, Pacific Marine Science Report 78-6, Institute of Ocean Sciences, Patricia Bay, Sidney, B.C., 57 pp. (2004 revision), 1978.

Gačić, M., Kovačević, V., Manca, B., Papageorgiou, E., Poulain, P.-M., Scarazzato, P., and Vetrano, A.: Thermohaline properties and circulation in the Strait of Otranto, in: Dynamics of Mediterranean Straits and Channels, Bull. Inst. Oceanogr., no. spécial 17, CIESM Science Series no. 2, edited by: Briand, F., Monaco, 117–145, 1996.

Guarnieri, A., Pinardi, N., Oddo, P., Bortoluzzi, G., and Ravaioli, M.: Impact of tides in a baroclinic circulation model of the Adriatic Sea, J. Geophys. Res. Oceans, 118, 166–183, doi:10.1029/2012JC007921, 2013.

Hendershott, M. C. and Speranza, A.: Co-oscillating tides in long, narrow bays; the Taylor problem revisited, Deep-Sea Res., 18, 959–980, 1971.

Huthance, J. M.: On coastal trapped waves: analysis and numerical calculation by inverse iteration, J. Phys. Ocean., 8, 74–92, 1978.

Janeković, I. and Kuzmić, M.: Numerical simulation of the Adriatic Sea principal tidal constituents, Ann. Geophys., 23, 3207–3218, doi:10.5194/angeo-23-3207-2005, 2005.

Janeković, I., Bobanović, J., and Kuzmić, M.: The Adriatic Sea M2 and K1 tides by 3D model and data assimilation, Estuar. Coast. Shelf S., 57, 873–885, 2003.

Klaić, Z. B., Pasarić, Z., and Tudor, M.: On the interplay between sea-land breezes and Etesian winds over the Adriatic, J. Mar. Sys., 78, S101–S118, 2009.

Kovačević, V., Gačić, M., and Poulain, P.-M.: Eulerian current measurements in the Strait of Otranto and in the Southern Adriatic, J. Mar. Sys., 20, 255–278, 1999.

Kovačević, V., Gačić, M., Mancero Mosquera, I., Mazzoldi, A., and Marinetti, S.: HF Radar Observations in the Northern Adriatic: Surface Current Field in Front of the Venetian Lagoon, J. Mar. Sys., 51, 95–122, 2004.

Kunze, E.: Near-inertial wave propagation in geostrophic shear, J. Phys. Oceanogr., 15, 544–565, 1985.

Leder, N. and Orlić, M.: Fundamental Adriatic seiche recorded by current meters, Ann. Geophys., 22, 1449–1464, doi:10.5194/angeo-22-1449-2004, 2004.

Leder, N., Smirčić, A., Ferenčak, M., and Vučak, Z.: Some results of current measurements in the area of the Otranto Strait, Acta Adriatica, 33, 3–16, 1992.

Lozano, C. and Candela, J.: The M2 tide in the Mediterranean Sea: dynamic analysis and data assimilation, Oceanol. Acta, 18, 419–441, 1995.

Malačič, V., Viezzoli, D., and Cushman-Roisin, B.: Tidal dynamics in the northern Adriatic Sea, J. Geophys. Res., 105, 26265–26280, doi:10.1029/2000JC900123, 2000.

Manca, B. B., Kovačević, V., Gačić, M., and Viezzoli, D.: Dense water formation in the Southern Adriatic Sea and spreading into the Ionian Sea in the period 1997–1999, J. Mar. Sys., 33–34, 133–154, 2002.

Martin, P. J., Book, J. W., Burrage, D. M., Rowley, C. D., and Tudor, M.: Comparison of model-simulated and observed currents in the Central Adriatic during DART, J. Geophys. Res., 114, C01S05, doi:10.1029/2008JC004842, 2009.

Michelato, A. and Kovačević, V.: Some dynamic features of the flow through the Otranto Strait, Boll. Oceanol. Teor. Appl., IX, 39–51, 1991.

Mihanović, H., Orlić, M., and Pasarić, Z.: Diurnal internal tides detected in the Adriatic, Ann. Geophys., 24, 2773–2780, doi:10.5194/angeo-24-2773-2006, 2006.

Mihanović, H., Orlić, M., and Pasarić, Z.: Diurnal thermocline oscillations driven by tidal flow around an island in the Middle Adriatic, J. Mar. Sys., 78, S157–S168, doi:10.1016/j.jmarsys.2009.01.021, 2009.

Mosetti, F. and Manca, B.: Le maree dell'Adriatico: Calcoli di nuove costanti armoniche per alcuni porti, Studi in onore di Giuseppina Aliverti, Istituto Universitario Navale di Napoli, Ist. di Meteorologia e Oceanografia, 163–177, 1972.

Orlić, M.: Anatomy of sea level variability – an example from the Adriatic, in: The Ocean Engineering Handbook, edited by: El-Hawary, F., CRC Press, Boca Raton (USA), 2001.

Orlić, M., Beg Paklar, G., Dadić, V., Leder, N., Mihanović, H., Pasarić, M., and Pasarić, Z.: Diurnal upwelling resonantly driven by sea breezes around an Adriatic island, J. Geophys. Res., 116, C09025, doi:10.1029/2011JC006955, 2011.

Pawlowicz, R., Beardsley, B., and Lentz, S.: Classical tidal harmonic analysis including error estimates in MATLAB using T_TIDE, Comput. Geosci., 28, 929–937, 2002.

Pereira, A. F., Beckmann, A., and Hellmer, H. H.: Tidal Mixing in the Southern Weddell Sea: Results from a Three-Dimensional Model, J. Phys. Oceanogr., 32, 2151–2170, 2002.

Polli, S.: La propagazione delle maree nell'Adriatico, Istituto Talassografico Sperimentale, Pubblicazione no. 370, Trieste, 11 pp., 1961.

Poulain, P.-M.: Tidal currents in the Adriatic as measured by surface drifters, J. Geophys. Res.-Oceans, 118, 1434–1444, doi:10.1002/jgrc.20147, 2013.

Taylor, L.: Tidal oscillations in gulfs and rectangular basins, Proc. London Math. Soc., 20, 93–204, 1921.

Thompson, R. E. and Crawford, W. R.: The generation of diurnal period shelf waves by tidal currents, J. Phys. Ocean., 12, 635–643, 1982.

Torrence, C. and Compo, G. P.: A practical Guide to Wavelet analysis, B. Am. Meteorol. Soc., 79, 61–78, 1998.

Torrence, C. and Webster, P. J.: Interdecadal changes in the ENSO-monsoon System, J. Climate, 12, 2679–2690, 1999.

Ursella, L. and Gačić, M.: Use of the Acoustic Doppler Current Profiler (ADCP) in the study of the circulation of the Adriatic Sea, Ann. Geophys., 19, 1183–1193, doi:10.5194/angeo-19-1183-2001, 2001.

Ursella, L., Kovačević, V., and Gačić, M.: Footprints of mesoscale eddy passages in the Strait of Otranto (Adriatic Sea), J. Geophys. Res., 116, C04005, doi:10.1029/2010JC006633, 2011.

Ursella, L., Gačić, M., Kovačević, V., and Deponte, D.: Low-frequency flow in the bottom layer of the Strait of Otranto, Cont. Shelf Res., 44, 5–19, doi:10.1016/j.csr.2011.04.014, 2012.

Vetrano, A., Gačić, M., and Kovačević, V.: Water fluxes through the Strait of Otranto, in: EC Ecosystems research report No. 32, The Adriatic Sea, Proceedings of the workshop "Physical and biogeochemical processes in the Adriatic Sea", Portonovo (Ancona), Italy, 23 to 27 April 1996, edited by: Hopkins, T. S., Artegiani, A., Cauwet, G., Degobbis, D., and Malej, A., 127–137, 1999.

Vilibić, I., Šepić, J., Dadić, V., and Mihanović, H.: Fortnightly oscillations observed in the Adriatic Sea, Ocean Dynam., 60, 57–63, doi:10.1007/s10236-009-0241-2, 2010.

Wunsch, C.: Internal tides in the ocaen, Rev. Geophys. Space Phys., 13, 167–182, 1975.

Yari, S., Kovačević, V., Cardin, V., Gačić, M., and Bryden, H. L.: Direct estimate of water, heat, and salt transport through the Strait of Otranto, J. Geophys. Res., 117, C09009, doi:10.1029/2012JC007936, 2012.

3

Comparative heat and gas exchange measurements in the Heidelberg Aeolotron, a large annular wind-wave tank

L. Nagel[1]**, K. E. Krall**[1]**, and B. Jähne**[1,2]

[1]Institute of Environmental Physics, University of Heidelberg, Im Neuenheimer Feld 229, 69120 Heidelberg, Germany
[2]Heidelberg Collaboratory for Image Processing, University of Heidelberg, Speyerer Straße 6, 69115 Heidelberg, Germany

Correspondence to: B. Jähne (bernd.jaehne@iwr.uni-heidelberg.de)

Abstract. A comparative study of simultaneous heat and gas exchange measurements was performed in the large annular Heidelberg Air–Sea Interaction Facility, the Aeolotron, under homogeneous water surface conditions. The use of two gas tracers, N_2O and C_2HF_5, resulted not only in gas transfer velocities, but also in the measurement of the Schmidt number exponent n with a precision of ± 0.025. The original controlled flux, or active thermographic, technique proposed by Jähne et al. (1989) was applied by heating a large patch at the water surface to measure heat transfer velocities. Heating a large patch, the active thermography technique is laterally homogeneous, and problems of lateral transport effects are avoided. Using the measured Schmidt number exponents, the ratio of the scaled heat transfer velocities to the measured gas transfer velocities is 1.046 ± 0.040, a good agreement within the limits of experimental uncertainties. This indicates the possibility to scale heat transfer velocities measured by active thermography to gas transfer velocities, provided that the Schmidt number exponent is known and that the heated patch is large enough to reach the thermal equilibrium.

1 Introduction

In 1989 Jähne et al. (1989) proposed to use heat as a proxy tracer for gas transfer velocities, then called the "controlled flux technique" (CFT). This technique provides transfer velocity measurements with high temporal resolution on the order of minutes and spatial resolution of less than a meter. However, using heat as a proxy for mass has one significant drawback. Because transfer velocities of two different tracers, including heat, scale with their diffusivity, the transfer

velocity of a gas k_{gas} can be extrapolated from the transfer velocity of heat k_{heat} by

$$k_{\text{gas}} = k_{\text{heat}}\left(\frac{D_{\text{gas}}}{D_{\text{heat}}}\right)^n = k_{\text{heat}}(Le)^{-n}. \tag{1}$$

Le denotes the Lewis number. To be able to use Eq. (1) the exponent n has to be known. The exponent n gradually decreases from $2/3$ for a smooth water surface to $1/2$ for a wavy surface (Jähne et al., 1987b; Richter and Jähne, 2011). If the water temperature is different as well, Eq. (1) generalizes to

$$k_{\text{gas}} = k_{\text{heat}}\left(\frac{Sc}{Pr}\right)^{-n}, \tag{2}$$

where the Schmidt number is $Sc = \nu/D_{\text{gas}}$ and the Prandtl number is $Pr = \nu/D_{\text{heat}}$. In the temperature range from 0 to 40 °C, the Schmidt numbers for volatile species range from 60 to 4000 and the Prandtl number from 4.3 to 14.5 (Jähne, 2009). Diffusion of heat is approximately 100 times faster than diffusion of mass in water. By performing simultaneous gas and heat transfer measurements in the Karlsruhe linear air–sea interaction facility, Jähne et al. (1989) validated this extrapolation.

The initial radiometer used by Jähne et al. (1989) was a point-measuring device. Once thermal imaging systems with a sufficiently low noise level became available, thermographic techniques evolved into an even more useful method to investigate small-scale air–sea interaction processes. With advanced imaging devices, it was not only possible to measure transfer velocities, but also to provide a direct "insight" into the small-scale processes taking place at the ocean surface. Therefore, these imaging devices were used to study the mechanisms determining the transfer of mass across the air–sea interface such as Langmuir circulation (Melville et al., 1998), microscale wave breaking (Jessup et al., 1997; Zappa et al., 2001, 2004), surface renewal processes (Zappa et al., 1998), and the surface velocity field (Garbe et al., 2003; Veron et al., 2008).

For field measurements, various modifications of the original CFT were applied (Haußecker et al., 2002). Haußecker et al. (1995) developed a method based on a surface renewal model to track the decay of a small heated spot and applied this technique during the Office of Naval Research (ONR) Marine Boundary Layer Accelerated Research Initiative (MBL ARI) cruise. However, this modification was not verified by independent laboratory measurements by directly comparing gas transfer and heat transfer velocities.

Schimpf et al. (1999) proposed not to apply an artificial infrared radiation to the surface, but to use the naturally occurring net heat flux instead. By an analysis of the temperature statistics, they estimated the temperature difference across the heat boundary layer at the sea surface. This approach was also based on a surface renewal model and was used in three field campaigns: CoOP 1995, CoOP1997, and GasEx1999 (Schimpf et al., 2004; Frew et al., 2004).

However, more recent experimental evidence suggests that extrapolating from heat transfer velocities to gas transfer velocities may lead to biased results. During the Fluxes, Air–sea Interactions, and Remote Sensing (FAIRS) experiment, Asher et al. (2004) measured heat transfer velocities by tracking heated spots and using a surface renewal model, i.e., a Schmidt number exponent $n = 1/2$. Simultaneously, gas transfer velocities were measured during FAIRS. Gas transfer velocities calculated by scaling the measured heat transfer velocities were found to be twice as large as the directly measured gas transfer velocities.

Atmane et al. (2004) performed simultaneous gas exchange measurements with He and SF_6 and heat transfer measurements in a 9.1 m long linear wind-wave tank. They found that heat and gas transfer velocities can be well matched using a modification of the surface renewal model, the random eddy model by Harriott (1962), which assumes that the boundary layer is only partly renewed. Later Jessup et al. (2009) provided further evidence for complete and partial surface renewal. The Hariott model also leads to varying Schmidt number exponents. However, this is not the only model with this property. Jähne (1985) showed that the experimentally found variation of the Schmidt number exponent n between 2/3 and 1/2 can be explained by different types of models, either the extended surface renewal model, where the probability for surface renewal depends on the distance to the surface, or the turbulent diffusion model with different assumptions about the increase of the turbulent diffusivity with the distance from the interface.

All previous comparisons, however, have one or both of the following two deficits. First, they were performed in linear facilities, where it is difficult to compare the locally measured heat transfer velocity with a gas transfer velocity, which is averaged over the whole facility. Second, all more recent comparisons include only the modification of active thermography with a small heated spot. With this technique, a three-dimensional modeling is actually required, because heat is also being transported horizontally by molecular diffusion, and by the shear current in the boundary layer. It is still unclear to which extend these effects influence the measured heat transfer velocity using the spot technique.

The purpose of this investigation is therefore a careful study with simultaneous heat and gas exchange measurements in a wind-wave tank with spatially more homogeneous conditions. Such a facility is the large annular Heidelberg Air–Sea Interaction Facility, the Aeolotron. The original controlled flux developed by Jähne et al. (1989) is applied, where a large patch at the water surface is heated. By heating a large patch, the active thermography technique is laterally homogeneous – provided the patch is large enough – and all problems with lateral transport effects are avoided. This investigation aims to answer the question of whether it is possible to scale heat transfer measurements performed with the CFT to gas transfer measurements without any model assumptions, provided the Schmidt number exponent n is known.

2 The wind-wave facility

2.1 The Heidelberg Aeolotron

The Heidelberg Aeolotron is an annular wind-wave facility with a diameter of 8.68 m at the inside wall and a width of 0.61 m (Fig. 1). The annular shape results in a quasi-stationary wave field with a virtually unlimited fetch. A detailed description of the facility is given in Krall (2013, chapter 4.1) and in Jähne (2001). For the conducted measurements, deionized water with a height of 1.0 m was used, which corresponds to a water volume of 17.9 m^3. The air part of the flume above the water has a height of 1.4 m. Wind is generated by two axial fans mounted diametrically in the ceiling of the air space of the flume. The maximum wind speed (scaled to the reference height of 10 m) that can be produced by the wind generator is $u_{10} = 20\,\mathrm{m\,s^{-1}}$.

2.2 Homogeneity of the wind field in the Aeolotron

Because of the geometry of the facility, a logarithmic wind profile is not formed in the Aeolotron. Centrifugal forces due to the curved walls generate secondary currents (Schlichting, 2000). Furthermore, the wind speed is not uniformly distributed throughout the whole facility due to the positions and type of the wind generators. Bopp (2014) showed that the wind speed, and therefore also the friction velocity at the measurement location of the thermography setup, is approximately 15 % higher than for the average of the whole flume. When comparing locally measured heat transfer velocities with gas transfer velocities, which are integrated over the

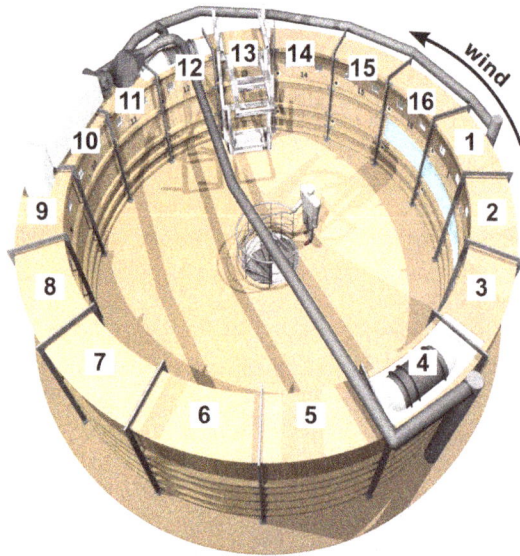

Figure 1. Rendered view of the Aeolotron, taken from Krall (2013). The 16 segments are numbered clockwise, while the wind direction is counterclockwise. The wind generating fans can be seen in segments 4 and 12. The heat transfer velocity is measured in segment 13, while the gas transfer measurements integrate over the whole water surface.

whole water area, this difference between the local and the averaged wind forcing has to be taken into account.

2.3   Measurement conditions

The measurements were conducted in spring 2010 using six different wind speed conditions between $u_{10} = 2.7$ and $12.7 \, \mathrm{m \, s^{-1}}$. Even at the highest wind speed, bubble formation by breaking waves was low. Therefore, bubble-induced gas transfer can be neglected.

Each condition was repeated two or three times. Table 1 summarizes wind speeds and friction velocities for each condition averaged over the three measurement days, as well as the mean water temperatures for each measuring day. Exact values for the wind speed and the friction velocity can be found in Appendix A for each condition. The wind speed conditions were chosen such that they were roughly equidistant in log space.

3   The controlled flux technique

The controlled flux technique inverts classical gas transfer measurements: a known flux density is forced to the water surface, and the resulting concentration difference is measured. For the heat exchange measurements, the setup tested in a pilot experiment under field conditions, as described in Schimpf et al. (2011), was used. A carbon dioxide laser (Evolution 100, Synrad Inc.) with an emitting wavelength of $\lambda = 10.6 \, \mu m$ creates a heat flux density, which is distributed ho-

mogeneously in wind direction over a rectangular area with a mirror scanning system (Micro Max 671, Cambridge Technology Inc.). The temperature response of the water surface is measured with an infrared camera (CMT256, Thermosensorik) with a resolution of 256×256 pixels in the wavelength regime of $\lambda = 3.4 - 5 \, \mu m$ and a noise equivalent temperature difference of less than $\Delta T = 20 \, mK$. During each condition, the laser is switched on and off with changing frequencies. This allows a system-theoretical approach for data analysis as proposed in Jähne et al. (1989). The thermal boundary layer acts like a low-pass filter to the laser forcing. For low forcing frequencies the surface reaches the equilibrium temperature of constant forcing. For higher frequencies the system can not reach the thermal equilibrium, the penetration depth is reduced, resulting in transport which is restricted to molecular diffusion and the temperature response is damped. From the measured temperature response of the system, the transfer function and therefore the cutoff frequency, which corresponds to the response time τ of the system, is determined in the Fourier domain.

From the assumption that right at the interface transport is only driven by molecular diffusion, the thickness of the mass boundary layer z_* can be defined and can be related to the transfer velocity k by

$$k = \frac{D}{z_*}. \tag{3}$$

as derived in Jähne et al. (1989). Then, the characteristic time constant for the transport across the mass boundary layer τ is given by the ratio of the boundary layer thickness to the transfer velocity as

$$\tau = \frac{z_*}{k}. \tag{4}$$

Substituting Eq. (3) into Eq. (4) gives then the relation between the transfer velocity k and the time constant τ:

$$k = \sqrt{\frac{D_{\mathrm{heat}}}{\tau}}. \tag{5}$$

The advantages of this data evaluation method are that it is independent of model assumption, that a flux density calibration is not required, and the low liability to reflections.

Furthermore, the temperature increase due to the periodic heating by the CO_2 laser is at most a few tenths of a degree centigrade. Therefore no significant buoyancy effect is introduced, which may influence the gas transfer process and thus also the parameters k, z_*, and τ.

4   Gas exchange

To measure gas exchange velocities k_{gas}, a box model method is employed. The wind-wave tank is interpreted as two well-mixed boxes. One of these boxes encompasses the

Table 1. Measurement conditions used in this study. Shown are the mean friction velocities u_* and the wind speeds u_{10} averaged over the three measurement days, as well as the mean water temperatures for each day, T_{mean}. The conditions at which measurements were conducted on each day are marked with an x.

Cond. no.	1	2	3	4	5	6	
u_* averaged [cm s^{-1}]	0.283	0.370	0.511	0.707	1.086	1.713	
u_{10} averaged [m s^{-1}]	2.74	3.51	4.69	6.24	8.90	12.66	T_{mean} [°C]
26 Apr 2010	x		x		x		20.1
28 Apr 2010	x	x	x	x	x	x	20.5
30 Apr 2010	x	x	x	x	x	x	20.8

air with a volume of V_a and a homogeneous trace gas concentration of c_a, and the other one the water with a volume of V_w and homogeneous tracer concentration c_w. Trace gases can be exchanged between both boxes through the water surface A. Allowing for the possibility of air leaks with a volume flux of $\dot{V}_a$, the transfer velocity k_{gas} can be calculated using the mass balance for the air box by

$$k_{gas} = \frac{V_a}{A} \cdot \frac{\dot{c}_a + \lambda_a c_a}{c_w} \cdot \frac{1}{1 - \alpha c_a/c_w}. \qquad (6)$$

The tracer's dimensionless solubility is denoted by α, and the leak rate is defined as $\lambda_a = \dot{V}_a/V_a$. The box model Eq. (6) is only applicable in this form when the concentration of the tracer ambient air is negligible and no water leaks exist. More thorough derivations of the box model equations can be found in Kräuter (2011), Krall (2013), and Mesarchaki et al. (2014).

Measuring time-resolved air and water-side concentrations allows the measurement of the transfer velocity of a gas using Eq. (6). Additionally, the geometry of the used wind-wave tank needs to be known, as well as the solubility of the trace gas used. Also, the leak rate λ_a needs to be known or measured.

The instrumentation used to measure concentrations in this study as well as the determination of the leak rate and a detailed analysis of the uncertainties of gas transfer velocity are described in Mesarchaki et al. (2014) and Krall (2013). In this study, the trace gases nitrous oxide (N$_2$O) and pentafluoroethane (C$_2$HF$_5$) were used. Their physico-chemical parameters are listed in Table 2.

To measure the Schmidt number exponent n, Schmidt number scaling (see Eq. 2) is applied to two gases. The transfer velocities of two trace gases, k_1 and k_2, with differing Schmidt numbers, Sc_1 and Sc_2, are measured simultaneously, and the Schmidt number exponent is then calculated as

$$n = -\frac{\ln(k_1/k_2)}{\ln(Sc_1/Sc_2)}. \qquad (7)$$

Table 2. Chemical and physical properties of the trace gases used, nitrous oxide (N$_2$O) and pentafluoroethane (C$_2$HF$_5$), as well as for carbon dioxide (CO$_2$) for comparison. All values are given at 20 °C for freshwater. Given Schmidt numbers are calculated using $Sc = \nu/D$, with $\nu = 0.010$ cm^2 s^{-1} (Kestin et al., 1978).

Formula	M g mol^{-1}	α	D 10^{-5} cm^2 s^{-1}	Sc
N$_2$O	44.01	0.59[a]	1.63[b]	613
C$_2$HF$_5$	120.02	0.45[c]	0.97[d]	1031
CO$_2$	44.01	0.94[e]	1.67[f]	599

[a] Young (1981), [b] Tamimi et al. (1994), [c] Krall (2013), [d] Yaws (2009), [e] Weiss (1974), [f] Jähne et al. (1987a).

5 Results and discussion

5.1 Measured heat transfer velocities

Heat and gas transfer velocities were measured during each wind speed condition with a closed air space. The determination of the heat transfer velocities and their uncertainties is described in detail in Nagel (2014). The friction velocity was measured under the same conditions, but at a different time. The measurement of the friction velocities is described in Bopp (2014). As Bopp (2014) used slightly different wind speeds in his measurements, a polynomial of 3rd order was fitted to the almost linearly related data points of friction velocities versus wind speed measured by Bopp (2014). Here, this relationship is used to determine the averaged friction velocity for each condition. For the local heat transfer measurements, the local friction velocity is assumed to be 15 % larger than the averaged one; see Sect. 2.2.

Figure 2 shows the heat transfer velocities plotted against the local friction velocity. For three conditions (no. 3 on 26 April 2010, no. 3 on 28 April 2010, and no. 1 on 30 April 2010), no transfer velocity could be determined, as the fit of the amplitude damping function did not converge due to large scatter in the measured amplitudes.

To use the system-theoretical approach, described in Sect. 3, for data analysis, it is necessary that a water parcel stays in the heated area for a time that is long enough to reach the thermal equilibrium. Therefore the response time

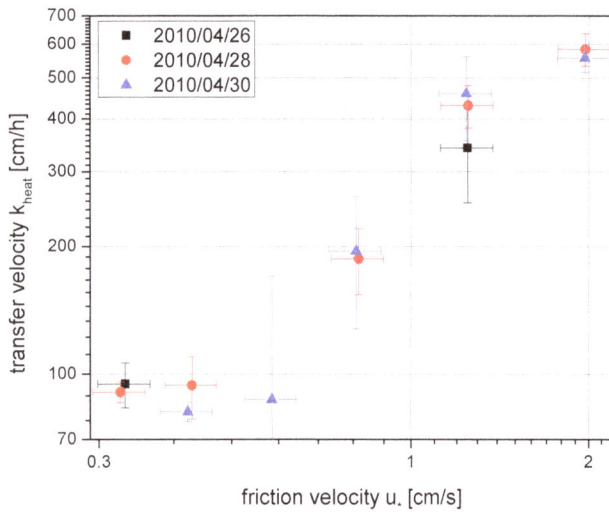

Figure 2. Measured heat transfer velocities plotted against the locally measured friction velocity. Dates are given in YYYY/MM/DD.

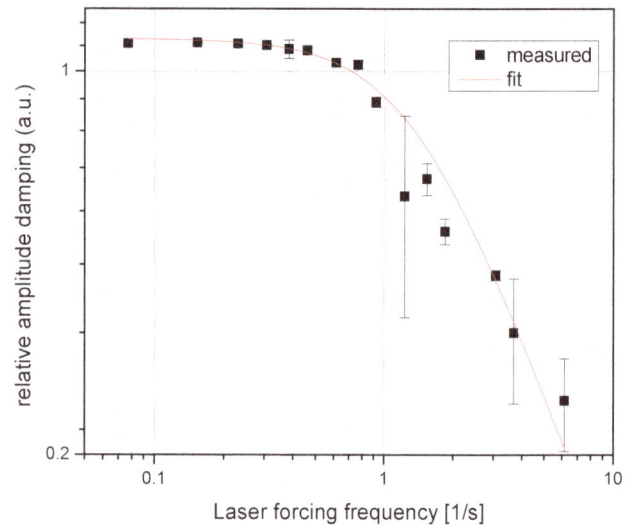

Figure 3. Example for measured amplitude damping plotted against the laser forcing frequency at a friction velocity of $u_* = 0.28 \, \mathrm{cm \, s^{-1}}$. At low forcing frequencies the thermal equilibrium was easily reached.

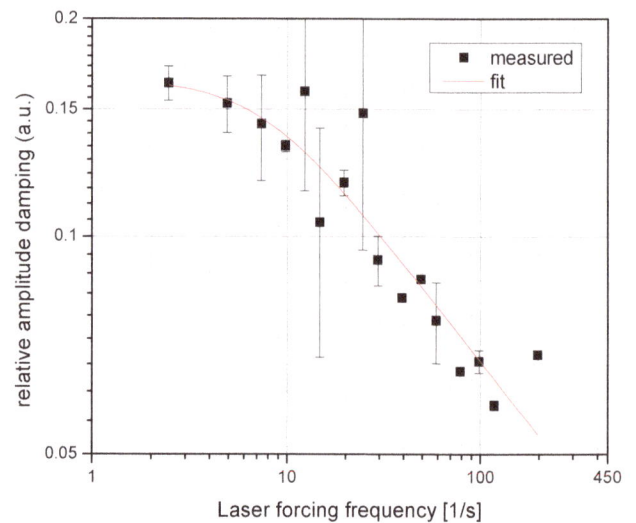

Figure 4. Example for measured amplitude damping plotted against that of the laser forcing frequency at a friction velocity of $u_* = 2.46 \, \mathrm{cm \, s^{-1}}$. At low forcing frequencies the thermal equilibrium was barely reached.

τ has to be smaller than the residence time of a water parcel in the heated patch. The size of the heated patch was 40 cm along-wind and 23 cm cross-wind. The required horizontal length scale can be estimated as follows. Gutsche (2014) measured the mean surface drift velocity in the Aeolotron. A good estimate is $0.036 \pm 6\%$ of the reference wind speed: $u_s = 0.036 \, U_{ref}$. The horizontal length scale x_* is then the product of the surface drift velocity u_s and the timescale τ: $x_* = u_s \tau$. Taking the values from Tables A1 and A2, the horizontal length scale is only 1.7 cm at the highest wind speed, but at the second-highest wind speed it is already 23 cm. For that reason it is likely that the transfer velocity of the lowest wind speed is overestimated, because the heated area was too small.

Because the expected response time of the system decreases with increasing wind speed, the used laser forcing frequencies were increased as well. Nevertheless, to determine the response time, it is necessary that the system can reach the thermal equilibrium at the lowest used forcing frequencies. How well the thermal equilibrium is reached at the lowest forcing frequencies influences the accuracy of the determination of the response time. Figures 3 and 4 show two examples of amplitude damping dependent upon the laser forcing frequency recorded under different wind speed conditions. The fit of the amplitude damping curve has a higher accuracy in Fig. 3, where the thermal equilibrium is reached at more than one of the low forcing frequencies. In contrast, the thermal equilibrium is barely reached in Fig. 4, leading to a larger uncertainty in the response time. Therefore, the accuracy of the calculated heat transfer velocities varies significantly.

5.2 Gas transfer velocities and Schmidt number exponent

The gas transfer velocities, which were measured simultaneously to the heat exchange velocities, are shown in Fig. 5 against the friction velocity averaged over the whole facility. The transfer velocity shown is scaled to a Schmidt number of $Sc = 600$ using Schmidt number scaling; see Eq. (2). The Schmidt number exponent n used for scaling was calculated using the measured transfer velocities of both gases,

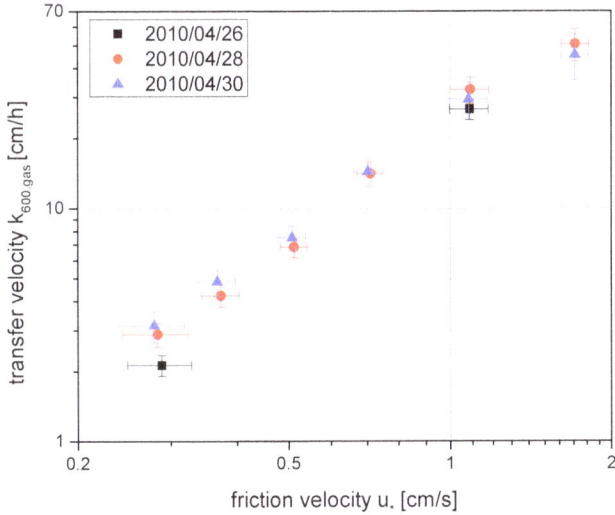

Figure 5. Measured gas transfer velocities scaled to a Schmidt number of $Sc = 600$ plotted against the global averaged friction velocity. Dates are given in YYYY/MM/DD.

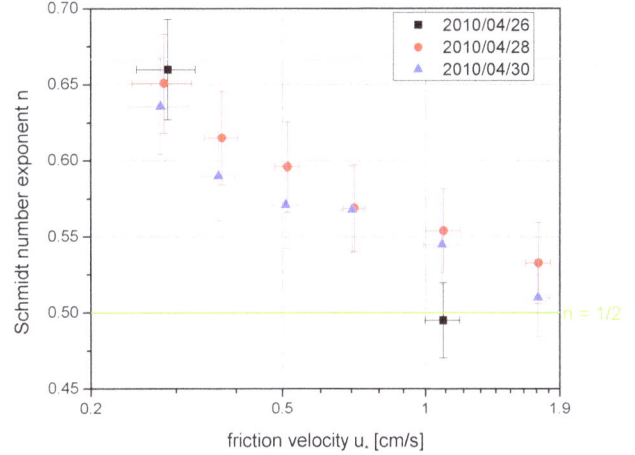

Figure 6. Measured Schmidt number exponents n with an error of 5 % plotted against the global averaged friction velocity u_*. Dates are given in YYYY/MM/DD.

N_2O as well as C_2HF_5, using Eq. (7). Therefore, scaling the measured gas transfer velocities of both gases to $Sc = 600$ yields the same transfer velocity k_{600}. That means that, even though the transfer velocities of two gases were measured in each condition, only one transfer velocity can be shown. For condition no. 3 on 26 April 2010 (see Table 1), which was measured as the first condition on that day, no gas transfer velocity could be calculated. During this condition the trace gases were insufficiently mixed into the water, leading to spatially varying water-side concentration, making the box model which requires homogeneous concentrations (see Sect. 4) no longer applicable.

The measured Schmidt number exponent n is shown in Fig. 6. It shows a smooth transition from $n = 2/3$ for the low wind speeds to $n = 1/2$ for the highest wind speed as described in Sect. 1. The error estimation of the Schmidt number exponent was done with a mean difference approach, resulting in an error of less than 0.025. The measured gas transfer velocities k_{600} as well as the transition of the Schmidt number exponent n from 2/3 to 1/2 are in good agreement with previous studies in wind-wave facilities (Jähne et al., 1987b; Zappa et al., 2001; Nielsen, 2004; Krall, 2013).

5.3 Comparison between measured gas and heat transfer velocities

To compare heat and gas transfer, all measured transfer velocities are scaled to a Schmidt number of $Sc = 600$ by Schmidt number scaling, as described in Sect. 1, Eq. (2), using the measured Schmidt number exponents.

Figure 7 shows the measured transfer velocities for heat and for a gas dependent upon the friction velocity. The shown gas transfer velocities are integrated over the whole facility,

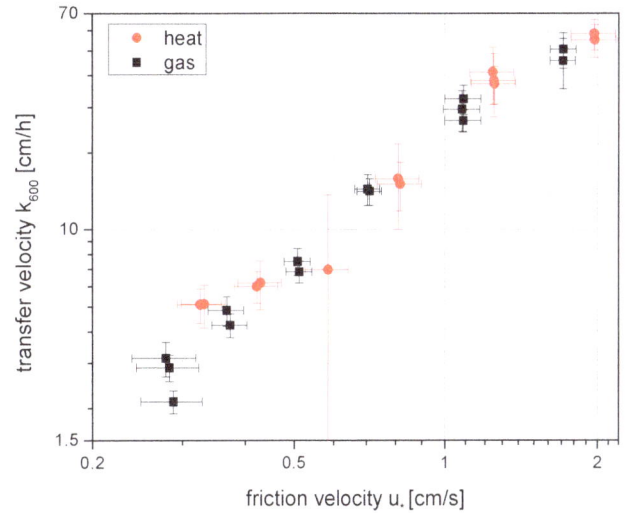

Figure 7. Simultaneously measured heat and gas transfer velocities, both scaled to a Schmidt number of $Sc = 600$ plotted against that of the friction velocity. The friction velocities for the local heat transfer measurements are 15 % enhanced in comparison to the global values for the friciton velocities for the gas transfer velocity measurements.

while the heat transfer velocities are measured locally. As described in Sect. 2.2, the wind speed and the friction velocity in the measurement region of the local heat transfer measurements are approximately 15 % higher than the averaged value of the whole facility. Therefore the transfer velocities are measured at different friction velocities, although they are measured simultaneously. The uncertainty for the local friction velocity is approximately 5 %, while uncertainties of the friction velocity averaged over the whole facility are taken directly from Bopp (2014).

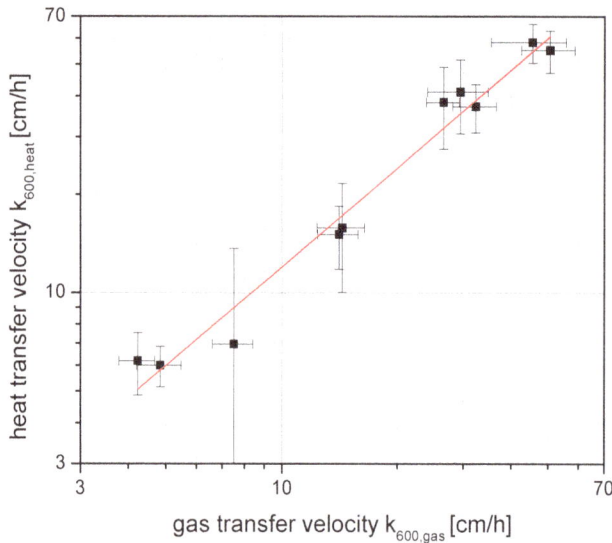

Figure 8. Heat transfer velocities plotted against simultaneously measured gas transfer velocities, both scaled to a Schmidt number of $Sc = 600$. The best-fit line has a slope of 1.20 ± 0.04.

Figure 7 shows the good agreement between the scaled heat and the scaled gas transfer velocities. To quantify the deviation of the scaled heat transfer velocities, Fig. 8 shows them against the simultaneous measured gas transfer velocities, also scaled to $Sc = 600$. The best-fit line shows a slope of 1.20 ± 0.04. That means that the heat transfer velocities are approximately 20 % higher than the simultaneously measured gas transfer velocities. As discussed above, 15 % can be attributed to the difference in the friction velocity of the local compared to the integrated measurements. Therefore the scaling factor between the heat transfer velocities and the values, which were expected from the gas transfer measurements, is just 1.046 ± 0.04. This is within the conservatively estimated error budget, which contains three different sources of errors.

First, the absolute uncertainty in the Schmidt number exponent n leads to a relative uncertainty for the heat transfer velocity scaled to a gas transfer velocity of

$$\frac{\sigma_k}{k_{gas}} = \ln\left(\frac{Sc}{Pr}\right)\sigma_n, \tag{8}$$

where σ_k and σ_n are the absolute uncertainties for the transfer velocity and the Schmidt number exponent, respectively. For $\sigma_n = 0.025$ (Fig. 6) and $Sc/Pr \approx 600/7.2$, the relative scaling for k_{gas} error is 11 %.

Second, the accuracy of the absolute value of the Schmidt number is less than 5 % (Jähne et al., 1987a). Third, the heat transfer velocities were measured at a local friction velocity, which is 15 ± 5 % higher than the friction velocity averaged over the whole facility. These three contributions lead, linearly added, to a total error of less than 21 %.

6   Conclusions and outlook

This study showed that it is possible to scale heat transfer velocities to gas transfer velocities. The mean deviation found experimentally is a factor of 1.046 ± 0.04. This is well below the possible maximum systematic deviation, conservatively estimated to be 21 %. This result was found by simultaneous gas transfer and heat transfer measurements using the original approach of Jähne et al. (1989), in the large annular Heidelberg Aeolotron wind wave tank. This approach does not depend on any model assumptions about the transfer processes; only the Schmidt number exponent n must be known. This opens up the opportunity to apply this technique to field measurements.

However, three issues must be addressed carefully.

First, water parcels at the surface must stay in the heated patch for a time that is longer than the response time τ of the heat transfer across the boundary layer (see Sect. 5.1). This condition is much harder to meet in the field than in a wind-wave facility and requires a platform that moves with the mean water surface drift velocity.

Second, the Schmidt number exponent n has to be known with a high level of certainty. All experimental and theoretical evidence suggests that it varyies between 2/3 and 1/2. With an unknown Schmidt number exponent in this range, the uncertainty of scaling from heat transfer to gas transfer is approximately a factor of 2 for a Schmidt number of 600. It is even larger for higher Schmidt numbers (tracers such as SF_6, DMS, and most organic volatiles) and lower for lower Schmidt numbers (e.g., He). Recent measurements by Krall (2013) showed that the exponent n does not simply decrease with increasing wind speed, but it also depends on the degree of contamination of the water surface with surface active material. Thus the relation between the Schmidt number exponent n and the surface conditions needs to be investigated carefully. Measurements by Zappa et al. (2004) and Jessup et al. (2009) indicate that it might be possible to infer the exponent from the infrared image sequences themselves, because it is possible to analyze microscale wave breaking and full/partial surface renewal events from them.

And third, active thermography does not see bubble-induced gas transfer. This is a clear disadvantage for scaling to gas transfer rates for high-wind-speed conditions. However, it can be turned into a clear advantage for combined gas transfer–heat transfer field campaigns to determine the bubble-induced portion of gas transfer.

Appendix A: Measured transfer velocities

Table A1 shows the environmental variables, and Table A2
the numerical results of the measurements described above.

Table A1. Environmental parameters at all measured conditions, including the date, the reference wind speed u_{ref}, the wind speed u_{10}, the friction velocity u_*, the water temperature T_{water}, the Schmidt number of N_2O, and the Schmidt number exponent n. The Prandtl number of heat was $Pr = 7.2$ under all measured conditions. The temperature-dependent Schmidt number of N_2O was taken from Degreif (2006).

Number	Date	u_{ref} $[m\,s^{-1}]$	u_{10} $[m\,s^{-1}]$ averaged	u_* $[cm\,s^{-1}]$ averaged	u_* $[m\,s^{-1}]$ local	T_{water} $[°C]$ averaged	Sc N_2O	n
1	26 Apr 2010	2.05	2.78	0.288	0.332	20.1	597	0.66
2	28 Apr 2010	2.02	2.73	0.283	0.326	20.4	586	0.65
3	30 Apr 2010	1.99	2.69	0.279	0.321	20.8	575	0.64
4	28 Apr 2010	2.66	3.53	0.373	0.429	20.5	585	0.62
5	30 Apr 2010	2.61	3.48	0.367	0.422	20.8	572	0.59
6	26 Apr 2010	3.64	4.70	0.512	0.589	20.1	598	–
7	28 Apr 2010	3.64	4.71	0.513	0.590	20.5	585	0.60
8	30 Apr 2010	3.61	4.68	0.509	0.585	20.9	573	0.57
9	28 Apr 2010	4.87	6.26	0.711	0.817	20.5	584	0.57
10	30 Apr 2010	4.83	6.21	0.704	0.809	20.9	573	0.57
11	26 Apr 2010	6.55	8.91	1.088	1.251	20.1	597	0.50
12	28 Apr 2010	6.56	8.92	1.089	1.253	20.5	584	0.55
13	30 Apr 2010	6.53	8.87	1.082	1.244	20.9	573	0.55
14	28 Apr 2010	8.34	12.67	1.715	1.972	20.5	584	0.53
15	30 Apr 2010	8.33	12.66	1.712	1.969	20.9	573	0.51

Table A2. Measured response times τ, heat transfer velocities k_{heat}, transfer velocities of N_2O (k_{N_2O}), and the k_{600} calculated from k_{heat} and k_{N_2O}.

Number	τ $[s]$	k_{heat} $[cm\,s^{-1}]$	k_{600} $[cm\,s^{-1}]$ heat	k_{N_2O} $[cm\,s^{-1}]$	k_{600} $[cm\,s^{-1}]$ N_2O
1	2.079 ± 0.520	94.74 ± 11.51	5.13 ± 0.50	2.13 ± 0.22	2.12 ± 0.22
2	2.270 ± 0.259	90.67 ± 5.03	5.17 ± 0.23	2.93 ± 0.35	2.88 ± 0.34
3	–	–	–	3.22 ± 0.49	3.13 ± 0.48
4	2.108 ± 0.726	94.08 ± 15.75	6.29 ± 0.97	4.29 ± 0.46	4.23 ± 0.45
5	2.807 ± 0.310	81.54 ± 4.37	6.16 ± 0.30	4.97 ± 0.64	4.84 ± 0.64
6	–	–	–	–	–
7	–	–	–	6.94 ± 0.67	6.84 ± 0.67
8	2.463 ± 4.834	87.05 ± 83.06	7.15 ± 6.79	7.70 ± 0.87	7.51 ± 0.30
9	0.535 ± 0.196	186.85 ± 33.28	15.32 ± 4.19	14.32 ± 1.61	14.10 ± 1.70
10	0.491 ± 0.348	194.97 ± 67.17	16.22 ± 8.87	14.77 ± 1.71	14.39 ± 2.01
11	0.160 ± 0.085	341.55 ± 88.06	38.34 ± 22.62	26.67 ± 2.64	26.61 ± 2.66
12	0.101 ± 0.024	430.15 ± 49.27	37.66 ± 10.79	32.79 ± 3.22	32.31 ± 4.22
13	0.089 ± 0.041	458.57 ± 102.8	42.20 ± 24.80	30.21 ± 3.21	29.47 ± 5.32
14	0.055 ± 0.010	583.12 ± 51.69	55.99 ± 14.81	51.38 ± 4.84	50.65 ± 8.08
15	0.060 ± 0.009	556.95 ± 42.42	59.75 ± 13.39	46.67 ± 4.40	45.60 ± 10.14

Acknowledgements. We would like to thank M. Bopp for fruitful discussions concerning the friction velocity in the Aeolotron. Financial support for this work by the German Federal Ministry of Education and Research (BMBF) joint project "Surface Ocean Processes in the Anthropocene" (SOPRAN, FKZ 03F0462F, and 03F0611F) within the international SOLAS project and of the Deutsche Forschungsgemeinschaft and Ruprecht-Karls-Universität Heidelberg within the funding program Open Access Publishing is gratefully acknowledged.

Edited by: J. Shutler

References

Asher, W. E., Jessup, A. T., and Atmane, M. A.: Oceanic application of the active controlled flux technique for measuring air-sea transfer velocities of heat and gases, J. Geophys. Res., 109, C08S12, doi:10.1029/2003JC001862, 2004.

Atmane, M. A., Asher, W., and Jessup, A. T.: On the use of the active infrared technique to infer heat and gas transfer velocities at the air-water free surface, J. Geophys. Res., 109, C08S14, doi:10.1029/2003JC001805, 2004.

Bopp, M.: Luft- und wasserseitige Strömungsverältnisse im ringförmigen Heidelberger Wind-Wellen-Kanal (Aeolotron), Masterarbeit, Institut für Umweltphysik, Universität Heidelberg, Germany, available at: http://www.ub.uni-heidelberg.de/archiv/17151 (last access: 23 January 2015), 2014.

Degreif, K.: Untersuchungen zum Gasaustausch – Entwicklung und Applikation eines zeitlich aufgelösten Massenbilanzverfahrens, Dissertation, Institut für Umweltphysik, Fakultät für Physik und Astronomie, Univ. Heidelberg, available at: http://www.ub.uni-heidelberg.de/archiv/6120 (last access: 23 January 2015), 2006.

Frew, N., Bock, E., Schimpf, U., Hara, T., Haussecker, H., Edson, J., McGillis, W., Nelson, R., McKenna, S., Uz, B., and Jähne, B.: Air-sea gas transfer: Its dependence on wind stress, small-scale roughness, and surface films, J. Geophys. Res., 109, C08S17, doi:10.1029/2003JC002131, 2004.

Garbe, C. S., Spies, H., and Jähne, B.: Estimation of surface flow and net heat flux from infrared image sequences, J. Math. Imaging Vis., 19, 159–174, doi:10.1023/A:1026233919766, 2003.

Gutsche, M.: Surface Velocity Measurements at the Aeolotron by Means of Active Thermography, Masterarbeit, Institut für Umweltphysik, Universität Heidelberg, Germany, available at: http://www.ub.uni-heidelberg.de/archiv/17431 (last access: 23 January 2015), 2014.

Harriott, P.: A Random Eddy Modification of the Penetration Theory, Chem. Eng. Sci., 17, 149–154, doi:10.1016/0009-2509(62)80026-8, 1962.

Haußecker, H., Reinelt, S., and Jähne, B.: Heat as a proxy tracer for gas exchange measurements in the field: principles and technical realization, in: Air–Water Gas Transfer: Selected Papers from the Third International Symposium on Air–Water Gas Transfer, edited by: Jähne, B. and Monahan, E. C., 405–413, AEON, Hanau, doi:10.5281/zenodo.10401, 1995.

Haußecker, H., Schimpf, U., Garbe, C. S., and Jähne, B.: Physics from IR image sequences: Quantitative analysis of transport models and parameters of air-sea gas transfer, in: Gas Transfer at Water Surfaces, edited by: Saltzman, E., Donelan, M., Drennan, W., and Wanninkhof, R., Vol. 127 of Geophysical Monograph, 103–108, American Geophysical Union, doi:10.1029/GM127p0103, invited, 2002.

Jessup, A. T., Zappa, C. J., and Yeh, H. H.: Defining and quantifying microscale wave breaking with infrared imagery, J. Geophys. Res., 102, 23145–23153, 1997.

Jessup, A. T., Asher, W. E., Atmane, M., Phadnis, K., Zappa, C. J., and Loewen, M. R.: Evidence for complete and partial surface renewal at an air-water interface, Geophys. Res. Lett., 36, 1–5, doi:10.1029/2009GL038986, 2009.

Jähne, B.: Transfer processes across the free water interface, Habilitation thesis, Institut für Umweltphysik, Fakultät für Physik und Astronomie, Univ. Heidelberg, doi:10.5281/zenodo.12202, 1985.

Jähne, B.: Aeolotron: the Heidelberg air sea interaction facility, CD-ROM, Hanau: AEON Verlag und Studio, doi:10.5281/zenodo.10281, available at: https://www.youtube.com/watch?v=UN0WLx9Ow9Q (last access: 23 January 2015), 2001.

Jähne, B.: Air-sea gas exchange, in: Encyclopedia Ocean Sciences, edited by: Steele, J. H., Turekian, K. K., and Thorpe, S. A., 147–156, Elsevier, doi:10.1016/B978-012374473-9.00642-1, invited, 2009.

Jähne, B., Heinz, G., and Dietrich, W.: Measurement of the diffusion coefficients of sparingly soluble gases in water, J. Geophys. Res., 92, 10767–10776, doi:10.1029/JC092iC10p10767, 1987a.

Jähne, B., Münnich, K. O., Bösinger, R., Dutzi, A., Huber, W., and Libner, P.: On the parameters influencing air-water gas exchange, J. Geophys. Res., 92, 1937–1950, doi:10.1029/JC092iC02p01937, 1987b.

Jähne, B., Libner, P., Fischer, R., Billen, T., and Plate, E. J.: Investigating the transfer process across the free aqueous boundary layer by the controlled flux method, Tellus, 41B, 177–195, doi:10.1111/j.1600-0889.1989.tb00135.x, 1989.

Kestin, J., Sokolov, M., and Wakeham, W. A.: Viscosity of Liquid Water in the Range −8 °C to 150 °C, J. Phys. Chem. Ref. Data, 7, 941–948, doi:10.1063/1.555581, 1978.

Krall, K. E.: Laboratory Investigations of Air-Sea Gas Transfer under a Wide Range of Water Surface Conditions, Dissertation, Institut für Umweltphysik, Fakultät für Physik und Astronomie, Univ. Heidelberg, available at: http://www.ub.uni-heidelberg.de/archiv/14392 (last access: 23 January 2015), 2013.

Kräuter, C.: Aufteilung des Transferwiderstands zwischen Luft und Wasser beim Austausch flüchtiger Substanzen mittlerer Löslichkeit zwischen Ozean und Atmosphäre, Diplomarbeit, Institut für Umweltphysik, Fakultät für Physik und Astronomie, Univ. Heidelberg, available at: http://www.ub.uni-heidelberg.de/archiv/13010 (last access: 23 January 2015), 2011.

Melville, W. K., Shear, R., and Veron, F.: Laboratory measurements of the generation and evolution of Langmuir circulations, J. Fluid Mech., 364, 31–58, doi:10.1017/S0022112098001098, 1998.

Mesarchaki, E., Kräuter, C., Krall, K. E., Bopp, M., Helleis, F., Williams, J., and Jähne, B.: Measuring air–sea gas exchange velocities in a large scale annular wind-wave tank, Ocean Sci. Discuss., 11, 1643–1689, doi:10.5194/osd-11-1643-2014, 2014.

Nagel, L.: Active Thermography to Investigate Small-Scale Air-Water Transport Processes in the Laboratory and the Field, Dissertation, Institut für Umweltphysik, Fakultät für Chemie und Geowissenschaften, Univ. Heidelberg, available at: http://www.ub.uni-heidelberg.de/archiv/16831 (last access: 23 January 2015), 2014.

Nielsen, R.: Gasaustausch – Entwicklung und Ergebnis eines schnellen Massenbilanzverfahrens zur Messung der Austauschparameter, Dissertation, Institut für Umweltphysik, Fakultät für Physik und Astronomie, Univ. Heidelberg, available at: http://www.ub.uni-heidelberg.de/archiv/5032 (last access: 23 January 2015), 2004.

Richter, K. and Jähne, B.: A laboratory study of the Schmidt number dependency of air-water gas transfer, in: Gas Transfer at Water Surfaces 2010, edited by: Komori, S., McGillis, W., and Kurose, R., 322–332, available at: http://hdl.handle.net/2433/156156 (last access: 23 January 2015), 2011.

Schimpf, U., Haußecker, H., and Jähne, B.: Studies of air-sea gas transfer and micro turbulence at the ocean surface using passive thermography, in: The Wind-Driven Air-Sea Interface: Electromagnetic and Acoustic Sensing, Wave Dynamics and Turbulent Fluxes, edited by: Banner, M. L., Sydney, Australia, 1999.

Schimpf, U., Garbe, C., and Jähne, B.: Investigation of transport processes across the sea surface microlayer by infrared imagery, J. Geophys. Res., 109, C08S13, doi:10.1029/2003JC001803, 2004.

Schimpf, U., Nagel, L., and Jähne, B.: The 2009 SOPRAN active thermography pilot experiment in the Baltic Sea, in: Gas Transfer at Water Surfaces 2010, edited by: Komori, S., McGillis, W., and Kurose, R., 358–367, available at: http://hdl.handle.net/2433/156156, 2011.

Schlichting, H.: Boundary Layer Theory, Springer, 8th Edn., 2000.

Tamimi, A., Rinker, E. B., and Sandall, O. C.: Diffusion Coefficients for Hydrogen Sulfide, Carbon Dioxide, and Nitrous Oxide in Water over the Temperature Range 293–368 K, J. Chem. Eng. Data, 39, 330–332, doi:10.1021/je00014a031, 1994.

Veron, F., Melville, W. K., and Lenain, L.: Wave-coherent air-sea heat flux, J. Phys. Oceanogr., 38, 788–802, doi:10.1175/2007JPO3682.1, 2008.

Weiss, R. F.: Carbon Dioxide in Water and Seawater: The Solubility of a Non-Ideal Gas, Mar. Chem., 2, 203–215, doi:10.1016/0304-4203(74)90015-2, 1974.

Yaws, C. L.: Diffusion coefficient in water – organic compounds, in: Transport Properties of Chemicals and Hydrocarbons, chap. 12, 502–593, William Andrew, doi:10.1016/B978-0-8155-2039-9.50017-X, 2009.

Young, C. L. (Ed.): IUPAC Solubility Data Series: Oxides of Nitrogen, Vol. 8, Pergamon Press, Oxford, England, 1981.

Zappa, C. J., Jessup, A. T., and Yen, H.: Skin layer recovery of free-surface wakes: Relation to surface renewal and dependence on heat flux and background turbulence, J. Geophys. Res., 103, 21711–21721, doi:10.1029/98JC01942, 1998.

Zappa, C. J., Asher, W. E., and Jessup, A. T.: Microscale wave breaking and air-water gas transfer, J. Geophys. Res., 106, 9385–9391, doi:10.1029/2000JC000262, 2001.

Zappa, C. J., Asher, W. E., Jessup, A. T., Klinke, J., and Long, S. R.: Microbreaking and the enhancement of air-water transfer velocity, J. Geophys. Res., 109, C08S16, doi:10.1029/2003JC001897, 2004.

4

Consequences of artificial deepwater ventilation in the Bornholm Basin for oxygen conditions, cod reproduction and benthic biomass – a model study

A. Stigebrandt[1], **R. Rosenberg**[2], **L. Råman Vinnå**[3], **and M. Ödalen**[4]

[1]Dept. of Earth Sciences, University of Gothenburg, Gothenburg, Sweden
[2]Dept. of Biological and Environmental Sciences – Kristineberg, University of Gothenburg, Kristineberg Fiskebäckskil, Sweden
[3]EPFL ENAC IIE APHYS, Lausanne, Switzerland
[4]Dept. of Meteorology, University of Stockholm, Stockholm, Sweden

Correspondence to: A. Stigebrandt (anst@gvc.gu.se)

Abstract. We develop and use a circulation model to estimate hydrographical and ecological changes in the isolated basin water of the Bornholm Basin. By pumping well-oxygenated so-called winter water to the greatest depth, where it is forced to mix with the resident water, the rate of deepwater density reduction increases as well as the frequency of intrusions of new oxygen-rich deepwater. We show that pumping $1000\,\text{m}^3\,\text{s}^{-1}$ should increase the rates of water exchange and oxygen supply by 2.5 and 3 times, respectively. The CRV (cod reproduction volume), the volume of water in the isolated basin meeting the requirements for successful cod reproduction ($S > 11$, $O_2 > 2\,\text{mL}\,\text{L}^{-1}$), should every year be greater than $54\,\text{km}^3$, which is an immense improvement, since it has been much less in certain years. Anoxic bottoms should no longer occur in the basin, and hypoxic events will become rare. This should permit extensive colonization of fauna on the earlier periodically anoxic bottoms. Increased biomass of benthic fauna should also mean increased food supply to economically valuable demersal fish like cod and flatfish. In addition, re-oxygenation of the sediments should lead to increased phosphorus retention by the sediments.

1 Introduction

The Baltic Sea is the second largest ($373\,000\,\text{km}^2$) brackish water system in the world, with an extensive drainage system that is 4 times larger than the sea surface area. The salinity varies from about 1 in the northern Bothnian Sea to about 17 in the deep basins of the southwestern Baltic. The Baltic is enclosed, with restricted water exchange with the adjacent seas: the Kattegat, the Skagerrak and the North Sea. Since the mid-1900s, large volumes of bottom water below the halocline have lacked oxygen concentrations high enough to support the respiration of aquatic organisms. During the last decade, areas of approximately $50\,000\,\text{km}^2$ (year 2011) of the seabed have been anoxic, and $27\,000\,\text{km}^2$ hypoxic ($< 2\,\text{mL}\,O_2\,\text{L}^{-1}$), thus lacking benthic animals (Karlson et al., 2002). The Baltic is the largest sea area with anthropogenically related low oxygen concentrations in the world (Diaz and Rosenberg, 2008) because of increased input of organic matter (Bonsdorff et al., 1997) that has changed the ecosystem properties (Elmgren, 1989; Rosenberg et al., 1990). The increased input of organic matter has been boosted by extensive leakage of phosphorus from anoxic bottoms that has increased the phosphorus content in the water column in spite of a halved external supply of phosphorus since the 1980s (Stigebrandt et al., 2014a). As a consequence, the structure and function of the Baltic ecosystem is affected by two main factors – horizontally by increasing salinity from north to south and vertically by decreasing oxygen concentrations by depth.

The annual fish catches increased in the Baltic in the middle of the last century, and the highest landings of cod, the most economically valuable species, peaked around 1990, with about 400 000 tons (MacKenzie et al., 2000). Cod is dependent on the deeper bottoms of the Baltic for reproduction and feeding, whereas herring and sprat are pelagic feeders. The cod stocks have declined since 1990 because of overly high fishing pressure and unfavorable spawning conditions. Other factors with a negative impact on the cod are the fact that herring and sprat are predators on egg and larvae, which deters the recovery of the cod stock.

The anoxic and hypoxic water in the deeper parts of the Baltic has great negative consequences for the ecosystem. Few or no animals are present at depths deeper than 70 m, which means that there are vast volumes that are lacking fish and fish food. The reproduction of cod is hampered as the eggs need an oxygen concentration greater than $2\,\mathrm{mL\,L^{-1}}$ and a salinity of at least 11 to be buoyant, but this varies somewhat with egg size and the age of the spawning cod (Hinrichsen et al., 2007; Vallin et al., 1999). Such a high salinity is found in the Bornholm Basin and, east of this basin, only close to the bottom and then frequently associated with toxic hydrogen sulfide, which is produced during anoxia. MacKenzie et al. (2002) summarized the volumes of water in the main cod spawning areas of the eastern Baltic and Bornholm basins where salinity and oxygen conditions could allow successful fertilization and hatching. From the mid-1980s, this was almost non-existent in the Gdansk and Gotland basins, and favorable conditions were only found in the Bornholm Basin.

Defaunated sediments appear laminated, which demonstrates that the biogeochemical processes that occur in oxidized sediments do not occur or are significantly changed (Jonsson et al., 1990). Thus, bioturbation (mixing of sediments) and irrigation (pumping of oxygenated water into the sediment) are lacking in laminated sediments. Lack of benthic animals means reduced nitrification and a reduced capacity to bind phosphorus in the top sediments, which may accelerate eutrophication and occasional blooms of cyanobacteria.

It has been suggested that the eutrophication of the Baltic Proper may be reduced by man-made oxygenation of the deepwater, which should decrease the phosphorus content in the water column, and that oxygenation should be done by pumping down well-oxygenated so-called winter water into the deepwater (Stigebrandt and Gustafsson, 2007). Gustafsson et al. (2008) and Conley et al. (2009b) discuss engineering methods that have been suggested to cure hypoxia in the Baltic Sea. Both papers consider below-halocline ventilation, the method applied in the present paper, to be the only realistic alternative.

Before considering oxygenation of the whole Baltic Proper, it may be wise to first look into oxygenation of a

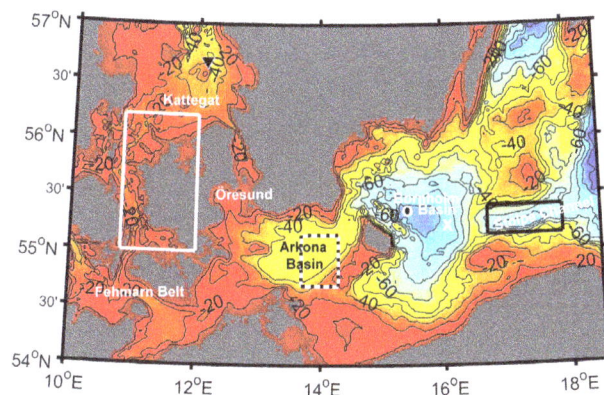

Figure 1. Map of the southern Baltic Sea area (hypsographic data from Seifert et al., 2001). Reference data for the Arkona Basin originate from the area within the white and black dashed rectangle, and reference data stations BY4 and BY5 in the Bornholm Basin are marked by a white o and x, respectively. Data from the Stolpe Channel used for forcing of the model originate from the area within the black rectangle. The forcing data for the freshwater pool model were obtained from within the white rectangle and from the Anholt E station marked by ▼.

smaller part of the Baltic. In this paper, we have chosen to consider oxygenation of the deepwater of the Bornholm Basin, beneath the level of the greatest depth (60 m) of the Stolpe Channel (Slupsk Furrow) in the southwestern Baltic Proper (Fig. 1). Very salty deepwater may stay in this basin for several years, whereby hypoxia and anoxia may develop in the lower parts.

Pumping winter water into the deepwater creates a mixture of winter water and deepwater that is buoyant and rises as an entraining plume until it reaches the level where the buoyancy is lost and interleaving occurs. Removal of deepwater by entraining plumes causes a compensatory downward vertical motion to replace the removed water. Pumping thus drives a vertical circulation cell between the levels of interleaving and pump outlet, respectively, with rising motion in the (narrow) buoyant plumes located at the pumps, and sinking motion everywhere outside the plumes. The sinking motion tends to reduce vertical concentration gradients of oxygen and other dissolved substances in the cell.

Stigebrandt and Kalén (2013) implemented pumping in a model of the Bornholm Basin. They showed that pumping $800\,\mathrm{m^3\,s^{-1}}$ of winter water from 40 to 90 m with a tenfold increase in the flow due to mixing at the outlet would have prevented the development of hypoxia and anoxia that occurred during a selected period starting in early May 2003 after a complete water exchange. The characteristics of the modeled pumped flow fit well with observations made in a comprehensive, long-term pilot experiment in the By Fjord, where $2\,\mathrm{m^3\,s^{-1}}$ of surface water were pumped into the usually anoxic deepwater (Stigebrandt et al., 2014b). The pumped water had a speed of $2\,\mathrm{m\,s^{-1}}$ through the outlet orifices. The total flux of deepwater carried by the plumes at a certain depth can readily be estimated from the speed of vertical migration of isohaline surfaces multiplied by the horizontal surface area of the fjord at that depth. In the By Fjord, the typical downward speed of isohaline surfaces was 2 m day^{-1}, implying an entrained flow of deepwater that was 30 times greater than the pumped flow. Preliminary estimates suggest a tenfold increase due to initial mixing and a further increase by a factor of 3 due to entrainment in the buoyant plume phase. Furthermore, the pumping in the By Fjord increased the rate of density reduction in the basin water by a factor of 100 and the frequency of water exchanges in the deeper parts of the basin by a factor of 10 (Stigebrandt et al., 2014b).

In the present paper we complement the model in Stigebrandt and Kalén (2013) with a time-dependent model of the inflow of new deepwater into the basin, which allows us to estimate how the rate of water exchange and the oxygen conditions in the deeper parts of the basin would be changed by pumping.

The goal of the present paper is to investigate hydrographical and ecological effects of enhanced deepwater circulation by pumping in the Bornholm Basin. First the pump is turned off (reference run), and the high quality of model predictions is demonstrated by comparisons between model results and hydrographic observations of salinity, temperature and oxygen in the deepwater of the Bornholm Basin for the period 1990–2010 (21 years). Then the pump is turned on to simulate how the vertical circulation of the Bornholm Basin should be affected with regard to residence time and changed salinity, temperature and oxygen conditions. Finally we compute how the changed hydrographical and oxygen conditions should affect the water volume available for successful cod reproduction, and describe the expected colonization of the oxygenated bottoms. We estimate the possible increase in benthic biomass and discuss the importance of this as fish feed and for biogeochemical sediment processes. We also briefly discuss possible effects on the Baltic Sea east of the Bornholm Basin due to changed properties of new deepwater entering from the Bornholm Basin.

2   Methods

2.1   The circulation model

We will use the vertical advection–diffusion model with high vertical resolution described for the Bornholm Basin by Stigebrandt and Kalén (2013). However, that model lacks a model of the inflow of new deepwater. Here we develop and use a relatively simple mechanistic model of the flow of new deepwater from Kattegat to the Bornholm Basin (for an area map, see Fig. 1). Since the model is used only in the hindcast mode, it may be tuned if necessary to improve the description of the inflow of new deepwater.

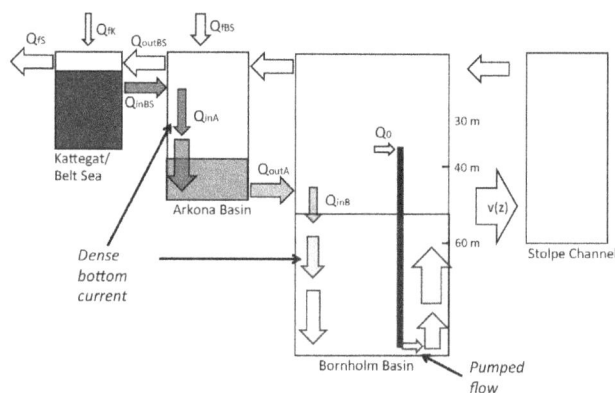

Figure 2. Schematic image of model flows. Grey is used to indicate salinity, with the saltiest water darkest. Dense bottom currents in the Arkona and Bornholm basins are shown by their vertical components. These flows increase with depth due to entrainment of ambient water. Pumping in the Bornholm Basin is shown by a vertical pipe transporting winter water into the deepwater. The pump-induced flow increases due to initial mixing at the pipe outlet and thereafter by plume mixing while rising. The flows are further described in Sect. 2.1.

A starting point for the construction of our inflow model is the fact that the salinity of the Baltic Sea is controlled essentially by two external factors, namely (i) the freshwater supply to the Baltic Sea and (ii) the sea level variability in Kattegat that drives the water exchange between Kattegat and the Baltic Sea inside the Fehmarn Belt and the Öresund (the Sound) (Stigebrandt, 1983). This implies that the inflow of new deepwater (and salt) to the Baltic Proper is independent of all oceanographic factors inside the entrance area other than the sea level. Freshwater from the Baltic Sea accumulates in the surface layer of the entrance area (Kattegat and the Belt Sea). Due to shallow sills, the presence of this layer partly blocks inflow of seawater from the deeper part of Kattegat to the Baltic, which strongly influences the salinity of new deepwater flowing into the Baltic. However, freshwater from the Baltic Sea is not only re-circulated back to the Baltic with the new deepwater, it is also permanently lost to Skagerrak by a flow that is driven by the excess pressure caused by the accumulated freshwater in the surface layer of Kattegat. Of course, the long-term freshwater export to Skagerrak equals the long-term supply to the whole Baltic Sea, including the entrance area. A schematic image of the model flows (further described below) is shown in Fig. 2.

The daily water exchange between the Baltic Sea and the entrance area is computed from observed changes in the Baltic Sea volume (sea level) and by accounting for freshwater supply inside the entrance area (Sect. 2.1.1). The amount of freshwater in the entrance area, computed by a simple budget model (Sect. 2.1.2), determines the salinity of the new deepwater flowing into the Baltic Sea, which is computed using an empirical function by Stigebrandt and Gustafsson (2003) (Sect. 2.1.3).

The inflowing new deepwater moves along the bottom of the Arkona Basin as a dense bottom current that entrains water from above, whereby the volume flow increases and the salinity decreases. The dense bottom current was observed, described and analyzed by Arneborg et al. (2007). This current replenishes the pool of new deepwater in the Arkona Basin. The flow and dynamics of new deepwater in the Arkona Basin are described in Sect. 2.1.4. A current transports water from the deepwater pool towards the Bornholm Basin (Liljebladh and Stigebrandt, 1996; Stigebrandt, 1987a). After having entered the Bornholm Basin, the current of new deepwater is transformed into a dense bottom current, which entrains water from above. When the new deepwater in the bottom current has got the same density as the ambient water, it is interleaved in the Bornholm Basin. Only the densest new deepwater makes it all the way through the halocline and down to the greatest depths; less dense new deepwater is interleaved in the halocline. Water exits the Bornholm Basin through the 59 m deep Stolpe Channel. For a description of the dynamics of the Bornholm Basin, see Sect. 2.1.5.

2.1.1   The flow through the entrance straits

The water exchange through the Öresund and the Fehmarn Belt changes the volume of the Baltic Sea and may thus be quantified from observations of the changing sea level. However, freshwater supply also changes the volume. From conservation of volume, one obtains the following equation for the volume change of the Baltic Sea:

$$A_{BS} \frac{dh}{dt} = Q_{BS} + Q_{fBS}. \tag{1}$$

Here dh/dt is the rate of change of the horizontal average sea level h, A_{BS} the horizontal surface area of the Baltic Sea, Q_{fBS} the freshwater supply to the Baltic Sea (inside the entrance area) and Q_{BS} the water exchange of the Baltic Sea with the entrance area. For use in the model, Q_{BS} is separated into two parts, Q_{inBS} and Q_{outBS}, describing inflow and outflow, respectively. The salinity and other properties of these flows are specified in Sect. 2.1.3.

The simple description of the inflowing water is complemented with a low-pass filter removing inflows resulting from high-frequency fluctuations. This is because the first water flowing towards the Baltic Proper after a period of outflow is just surface water from the Baltic. This can be handled by removing certain volumes of the first coming transport, both outgoing and ingoing, using a buffer volume. Such buffer volumes were introduced by Stigebrandt (1983) and discussed by, for instance, Stigebrandt and Gustafsson (2003). The presence of a buffer volume delays the onset of flow events and reduces the importance of high-frequency fluctuations. An appropriate value of the buffer volume will be determined by tuning in Sect. 3.2.

An occasional shift in flow, which is too small for the buffer volume to be completely filled (emptied), will only pause the flow event. If the pause is too long, a new flow event starts when the buffer volume is restored. This buffer restoration time limit will be tuned so that inflow event lengths match the observed inflows; see Sect. 3.2. Finally, for each flow event, an average flow rate is calculated from the cumulated sum of water flow during the event.

2.1.2   The freshwater pool in the entrance area

To compute the salinity S_{in} of the inflowing new deepwater, we need to know the mixture of freshwater and seawater. The latter is the deepwater in Kattegat, which is modified Skagerrak water. This is computed from the amount of freshwater in the entrance area, as shown in Sect. 2.1.3 below. Therefore, we construct a model of the freshwater pool in the entrance area. The thickness of the freshwater pool, H_{fK}, is controlled by Q_{BS} and the salinity of this flow, the freshwater supply to the entrance area, Q_{fK}, and freshwater outflow from the entrance area to Skagerrak, Q_{fS}.

The rate of change of freshwater thickness, dH_{fK}/dt in the entrance area, can be calculated from the following continuity equation for freshwater:

$$A_K \frac{dH_{fK}}{dt} = Q_{fK} + Q_{outBS} \frac{S_0 - S_B}{S_0}$$
$$- Q_{inBS} \frac{S_0 - S_{in}}{S_0} - Q_{fS}. \tag{2}$$

Here A_K is the horizontal area of the freshwater pool, S_B the salinity of the water flowing out from the Baltic Sea and S_0 the salinity of the seawater in Kattegat. The exact value of A_K is determined by model tuning in Sect. 3.2.

The freshwater outflow from the entrance area to Skagerrak can be described as in Stigebrandt and Gustafsson (2003) by

$$Q_{fS} = \frac{g \beta S_0 H_{fk}^2}{2f}. \tag{3}$$

Here g is the acceleration of gravity, f the Coriolis parameter and $\beta \approx 8 \times 10^{-4}$ the salt contraction coefficient, defined by the following approximate equation of the state of seawater:

$$\rho = \rho_f (1 + \beta S), \tag{4}$$

where ρ_f is the density of freshwater and S is the salinity. This equation is applicable when density variations are essentially due to salinity variations, which is the case in the entrance area.

2.1.3   Properties of new deepwater from the entrance area

The new deepwater flowing into the Baltic Sea is a mixture of freshwater and deepwater from the entrance area. Its salinity

S_{in} can be calculated from the following empirical equation (Stigebrandt and Gustafsson, 2003):

$$S_{in} = S_0 \frac{H_K - H_{fK}}{H_K}. \tag{5}$$

The thickness H_K is an empirical constant, which is tuned in order to achieve conservation of salt in the Baltic Sea system; see Sect. 3.2. Equation (5) is valid as long as $S_{in} > S_B$ because the inflowing new deepwater must be saltier than the surface water in the Arkona Basin.

Other state variables of the new deepwater, e.g., oxygen concentration and temperature, represented here by the dummy c_{in}, are calculated according to

$$c_{in} = c_{surf} \frac{S_0 - S_{in}}{S_0} + c_0 \frac{S_{in}}{S_0}. \tag{6}$$

Here, c_0 and c_{surf} are approximated by monthly means of the seawater and surface values in the entrance area as described in Sect. 3.1.

In the model we use the event average of Q_{inBS} and Q_{outBS} for each time step during a flow event, as described in Sect. 2.1.1. However, the salinity, temperature and oxygen concentration of inflowing new deepwater, computed using Eqs. (5) and (6), change for each time step.

2.1.4   The deepwater pool in the Arkona Basin

Because new deepwater from the entrance area is denser than the surface water in the Arkona Basin, it descends along the bottom as a dense bottom current. By mixing with overlying water, the volume flow increases and the salinity (density) decreases. The entrainment of ambient water into the dense bottom current is computed using the plume model in Stigebrandt and Kalén (2013), but with a reversed, i.e., downward, travelling direction and with a value of the empirical entrainment coefficient, EA, that will be tuned, see Sect. 3.2, in order to achieve a good fit with observations. The properties of the entrained water are described by monthly averages of observations from 15 m depth in the Arkona Basin; see Sect. 3.1.

After entrainment, the dense bottom current has the flow rate Q_{inA} and salinity S_{inA} when it reaches the halocline in the Arkona Basin. Below this is a thoroughly mixed pool of dense deepwater that is replenished by the inflowing new deepwater. The pool loses water by the geostrophic flow of deepwater out from the Arkona Basin. The thickness of the Arkona pool, H_A, changes according to the following equation:

$$\frac{d[A_A \times H_A]}{dt} = Q_{inA} - Q_{outA}. \tag{7}$$

Here A_A is the depth-dependent area of the pool and Q_{outA} is the baroclinic geostrophic outflow from the pool, which can be computed as in Stigebrandt (1987a); thus,

$$Q_{outA} = \frac{g' H_A^2}{2 f}. \tag{8}$$

In this equation, $g' = g \, \Delta\rho / \rho_f$ where $\Delta\rho$ is the density difference between the pool water, salinity S_A, and the water overlying it, salinity S_B. Using Eq. (4) gives $\Delta\rho / \rho_f = \beta(S_A - S_B)$. The salinity S_A of the pool water changes according to the following equation:

$$\frac{d[S_A \times A_A \times H_A]}{dt} = Q_{inA} \, S_{inA} - Q_{outA} \, S_A. \tag{9}$$

Temperature and oxygen concentrations are computed analogously. Q_{outA} may empty the pool within a few weeks. This allows us to assume that the residence time of the pool water in the Arkona Basin is so short that oxygen consumption in the pool may be neglected for simplicity. The validity of this assumption is discussed in Sect. 4.2 below.

2.1.5   The Bornholm Basin model

The water flowing out of the Arkona Basin is feeding a dense bottom current in the Bornholm Basin that, after entrainment of ambient water, is interleaved in the stratified basin. It then has the volume flow Q_{inB} and the salinity S_{inB}. The computations of the flow in the dense bottom current are performed in the same way as described in Sect. 2.1.4 above. As described in Sect. 3.2 below, the empirical entrainment coefficient for the Bornholm Basin, EB, is tuned in order to achieve a good fit between model results and observations.

The outflow from the Bornholm Basin is geostrophic, driven by the horizontal west–east baroclinic pressure gradient between the Bornholm Basin and the Stolpe Channel. To resolve the flow vertically, we assume that the current speed, $v(z)$, is proportional to the horizontal baroclinic pressure gradient $dP(z)$ at each depth z; thus,

$$v(z) \sim dP(z). \tag{10}$$

The difference in baroclinic pressure between the two basins, $dP(z)$ equals $P_{BB}(z) - P_{SC}(z)$. It is calculated using daily time series, interpolated from observed profiles, from the Stolpe Channel (pressure $P_{SC}(z)$) and modeled profiles from the Bornholm Basin (pressure $P_{BB}(z)$).

We then normalize $v(z)$ so that the total outflow from the Bornholm Basin, i.e., the sum of the flows in all flowing layers, equals the flow we obtain from Eq. (8) modified for use in the Bornholm Basin. We use the salinity S_{SC} in the topmost 30 m in the Stolpe Channel to compute the reference salinity corresponding to S_B used in Sect. 2.1.4 above.

Our goal is to achieve a good model of hydrographic and oxygen conditions below 60 m in the Bornholm Basin. Since the new deepwater is not in contact with the sea surface after having entered the Arkona Basin, we choose not to model the upper layers driven by processes at the sea surface. Instead, we use interpolated observational data to describe the upper layers.

The rate of oxygen consumption in the basin, OXC, should vary with the export production from the euphotic zone, which should be proportional to the total phosphorus content,

TP, in the 0–40 m layer in winter. The reference oxygen consumption rate from Stigebrandt and Kalén (2013) (OXC_{SK}) was accordingly multiplied by the factor TP/TP_n, where TP_n is the average phosphorus content in the period 1990–2010.

$$OXC = \begin{cases} OXC_{SK}\dfrac{TP}{TP_n} & O_2 \geq 1.5\,mL\,L^{-1} \\[2mm] B \times OXC_{SK}\dfrac{TP}{TP_n}, & 1.5\,mL\,L^{-1} > O_2 > 0\,mL\,L^{-1} \\[2mm] OXC_{SK}\dfrac{TP}{TP_n}, & O_2 < 0\,mL\,L^{-1} \end{cases} \quad (11)$$

To adapt consumption to observations, we need to decrease the consumption rate by a factor B in the interval $1.5 > O_2 > 0$ mL L^{-1}; see Sect. 3.2. The reason for this is oxidation due to reduction of nitrate, nitrite and iron and manganese oxides/hydroxides. Equation (11) is used during stagnant conditions without advective water exchange. However, when the bottom water is oxidized by inflowing new deepwater, it is assumed that both hydrogen sulfide and ammonia, if present in the water column, are oxidized as described in Sect. 3.1. To oxidize 1 mole of hydrogen sulfide to sulfate, 2 moles of oxygen are needed, and to oxidize 1 mole of ammonium to nitrate, 1.5 moles of oxygen are needed.

2.2 Biological issues

2.2.1 Critical oxygen concentrations and re-colonization

In their reviews of marine benthic hypoxia, Diaz and Rosenberg (1995) and Karlson et al. (2002) concluded that critical oxygen concentrations for marine benthic animals varied between species. Mass mortality occurred when the oxygen concentration dropped below the range 0.5 to 1 mL L^{-1}. Rosenberg et al. (2002) studied the reactions of benthic animals in the field during successively declining oxygen concentrations and found that mass mortality occurred at 0.7 mL L^{-1}. However, behavioral responses and effects on reproduction occur in higher concentrations – commonly around 2 mL L^{-1}.

The value of 2 mL L^{-1} has been used as the limit below which hypoxia is defined and hypoxic waters are considered to harbor only a reduced benthic macrofauna under stress with adverse behavior and lack of reproduction ability (Conley et al., 2009a; Diaz and Rosenberg, 1995). In hypoxic areas where the sediments also contain toxic hydrogen sulfide close to the sediment surface, the conditions will be particularly unsuitable for benthic animals. Where oxygen minimum zones occur naturally, like in the eastern Pacific, animals are adapted to oxygen deficiency and tolerate oxygen levels close to zero (Levin, 2003). Such adaptations have not occurred in the Baltic, which is a young sea established after the last glaciation (Elmgren, 2001). Fish species are generally less tolerant to low oxygen concentrations and show avoidance reactions and reproductive failure in oxygen saturations less than 50 % (≈ 4 mL L^{-1}) (Breitburg, 2002).

The recovery process of a formerly hypoxic/anoxic system that is re-oxygenated depends on the severity and extent to which the system was disturbed. The succession of benthic animals will most probably follow the Pearson and Rosenberg (1978) paradigm, which shows that the colonization begins with small, opportunistic species in high numbers, which will facilitate the sedimentary habitat for the later successional stages. Transitory immigrants will follow the pioneering stages and, finally, a more mature benthic community will establish itself. In areas where the sediments have a high organic content because of heavy loadings, the time of recovery to pre-pollution levels could be 6 to 8 years (Rosenberg, 1976). In a Swedish fjord, where the low oxygen concentrations caused mass mortality, but the near-bottom water was soon re-oxygenated, the recovery of the benthic fauna to pre-hypoxic compositions in this case took about 2 years (Rosenberg et al., 2002), the reason being that the sediment was not over-enriched by organic matter. In an experiment with oxygenation of the basin water in the usually anoxic By Fjord, on the western coast of Sweden, colonization of the bottoms started after about 1 year, when the sediment surface had been oxidized (Stigebrandt et al., 2014b).

2.2.2 The cod reproduction volume (CRV)

The cod reproduction volume (CRV) is defined as the water volume where the physical properties allow the survival of cod eggs, a prerequisite for successful cod spawning. The criteria that need to be fulfilled are salinity $S > 11$, oxygen concentration $O_2 > 2$ mL L^{-1}, and temperature $T > 1.5\,°C$. Vallin et al. (1999) conclude in their review that the combination of high salinity and high availability of oxygen is necessary for large recruitment of cod in the Baltic Sea. Thus, large recruitment is only possible after major inflows of new deepwater. Since inflow of new deepwater in general occurs during the winter months (Matthäus and Schinke, 1994), and oxygen consumption causes a reduction in the oxygen concentration over time, the CRV will be smaller later in the spawning season (Vallin et al., 1999). Under present environmental conditions, the Bornholm Basin has been shown to be the most favorable area for successful cod reproduction in the Baltic Sea (MacKenzie et al., 2000). The season for cod spawning has shifted from the spring months towards the summer months (e.g., Hinrichsen et al., 2007; Vallin et al., 1999). To reveal seasonal effects, CRV in the second and third quarters of the year should be presented separately.

3 Data and tuning

3.1 Data

To calculate the inflow Q_{BS}, we use Eq. (1), which needs data on the horizontal mean sea level h and the freshwater supply. For a first estimate of Q_{BS} with the temporal resolution of 1 day, we use 5 day running averages of the observed sea level in Stockholm. Strictly speaking, one should use the horizontal mean sea level of the Baltic Sea h for the computations of Q_{BS}, but since Stockholm is situated in the nodal line of the first mode of sea surface oscillations, we use the sea level in Stockholm as a proxy for horizontal mean sea level (e.g., Samuelsson and Stigebrandt, 1996). For Q_{fBS}, we use a time series of monthly values of river runoff to the Baltic Sea, with the Arkona Basin as the western limit. To this, we add the average annual cycle for precipitation minus evaporation from the sea surface (Rutgersson et al., 2002). The average of Q_{fBS} in the period 1990–2010 was calculated to be $15\,500\,\mathrm{m}^3\,\mathrm{s}^{-1}$.

The start value for H_{fK}, the thickness of the freshwater pool in the entrance area, is estimated from salinity observations from Anholt E ($56°40'$ N, $12°7'$ E, in Fig. 1) from the end of November 1989 to the end of January 1990. Of 1636 available vertical hydrographic profiles from the southern Kattegat ($55°0'–56°12'$ N, $10°48'–12°0'$ E, white box in Fig. 1), all temperature and oxygen data from 5 m depth (1635 and 782, respectively) are used to compute monthly averages to represent the annual cycle of the surface layer in the entrance area. All temperature and oxygen data measured in deepwater of salinity ≥ 32 are used similarly to compute monthly averages representing the seawater annual cycle. The properties of the inflowing new deepwater are then given by the monthly averages and the proportions of surface and deepwater in the specific inflow, as described by Eq. (6).

We use all available data (historic vertical hydrographic profiles) at 15 m depth from 1990 to 2010 from the eastern half of the Arkona Basin ($54°42.0'–55°9.0'$ N, $13°41.4'–14°18.0'$ E, dashed box in Fig. 1) (455 observations) to compute monthly averages of salinity, temperature and oxygen describing the properties of the upper layer in the Arkona Basin. Water with these properties is entrained into the dense bottom current with new deepwater. 421 observed vertical salinity profiles from the same area, going down to at least 35 m depth and with bottom salinity greater than 11.6, were used to calculate the baroclinic outflow from the dense bottom pool according to Stigebrandt (1987a). The observed outflow is compared to model estimates for different values of the entrainment coefficient EA; see Sect. 3.2 below.

In the Bornholm Basin, we use 361 profiles from station BY4 ($55°23.0'$ N, $15°20.0'$ E, white o in Fig. 1) and 247 observed vertical hydrographic profiles from station BY5 ($55°15.0'$ N, $15°59.0'$ E, white x in Fig. 1) from 1990 to 2010 for the following computations. An average of BY5 profiles from 17 November 1989 and 24 January 1990 is used for the initial water column properties of the model on the start date (3 January 1990). The vertical resolution varies throughout the profiles (decreases with depth) and also somewhat between profiles, due to missing data values. The most common resolution is 5 m between 0 and 20 m depth, and then a 10 m resolution down to 90 m. Data gaps are more common in the winter, when storms and ice cover stop the observation vessels from reaching the stations. The observed salinity, tem-

perature and oxygen concentration profiles from BY4 and BY5 are interpolated to form daily time series that are then compared to the corresponding time series resulting from the model. These vertical profiles are also used to compute the "observed" cod reproduction volume during the second and third quarters of each year from 1990 to 2010. This is then compared to the modeled cod reproduction volume. The interpolated daily time series from BY4 are chosen to represent the state variables (salinity, temperature and oxygen) above 30 m depth in the Bornholm Basin.

From observations obtained at BY5, we construct time series of the annual winter content of total phosphorus in the uppermost 40 m, TP, to estimate the year-to-year variation of the export production supplying organic matter below 60 m depth in the Bornholm Basin as described in connection with Eq. (11).

From observations from depths greater than or equal to 85 m at BY5 (43 observations), we have estimated that changes in the mole ratio between ammonium and hydrogen sulfide equals $1/5$ under anoxic conditions. Furthermore, we estimated that the concentration of ammonium when the water turns from oxic to anoxic equals $4\,\mathrm{mmol\,m}^{-3}$. Ammonium and hydrogen sulfide are oxidized when water exchange mixes old oxygen-free deepwater and new deepwater as described in Sect. 2.1.5.

219 vertical salinity and temperature profiles from the Stolpe Channel area ($55°9'–55°24'$ N, $16°42'–17°54'$ E, black box in Fig. 1) are used to form the interpolated daily time series used to calculate P_{SC} for the calculation of the geostrophic outflow from the Bornholm Basin into the Baltic Sea. The same data set is used to calculate a time series of S_{SC}, the mean salinity of the upper 30 m in Stolpe Channel.

Finally, hypsographic data for the studied sub-basins of the Baltic Sea were obtained from Seifert et al. (2001). The horizontal surface area of the Baltic Sea, A_{BS}, used in Eq. (1), equals $3.7 \times 10^{11}\,\mathrm{m}^2$. For other data sources, see the Acknowledgements.

3.2 Tuning

Model coefficients have been tuned in order to improve the match between computed and observed data. Accordingly, the buffer volume located at the entrance sills equals $30\,\mathrm{km}^3$, with a restoration time limit of 7 days. When the model run begins, the buffer volume is empty. This approach gives long-term average flow rates to/from the Baltic of $16\,400/32\,100\,\mathrm{m}^3\,\mathrm{s}^{-1}$, respectively, for the investigated period 1990–2010. Observations from 15 m depth in the Arkona Basin for the same period give an average outgoing salinity of 7.9. If salt content and volume of the Baltic are assumed to be preserved over the period 1990–2010, new deepwater needs to have an average salinity of 15.5. This is achieved in the model by setting S_0 equal to 33 and H_K in Eq. (5) equal to 6.7 m. The freshwater pool area, A_K, was adjusted to avoid overly salty inflows. The value

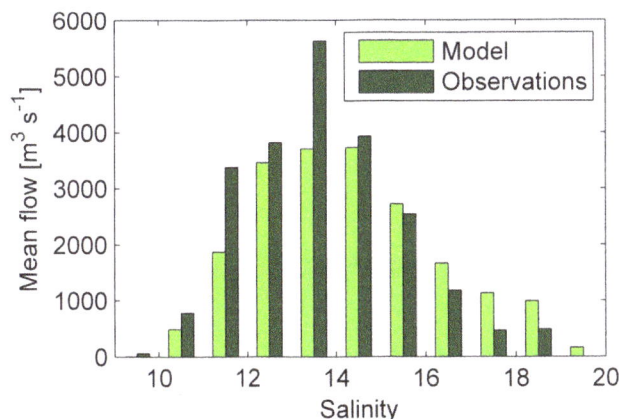

Figure 3. Distribution of observed and modeled average flow rates among different salinities for the baroclinic outflow from the Arkona Basin.

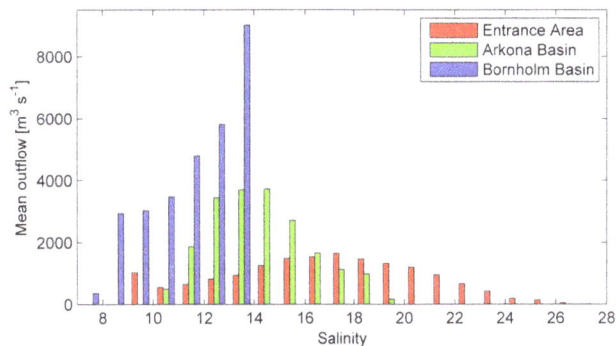

Figure 4. Distribution of the average volume flow among different salinities for the modeled deepwater flows from the entrance area, the Arkona Basin and the Bornholm Basin.

4   Results and discussion

The modeled flows of new deepwater in different salinity intervals from the Baltic entrance area, out from the Arkona Basin and out from the Bornholm Basin, respectively, are shown in Fig. 4. As expected, the salinity decreases and the volume flow increases downstream due to mixing in the basins.

4.1   Arkona Basin – deepwater flow estimates using historic vertical hydrographic profiles versus the pool model

The distributions of average flow rates among different salinities according to the Arkona Basin pool model and computed from hydrographical observations are shown in Fig. 3. We find that hydrographical observations for the period 1990–2010 give an ensemble average flow rate from the dense pool of $22\,300\,\mathrm{m^3\,s^{-1}}$ with an average salinity of 13.7, which is similar to the results estimated from hydrographical observations for an earlier period in Stigebrandt (1987a). The corresponding values according to our pool model are $19\,900\,\mathrm{m^3\,s^{-1}}$ and 14.4, respectively. The modeled outflow of deepwater from the Arkona Basin thus follows the observations rather well.

It was shown in Stigebrandt (1987a) that, due to the large variability of the flow of new deepwater in the Arkona Basin, about 500 independent observations are needed to estimate the mean inflow with an error of less than 10 %. Thus, from the number of observations available (451, see Sect. 3.1), we expect an error within the range ±10 %. This means that the computed flow might be closer to reality than statistics produced by the observations. The evolution of observed and modeled vertical distributions of salinity, temperature and oxygen shown below may deviate rather much due to the combination of large variability and sparse observations. Another reason for deviations is the fact that model results are from a horizontally integrated model with high temporal res-

$42\,000\,\mathrm{km^2}$, proposed by Stigebrandt (2001), gives good correspondence between the model and observations. The thickness H_A of the dense Arkona pool is measured upwards from 41 m depth, which is the sill depth of the Arkona Basin (Stigebrandt, 1987a).

The empirical entrainment coefficient of dense bottom currents, which determines the rate of increase in the volume transport per unit depth, was tuned to give a good fit between model results and observations. For the Arkona Basin, $EA = 0.02\,\mathrm{m^{-1}}$ gives a satisfactory relationship between salinity and volume flowing out from the Arkona Basin, as seen in Fig. 3. For the Bornholm Basin, $EB = 0.01\,\mathrm{m^{-1}}$ gave a good fit between observations and modeled values of salinity, oxygen and temperature. A wide and thin bottom gravity current, flowing on a sloping plane, entrains more ambient water than a thicker and narrower current that runs in a canyon (Stigebrandt, 1987b). This qualitatively explains the difference between EA (plane) and EB (canyon).

The salinities S_in of seven large modeled inflows into the Bornholm Basin were adjusted manually in order to get a better fit between modeled and observed salinity in the Bornholm Basin. The salinity was slightly increased in the inflows that occurred in 1994, 2003 and 2010 and slightly decreased for the inflows in 1997 and 2007. The salinity in two subsequent inflows during 1998 was increased and decreased, respectively.

To conform to observations in the Bornholm Basin, we halved the oxygen consumption rate for the interval $1.5 > \mathrm{O_2} > 0\,\mathrm{mL\,L^{-1}}$, thus using $B = 0.5$ in Eq. (11). Although interesting in itself, we found no strong reason for the present investigation to fine-tune the parameterization of oxygen consumption by, for instance, letting B vary in the interval $0–1.5\,\mathrm{mL\,L^{-1}}$ or using a B value different from 1 in the anoxic case.

olution, while observations are obtained at a few depths of a sparsely visited location (BY4).

4.2 Bornholm Basin – observed and modeled vertical hydrographic profiles

The observed salinity at station BY4 (Fig. 5a) can be compared to the modeled salinity (Fig. 5b). The modeled salinity is well correlated with observations below 40 m. The evolution of the salinity is clearly governed by three different physical processes that are included in the model: (i) during deepwater inflows, salinity increases and the halocline rises; (ii) between inflows, vertical diffusion dominates, whereby salinity decreases with time; and (iii) outflows through the Stolpe Channel lower the halocline. The observed and modeled temperatures are shown in Fig. 6.

The observed oxygen concentration at BY4 (Fig. 7a) is compared to modeled levels in Fig. 7b. In the plots, hydrogen sulfide is expressed as negative oxygen such that 2 moles of negative oxygen correspond to 1 mole of hydrogen sulfide. The actions of the three advection–diffusion processes seen in Fig. 5 can also be seen in Fig. 7. However, in addition, oxygen is influenced by oxygen consumption (also included in the model) that decreases the oxygen concentration. Vertical diffusion of oxygen from above tends to compensate for oxygen consumption in the enclosed basin. The modeled oxygen concentration largely follows the observations from station BY4, but also fits data from station BY5. In Fig. 7a we do not see any traces of small volumes of water with relatively low oxygen concentrations that might come from the Arkona Basin in the fall. This is discussed further below.

The comparison between observations and the reference model run (Figs. 5a and b, 7a and b) shows that the timing of the modeled inflows corresponds well to the observations in the Bornholm Basin. However, the observed vertical hydrographic profiles used for validating the model generally have a monthly time resolution, usually with still lower resolution during winter months. The resolution of observations is thus very low compared to the daily resolution of the model results. It is therefore possible that the observations may miss the precise timing of the inflow and the exact water properties just after a deepwater inflow. Sometimes the observed bottom salinities appear to be higher than the modeled bottom salinities. This suggests that the model produces a more homogenous water column in the deepest parts of the basin, but with a similar salt content to that observed in the whole water column.

Observations show that the halocline uplift after larger inflows, such as the one in early 2003, can persist for up to 1 year. In the model, such uplifts seem to decay more quickly, suggesting that the modeled outflow might be too fast. However, another explanation may be that the vertical resolution of observations is typically 10 m, while the model resolution equals 1 m. Observations would then give the im-

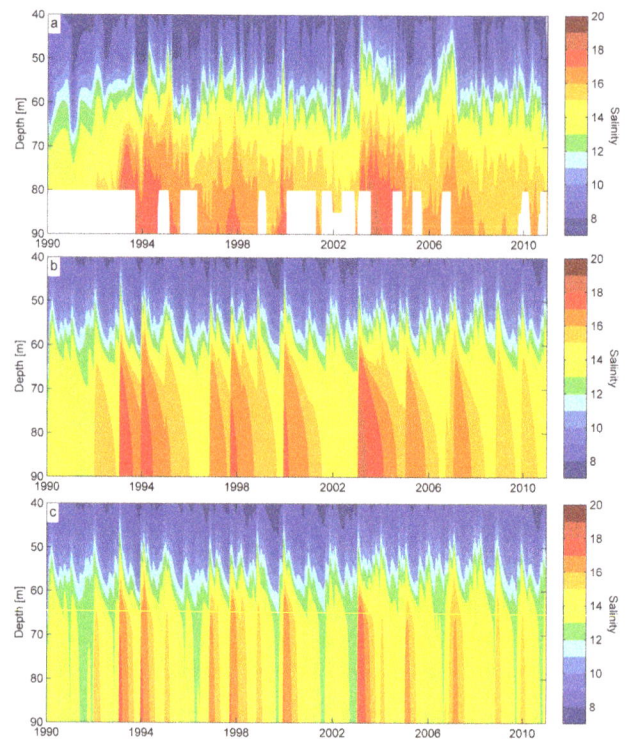

Figure 5. (**a**) The observed salinity from BY4, (**b**) the modeled salinity (reference run) and (**c**) the modeled salinity when pumping 1000 m^3 s^{-1} from 30 m depth down to 90 m.

pression of a persistent halocline in the depth interval 50 to 60 m. Since the modeled conditions in the basin water below sill depth (60 m) appear to follow observations quite well, the possibly slightly too rapid outflow (above 60 m depth) should not be critical for the results in the basin water (below 60 m depth).

It is obvious that the modeled temperature (Fig. 6b) in the halocline (50–65 m depth) is often too low in summer and too high in winter compared to the observed temperature (Fig. 6a). One reason for the deviation in winter may be that our model lacks surface dynamics and deep vertical convection occasionally reaching down to the sill level. Deviations all around the year may be due to overly sparse data in the entrance area, where water temperatures are given by monthly means, meaning that inflows with extreme temperatures are missed. However, since density variations in the Bornholm Basin are dominated by salinity variations, it is of little dynamical importance that the modeled temperature is smoothed.

Our model assumes that there is no oxygen consumption in the new deepwater when passing the Arkona Basin. We will investigate whether or not this assumption may be violated in periods. The volume of the basin water, below the sill level at 41 m depth, in the Arkona Basin is about 12 km^3, estimated using the topographic database in Seifert et al. (2001), which is only 8 % of the volume of the basin

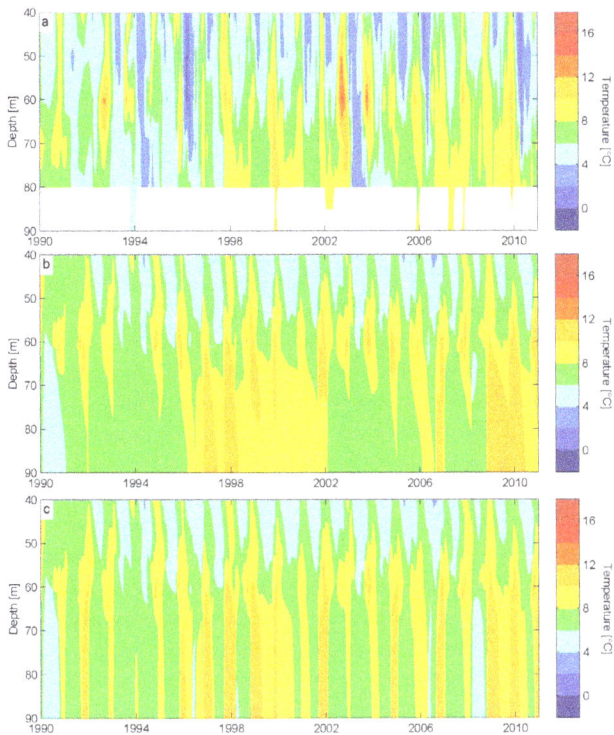

Figure 6. (a) The measured temperature from BY4, **(b)** the modeled temperature (reference run) and **(c)** the modeled temperature when pumping $1000\,\mathrm{m}^3\,\mathrm{s}^{-1}$ from 30 m depth down to 90 m.

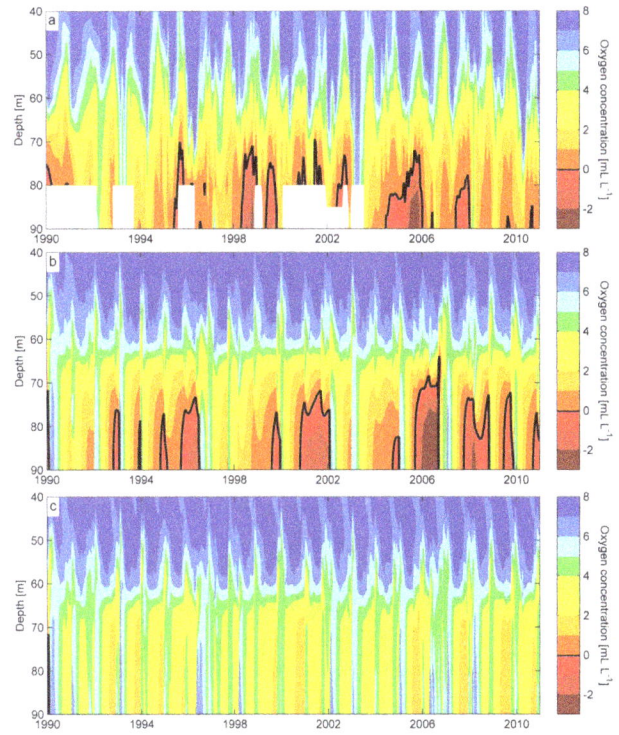

Figure 7. (a) The measured oxygen content from BY4, **(b)** the modeled oxygen content (reference run) and **(c)** the modeled oxygen content when pumping $1000\,\mathrm{m}^3\,\mathrm{s}^{-1}$ from 30 m depth down to 90 m.

water in the Bornholm Basin. If initially of high salinity, water may stay in the basin from spring to fall, whereby it may obtain a low oxygen concentration due to local consumption. The lowest record of oxygen concentration in the Arkona Basin, according to the data available for this study, equals $0.33\,\mathrm{mL}\,\mathrm{L}^{-1}$ $(0.47\,\mathrm{mg}\,\mathrm{L}^{-1})$ (December 1997) $(1\,\mathrm{mL}\,\mathrm{L}^{-1} \approx 1.4\,\mathrm{mg}\,\mathrm{L}^{-1})$.

In Table 1 are given volume and mean salinity and mean oxygen concentration of all 10 cases of hypoxic water with volumes greater than $3\,\mathrm{km}^3$ estimated from hydrographical observations in the Arkona Basin during the period 1990–2011 (22 years). When the basin water resting in the Arkona Basin is eventually flushed by a new denser deepwater, parts of the flushed basin water will be entrained into the new deepwater, while the remainder is lifted above the sill level; cf. the process of water exchange in the By Fjord (Stigebrandt et al., 2014b). After the uplift, the flushed deepwater will be contained in a rather thin layer for topographical reasons. For example, a layer of volume $10\,\mathrm{km}^3$ centered at 35 m depth will be only 2 m thick. This layer has a lower oxygen concentration than the surrounding water, which is why oxygen will diffuse efficiently into the layer from both above and below. Hereby, the oxygen content is expected to increase already before the water in this layer exits through the Bornholm Channel.

In the Bornholm Basin, the flushed basin water from the Arkona Basin will entrain ambient water of low salinity (average 9) and high oxygen content (average $6.7\,\mathrm{mL}\,\mathrm{L}^{-1}$ $(9.8\,\mathrm{mg}\,\mathrm{L}^{-1})$) at the rate EB, meaning that the volume has increased by 30 % when reaching 60 m depth. The estimated salinity and oxygen concentration of the water when reaching the depth of 60 m in the Bornholm Basin are given in Table 1. As can be seen, the water is in none of the cases hypoxic when reaching 60 m, and the salinity is usually too low to permit penetration down into the deep basin. The figures for oxygen concentration given in Table 1 should be conservative, because the probably substantial diffusive supply of oxygen by vertical mixing in the Arkona Basin after the uplift discussed above is not accounted for.

4.3 Hydrographical changes of pumping in the Bornholm Basin

The model was run with the pumping turned on with an initial (jet) mixing rate $\alpha = 10$, meaning an eleven-fold increase in the volume flow, and plume entrainment $E = 0.05$, meaning a 5 % increase in the volume flow when the buoyant plume rises by 1 m. This fits well with the observed mixing of the pumped flow in the By Fjord experiment (Stigebrandt et al., 2014b). The natural vertical mixing in the basin is modeled as specified in Stigebrandt and Kalén (2013). The

Table 1. Volume, $V(A)$, mean salinity, $S(A)$, and mean oxygen concentration, $O_2(A)$, of hypoxic water observed in the Arkona Basin during the period 1990–2011. Also shown are estimated salinity, $S(B)$, and oxygen concentration, $O_2(B)$, when the water reaches 60 m depth in the Bornholm Basin.

Day	Month	Year	$V(A)$[a]	$S(A)$	$O_2(A)$[b]	$S(B)$	$O_2(B)$[b]
31	08	1992	23.35	14.33	0.90 (1.29)	13.1	2.28 (3.25)
1	09	1994	15.13	15.07	1.57 (2.24)	13.82	2.80 (3.99)
15	12	1997	15.13	16.62	1.09 (1.55)	14.86	2.42 (3.46)
23	09	1998	19.09	14.12	1.50 (2.14)	12.94	2.74 (3.91)
25	09	2002	15.13	16.66	1.69 (2.41)	14.89	2.89 (4.12)
30	10	2007	19.09	14.53	1.20 (1.71)	13.25	2.50 (3.57)
17	09	2008	19.09	14.42	1.46 (2.09)	13.17	2.70 (3.86)
16	09	2009	11.50	15.30	1.82 (2.61)	13.85	2.99 (4.27)
15	09	2010	15.13	13.64	1.53 (2.18)	12.57	2.77 (3.95)
14	09	2011	8.26	13.01	1.33 (1.90)	12.08	2.61 (3.72)

[a] Volume is given in km^3. [b] Oxygen values are given in $mL\,L^{-1}$ ($mg\,L^{-1}$).

results with pumping turned on are compared to the model results with pumping turned off presented in Sect. 4.2 above. We are primarily interested in finding out the volume transport by pumping required to maintain oxic conditions in the deepwater. The water intake depth was set to 30 m, but intake depths at 0 and 40 m were also tested for comparison. In all model runs, the horizontal pump outlet is at 90 m depth. The investigated pumping rates were 800, 1000, 1200, 1500 and $4000\,m^3\,s^{-1}$. At the present stage, it is not necessary to specify the number of pumps, their geographical location and the outlet velocity needed in practice to achieve the properties of the modeled pumping.

The salinity and oxygen content in the water column with a pumped volume flow of $1000\,m^3\,s^{-1}$ can be seen in Figs. 5c and 7c, respectively. Notice that pumping leads to an increased number of inflow events in the basin water and that the salinity never falls below 11 beneath 65 m depth. Notice also that the oxygen content for this pumping rate only rarely drops below $1.5\,mL\,L^{-1}$ and never below $1\,mL\,L^{-1}$. It is also to be noted that, on all occasions when the oxygen content in the pumping scenario (Fig. 7c) drops below $2\,mL\,L^{-1}$, the oxygen content in the model reference run (Fig. 7b) is slightly too low compared to observations (Fig. 7a). The bottoms deeper than 75 m, with an area of about $5000\,km^2$, will be habitable for benthic biomass, as further discussed in Sect. 4.6. The oxygen conditions produced by the model in the basin water are discussed further below.

The effects of different pumping rates on the mean flow rate, the number and properties of inflow events and on the basin salinity, temperature and oxygen levels, represented by values from 75 m depth, are shown together with the effects on the CRV in Table 2. It can be seen that an increased pumping rate leads to an increased rate of water exchange at the greatest depths. For the highest pumping rates, we get the highest average inflow rates, but notice fewer inflow events, since inflow events merge; compare Table 2 for pumping rates 1500 and $4000\,m^3\,s^{-1}$. Like in the

By Fjord experiment (Stigebrandt et al., 2014b), the average salinity in the basin is decreased by pumping. This explains why pumping allows for increased basin ventilation by inflows that previously were not able to penetrate to the bottom of the basin. The minimum basin salinity decreases approximately linearly with increased pumping rate (approximately -0.1 per $100\,m^3\,s^{-1}$); thus, if the pumping rate exceeds $2200\,m^3\,s^{-1}$, the minimum salinity may occasionally drop below 11, which affects, e.g., the minimum CRV for the highest investigated pumping rate $4000\,m^3\,s^{-1}$ (see Table 2).

Inflows contribute 3 to 5 times more oxygen than the pumped water, dependent on the magnitude of the pumped volume flow (Table 2). By pumping $1000\,m^3\,s^{-1}$, the inflow of oxygen by new deepwater is increased by 144 %. An additional increase in the oxygen inflow by about 56 % is contributed by the pumped water. Both average and minimum oxygen levels in the basin water increase significantly with increased pumping. A pumped volume flow of $1000\,m^3\,s^{-1}$ increases the average oxygen level at 75 m from 1.9 to $3.6\,mL\,L^{-1}$. The minimum (average of the 2 % lowest values) level increases from -2.4 (anoxic) to $1.4\,mL\,L^{-1}$, respectively. The pumping makes particularly large difference in the deepest parts of the basin because the vertical downward motion in the whole basin, induced by the pumping, transports oxygen and other substances from higher levels much more efficiently than the natural vertical diffusion does; cf. Fig. 7c and an example in Stigebrandt and Kalén (2013).

In the fall, hypoxic basin water from the Arkona Basin may be exchanged and flow into the Bornholm Basin, as discussed at the end of Sect. 4.2. The oxygen deficit of a package of $20\,km^3$ of water with an O_2 concentration of $2.7\,mL\,L^{-1}$ ($3.9\,mg\,L^{-1}$) instead of $6.7\,mL\,L^{-1}$ ($9.8\,mg\,L^{-1}$) (oxygen deficit 6 mg per kg water) equals 120 kton of oxygen. The total oxygen supply in the reference case without pumping equals $647\,kton\,yr^{-1}$ (Table 2). For pumping $1000\,m^3\,s^{-1}$, the supply is estimated at 1940 kton

Table 2. Model results for 1990–2010 with the pump intake placed at 30 m depth and for different pumping rates.

Pumped volume flow ($m^3 \, s^{-1}$)	Ref.[a]	800	1000	1200	1500	4000
Average inflow rate ($km^3 \, year^{-1}$)[b]	66.8	141.0	160.0	177.7	205.2	587.5
Number of inflows[c]	17	38	45	50	47	37
Average inflow salinity[d]	15.89	15.05	14.90	14.79	14.69	14.08
Average inflow of O_2 ($kton \, year^{-1}$)[b]	647	1385	1579	1769	2051	6076
Flow of O_2 via pump ($kton \, year^{-1}$)	0	288	360	431	536	1438
Average basin O_2 ($mL \, L^{-1}$)[e]	1.92	3.44	3.64	3.81	4.04	5.10
Minimum basin O_2 ($mL \, L^{-1}$)[f]	−2.41	1.26	1.38	1.45	1.68	3.78
Average basin salinity[e]	15.41	14.52	14.38	14.25	14.07	13.11
Minimum basin salinity[f]	13.34	12.57	12.46	12.32	12.09	10.80
Average basin temp (°C)[e]	8.16	8.12	8.11	8.11	8.09	7.92
Q1 of CRV (km^3)[g]	41.9	57.2	65.7	81.1	100.6	112.2
Minimum CRV (km^3)[h]	15.2	54.0	54.0	54.0	54.0	0

[a] Model reference values (pumping rate set to 0). [b] Volumes reaching 90 m depth. [c] The number of inflows is defined as the number of new inflow events reaching 90 m and lasting at least 24 h after a minimum of 1 week of stagnation. [d] Volume-weighted average for salinities of the inflow of new deepwater. [e] The average of values from 75 m depth. [f] The average of the 2 % lowest values at 75 m depth. [g] The average of the 25 % smallest CRVs calculated for the cod spawning period, April to June and June to September. [h] Calculated for the cod spawning period, April to June and June to September.

(360 kton from pumping and 1580 kton from inflows). The oxygen supply has thus increased by 200 %. If we include the package with low oxygen content, the total supply decreases to 1820 kton. This gives an increase in the oxygen supply by 181 %. This should be a conservative estimate, because most of the packages in Table 1 should not reach the greatest depth of the Bornholm Basin at the suggested rate of pumping, as their salinity is too low. In addition, the packages from the Arkona Basin are released in September or later, which is too late in the year to affect the CRV, because the spawning occurs in the time period April–September. It would be relatively easy to complement the oxygenation system in the Bornholm Basin by a system in the Arkona Basin to keep the small deepwater volume there oxygenated and in this way to eliminate the possibility of low oxygen concentrations developing in the Arkona Basin. This would also be quite beneficial for the bottom fauna in the deepest part of the Arkona Basin.

The average basin temperature is 8.1 °C in the reference case and does not change significantly with any of the tested pumping rates. Hence, the long-term basin average temperature does not appear to be affected by pumping when the pump intake depth is in the interval 30–40 m. However, the average and maximum temperatures at 75 m depth increase by approximately 0.2 °C if the water intake is located at the sea surface. The salinity changes very little if the intake depth is changed to 0 or 40 m.

The changes in salinity due to pumping can easily be seen in Fig. 6 in Stigebrandt and Kalén (2013), which shows that the pumped winter water, of salinity about 8, is interleaved in the lower part of the halocline, well below 60 m depth, due to strong mixing with the ambient deepwater. When interleaved, the volume flow of the pumped plumes has increased by a factor of 20 to 30. Equally strong mixing of the pumped flow was observed in the By Fjord experiment, where the volume flow of pumped plumes increased by a factor of 30 before they were interleaved (Stigebrandt et al., 2014b). Because the pumped water is interleaved in the lower part of the halocline, the dynamics of the surface layer, including the local supply of nutrients and the production of organic mat-

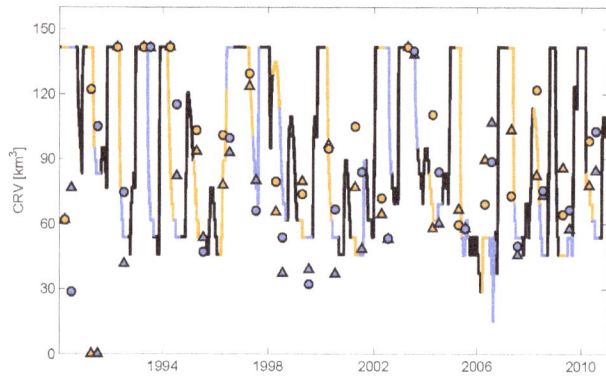

Figure 8. The modeled cod reproduction volume (CRV) (black line). The second and third quarter periods are highlighted by yellow and blue, respectively. The second and third quarterly average cod reproduction volumes, calculated from BY4 and BY5 data, are represented by triangles and circles, respectively.

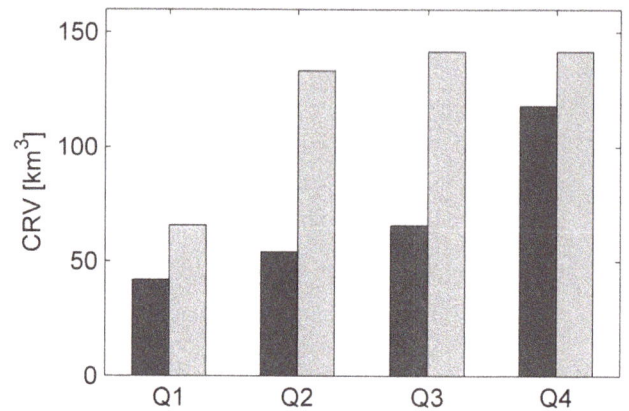

Figure 9. Model average cod reproduction volumes (CRV) from the second and third quarters of the year separated by quartiles and averaged, here presented in rising order (Q1–Q4). Dark grey bars represent the model reference run (no pump) and the light grey bars represent the model with a pump volume flow of $1000 \, \mathrm{m}^3 \, \mathrm{s}^{-1}$.

ter, should be negligibly influenced by the pumping. However, since the oxygenation reduces the leakage of phosphorus into the deepwater of the Bornholm Basin (Stigebrandt et al., 2014a), this water will carry less phosphorus with it when it is flushed and further transported into the basins east of the Bornholm Basin. Consequently, there will be a decreased upwelling of P in the basins east of the Bornholm Basin. The effect of decreased P loading of the Baltic Proper will eventually also be felt in the surface layer of the Bornholm Basin, as an indirect effect of the pumping, and lead to a decreased production of organic matter sinking down into the deep basin. This will be beneficiary to cod recruitment, since the oxygen consumption decreases in the deepwater. The benthic biomass will get less feed falling down from the surface layers.

4.4 The effect on cod reproduction volume by artificial mixing

The CRV was calculated as the volume below 61 m depth that fulfills the definition in Sect. 2.2.2. These conditions may also be met above 61 m depth, but cod eggs present above this depth might be transported out from the Bornholm Basin by the outflow and, during that process, salinity will likely decrease due to entrainment. Therefore, the water volume above this depth is not included in the calculation of CRV. The total water volume below 61 m depth is $142 \, \mathrm{km}^3$, which is thus the upper limit for the CRV in our approach.

The quarterly CRV from the model (no pumping) is compared to the CRV calculated from observations from BY4 and BY5 (Fig. 8). The second (April to June) and third (July to September) quarters are the periods when cod spawning occurs, and these periods are highlighted in Fig. 8. The minimum and average CRV given by the model are 16 and $88 \, \mathrm{km}^3$, respectively. The corresponding values from observations were calculated to 0 and $78 \, \mathrm{km}^3$ for BY4 and 29 and

$89 \, \mathrm{km}^3$ for BY5. It is apparent from these values and from Fig. 8 that the modeled CRV generally follows the observed CRV well.

There are apparently observations during the studied period when the CRV is small and even vanishes (Fig. 8). We have partitioned the CRV data by quartiles, Q1–Q4, and calculated the average for each partition. The average of the lowest 25 % of the CRV values (Q1) based on spring and summer model values equals approximately $40 \, \mathrm{km}^3$ (Fig. 9). By pumping $1000 \, \mathrm{m}^3 \, \mathrm{s}^{-1}$, this value would increase to more than $65 \, \mathrm{km}^3$ (see also Table 2 for results of other tested pumping rates). Figure 9 shows that it is in particular during years with small CRVs that pumping might make a big difference. Thus, by pumping, there will be a large CRV each year, and cod recruitment should not fail in any single year due to lacking CRV.

The model runs with the pump intake at 0 and 40 m depth showed a similar variation as the runs in Table 2, but with a somewhat smaller CRV for the 40 m intake. For example, the Q1 and minimum CRV with a pump volume flow of $1000 \, \mathrm{m}^3 \, \mathrm{s}^{-1}$ were 66 and $54 \, \mathrm{km}^3$ for the 30 m intake depth. For an intake placed at 40 m, these values were 60 and $46 \, \mathrm{km}^3$. The corresponding values with the intake located at the sea surface are 70 and $54 \, \mathrm{km}^3$.

Gustafsson et al. (2008) performed computations showing that halocline ventilation might not increase CRV. This is opposite to the conclusion reached in the present paper. The reason for the diverging opinions is that Gustafsson et al. (2008) reached their result without studying the conditions in the Bornholm Basin explicitly, the only basin with successful cod recruitment during the last decades.

The oxygenation of the deepwater in the Bornholm Basin may only marginally change the supply of organic matter and the oxygen consumption in the basin water beneath the sill depth, as explained above. The changes in hydrography and oxygen conditions induced by pumping will therefore be lost when the pumps are turned off for a longer period or permanently.

4.5   Outflow of deepwater from the Bornholm Basin

The effects of artificial mixing on the water flowing from the Bornholm Basin to the Stolpe Channel were investigated for the pumping rate of $1000 \, \mathrm{m^3 \, s^{-1}}$. The total geostrophic transport, reference flow ca. $30\,000 \, \mathrm{m^3 \, s^{-1}}$, increases by approximately 2 %. Transports with the highest salinities (ref. ca. $9000 \, \mathrm{m^3 \, s^{-1}}$; $S > 13$) decrease by 17 %. Flows with oxygen content less than $2 \, \mathrm{mL \, L^{-1}}$ disappear completely, while flows in general are shifted towards higher oxygen content. For instance, volume transports with very high oxygen content (ref. ca. $6300 \, \mathrm{m^3 \, s^{-1}}$; $O_2 > 7 \, \mathrm{mL \, L^{-1}}$) increase by 10 %. The model thus shows that the salinity distribution of the outflow to the Stolpe Channel becomes narrower and the oxygen concentration becomes higher when artificial mixing is applied. This should decrease the vertical stratification, as shown by a numerical experiment in Stigebrandt (1987b), and improve the oxygen conditions in the deeper parts of the Baltic Proper east of the Bornholm Basin.

4.6   Benthic fauna, potential new biomass and fish feed

The benthic fauna in the Bornholm Basin was studied in the 1950s and 1960s by Demel and Mulicki (1954) and Leppäkoski (1969), respectively, in relation to changes in salinity (13.5–17) and oxygen concentrations (0–$5 \, \mathrm{mL \, L^{-1}}$). Up to seven species were recorded in the area deeper than 70 m, and in the 1950s, the bivalve *Astarte borealis* was the most common and had the highest biomass. Among potential colonizers in the deeper Bornholm Basin, should the oxygen be high over several years, are the bivalves *Astarte borealis*, *Macoma balthica* and *Macoma calcarea*; the polychaetes *Halicryptus spinulosus*, *Scoloplos armiger*, *Capitella capitata* and *Harmothoe sarsi*; and the crustaceans *Diasylis rathkei*, *Pontoporeia affinis* and *Pontoporeia femorata*. In the 1960s, bivalves were rare and, instead, polychaetes, particularly *Scoloplos armiger*, colonized the bottoms during periods of re-oxygenation. However, the oxygen could drop from 4–$5 \, \mathrm{mL \, L^{-1}}$ to less than $1 \, \mathrm{mL \, L^{-1}}$ during a few months.

The biomasses of the benthic fauna in the 1960s were much lower at depths greater than 70 m than in the 1950s, probably as a result of overly short periods of suitable oxygen conditions to let a healthy community establish itself. The average biomass at seven stations greater than 70 m in depth in the 1950s was $17.5 \, \mathrm{g \, m^{-2}}$ wet weight. Karlson et al. (2002) estimated that about 50 000 ton of benthic biomass could be missing in the Bornholm Basin during hypoxic/anoxic conditions in the area. Should re-oxygenation of the Bornholm Basin also have positive secondary effects on the oxygen conditions in the Baltic Proper, an additional biomass on the

order of 1 400 000 ton could be the result, as was estimated by Karlson et al. (2002).

The recruitment of benthic animals into pre-hypoxic areas with a low diversity may be facilitated by the fact that only a few niches are preoccupied and competition is low. Initially, the new recruits are likely to have a high production, and demersal fish such as cod could benefit from the new food source.

A local example of re-colonization of benthic fauna of formerly dead bottoms is from the rather enclosed inner Stockholm archipelago. After reducing the input of phosphorus from the sewage of the city, the oxygen concentrations in the deeper waters improved. In 1996 to 1998, all sediment samples at 20 to 50 m were black and reduced (Rosenberg and Diaz, 1993), but, in 2008, seven of these eight stations had an oxidized sediment surface (Karlsson et al., 2010). The authors recorded abundances of between 2300 and 5600 ind. $\mathrm{m^{-2}}$ and biomasses of between 6 and 65 $\mathrm{g \, m^{-2}}$ (wet weight, excluding the reduced station). The invasive polychaete *Marenzelleria neglecta* was totally dominant at all stations, and is a fast colonizer in great numbers. *Marenzelleria* spp. could possibly also colonize the deeper parts of the presently anoxic/hypoxic Bornholm Basin and the Baltic Proper if these areas were oxidized. Should this worm establish itself in the present defaunated area of ca. 80 000 $\mathrm{km^2}$ in the Baltic Proper with biomasses of the same weights as in the Stockholm archipelago, the total biomass could be on the order of 0.5 to 5.0 million ton, i.e., on the same order as calculated above. As a comparison, Cederwall et al. (1999) found *Marenzelleria viridis* invading the Gulf of Riga in 1993 to 1996 in average biomasses of 11.8 $\mathrm{g \, m^{-2}}$. The species *Marenzelleria* spp. is described in more detail in Sect. 4.7.

4.7   Bioturbation and related benthic biogeochemical processes

Benthic faunal activities have a significant impact on biogeochemical processes in the sediment globally through bioturbation, i.e., reworking the sediment by digging and burrowing; oxygenation of burrows and voids; and biodeposition of particles on the sediment surface and into the water column. Different species have different activity patterns and, in general, larger and more active species have a larger impact on sediment and irrigation processes. As the number of species in the brackish Baltic Sea are reduced compared to in more saline areas, the number of functions are also comparatively lower. Based on if animals are mobile or sessile, and where and how they feed, the number of "functional groups" can be calculated (Pearson and Rosenberg, 1987). Based on such a scheme, the number of functional groups in the Skagerrak are above 20 but only 8 in the southern Baltic (Bonsdorff and Pearson, 1999). Species missing in the Baltic (until recently) were those that were digging deep and feeding within the sediment (sub-surface deposit feeders) and animals using tentacles as feeding appendages. Thus, commonly only the upper few centimeters were bioturbated in the Baltic compared to several tens of centimeters in oceanic waters.

The spionid polychaete *Marenzelleria* spp., a recent invasive species first recorded in 1985, is a deep burrowing worm with consequences for the biogeochemical processes. *Marenzelleria* spp. consists of three different species that look alike and are difficult to separate into species. These species are nowadays common from shallow to deep waters and often dominant in many oxygenated areas of the Baltic. *Marenzelleria* spp. are both surface and subsurface deposit feeders and burrow deep (10–30 cm) into the sediment. They can locally reach numbers as high as 30 000 ind. m^{-2} (Zettler et al., 2002).

By their intrusion into the Baltic, *Marenzelleria* spp. have added a new dimension to the Baltic sedimentary system. This genus constructs galleries in the sediment down to more than 10 cm and adds new niches and complex functions to the ecosystem services. The three-species complex of *Marenzelleria* is spread from shallow waters down to deeper oxygenated bottoms from the southern Baltic up to the Bothnian Sea and the Gulf of Finland. The genera are dominant in many areas and occur, with great variations, frequently in numbers of between 500 and 5000 ind. m^{-2}, but with a somewhat heterogeneous distribution. The *Marenzelleria* spp. are potential invaders of the deeper Baltic sediments if the oxygenated areas are extended. It can colonize new areas by settlement of pelagic larvae or by actively swimming or crawling adults (Norkko et al., 2011). However, to what extent they will colonize these presently defaunated areas, and in which densities, remains unknown. Nonetheless, this is most likely to happen and the new invaders will, in addition to other colonizers, have a significant impact on the fluxes of nutrient in the sediment through their activities in the deep galleries.

The pool of inorganic phosphate in the sediment in the Baltic is clearly related to the oxygen concentration in the bottom-near water; phosphorus is bound to the sediment during oxic conditions and released under anoxic conditions (e.g., Conley et al., 2009a; Stigebrandt, et al., 2014a). Thus, when animals are present in the sediment, their bioturbation and ventilation of the sediment will increase the amount of phosphorus bound in the sediment. When *Marenzelleria* spp. occur in high densities, they can, through their deep burrowing activity, have a significant impact on these biogeochemical cycles. Quintana et al. (2011) found that *Marenzelleria viridis* had two types of ventilation – a muscular pumping of water out of the burrow and a ciliar pumping of water into the burrow. Significant amounts of water percolate upwards to the sediment surface, which will also have a significant effect on the reduction of sulfate in the sediment (Kristensen et al., 2011). Norkko et al. (2011) demonstrated in a model that *Marenzelleria* spp. had a significant density-dependent impact on the phosphorus cycling in the Baltic, where increasingly more phosphorus is bound to the sediment when the density increases beyond 3000 ind. m^{-2}. The great capacity of *Marenzelleria*-bioturbated sediments to store phosphorus was demonstrated by the fact that twice as much was bound annually to the present-day oxic sediments in the Stockholm archipelago as annually could be removed by the sewage treatment plants in Stockholm.

Another possibility is that changed redox conditions in the sediment and activity of *Marenzelleria* spp. could increase the flux of contaminants such as PCBs compared to if the sediment remains anoxic (Granberg et al., 2008). Hedman et al. (2008) showed that contaminants and cadmium could be either buried or remobilized in the sediment because of the activity of different infaunal species. The authors emphasized the importance of understanding the complex interactions between ecological and physiochemical processes when assessing the fate of contaminants in aquatic ecosystems. Thus, more research is needed before any general and firm conclusions could be made regarding the fate of contaminants and metals in the Baltic if the seabed is re-oxygenated. However, the By Fjord oxygenation experiment did not lead to increased leakage of toxic organic compounds or toxic metals from the sediments when these became oxidized (Stigebrandt et al., 2014b).

Oxygenation of the deepwater should tend to decrease the oxygen consumption through the coupling between primary production and decreased leakage of phosphorus from deep bottoms and the following decreased phosphorus content in the surface layers. As in eutrophic lakes, a substantial fraction of the oxygen consumption might be due to fluxes of reduced substances from the sediments (Müller et al., 2012), which is why the decrease in oxygen consumption initially might be less than expected. However, provided the eutrophication of the Baltic Proper does not increase, there is no reason to expect that oxygen consumption should increase when artificial oxygenation is turned on in the Bornholm Basin at the suggested pumping rate.

5 Concluding remarks

From simulations of the vertical circulation in the Bornholm Basin in the period 1990 to 2010, we have shown that the deep Bornholm Basin may be kept oxic by pumping down about 1000 m^3 s^{-1} of water from around 30 m depth. The average oxygen concentration in the basin will then be 3.6 mL L^{-1}, and the minimum concentration is estimated to 1.3 mL L^{-1}, compared to 1.9 and -2.4 mL L^{-1} without pumping. The rate of water exchange at 90 m depth increases from 66 to 160 km^3 year^{-1}. Like in the By Fjord experiment (Stigebrandt et al., 2014b), increased water exchange in the deepwater is the main reason for why the supply of oxygen increases. The average salinity of the basin decreases from 15.4 to 14.4, while the average temperature of 8.1 °C remains almost unchanged.

The small CRV in certain years, endangering the recruitment of cod, would increase with oxygenation. The average of the smallest 25 % of CRV should, by pumping $1000 \, \text{m}^3 \, \text{s}^{-1}$, increase from about 40 to more than $65 \, \text{km}^3$. Thus, oxygenation by increased deepwater ventilation might be a powerful method to strengthen cod reproduction in the Baltic Sea in years when the CRV otherwise would be small and even vanishing.

Benthic biomass will be established by colonization of sea bottoms, which means a new food source for demersal fish like cod and flatfish. Re-oxygenation of the deeper basins in the Baltic will undoubtedly have a positive effect on the recruitment and food availability for cod, which would be beneficial for the fisheries in the long run. Today the total annual catches of cod are only on the order of 50 000 tons, whereas the potential might be 400 000 tons, as it was in the 1990s (MacKenzie et al., 2002).

Reduced leakage of phosphorus from the sea bed in the Bornholm Basin by about 7500 ton per year (Stigebrandt et al., 2014a) will reduce the phosphorus content in the surface layer and thereby contribute to decreasing the eutrophication of the Baltic Sea. It will also reduce the magnitude and spatial extent of toxic cyanobacteria blooms in summer, with large socioeconomic and economic benefits for tourism. The present investigation also indicates that an isolated effort in the Bornholm Basin might give rise to positive effects in the East Gotland Basin, because the deepwater of the Bornholm Basin, which is exported to the East Gotland Basin, would be of lower salinity and contain less phosphorus and more oxygen. This should have positive impacts for the functioning of the ecosystem of the Baltic Proper.

The increase in biological production since the 1950s is the main reason for present-day hypoxia and anoxia. Oxygenation of the Bornholm Basin can contribute to decreased biological production, but the primary goal is to improve the oxygen conditions in the deep part of the basin and thereby also make the deepest parts habitable. However, as long as the biological production is not reduced to the level of the 1950s, the oxygen conditions in the Bornholm Basin will return to present-time conditions if the pumping is shut off. A reduction in the large-scale eutrophication might possibly be achieved by restoration of the Baltic Proper, which would require simultaneous oxygenation of all anoxic bottoms in the Baltic Proper. This will be discussed in a forthcoming paper.

The present paper contributes an important part of an analysis of the possible consequences of artificial oxygenation of the deepwater in the Bornholm Basin. Since the consequences are suggested to be positive, artificial oxygenation by pumping is recommended to be undertaken in the Bornholm Basin. It would be positive for the deepest bottom fauna also to oxygenate the Arkona Basin, which only requires a smaller effort. A carefully constructed observational control program, like the one in the By Fjord experiment (Stigebrandt et al., 2014b), should be run to study the various ecological and biogeochemical effects of such an oxygenation.

Acknowledgements. This work was supported by the Swedish EPA through contract number NV 08/307 F-255-08 (the BOX project) and by the Swedish Agency for Marine and Water Management through contract number HaV dnr 2415-11 and by the BSAP fund through the Nordic Investment Bank (NIB) (the BOX-WIN project). Stockholm sea level data and hydrographic data from stations BY4, BY5 and Anholt E were obtained through the Swedish Meteorological and Hydrological Institute (SMHI). Data from the entrance area, the Arkona Basin and the Gdansk Deep have been retrieved through the Baltic Nest Institute DAS application, with the assistance of Miguel Rodriguez Medina at the Baltic Sea Centre of Stockholm University, and the number of hydrographic profiles available for this study was increased by permission to access data from the Landesamt f. Umwelt, Naturschutz und Geologie M-V (LUNG M-V) and the NMFRI (former Sea Fisheries Institute in Gdynia, Poland). Anders Omstedt at the University of Gothenburg (Department of Earth Sciences) kindly provided data of river runoff to the Baltic Sea. Suggestions from the referees helped to improve the paper.

Edited by: J. A. Johnson

References

Arneborg, L., Fiekas, V., Umlauf, L., and Burchard, H.: Gravity current dynamics and entrainment – a process study based on observations in the Arkona Basin, J. Phys. Oceanogr., 37, 2094–2113, 2007.

Bonsdorff, E. and Pearson, Y. H.: Variation in the sublittoral macrozoobenthos of the Baltic Sea along environmental gradients: A functional-group approach, Aust. J. Ecol., 24, 312–326, 1999.

Bonsdorff, E., Blomqvist, E. M., Mattila, J., and Norkko, A.: Coastal eutrophication: causes, consequences and perspectives in the archipelago area of the northern Baltic Sea, Estuar. Coast. Shelf Sci., 44, 63–72, 1997.

Breitburg, D. L.: Effects of hypoxia, and the balance between hypoxia and enrichment, on coastal fishes and fisheries, Estuaries, 25, 767–781, 2002.

Cederwall, H., Jermakovs, V., and Lagzdins, G.: Long-term changes in the soft-bottom macrofauna of the Gulf of Riga, ICES J. Mar. Sci., 56, 41–48, 1999.

Conley, D. J., Björck, S., Bonsdorff, E., Carstensen, J., Destouni, G., Gustafsson, B. G., Hietanen, S., Kortekaas, M., Kuosa, H., Meier, H. E. M., Müller-Karulis, B., Nordberg, K., Norkko, A., Nürnberg, G., Pitkänen, H., Rabalais, N. N., Rosenberg, R., Savchuk, O. P., Slomp, C. P., Voss, M., Wulff, F., and Zillén, L.: Hypoxia-related processes in the Baltic Sea, Environ. Sci. Technol., 43, 3412–3420, doi:10.1021/es802762a, 2009a.

Conley, D. J., Bonsdorff, E., Carstensen, J., Destouni, G., Gustafsson, B. G., Hansson, L.-A., Rabalais, N. N., Voss, M., and Zillén, L.: Tackling hypoxia in the Baltic Sea: Is engineering a solution?, Environ. Sci. Technol., 43, 3407–3411, doi:10.1021/es8027633, 2009b.

Demel, K. and Mulicki, Z.: Quantitative investigations on the biological bottom productivity of the South Baltic (in Polish with English summary), Prace Morskiego Instytutu Rybackiego w Gdyni, 7, 75–126, 1954.

Diaz, R. J. and Rosenberg, R.: Marine benthic hypoxia: a review of its ecological effects and behavioural responses of marine macrofauna, Oceanogr. Mar. Biol., 33, 245–303, 1995.

Diaz, R. J. and Rosenberg, R.: Spreading dead zones and consequences for marine ecosystems, Science, 321, 926–929, 2008.

Elmgren, R.: The eutrophication status of the Baltic Sea: Input of nitrogen and phosphorus, their availability for plant production, and some management implications, Baltic Sea Environ. Proc. Helsinki, 30, 12–31, 1989.

Elmgren, R.: Understanding human impact on the Baltic ecosystem: changing views in recent decades, AMBIO, 30, 222–231, 2001.

Granberg, M., Gunnarsson, J., Hedman, J., Rosenberg, R., and Jonsson, P.: Bioturbation-driven release of sediment-associated contaminants from Baltic Sea sediments mediated by the invading polychaete *Marenzelleria neglecta*, Environ. Sci. Technol., 42, 1058–1065, 2008.

Gustafsson, B. G., Meier, H. E. M., Savchuk, O. P., Eilola, K., Axell, L., and Almroth, E.: Simulation of some engineering measures aiming at reducing effects from eutrophication of the Baltic Sea, Earth Sciences Centre, University of Gothenburg, Earth Sciences Report Series, C82, ISSN 1400-383X, 59 pp., 2008.

Hedman, J. E., Bradshaw, C., Thorsson, M. H., Gilek, M., and Gunnarsson, J. S.: Fate of contaminants in Baltic Sea sediments: role of bioturbation and settling organic matter, Mar. Ecol. Prog. Ser., 356, 25–38, 2008.

Hinrichsen, H.-H., Voss, R., Wieland, K., Köster, F., Andersen, K. H., and Margonski, P.: Spatial and temporal heterogeneity of cod spawning environment in the Bornholm Basin, Baltic Sea, Mar. Ecol. Prog. Ser., 345, 245–254, doi:10.3354/meps06989, 2007.

Jonsson, P., Carman, R., and Wulff, F.: Laminated sediments in the Baltic – a tool for evaluating nutrient mass balances, AMBIO, 19, 152–158, 1990.

Karlson, K., Rosenberg, R., and Bonsdorff, E.: Temporal and spatial large-scale effects of eutrophication and oxygen deficiency on benthic fauna in Scandinavian and Baltic waters – a review, Oceanogr. Mar. Biol., 40, 427–489, 2002.

Karlsson, M., Jonsson, P. O., Lindgren, D., Malmaeus, J. M., and Stehn, A.: Indicators of recovery from hypoxia in the inner Stockholm archipelago, AMBIO, 39, 486–495 2010.

Kristensen, E., Hansen, T., Delefosse, M., Banta, G., and Quintana, C. O.: Contrasting effects of the polychaetes *Marenzelleria viridis* and *Nereis diversicolor* on benthic metabolism and solute transport in sandy coastal sediment, Mar. Ecol.-Prog. Ser., 425, 125–139, 2011.

Leppäkoski, E.: Transitory return of the benthic fauna of the Bornholm Basin, after extermination by oxygen insufficiency, Cah. Biol. Mar. ,10, 163–172 1969.

Levin, L. A.: Oxygen minimum zone benthos: adaptation and community response to hypoxia, Oceanogr. Mar. Biol., 41, 1–45, 2003.

Liljebladh, B. and Stigebrandt, A.: Observations of the deepwater flow into the Baltic Sea, J. Geophys. Res., 101, 8895–8911, 1996.

MacKenzie, B. R., Hinrichsen, H.-H., Plikshs, M., Wieland, K., and Zezera, A. S.: Quantifying environmental heterogeneity: habitat size necessary for successful development of cod *Gadus morhua* eggs in the Baltic Sea, Mar. Ecol.-Prog. Ser., 193, 143–156, 2000.

MacKenzie, B. R., Alheit, J., Conley, D. J., Holm, P., and Kinzie, C. C.: Ecological hypotheses for a historical reconstruction of upper trophic level biomass in the Baltic Sea and Skagerrak, Can. J. Fish. Aquat. Sci., 59, 173–190, 2002.

Matthäus, W. and Schinke, H.: Mean atmospheric circulation patterns associated with major Baltic inflows, Dtsch. Hydrogr. Z., 46, 321–339, 1994.

Müller, B., Bryant, L.D., Matzinger, A., and Wüest, A.: Hypolimnetic oxygen depletion in eutrophic lakes, Environ. Sci. Technol., 46, 9964–9971, 2012.

Norkko, J., Reed, D. C., Timmermann, K., Norkko, A., Gustafsson, B. G., Bonsdorff, E., Slomp, C. P., Carstensen, J., and Conley, D. J.: A welcome can of worms? Hypoxia mitigation by an invasive species, Glob. Change Biol., 18, 422–434, doi:10.1111/j.1365-2486.2011.02513.x, 2011.

Pearson, T. H. and Rosenberg, R.: Macrobenthic succession in relation to organic enrichment and pollution of the marine environment, Oceanogr. Mar. Biol., 16, 229–311, 1978.

Pearson, T. H. and Rosenberg, R.: Feast and famine: Structuring factors in marine benthic communities, in: 27th Symposium of the British Ecological Society Aberystwyth 1986, edited by: Gee, J .H. R. and Giller, P. S., Blackwell Scientific Publications, Oxford, 373–395, 1987.

Quintana, C. O., Hansen, T., Delefosse, M., Banta, G., and Kristensen, E.: Burrow ventilation and associated porewater irrigation by the polychaete *Marenzelleria viridis*, J. Exp. Mar. Biol. Ecol., 397, 179–187, 2011.

Rosenberg, R.: Benthic faunal dynamics during succession following pollution abatement in a Swedish estuary, Oikos, 27, 414–427, 1976.

Rosenberg, R. and Diaz, R. J.: Sulfur bacteria (*Beggiatoa* spp.) mats indicate hypoxic conditions in the inner Stockholm Archipelago, AMBIO, 22, 32–36, 1993.

Rosenberg, R., Elmgren, R., Fleischer, S., Jonsson, P., Persson, G., and Dahlin, H.: Marine eutrophication case studies in Sweden, AMBIO, 19, 102–108, 1990.

Rosenberg, R., Agrenius, S., Hellman, B., Nilsson, H. C., and Norling, K.: Recovery of benthic habitats and fauna in a Swedish fjord following improved oxygen conditions, Mar. Ecol.-Prog. Ser., 234, 43–53, 2002.

Rutgersson, A., Omstedt, A., and Räisänen, J.: Net precipitation over the Baltic Sea during present and future climate conditions, Clim. Res., 22, 27–39, 2002.

Samuelsson, M. and Stigebrandt, A.: Main characteristics of the long-term sea level variability in the Baltic Sea, Tellus, 48A, 672–683, 1996.

Seifert, T., Tauber, F., and Kayser, B.: A high resolution spherical grid topography of the Baltic Sea – 2nd edition, Baltic Sea Science Congress, Stockholm, 25–29 November 2001, Poster #147, available at: http://www.io-warnemuende.de/topography-of-the-baltic-sea.html (last access: 4 July 2014), 2001.

Stigebrandt, A.: A model for the exchange of water and salt between the Baltic and the Skagerrak, J. Phys. Ocean., 13, 411–427, 1983.

Stigebrandt, A.: Computations of the flow of dense water into the Baltic Sea from hydrographical measurements in the Arkona Basin, Tellus A, 39, 170–177 1987a.

Stigebrandt, A.: A model for the vertical circulation of the Baltic deep water, J. Phys. Oceanogr., 17, 1772–1785 1987b.

Stigebrandt, A.: Physical oceanography of the Baltic Sea, in: A systems analysis of the Baltic Sea, edited by: Wulff, F., Rahm, L., and Larsson, P., Springer-Verlag, Berlin, Heidelberg, New York, 19–74, 2001.

Stigebrandt, A. and Gustafsson, B.: Response of the Baltic Sea to climate change – theory and observations, J. Sea Res., 49, 243–256, 2003.

Stigebrandt, A. and Gustafsson, B. G.: Improvement of Baltic proper water quality using large-scale ecological engineering, AMBIO, 36, 280–286, 2007.

Stigebrandt, A. and Kalén, O.: Improving oxygen conditions in the deeper parts of Bornholm Sea by pumped injection of winter water, AMBIO, 42, 587–595, doi:10.1007/s13280-012-0356-4, 2013.

Stigebrandt, A., Rahm, L., Viktorsson, L., Ödalen, M., Hall, P. O. J., and Liljebladh, B.: A new phosphorus paradigm for the Baltic proper, AMBIO, 43, 634–643, doi:10.1007/s13280-013-0441-3, 2014a.

Stigebrandt, A., Liljebladh, B., De Brabandere, L. Forth, M., Ganmo, Å., Hall, P. O. J., Hammar, J., Hansson, D., Kononets, M., Magnusson, M., Norén, F., Rahm, L., Treusch, A., and Viktorsson, L.: An experiment with forced oxygenation of the deepwater of the anoxic By Fjord, western Sweden, AMBIO, doi:10.1007/s13280-014-0524-9, 2014b.

Vallin, L., Nissling, A., and Westin, L.: Potential factors influencing reproductive success of Baltic cod, *Gadus morhua*: a review, AMBIO, 28, 92–99, 1999.

Zettler, M. L., Daunys, D., Kotta, J., and Bick, A.: History and success of an invasion into the Baltic Sea: the polychaete *Marenzelleria* cf. *viridis*, development and strategies, in: Invasive aquatic species of Europe: distribution, impacts and management, edited by: Leppäkoski, E., Gollasch, S., and Olenin, S., Kluwer Academic Publishers, Dordrecht, 66–76, 2002.

5

Measuring air–sea gas-exchange velocities in a large-scale annular wind–wave tank

E. Mesarchaki[1], C. Kräuter[2], K. E. Krall[2], M. Bopp[2], F. Helleis[1], J. Williams[1], and B. Jähne[2,3]

[1]Max-Planck-Institut für Chemie (Otto-Hahn-Institut) Hahn-Meitner-Weg 1, 55128 Mainz, Germany
[2]Institut für Umweltphysik Universität Heidelberg, Im Neuenheimer Feld 229, 69120 Heidelberg, Germany
[3]Heidelberg Collaboratory for Image Processing (HCI), Universität Heidelberg, Speyerer Straße 6, 69115 Heidelberg, Germany

Correspondence to: E. Mesarchaki (evridiki.mesarchaki@mpic.de)

Abstract. In this study we present gas-exchange measurements conducted in a large-scale wind–wave tank. Fourteen chemical species spanning a wide range of solubility (dimensionless solubility, $\alpha = 0.4$ to 5470) and diffusivity (Schmidt number in water, $Sc_w = 594$ to 1194) were examined under various turbulent ($u_{10} = 0.73$ to $13.2 \, \mathrm{m \, s^{-1}}$) conditions. Additional experiments were performed under different surfactant modulated (two different concentration levels of Triton X-100) surface states. This paper details the complete methodology, experimental procedure and instrumentation used to derive the total transfer velocity for all examined tracers. The results presented here demonstrate the efficacy of the proposed method, and the derived gas-exchange velocities are shown to be comparable to previous investigations. The gas transfer behaviour is exemplified by contrasting two species at the two solubility extremes, namely nitrous oxide (N_2O) and methanol (CH_3OH). Interestingly, a strong transfer velocity reduction (up to a factor of 3) was observed for the relatively insoluble N_2O under a surfactant covered water surface. In contrast, the surfactant effect for CH_3OH, the high solubility tracer, was significantly weaker.

1 Introduction

The world's oceans are key sources and sinks in the global budgets of numerous atmospherically important trace gases, in particular CO_2, N_2O and volatile organic compounds (VOCs) (Field et al., 1998; Williams et al., 2004; Millet et al., 2008, 2010; Carpenter et al., 2012). Gas exchange between the ocean and the atmosphere is therefore a significant conduit within global biogeochemical cycles. Air–sea gas fluxes, provided by either direct flux measurements or accurate gas transfer parameterisations, are a prerequisite for global climate models tasked to deliver accurate future predictions (Pozzer et al., 2006; Saltzman, 2009).

The principles behind gas exchange at the air–sea interface have been reported in detail within previous reviews (Jähne and Haußecker, 1998; Donelan and Wanninkhof, 2002; Wanninkhof et al., 2009; Jähne, 2009; Nightingale, 2009). A simplified conceptual two layer model is generally accepted. The model assumes that close to the interface turbulent motion is suppressed and that the transfer of gases is controlled by molecular motion (expressed by the diffusion coefficient D). This leads to the formation of two mass boundary layers on both sides of the interface. In the upper part of the air-side mass boundary layer, turbulent transport becomes significant. Further away from the interface the significance of the air-side turbulent transport increases. In the water-side, due to lower diffusivities, molecular transport remains the controlling factor of the transfer. Depending on the solubility of the gas in question, its transfer could be restricted by one or both sides of the interface (i.e air-side and water-side controlled).

The transfer velocity, k (in $\mathrm{cm \, h^{-1}}$), of a gas across the surface is defined as the gas flux density, F, divided by the concentration difference, Δc, between air and water (henceforth named k_t expressing the transfer through both boundary layers against the single air and water layer transfer, k_a and k_w, accordingly). Wind-driven turbulence near the water surface as well as the resultant processes (surface stress and roughness, waves, breaking waves, bubbles, spray, etc.) influences the thickness of the mass boundary layers. Thus, the transfer velocity is related to the degree of turbulence on both sides close to the interface as well as the tracer characteristics, i.e. their solubility and diffusion coefficients (Danckwerts, 1951; Liss and Slater, 1974). Surface films are also known to have a strong influence on the transfer velocity by inhibiting waves and decreasing the near-surface turbulence (e.g Frew et al., 1990; Jähne and Haußecker, 1998; Zappa et al., 2004; Salter et al., 2011). To date, research on surface films (using different film thicknesses and types) and their effect on transfer velocity is only in the very early stages. The impact of wind-driven mechanisms, surface films and diverse physiochemical tracer characteristics on the gas-exchange rates can be studied in detail through transfer velocity measurements of individual species provided by the method proposed here. Such studies aim to improve our understanding of air–sea gas transfer and provide new insights into the theoretical background.

Gas transfer velocities have been determined in both field studies (using mass balance, eddy correlation or controlled flux techniques) and laboratory experiments described in previous gas-exchange reviews (Jähne and Haußecker, 1998; Donelan and Wanninkhof, 2002; Wanninkhof et al., 2009; Jähne, 2009; Nightingale, 2009) and references therein. Wind–wave tanks, in contrast to the open ocean, offer a unique environment for the investigation of individual mechanisms related to the air–sea gas exchange under controlled conditions.

Mass balance methods have been applied in the field using geochemical tracers (O_2, ^{14}C, Radon, for instance in Broecker et al., 1985) and dual tracer (SF_6, ^3He, for instance in Watson et al., 1991; Wanninkhof et al., 1993) techniques. The main drawback of these approaches was the relatively low temporal resolution (Jähne and Haußecker, 1998). Furthermore, the transfer velocity measurements were based primarily on sparingly soluble tracers, and very few experimental results of highly soluble trace gas transfer velocities are available.

In this study, gas-exchange experiments were performed in a state-of-the-art large-scale annular wind–wave tank. An experimental approach based on mass balance has been developed, whereby air- and water-side concentrations of various tracers are monitored using instrumentation capable of online measurement. For the first time, parallel measurements of total air and water-side transfer velocities for 14 individual gases within a wide range of solubility, have been achieved. Wind speed conditions (reported at 10 metres height, u_{10}) as low as 0.73 and reaching up to $13.2 \, \text{m s}^{-1}$ were investigated. Supplementary parameters directly linked with the gas-exchange velocities, such as friction velocity and mean square slope of the water surface, were additionally measured under the same conditions. This paper details the entire instrumental set-up and provides a validation of the over-

all operation and concept through transfer velocity measurements of nitrous oxide and methanol. These species are chosen as they bracket the wide range of solubilities among the investigated tracers and clearly show different gas-exchange behaviours. Transfer velocity measurements of the remaining examined tracers are going to be presented in a follow-up publication.

2 Method

In this study, total transfer velocities for low as well as medium to highly soluble tracers were determined using a mass balance approach. The wind–wave tank is interpreted in terms of a box model.

2.1 The box model

The basic idea of the box model method is the development of a direct correlation between the air- and water-phase concentrations, c_a and c_w, and the desired transfer rates, k_t, of various inert tracers.

Figure 1 shows a schematic representation of the wind–wave tank in a box model (Kräuter, 2011; Krall, 2013). Water and air spaces are assumed to be two well-mixed separate boxes with volumes, V_w and V_a, between which tracers can be exchanged only through the water surface, A. Further possible pathways of tracers entering or leaving the box are also shown in Fig. 1. Assuming constant volumes, temperature and pressure conditions, the mass balance for the air and the water phases of the box yields for a water-side perspective:

$$V_a \dot{c}_a = Ak_{tw}(c_w - \alpha c_a) + \dot{V}_a^i c_a^i - \dot{V}_a c_a + \dot{V}_a c_a^0 \quad (1)$$

$$V_w \dot{c}_w = -Ak_{tw}(c_w - \alpha c_a) \quad (2)$$

and an air-side perspective:

$$V_a \dot{c}_a = -Ak_{ta}\left(c_a - \frac{c_w}{\alpha}\right) + \dot{V}_a^i c_a^i - \dot{V}_a c_a + \dot{V}_a c_a^0 \quad (3)$$

$$V_w \dot{c}_w = Ak_{ta}\left(c_a - \frac{c_w}{\alpha}\right), \quad (4)$$

where $\alpha = c_w/c_a$ denotes the dimensionless solubility and k_{tw}, k_{ta} the total transfer velocities for a water- and an air-sided viewer, respectively. The two transfer velocities differ by the solubility factor of the tracer (see Eq. A1 in Appendix A). The dotting stands for the time derivative of the related symbol.

The first term on the right hand side of each equation represents the exchange of a tracer from one phase to the other due to a concentration gradient. The second term stands for possible tracer input ($\dot{V}_a^i c_a^i$), the third term for possible tracer output (flushing/leaking term: $\dot{V}_a c_a$) and the fourth term for a possible tracer coming in through leaks from the surrounding room or through the flushing ($\dot{V}_a c_a^0$).

In the following sections, two different box model solutions, as used in this work for the low solubility tracers (water-side controlled: Sect. 2.2) and for the medium to

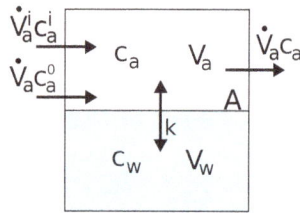

Figure 1. Mass balances for the air and water side. Naming convention is as follows: A: water surface area; V_a: air volume; V_w: water volume; k: gas transfer velocity; c_a: air-side concentration; c_w: water-side concentration; c_a^i: input tracer concentration; c_a^0: tracer concentration in the ambient air. The dotting denotes the time derivative of the related symbol.

high solubility tracers (air-side controlled: Sect. 2.3), are presented in detail. The simulated air/water concentration time series derived for a water-side (b and c in purple) and an air-side controlled (d and e in orange) tracer are presented in Fig. 2.

2.2 Water-side controlled tracers

The following approach was used for tracers with relatively low solubility ($\alpha < 100$) for which the transfer velocity, k_t, is mainly restricted due to the water-side resistance. Here, a low solubility tracer is dissolved in the water volume which is considered well mixed. High tracer concentration in the water and very low concentrations in the air ($c_a \simeq 0$) direct the flux from the water to the air (evasion).

Figure 2b shows the simulated air-phase concentration time series of an example water-side controlled tracer at three example wind speed conditions (as seen in Fig. 2a, change of wind speed is denoted with grey dashed lines). Each condition starts with a closed air-space tank configuration (closed box – no flushing; see more details in Sect. 3.2.5), where the air-side concentration, starting from circa zero, increases linearly with time, due to the water-to-air gas exchange. At time t_1, the air space is opened (open box – flushing on; flushing time is denoted with grey background) and a drastic decrease is observed due to dilution of the air-space concentration with the relatively clean ambient air entering the facility. As indicated in the figure, the higher the wind speed the faster the concentration increase. Figure 2c presents the water-phase concentration of the same tracer which in parallel starts from the highest concentration point and gradually decreases during the course of the experiment as more and more molecules escape the water to enter the gas phase.

The ambient tracer concentration in the air entering the air space through leaks or during flushing can be safely assumed as negligible in comparison to the levels used for all examined tracers. Omitting parameter c_a^0, simplifies the box model Eq. (1), which can be subsequently solved for k_{tw} as follows:

$$k_{tw} = \frac{V_a}{A} \cdot \frac{\dot{c}_a + \lambda_{f,x} c_a}{c_w} \cdot \frac{1}{1 - \alpha c_a / c_w}, \tag{5}$$

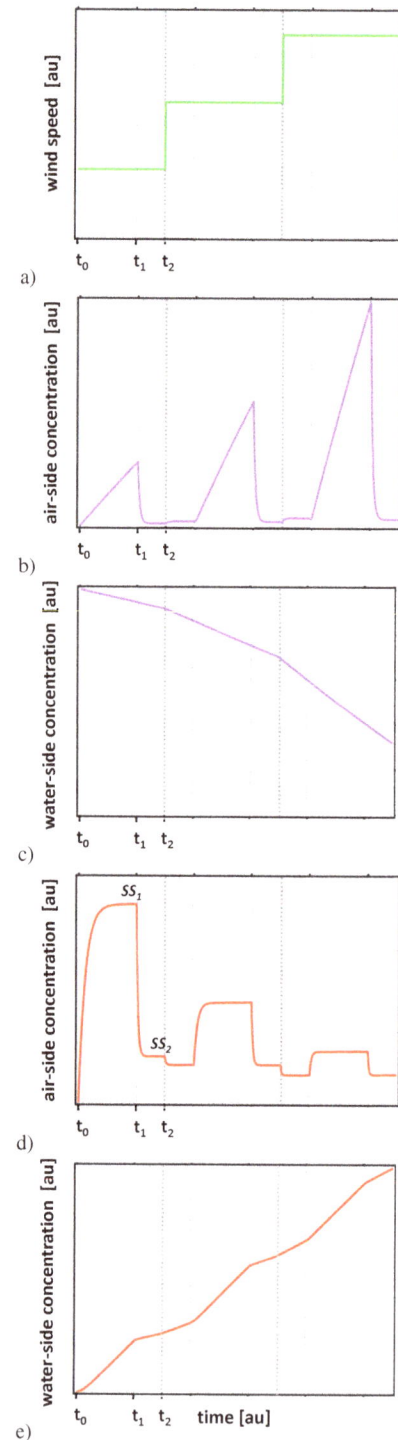

Figure 2. Simulated concentration time series for a water (**b** and **c**) and an air-side controlled (**d** and **e**) tracer in both air and water phase, at three example wind speed conditions. The grey background denotes the air-phase flushing periods and the dashed lines the change of the wind speed condition. SS_1 and SS_2 mark the developed steady states.

where $\lambda_{f,x} = \dot{V}_a / V_a$ is the leak or flush rate for x being 1 or 2, respectively.

Applying Eq. (5), the instantaneous total transfer velocities (k_{tw}) can be calculated from time-resolved measurements of air- and water-side concentrations.

2.3 Air-side controlled tracers

In this approach, tracers with relatively high solubility ($\alpha > 100$) for which k_t is expected to be controlled mainly by air-side processes, were used. Here, a relatively high solubility tracer is introduced with a constant flow to the air volume, continuously during the experiment. Due to low concentrations in the water volume, the net gas-exchange flux is directed from the air to the water (invasion).

In Fig. 2d the air-phase concentration of an example air-side controlled tracer is shown. During the closed air-space period (t_0 to t_1), the concentration increases exponentially, as a fraction of the air-space molecules transmit into the water due to the air–water gas exchange. At t_1, the concentration reaches a steady state, SS_1, where the input rate of the tracer is equal to the exchange rate between the two phases and the leak/flush rate.

At an equilibrium point, the concentration time derivative $\dot{c}_a$ is approximately zero so that Eq. (3) can be written as

$$c_a = \frac{\lambda_{ta} \frac{c_w}{\alpha} + \lambda^i c_a^i}{\lambda_{ta} + \lambda_{f,x}}, \tag{6}$$

where $\lambda_{ta} = \frac{A}{V_a} k_{ta}$ is the exchange rate and $\lambda^i = \frac{\dot{V}_a^i}{V_a}$ the input rate.

After SS_1, the facility is flushed with ambient air (open air space) and the concentration decreases abruptly. Under these conditions (t_2), a second steady state, SS_2, is developed at a lower concentration range. In SS_1, a very small leak rate is present ($\lambda_{f,1} \approx 0$, leak rate) while in SS_2 the leak rate is much larger due to the open air space ($\lambda_{f,2}$, flush rate). Dividing the air-side concentrations of the two steady states $\frac{c_{a,1}}{c_{a,2}}$ (as given in Eq. 6) and solving it with respect to the exchange rate yields

$$\lambda_{ta} = \frac{\lambda_{f,2} c_{a,2} - \lambda_{f,1} c_{a,1}}{(c_{a,1} - c_{a,2})}. \tag{7}$$

The total transfer velocities in the wind–wave tank box are calculated from

$$k_{ta} = \frac{\lambda_{ta} V_a}{A}. \tag{8}$$

2.3.1 Leak and flush rate

In most wind–wave facilities, small air leaks are inevitable. The amount of tracer escaping the air space of the facility needs to be monitored and corrected for, as described in Sects. 2.2 and 2.3. To measure the leak/flush rate $\lambda_{f,x}$ for the open and closed configuration of the wind–wave tank, a non-soluble tracer (here CF_4), called a leak test gas, is used. Directly after closing the wind–wave tank, a small amount of the leak test gas is injected rapidly into the air space. As the leak test gas is non-soluble, the water-side concentration, c_w, as well as the gas-exchange velocity, k_{ta}, in Eq. (3) are equal to zero, reducing the air-side mass balance equation to

$$V_a \dot{c}_a = \dot{V}_a^i c_a^i - \dot{V}_a c_a. \tag{9}$$

After the initial injection, the input term $\dot{V}_a^i c_a^i$ in Eq. (9) vanishes, yielding $V_a \dot{c}_a = -\dot{V}_a c_a$. This simple differential equation can be solved easily with

$$c_a(t) = c_a(0) \cdot \exp(-\lambda_{f,x} \cdot t), \tag{10}$$

where $c_a(0)$ is the concentration directly after the input of the leak test gas. Monitoring the concentration of the leak test gas over time and fitting an exponentially decreasing curve to this concentration time series yields the leak/flush rate $\lambda_{f,x}$ of the system. In the Aeolotron facility, typical leak and flush rates were of the order of 0.05 to $0.4\,h^{-1}$ and 20 to $50\,h^{-1}$, respectively.

3 Experiments

3.1 The Aeolotron wind–wave tank

The air–water gas-exchange experiments were conducted in the large-scale annular Aeolotron wind–wave tank at the University of Heidelberg, Germany (Fig. 3). With an outer diameter of 10 m, a total height of 2.4 m and a typical water volume of 18 000 L, the Aeolotron represents the world's largest operational ring shaped facility (much larger than annular facilities used in previous investigations, Jähne et al., 1979, 1987). The chamber is mostly gastight, thermally isolated, chemically clean and inert. In Fig. 3, a list of the main dimensions along with an aerial illustration of the facility are given. The tank is divided in 16 segments and an inner window extending through segments 16 to 4 allows visual access to the wind formed waves. The facility ventilation system consists of two pipes through which the air space can be flushed with ambient air at a rate of up to $50\,h^{-1}$. Two diametrically positioned ceiling mounted axial ventilators (segment 4 and 12) are used to generate wind velocities of up to $u_{ref} = 12\,m\,s^{-1}$.

In the facility, several ambient parameters are monitored. Temperature measurements are provided by two temperature sensors (PT-100) installed in the water and air phase of segment 15 (at heights of 0.5 and 2.3 m, respectively). On the ceiling of the same segment a fan anemometer (STS 020 by Greisinger electronic GmbH) installed in the centre line, determines the wind velocity. Two humidity sensors are mounted in segments 2 and 13. An optical ruler provides the water height using the principle of communicating vessels. Segments 1 and 11 contain the tracer inlets for the air and water phase, respectively. The leak test gas is introduced in segment 11.

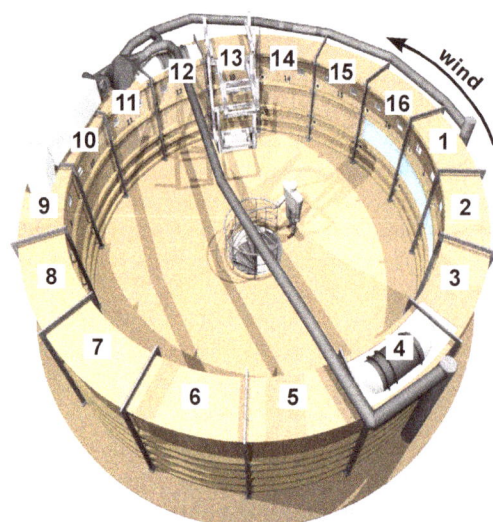

Main Features	
mean circumference	29.2 m
mean diameter	9.3 m
mean width	61 cm
mean total height	241 cm
typical water depth	100 cm
surface area	17.9 m^2
typical water volume	17.9 m^3
typical air volume	24.4 m^3

Figure 3. An aerial illustration of the Aeolotron tank and its main features. The numbers denote the segments. The axial fans producing the wind can be seen in the roof of segments 4 and 12. The air pipes supplying fresh air and removing waste air are shown in grey; figure adapted from (Krall, 2013)).

The annular geometry of the wind–wave tank, contrary to a linear geometry, permits homogeneous wave fields and unlimited fetch. The well-mixed air space (at few centimetres height above the surface) ensures no concentration gradients and therefore concentration measurements independent of the sampling height. On the other hand, the restricted size of the facility which leads to waves reflecting off the walls, results with a different wave field to that found in the open ocean.

3.2   Tracers and instrumentation

A series of 14 tracers covering a wide solubility ($\alpha = 0.4$ to 5470) and diffusivity ($Sc_w = 594$ to 1194) range, were selected for this study. Many of these tracers are very common in the ocean environment, while the rest are used to extend the solubility and diffusivity ranges, a significant criterion for further physical investigations of the gas-exchange mechanisms. Table 1 gives an overview of the examined tracers, with their respective molecular masses, solubility and Schmidt numbers, Sc (the dimensionless ratio of the kinematic viscosity of water v and the diffusivity of the tracer D, $Sc = v/D$) at 20 °C.

All tracers were monitored on-line in both the air and the water phase. The VOC measurements were performed using proton reaction mass spectrometry (PTR-MS) from Ionicon Analytik GmbH (Innsbruck, Austria), while for the halocarbons and N$_2$O, two Fourier transform infrared (FT-IR) Spectrometers (Thermo Nicolet iS10) were used. As leak test gas, carbon tetrafluoride (CF$_4$) was used; it was also measured by FT-IR spectrometry.

For the surfactant experiments, the soluble substance Triton X-100, C$_{14}$H$_{22}$O(C$_2$H$_4$O)$_{9.5}$ (Dow Chemicals, listed

$M_r = 647$ g mol^{-1}) was used to cover the water surface. Triton X-100 was chosen because of its common use as a reference substance to quantify the surface activity of unknown surfactant mixtures found in the open ocean (Frew et al., 1995; Cosovic and Vojvodic, 1998; Wurl et al., 2011).

The operation and sampling conditions for both air and water phases are briefly described below. Additional instrumentation for substantial supplementary measurements follows.

3.2.1   Water-phase measurements

In the water phase a PTR-Quadrupole-MS (PTRQ-MS) (water inlet in segment 3) and a FT-IR spectrometer (water inlet in segment 6) were used to measure the concentration levels of the VOCs and the halocarbons and N$_2$O, respectively. Our instrumentation, which is normally suited only for air sampling, was combined with an external membrane equilibrator (the oxygenator Quadrox manufactured by Maquet GmbH, Rastatt, Germany) to establish equilibrium between the water concentration and the gas stream to be measured. In this way, water-side concentrations could be obtained and used for the calculation of the transfer velocities for the low solubility tracers (see Sect. 2.2).

Membrane equilibrator configuration

The membrane equilibrator device includes a thin gas permeable membrane capable of separating the gas from the liquid phase (commercially available and often used in medicine as a human lung replacement to oxygenate blood). Water from the Aeolotron is constantly pumped through the membrane device. Inside the equilibrator gas exchange occurs, due to

Table 1. Molecular masses (M in $g\,mol^{-1}$), dimensionless solubility (α) and Schmidt numbers in air (Sc_a) and water (Sc_w) for the investigated tracers at $20\,^\circ C$.

Gas	Formula	M	α	Sc_a^l	Sc_w^l
methanol	CH_3OH	32.04	5293[a]	1.0268	671.04
1-butanol	C_4H_9OH	74.12	4712[b]	1.8198	1141.7
acetonitrile	CH_3CN	41.05	1609[c]	1.2957	832.07
acetone	$(CH_3)_2CO$	58.08	878.0[c]	1.4921	880.53
2-butanone	$C_2H_5COCH_3$	72.11	598.9[b]	1.7344	1159.8
acetaldehyde	CH_3CHO	44.05	378.7[d]	1.0786	824.90
ethyl acetate	$CH_3C(O)OC_2H_5$	88.10	156.4[e]	1.8183	997.46
dms	CH_3SCH_3	62.13	16.62[f]	1.4484	979.40
benzene	C_6H_6	78.11	5.672[g]	1.6785	980.46
toluene	$C_6H_5CH_3$	92.14	4.529[g]	1.8409	1176.3
trifluoromethane	CHF_3	70.01	0.760[h]	1.2132	747.50
nitrous oxide	N_2O	44.01	0.676[i]	1.0007	593.90
isoprene	C_5H_8	68.12	0.31[j]–0.69[k,*]	1.6617	1193.9
pentafluoroethane	CF_3CHF_2	120.0	0.415[h]	1.5106	1027.0

[a] Schaffer and Daubert (1969), [b] Snider and Dawson (1985), [c] Benkelberg et al. (1995), [d] Betterton and Hoffmann (1988), [e] Janini and Quaddora (1986), [f] Dacey et al. (1984), [g] Robbins et al. (1993), [h] Krall (2013), [i] Weiss and Price (1980), [j] Yaws and Pan (1992), [k] Sander (1999), [l] Yaws (1995), * only available values at $25\,^\circ C$.

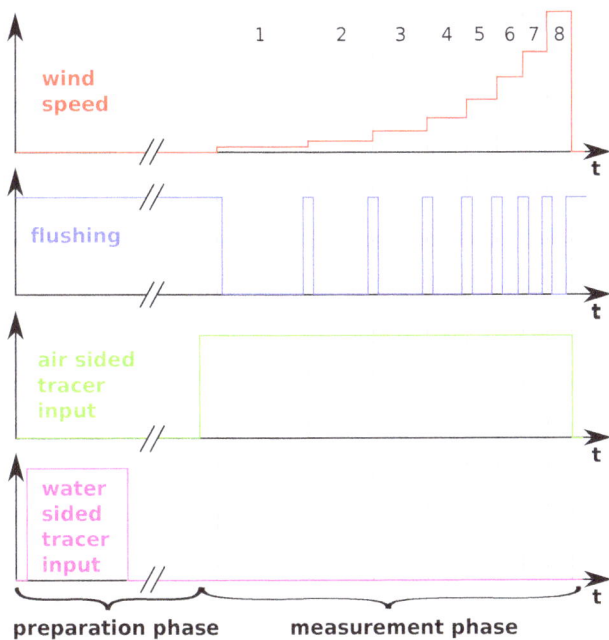

Figure 4. Schematic time series of the wind speed, flushing periods and air/water tracer inputs.

the partial pressure difference of the gases involved, until equilibrium between air and water is achieved (Henry's law at constant temperature).

A detailed configuration of the membrane set-up in conjunction with the PTR-MS is shown in Fig. 6. The system consists of a water and an air loop, both constantly in contact with the membrane equilibrator. The dark blue lines represent the water loop where water was being pumped from the Aeolotron through the membrane and back into the facility, with a constant flow of $3.4\,L\,min^{-1}$. The light orange coloured lines represent the air loop which has a link to the PTRQ-MS instrument. A synthetic air inlet and an excess flow exhaust are used to regulate the flow inside the air loop constant at $1\,L\,min^{-1}$ and the systems pressure at 1013 hPa. Part of the air that comes out of the equilibrator is driven to the PTRQ-MS for analysis, while the rest remains in the loop. The relative humidity in the equilibrated air increases after passing through the equilibrator; therefore, the air tubing was heated to a few degrees above room temperature to avoid water condensation.

A similar set-up using a second membrane equilibrator was connected to the FT-IR instrument. The water flow was kept at a rate of about $3\,L\,min^{-1}$. Here the instrument's measuring cell was integrated into the air loop, removing the need for sample extraction, a synthetic air inlet and an exhaust. The air was circulated in the closed loop at a rate of approximately $150\,mL\,min^{-1}$. Between the equilibrator and the measuring cell, a dehumidifying unit containing phosphorous pentoxide was used to remove water from the air stream and in this way protect the optical windows of the IR measuring cell.

The time constant of the membrane equilibrator was evaluated as described in Krall and Jähne (2014), providing a very fast response of $\simeq 1\,min$.

PTRQ-MS configuration

The PTR-MS detection technique has been described in detail elsewhere (Lindinger et al., 1998). Here, a PTRQ-MS

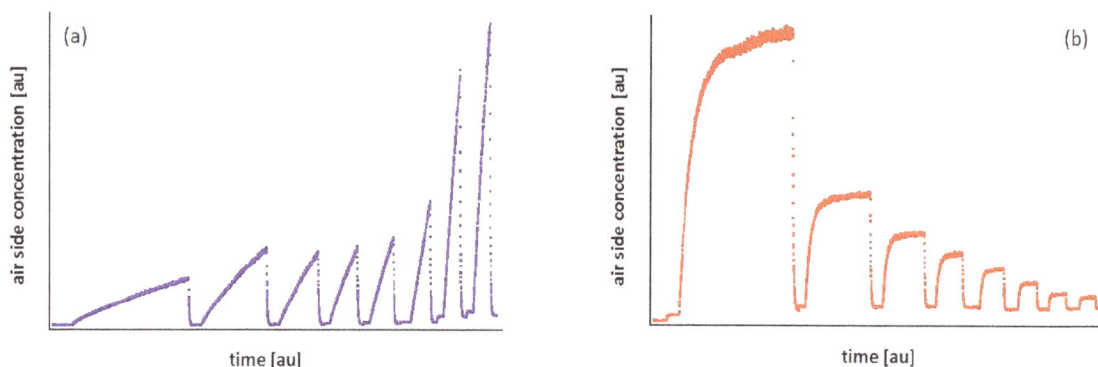

Figure 5. Air-side concentrations obtained for example water-side (**a**) and air-side (**b**) controlled tracers throughout the experimental procedure.

was operated under $30\,\mathrm{mL\,min^{-1}}$ sampling flow, $2.1\,\mathrm{mbar}$ drift pressure, and $600\,\mathrm{V}$ drift voltage ($E/N = 130\,\mathrm{Td}$, $\mathrm{Td} = 10^{-17}\,\mathrm{cm^2\,V\,molecule^{-1}}$). A total of 30 masses were measured sequentially with a dwell time of $1\,\mathrm{s}$ (time resolution is $30\,\mathrm{s}$). Possible mass overlapping was prevented by the careful reselection of the analysed compounds based on the initial mass scan.

Water-phase calibrations were performed in conjunction with the membrane equilibrator set-up (see Fig. 6). Known VOC concentrations were diluted in deionised water and then introduced into the water phase of the facility, in precise volume quantities. To avoid losses of the investigated tracers into the air phase due to air–water gas exchange, the water surface was covered with a large amount of an organic surfactant ($0.446\,\mathrm{mg\,L^{-1}}$, Triton X-100) and calm wind conditions were used to gently mix the air space. Under such conditions gas-exchange velocities were estimated to be negligible. Linear behaviour was established for all examined tracers at concentration levels embracing the characteristic water-phase concentration ranges detected during the experiments.

FT-IR configuration

The key aspects of FT-IR spectroscopy are described in detail in Griffiths (2007). In this study, a Nicolet iS10 (manufactured by Thermo Fischer Scientific Inc., Waltham, MA., USA) FT-IR spectrometer with a custom made measuring cell of approximately $5\,\mathrm{cm}$ length was used. About every $5\,\mathrm{s}$, one infrared absorbance spectrum with wave numbers between 4000 and $650\,\mathrm{cm^{-1}}$ with a resolution of $0.214\,\mathrm{cm^{-1}}$ was acquired. Six of these single spectra were averaged to minimise noise and stored for further evaluation, leading to a time resolution of about 1 spectrum every $30\,\mathrm{s}$. Signal conversion to water-phase concentration, calibration and uncertainty estimation are described in detail in Krall (2013).

3.2.2 Air-phase measurements

In the air phase, a PTR-time of flight (ToF)-MS (inlet in segment 3) with a time resolution of $10\,\mathrm{s}$ provided very fast on-line measurements for the VOCs while a FT-IR (inlet in segment 2) with a time resolution of $30\,\mathrm{s}$ was in parallel monitoring the halocarbons and N_2O. High time resolution measurements enabled a fast experimental procedure and at the same time high accuracy data analysis. Additionally, due to the fast on-line measurements, the transient response of the system could be followed very efficiently throughout the experimental procedure. Example air-side measurements are shown in Fig. 5 for a water and an air-side controlled tracer.

PTR-ToF-MS configuration

The ionisation principle of the PTR-ToF-MS is the same as the PTRQ-MS; however, here a time-of-flight mass spectrometer is used. Throughout the measurements, the PTR-ToF-MS was configured in the standard V mode with a mass resolution of approximately $3700\,\mathrm{m}\,\Delta\mathrm{m^{-1}}$. The drift voltage was maintained at $600\,\mathrm{V}$ and the drift pressure at $2.20\,\mathrm{mbar}$ (E/N $140\,\mathrm{Td}$). Mass spectra were collected over the range 10–$200\,m/z$ (mass-to-charge ratio) and averaged every $10\,\mathrm{s}$, providing a mean internal signal for each compound. After acquisition all spectrum files were mass calibrated using $(H_2O)H^+$, NO^+ and $(C_3H_6O)H^+$ ions to correct for mass peak shifting.

Calibrations in the air phase were conducted under high humidity conditions equivalent to the sampling conditions during the experiments (85–90 % RH). The desired mixing ratios (1–600 ppbv) were obtained by appropriate dilution of the multi-component VOC gas standard with synthetic air. Linear response was established for all examined tracers.

FT-IR configuration

A second Thermo Scientific Nicolet iS10 FT-IR spectrometer was used to measure the air-side concentrations. The measuring cell with a folded light path of a total length of $2\,\mathrm{m}$, was kept at a constant temperature of 35 °C using a Thermo Nicolet cell cover. Air from the Aeolotron was sampled at a rate of 150 min^{-1} at segment 13. As with the water-side instrumentation, water vapour was removed before entering the measuring cell using phosphorous pentoxide. The spectrometer settings, data acquisition as well as data processing, were identical to the water-side instrumentation, see Sect. 3.2.1.

3.2.3 Error analysis of k_t

The individual total transfer velocity uncertainties were calculated applying the propagation of error for uncertainties independent from each other to Eq. (5) for the k_{tw}, and Eqs. (7) and (8) for the k_{ta}.

The concentration uncertainties for the PTR-MS measurements were calculated using the background noise and the calibration uncertainty of each examined tracer. Relatively low uncertainties were obtained for the air-phase concentration levels SS$_1$ and SS$_2$ ranging between 1–1.5 % and 1.5–2.5 %, respectively. The water-phase concentration uncertainty, Δc_w, was estimated the same way and the uncertainties were between 6.5 and 8 % for the concentration ranges used.

The uncertainty of the concentration measurement with the FT-IR spectrometers was found to be concentration dependent. All concentration uncertainties lie below 4 % for the typical concentrations measured in the described experiments.

The individual uncertainties for the leak and flush rates of all conditions were of the order of 0.5 and 1 %, respectively. Based on the geometrical parameters of the facility the surface area uncertainty was calculated to be approximately 2 %, while a maximum of 3 % is estimated for the volume uncertainty. For the solubility values provided by literature, accurate uncertainty estimations are difficult. Here we assume a maximum uncertainty of 10 % for all literature sources.

The overall estimated total transfer velocity uncertainties therefore ranged between 6–12 and 6–20 %, respectively, for the k_{ta} and k_{tw} values of all examined tracers.

3.2.4 Additional instrumentation

Supplementary measurements of wind driven, surface associated, physical parameters, such as the mean square slope and the water-sided friction velocity, were additionally made in the Aeolotron wind–wave tank to enable further investigations of the physical mechanisms of air–water gas exchange.

The mean square slope measurements, reflecting the surface roughness conditions, were performed in parallel with the gas-exchange measurements using a colour imaging slope gauge (CISG) installed in segment 13. The CISG device uses the refraction properties of light at the air–water boundary. A colour coded light source was placed below the water while a camera observed the water surface from above. Using lenses to achieve a telecentric set-up, a relationship

between surface slope and the registered colour can be determined. Errors are calculated from the statistical fluctuations of the individually measured mean square slope values. A more detailed description can be found in Rocholz (2008).

The water-sided friction velocity, $u_{*,w}$, measurements, expressing the shear stress created on the water interface, were accomplished at a later stage using the same setting of the wind generator and the same surfactant coverage of the water surface. The momentum balance method was used as described in Bopp (2014) and Nielsen (2004). To apply this method, the friction between the water and the walls needs to be measured first. This is done by monitoring the decrease of the velocity of the bulk water after switching off the wind. In a stationary equilibrium, that is characterised by an equality of the momentum input into the water by the wind and the momentum loss due to friction at the walls, the friction velocity, $u_{*,w}$, can be calculated from the mean water velocity. The water velocity was measured using a three-axis Modular Acoustic Velocity Sensor (MAVS-3 manufactured by NOBSKA, Falmouth, MA, USA) installed in the centre of the water channel in segment 4 of the Aeolotron at a water depth of around 50 cm. The uncertainty of the friction velocity measurements is calculated from the statistical fluctuations of the bulk water velocity measurement as well as the uncertainty in the friction parameter used in the momentum balance method. Both sources of error are described in detail in Bopp (2014). Subsequently, simple error propagation was used to derive the wind speed (u_{10}) uncertainty from the Smith and Banke (1975) empirical relationship (see Appendix B), the error of which is assumed to be negligible.

3.2.5 Experimental arrangement

The Aeolotron facility was filled to 1 m height ($\sim$ 18 m^3 water volume) with clean deionised water. Diluted aqueous mixtures of low solubility tracers were introduced into the water phase of the facility a day prior to an experiment and homogeneity was achieved using two circulating pumps. Before the beginning of each experiment (for the clean water surface cases), the water surface was skimmed to clean off any possible surface contamination. To do this, a small barrier with a channel is mounted between the walls of the tank, perpendicular to the wind direction while the wind is turned on at a low wind speed ($u_{ref} \approx 3$ m s^{-1}). The wind pushes the water surface over the barrier into the channel removing any surfactant. A pump continuously empties the channel and drains the water contaminated with surface active materials.

Individual gas-washing bottles containing highly solubility tracers in liquid form were purged with a controlled flow of clean air that swept the air-tracer gas mixture into the air phase of the facility. The bottles were kept in a thermostatic bath at 20 °C throughout the experimental procedure.

At the beginning of each experiment, the first wind speed condition was applied while the flushing of the air space was turned on (open air space) in order to achieve a background

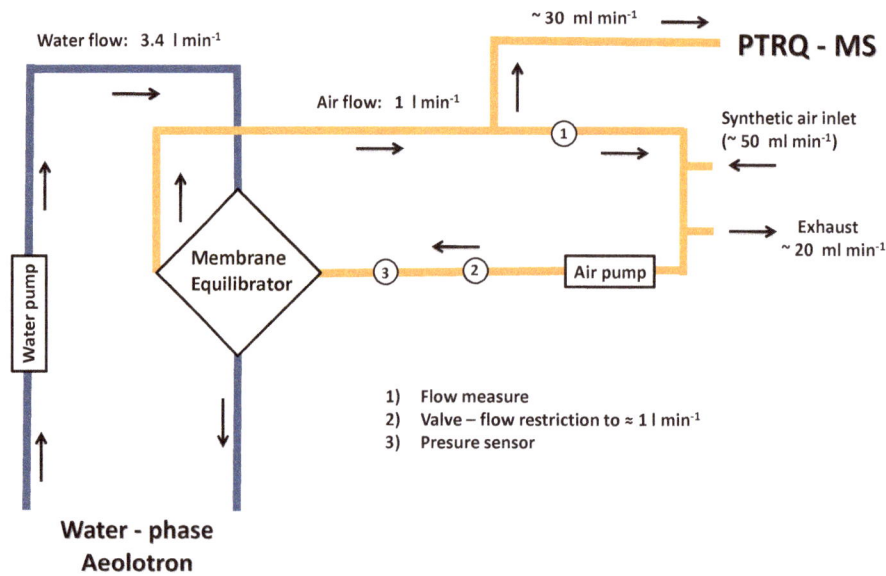

Figure 6. Membrane equilibrator – PTRQ-MS set-up schematic. The dark blue and orange lines represent the water and air loops of the system, accordingly.

point for all tracers. Thereafter, the flushing was turned off (closed air space) and the tracer concentration (air and water-side controlled) started to increase (see more in Sect. 2.3). Immediately after turning off the flushing, the leak test gas was introduced into the air space. After the steady-state point (SS_1) for the air-side controlled tracers was approached, the air space was flushed once more with ambient air and an abrupt concentration decrease was observed. The same process was repeated for eight different wind speed conditions, progressing from lower to higher values. In Fig. 4 a time series of the experimental conditions (wind speed, flushing periods and air and water tracer inputs) are schematically represented. The obtained air-sided concentration time series over the eight wind speed conditions for a water- and an air-sided example tracer are given in Fig. 5.

The wind speed varied from very low values ($u_{ref} = 0.74\,m\,s^{-1}$, equivalent to $u_{10} = 0.73\,m\,s^{-1}$) up to higher ones ($u_{ref} = 8.26\,m\,s^{-1}$, equiv. $u_{10} = 13.2\,m\,s^{-1}$). At the very beginning of the experiment, hardly any surface movement was seen. As the experiment progressed, the first capillary waves became apparent and started breaking above $u_{ref} = 4.8\,m\,s^{-1}$, equiv. $u_{10} = 6.6\,m\,s^{-1}$. Reaching larger wavelengths, wave braking and bubble formation was observable only at the highest wind speed condition.

The experimental procedure described above was repeated four times at clean surface conditions for all tracers listed in Table 1. Three further repetitions were accomplished with a surfactant (Triton X-100) covered water surface. The surfactant concentration in the fifth repetition was $0.033\,mg\,L^{-1}$ while in the last two a larger amount of $0.167\,mg\,L^{-1}$ was used.

Despite the well-reproduced experimental conditions, small variations between the repetitions were observed. Table 2 displays a mean value of the main measured parameters along with the standard deviation, expressing the extent of variability between the repetitions, of each case. For the highest wind speed condition of the clean case, only three repetitions were performed. Also in repetition two, the σ_s^2 values observed in conditions 4, 5 and 6 were significantly lower and therefore omitted from the averaging. Here we assume that the water surface was probably insufficiently skimmed before the experiment or that surfactant material might have entered the facility during the flushing phases. In the higher surfactant case (case 3), the first condition was omitted for reasons of experimental simplicity while σ_s^2 are available only for one repetition.

4 Results

In this work, total transfer velocities of two contrasting tracers at opposite ends of the solubility spectrum, N_2O ($\alpha = 0.67$) anticipated as only water-side controlled (i.e $k_{tw} \sim k_w$) and CH_3OH ($\alpha = 5293$) similarly anticipated as only air-side controlled (i.e. $k_{ta} \sim k_a$), are presented. In this way, we intend to validate the above described method and apparatus and subsequently to compare with previous air–sea gas-exchange studies. A full investigation of the mechanisms influencing the gas-exchange transfer and their relationship to individual gases over a broad solubility range will be presented in separate publications.

Table 2. Reference velocities, u_{ref} (m s^{-1}), friction velocities, $u_{*,w}$ (cm s^{-1}), mean square slope, σ_s^2, air temperature, t_a (°C), water temperature, t_w (°C) mean values and % standard deviations as quantified in the Aeolotron facility for case 1: clean surface experiments; case 2: surface covered with 0.033 mg L^{-1} Triton X-100; and case 3: surface covered with 0.167 mg L^{-1} Triton X-100, at eight different wind speed conditions. The number of replicates used for each case is given in brackets.

Case	Parameter	Cond.1	Cond.2	Cond.3	Cond.4	Cond.5	Cond.6	Cond.7	Cond.8
1 ($\times$ 4)	u_{ref} ($\pm$%)	0.744 (1.3)	1.421 (0.5)	2.052 (0.3)	2.674 (0.5)	3.621 (0.1)	4.805 (0.3)	6.465 (0.2)	8.256 (0.1)
	$u_{*,w}$ ($\pm$%)	0.063 (1.4)	0.135 (0.6)	0.216 (0.4)	0.309 (0.6)	0.473 (0.2)	0.720 (0.5)	1.141 (0.3)	1.967 (0.1)
	σ_s^2 ($\pm$%)	0.002 (1.7)	0.007 (1.4)	0.013 (1.2)	0.016 (0.5)	0.024 (2.2)	0.046 (2.0)	0.078 (3.0)	0.118 (6.3)
	t_a ($\pm$%)	21.29 (1.5)	21.18 (1.5)	21.09 (1.6)	20.99 (1.6)	20.88 (1.6)	20.75 (1.6)	20.59 (1.7)	20.50 (1.8)
	t_w ($\pm$%)	19.30 (1.7)	19.37 (1.8)	19.41 (1.8)	19.41 (1.8)	19.39 (1.8)	19.34 (1.9)	19.29 (2.0)	19.26 (2.1)
2 ($\times$ 1)	u_{ref} ($\pm$%)	0.800 ($-$)	1.460 ($-$)	2.091 ($-$)	2.717 ($-$)	3.650 ($-$)	4.851 ($-$)	6.502 ($-$)	8.288 ($-$)
	$u_{*,w}$ ($\pm$%)	$-$	0.163 ($-$)	0.246 ($-$)	0.336 ($-$)	0.484 ($-$)	0.698 ($-$)	1.038 ($-$)	1.464 ($-$)
	σ_s^2 ($\pm$%)	0.002 ($-$)	0.002 ($-$)	0.002 ($-$)	0.008 ($-$)	0.010 ($-$)	0.020 ($-$)	0.071 ($-$)	0.111 ($-$)
	t_a ($\pm$%)	22.11 ($-$)	22.04 ($-$)	21.94 ($-$)	21.77 ($-$)	21.64 ($-$)	21.44 ($-$)	21.19 ($-$)	21.11 ($-$)
	t_w ($\pm$%)	19.80 ($-$)	19.88 ($-$)	19.93 ($-$)	19.93 ($-$)	19.93 ($-$)	19.90 ($-$)	19.88 ($-$)	19.85 ($-$)
3 ($\times$ 2)	u_{ref} ($\pm$%)	$-$	1.451 (1.2)	2.075 (0.0)	2.707 (0.6)	3.667 (0.3)	4.913 (0.4)	6.615 (0.2)	8.371 (0.0)
	$u_{*,w}$ ($\pm$%)	$-$	0.18 (1.0)	0.239 (0.0)	0.295 (0.5)	0.381 (0.3)	0.524 (0.7)	0.84 (0.4)	1.407 (0.1)
	σ_s^2 ($\pm$%)	$-$	0.002 ($-$)	0.002 ($-$)	0.002 ($-$)	0.005 ($-$)	0.007 ($-$)	0.040 ($-$)	0.096 ($-$)
	t_a ($\pm$%)	$-$	21.51 (1.4)	21.59 (1.3)	21.60 (1.0)	21.53 (0.9)	21.51 (0.6)	21.34 (0.7)	21.22 (0.9)
	t_w ($\pm$%)	$-$	19.83 (0.5)	19.86 (0.6)	19.90 (0.7)	19.94 (0.6)	19.95 (0.7)	19.95 (0.7)	19.95 (0.7)

4.1 Gas-exchange transfer velocities

In Figs. 7 and 8, we present the experimentally obtained k_{tw} for N$_2$O and k_{ta} for CH$_3$OH as a function of $u_{*,w}$ for the clean water surface experiments. In both figures, the experimental results of all repetitions are nicely reproduced. Occasionally, the variation between the transfer velocity values exceeded the given uncertainty bars. A more apparent example is provided by the lower transfer velocity points (circles) at conditions 4, 5 and 6 which arise as a result of the lower σ_s^2 values observed in repetition 2 (as described in Sect. 3.2.5). This effect could be taken as an indication that only one physical parameter is not enough to effectively describe the complicated process of the air–sea gas exchange. As the experimental conditions used in the four repetitions were similar but not identical (see Table 2), a four replicate mean value calculation was avoided and instead a fit through all points is chosen (black dashed line).

As indicated in Fig. 7, the k_{tw} increases non-linearly with $u_{*,w}$. The correlation could be described as linear up to $u_{*,w} = 0.72$ cm s^{-1} (equiv. $u_{10} = 6.6$ m s^{-1}) while above this point, a faster increase is observed. This sudden increase in the so far linear tendency can be attributed to various water surface effects (e.g. initiation of capillary wave braking), which are not going to be discussed here.

The air-sided transfer velocities k_{ta} (Fig. 8) in contrast, increase linearly ($R^2 = 0.99$) with $u_{*,w}$ throughout the examined velocity range ($u_{*,w} = 0.063$–1.7 cm s^{-1} equiv. $u_{10} = 0.73$–13.2 m s^{-1}). As it appears from Fig. 8, the first transfer velocity values of CH$_3$OH (i.e. those at the lowest turbulent condition) are slightly underestimated relative to

the linear trend ($\simeq 10$%). This could be explained as being due to the inefficiently mixed air space caused by the lower turbulence conditions applied.

Overall, the observed trends and transfer velocity magnitudes of both k_{tw} and k_{ta} are in good agreement with observations made by previous studies. A more detailed comparison with literature follows in Sect. 4.3.

4.2 Effect of surfactants

After obtaining clear, reproducible transfer velocity trends for a clean water surface, the effect of a surfactant was evaluated using two different surfactant (Triton X-100) concentrations. As expected, the surfactant suppressed the transfer velocity as well as the friction velocity, $u_{*,w}$, and mean square slope, σ_s^2 (see Table 2). In Fig. 9, the transfer velocities of all seven experiments for N$_2$O and CH$_3$OH are presented against the reference wind speed, u_{ref}; a parameter which is not affect by the surfactant layers.

As indicated in Fig. 9, the surfactant effect shows significant differences between the two contrasting tracers. A generally stronger suppression is observed for N$_2$O and a significantly weaker for CH$_3$OH where the transfer is mainly controlled by the air-side boundary layer. Under low turbulence conditions, the surfactant diminished the transfer velocity of N$_2$O by a factor of 3 for both examined concentrations (0.033 and 0.167 mg L^{-1} Triton X-100) (Fig. 9a). In the case of CH$_3$OH, the effect was weaker demonstrated by a factor of 1.5 (Fig. 9b). At slightly higher wind speeds (i.e $\geq u_{ref}$ 3 m s^{-1}; see Table 2 for the equivalent $u_{*,w}$ of each case), different trends are observed between low and high

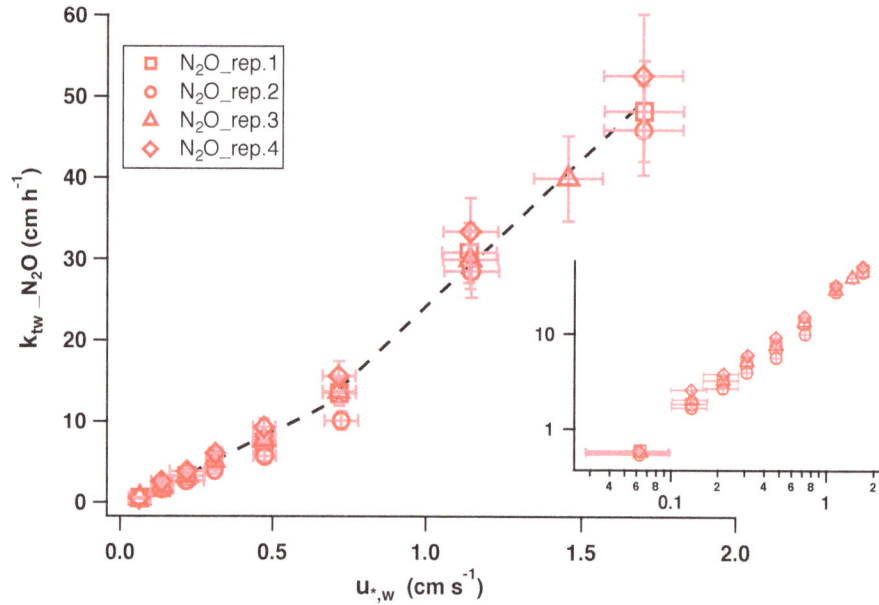

Figure 7. Total transfer velocity of N_2O of four clean case repetitions plotted against $u_{*,w}$. Squares correspond to repetition 1, circles to repetition 2, triangles to repetition 3 and diamonds to repetition 4. Vertical bars in light red give the individual transfer velocity uncertainty (see Sect. 3.2.3) and the horizontal bars the uncertainty of the $u_{*,w}$ measurements.

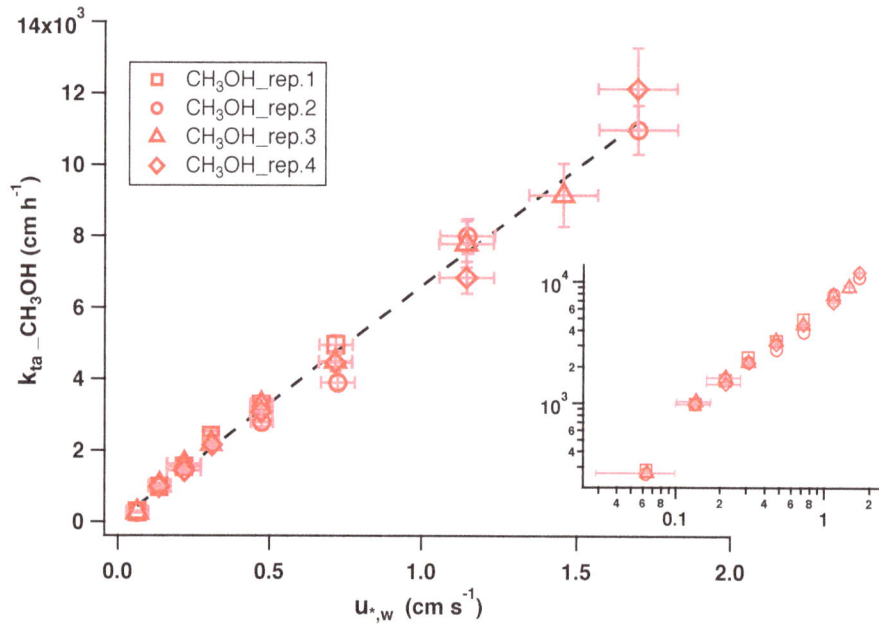

Figure 8. Total transfer velocity of CH_3OH of four clean case repetitions plotted against $u_{*,w}$. Symbols and bars are the same as in Fig. 7.

surfactant concentrations with the higher concentration causing a more prominent suppression. Reaching the highest examined wind speed where waves, wave breaking and bubbles are present, the surfactant effect is significantly weaker. In the case of CH_3OH, no further impact is observed (values within the error bars), while a still significant suppression (around a factor of 1.5) is apparent for the k_{tw} of N_2O under a $0.167\,\mathrm{mg\,L^{-1}}$ Triton X-100 surfactant film.

Reduced surface stress and roughness change the hydrodynamic properties of the water surface and consequently affect the gas transfer. As given in Table 2, the suppression caused by a surfactant seems to be relatively strong for σ_s^2 (like in the case of N_2O) and rather weak for $u_{*,w}$ (here, note the higher uncertainties at the low wind speed regime). The trend of the reduction for both $u_{*,w}$ and σ_s^2 was similar to the one of k_t, more apparent at low wind speeds and weaker under higher

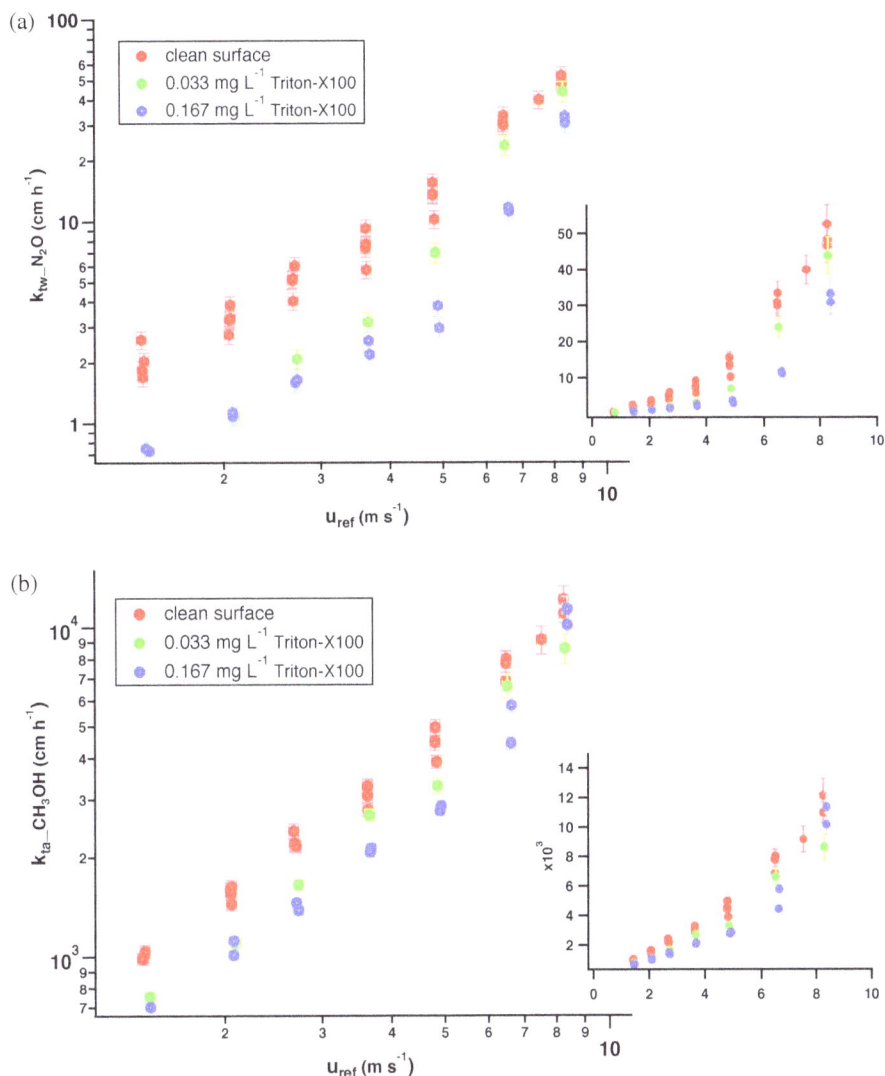

Figure 9. Effect of the two different surfactant concentrations on the total transfer velocities of (**a**) N_2O and (**b**) CH_3OH. The clean water surface results are given in red, the results obtained using $0.033\,\text{mg}\,L^{-1}$ Triton X-100 in green and the ones using $0.167\,\text{mg}\,L^{-1}$ Triton X-100 in blue.

turbulence. The way in which each of these parameters affect the gas transfer needs further investigation and is going to be presented in a following publication.

4.3 Comparison with previous studies

The gas transfer velocities of weakly soluble tracers has been extensively studied over the previous years. Numerous k_w parameterisations are available, derived from experimental (laboratory and field) measurements as well as physical models. In Fig. 10, a selection of some representative, experimentally derived parameterisations (coloured lines), are used for comparison with the k_{tw} (here $k_{tw} \sim k_w$) measurements of N_2O (red points). The transfer velocities are plotted against the wind speed at 10 m height, u_{10}.

Looking at the lower wind speed range (0.7 to $4\,\text{m s}^{-1}$) an obvious spread between the various k_w predictions can be observed extending through more than 2 orders of magnitude. Transfer velocity measurements at the very low wind speed regions are difficult to conduct and the extended fits based on higher wind speed ranges can lead to incorrect estimations. One target of this study was to tackle the challenging measurements at the low wind speed end. The transfer velocites obtained in this study are characterised by low uncertainties and definitely provide a better indication of the low wind speed end gas-exchange behaviour. The results presented here are obtained under slightly stable atmosphere conditions (water temperature lower that the air temperature). No gas exchange due to convection is apparent. In the open ocean, depending on the temperature and the light intensity, convec-

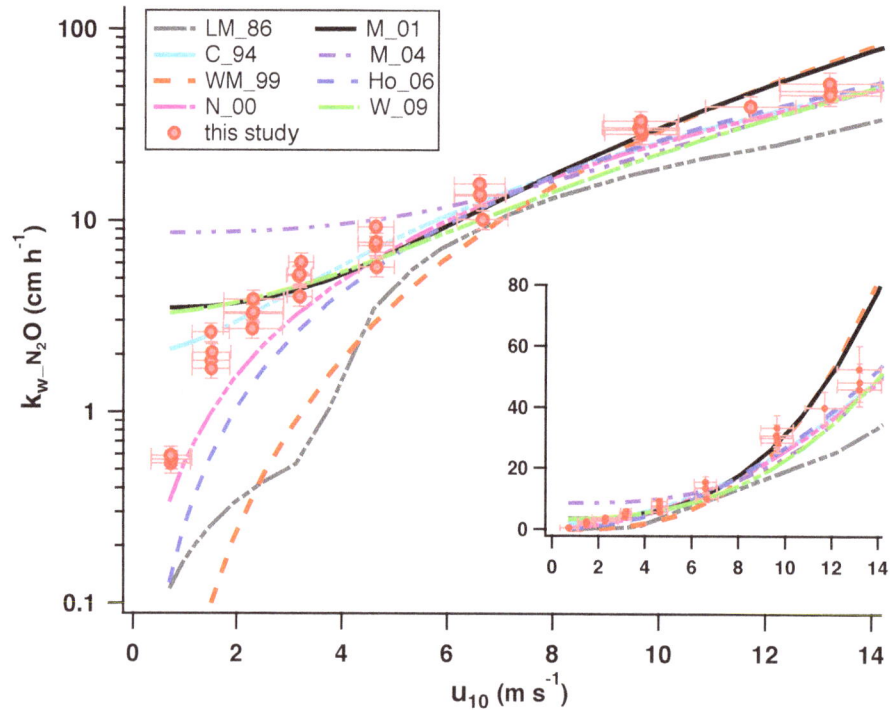

Figure 10. Comparison between the N_2O measurements of this study (red circles) and previous k_w parameterisations. The coloured lines correspond to LM_86: Liss and Merlivat (1986); C_94: Clark et al. (1994); WM_99: Wanninkhof and McGillis (1999); N_00: Nightingale et al. (2000); M_01: McGillis et al. (2001); M_04: McGillis et al. (2004); Ho_06: Ho et al. (2006); and W_09: Wanninkhof et al. (2009).

tion can provoke an increase in the gas exchange which at very low wind speeds is considered significant. Such environmental differences could provide an additional explanation for the huge disagreement observed. The projected absolute quantity differences to the atmospheric budgets, however, are estimated to be small since the fluxes themselves are small. The middle wind speed range seems to be well represented in all studies. At the higher wind speed range, a smaller spread was observed especially above $10\,\mathrm{m\,s^{-1}}$. This spread, a little more than an order of 2, can lead to great discrepancies in the atmospheric budgets of the related tracers as the corresponding transfer velocities are much larger there.

The transfer velocities obtained in this study, show a closer agreement with the Clark et al. (1994) parameterisation apart from the lowest wind speed. There, our results agree better with Nightingale et al. (2000).

In contrast to the weakly soluble gases, high solubility tracers have received much less attention. In Fig. 11, the total transfer velocity measurements of CH_3OH (here $k_{ta} \sim k_a$) are compared with some available tunnel (Liss, 1973; Mackay and Yeun, 1983, coloured lines) and model (Duce et al., 1991; Jeffery et al., 2010, from COARE Algorithm; Fairall et al., 2003, grey lines) k_a parameterisations as well as a recent CH_3OH field study (Yang, 2013, black line with triangles).

As indicated in Fig. 11, our results agree very well with the previous laboratory parameterisations lying nearer to Mackay and Yeun (1983). Here again the first transfer veloc-

ity point deviates, though an increase by the estimated 10 % would still not change this trend. We note that also in the case of k_a, the lower wind speed range of the other experimental studies is covered by a fit extension based on transfer velocities obtained at higher wind speeds.

The transfer velocity values provided by model and field studies are about 1.5 to 2 times lower than the ones derived from the laboratory measurements. This is to be expected as model and field studies include an extra turbulent resistance in the air space at 10 m height.

5 Conclusions

This study has demonstrated that the Aeolotron wind–wave tank in combination with the adopted box model methodology, experimental procedure and instrumentation are capable of generating reliable and reproducible gas transfer velocities for species spanning a wide range of solubilities. The molecules nitrous oxide and methanol have been used to exemplify the behaviour of sparingly soluble and highly soluble species. These represent cases of a water-side and an air-side layer control, as described in Sect. 2. Small differences between the obtained transfer velocity values of the four repetitions of the clean case indicate that various physical parameters should be taken into account in future parameterisations in order to produce better transfer velocity estimations. The

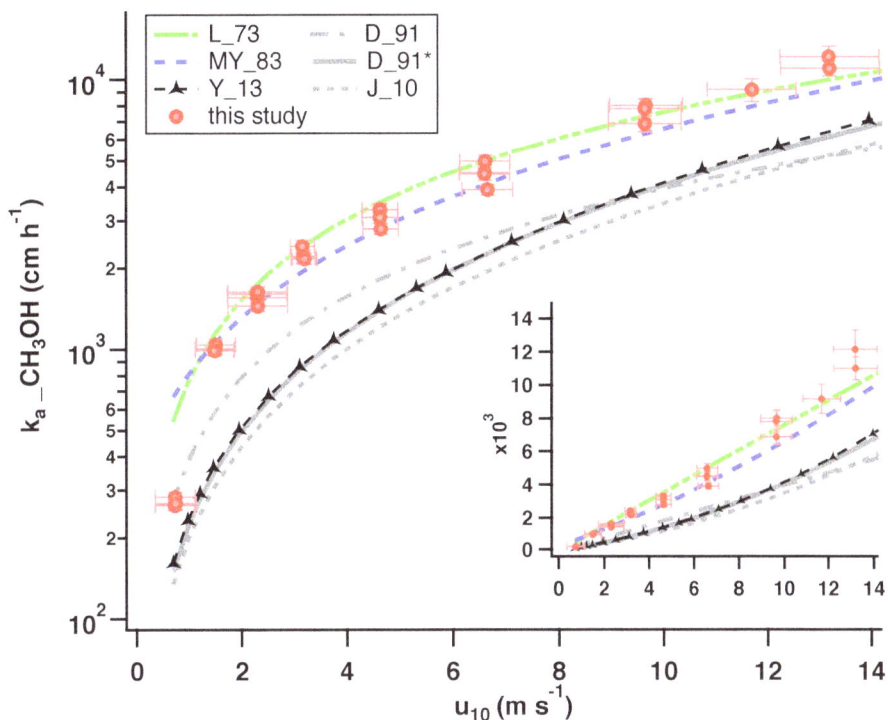

Figure 11. Comparison between the CH_3OH measurements of this study (red circles) and previous k_a parameterisations. Experimental studies are presented with coloured lines: L_73: Liss (1973); MY_83: Mackay and Yeun (1983); and Y_13: Yang (2013), while model studies are given with grey lines: D_91 using the M; D_91* using the Sc_a: Duce et al. (1991); and J_10: Jeffery et al. (2010) using a Smith and Banke (1975) derived drag coefficient term.

complete data set for all species (including the intermediate cases of both layer control) along with the available micro-scale surface property parameters, extending over a low to medium wind speed regime, can be used to generate a generalised parameterisation for the total transfer velocity. The derivation of this expression, which will be invaluable to future modelling efforts, will be presented in a separate publication.

Particularly interesting are the effects on the gas transfer velocity induced by the addition of a surfactant. Despite the surface micro-layer being commonly present on the ocean, its effect on air–sea gas transfer is poorly understood and there is a paucity of data both from the laboratory and the field. The impact of the surfactant is markedly different on the two tracers shown here. A strong reducing effect (up to a factor of 3) was observed for the water-side controlled tracer, N_2O while in the case of CH_3OH, the surfactant showed a quite weaker impact.

We maintain that it is important to monitor the transfer process in both the water-phase (using water-side controlled tracers) and the air-phase layer (using air-side controlled tracers) in order to develop a true enduring and generally applicable model for air–sea gas transfer. The results produced here correspond reasonably well with previous expressions for k_a and k_w. In case of k_w, at low wind speeds there is a wide spread in the literature values, with this study corre-

sponding most closely with those of Clark et al. (1994) and Nightingale et al. (2000). At high wind speeds the previous parameterisations are divided into three groups and this study lies in the central group. Despite the relatively small number of investigations, in case of k_a, the literature spread is much smaller with our results nicely corresponding to the previous laboratory parameterisations (i.e. Liss, 1973; Mackay and Yeun, 1983).

This study, based on data from the world's largest operational annular wind–wave facility, derived from advanced analytical technology which has been set-up to monitor the gas concentration changes in both the air and the water phase simultaneously at unprecedented measurement frequency, has proven to produce high quality transfer velocity measurements. On the basis of our results, we recommend the proposed methodology for future air–water gas-exchange measurements.

Appendix A: Relationship between air-sided and water-sided variables

The experimentally calculated water-sided total transfer velocities, k_{tw}, were converted to the equivalent air-sided total transfer velocities, k_{ta}, using

$$k_{ta} = \alpha k_{tw}. \tag{A1}$$

Air-sided friction velocities can be converted to water-sided friction velocities by

$$u_{*,w} = \sqrt{\frac{\rho_a}{\rho_w}} u_{*,a}. \tag{A2}$$

Appendix B: u_{10} derivation using $u_{*,a}$

The wind speed at a height of 10 m, u_{10}, is calculated from its relationship with the air-sided friction velocity and the drag coefficient, C_d, using

$$C_d = \frac{u_{*,a}^2}{u_{10}^2}. \tag{B1}$$

Here, the Smith and Banke (1975) empirical relationship between the drag coefficient and the wind speed was used

$$10^3 C_d = 0.63 + 0.066 u_{10}. \tag{B2}$$

Appendix C: Naming conventions

k_w	transfer velocity in water for a water-sided viewer
k_a	transfer velocity in air for an air-sided viewer
k_{tw}	total transfer velocity for a water-sided viewer
k_{ta}	total transfer velocity for an air-sided viewer
c_w	water-side concentration
c_a	air-side concentration
V_w	water volume
V_a	air volume
A	water surface
$\lambda_{f,1}$	leak rate
$\lambda_{f,2}$	flush rate
$\alpha = \frac{c_w}{c_a}$	dimensionless solubility
Sc_w	Schmidt number in water
Sc_a	Schmidt number in air
M	molecular mass
$u_{*,w}$	water-sided friction velocity
$u_{*,a}$	air-sided friction velocity
ρ_w	density of water
ρ_a	density of air
u_{ref}	reference wind speed
σ_s^2	mean square slope
u_{10}	wind speed at 10 m height

Acknowledgements. We own a special thank to J. Auld and T. Klüpfel for their valuable assistance and support during the gas exchange experiments. The mean square slope values were kindly provided by R. Rocholz. R. Sander is thanked for the helpful and insightful discussions on solubility matters considering this manuscript. Furthermore, we thank all members of B. Jähne's group for their understanding and support during the measurements. We acknowledge the financial support of the BMBF Verbundprojekt SOPRAN (www.sopran.pangaea.de; SOPRAN grant 03F0611A, 03F0611K, 03F0611F and 03F0662F)

The service charges for this open access publication have been covered by the Max Planck Society.

Edited by: J. Shutler

References

Benkelberg, H. J., Hamm, S., and Warneck, P.: Henry's law coefficients for aqueous solutions of acetone, acetaldehyde and acetonitrile, and equilibrium constants for the addition compounds of acetone and acetaldehyde with bisulfite, J. Atmos. Chem., 20, 17–34, doi:10.1007/Bf01099916, 1995.

Betterton, E. A. and Hoffmann, R. M.: Henry's law constants of some environmentally important aldehydes., Environ. Sci. Technol., 22, 1415–1418, doi:10.1021/Es00177a004, 1988.

Bopp, M.: Luft und wasserseitige Strömungsverhältnisse im ringförmigen Heidelberger Wind-Wellen-Kanal (Aeolotron), Master Thesis, University of Heidelberg, available at: http://www.ub.uni-heidelberg.de/archiv/16962 (last access: 18 June 2014), 2014.

Broecker, W. S., Peng, T. H., Ostlund, G., and Stuiver, M.: The Distribution of Bomb Radiocarbon in the Ocean, J. Geophys. Res.-Oceans, 90, 6953–6970, doi:10.1029/Jc090ic04p06953, 1985.

Carpenter, L. J., Archer, S. D., and Beale, R.: Ocean-atmosphere trace gas exchange, Chem. Soc. Rev., 41, 6473–6506, 2012.

Clark, J. F., Wanninkhof, R., Schlosser, P., and Simpson, H. J.: Gas-Exchange Rates in the Tidal Hudson River Using a Dual Tracer Technique, Tellus B, 46, 274–285, 1994.

Cosovic, B. and Vojvodic, V.: Voltammetric analysis of surface active substances in natural seawater, Electroanalysis, 10, 429–434, doi:10.1002/(SICI)1521-4109(199805)10:6<429::AID-ELAN429>3.0.CO;2-7, 1998.

Dacey, J. W. H., Wakeham, G. S., and Howes, L. B.: Henry's law constants for dimethylsulfide in freshwater and seawater, Geophys. Res. Lett., 11, 991–994, doi:10.1029/Gl011i010p00991, 1984.

Danckwerts, P. V.: Significance of Liquid-Film Coefficients in Gas Absorption, Ind. Eng. Chem., 43, 1460–1467, 1951.

Donelan, M. A. and Wanninkhof, R.: Concepts and Issues, in: Gas Transfer at Water Surfaces (127), American Geophysical Union, edited by: Donelan, M. A., Drennan, W. M., Saltzman, E. S., and Wanninkhof, R., 1–10, 2002.

Duce, R. A., Liss, P. S., Merrill, J. T., Atlas, E. L., Buat-Menard, P., Hicks, B. B., Miller, J. M., Prospero, J. M., Arimoto, R., Church, T. M., Ellis, W., Galloway, J. N., Hansen, L., Jickells,

T. D., Knap, A. H., Reinhardt, K. H., Schneider, B., Soudine, A., Tokos, J. J., Tsunogai, S., Wollast, R., and Zhou, M.: The atmospheric input of trace species to the world ocean, Geophys. Res. Lett., 5, 193–259, doi:10.1029/91GB01778, 1991.

Fairall, C. W., Bradley, E. F., Hare, J. E., Grachev, A. A., and Edson, J. B.: Bulk Parameterization of Air–Sea Fluxes: Updates and Verification for the COARE Algorithm, J. Climate, 16, 571–591, doi:10.1175/1520-0442(2003)016<0571:BPOASF>2.0.CO;2, 2003.

Field, C. B., Behrenfeld, M. J., Randerson, J. T., and Falkowski P.: Primary production of the biosphere, Science, 281, 237–240, 1998.

Frew, N. M., Goldman, J. C., Denett, M. R., and Johnson, A. S.: Impact of phytoplankton-generated surfactants on air-sea gas exchange, J. Geophys. Res.-Oceans, 95, 3337–3352, doi:10.1029/JC095iC03p03337, 1990.

Frew, N. M., Bock, E. J., McGillis, W. R., Karachintsev, A. V., Hara, T., Münsterer, T., and Jähne, B.: Variation of air-water gas transfer with wind stress and surface viscoelasticity, in: Air-water Gas Transfer, Selected Papers from the Third International Symposium on Air-Water Gas Transfer, edited by: Jähne, B. and Monahan, E. C., Aeon, Hanau, 529–541, 1995.

Griffiths, P. R.: Fourier Transform Infrared Spectrometry, Wiley Interscience, 2nd Edn., 2007.

Ho, D. T., Law, C. S., Smith, M. J., Schlosser, P., Harvey, M., and Hill, P.: Measurements of air-sea gas exchange at high wind speeds in the Southern Ocean: Implications for global parameterizations, Geophys. Res. Lett., 33, L16611, doi:10.1029/2006GL026817, 2006.

Jähne, B.: Air-sea gas exchange, in: Encyclopedia Ocean Sciences, Elsevier, 3434–3444, 2009.

Jähne, B. and Haußecker, H.: Air-water gas exchange, Annu. Rev. Fluid Mech., 30, 443–468, 1998.

Jähne, B., Münnich, K. O., and Siegenthaler, U.: Measurements of Gas-Exchange and Momentum-Transfer in a Circular Wind-Water Tunnel, Tellus, 31, 321–329, 1979.

Jähne, B., Münnich, K. O., Bösinger, R., Dutzi, A., Huber, W., and Libner, P.: On the Parameters Influencing Air-Water Gas-Exchange, J. Geophys. Res.-Oceans, 92, 1937–1949, 1987.

Janini, G. M. and Quaddora, A. L.: Determination of activity coefficients of oxygenated hydrocarbons by liquid-liquid chromatography, J. Liq. Chromatogr., 9, 39–53, doi:10.1080/01483918608076621, 1986.

Jeffery, C. D., Robinson, I. S., and Woolf, D. K.: Tuning a physically-based model of the air-sea gas transfer velocity, Ocean Model., 31, 28–35, doi:10.1016/j.ocemod.2009.09.001, 2010.

Krall, K. E.: Laboratory Investigations of Air-Sea Gas Transfer under a Wide Range of Water Surface Conditions, Dissertation, University of Heidelberg, available at: http://www. ub.uni-heidelberg.de/archiv/14392 (last access: 18 June 2014), 2013.

Krall, K. E. and Jähne, B.: First laboratory study of air-sea gas exchange at hurricane wind speeds, Ocean Sci., 10, 257–265, doi:10.5194/os-10-257-2014, 2014.

Kräuter, C.: Aufteilung des Transferwiderstandes zwischen Luft und Wasser beim Austausch flüchtiger Substanzen mittlerer Löslichkeit zwischen Ozean und Atmosphäre, Diploma, University of Heidelberg, Germany, available at: http://www.ub.

uni-heidelberg.de/archiv/13010 (last access: 18 June 2014), 2011.

Lindinger, W., Hansel, A., and Jordan, A.: On-line monitoring of volatile organic compounds at pptv levels by means of proton-transfer-reaction mass spectrometry (PTR-MS) – Medical applications, food control and environmental research, Int. J. Mass. Spectrom., 173, 191–241, doi:10.1016/S0168-1176(97)00281-4, 1998.

Liss, P. S.: Processes of Gas-Exchange across an Air-Water Interface, Deep-Sea Res., 20, 221–238, 1973.

Liss, P. S. and Merlivat, L.: Air-sea gas exchange rates: Introduction and synthesis, in The role of air-sea exchange in geochemical cycling, Reidel, Boston, MA, 113–129, 1986.

Liss, P. S. and Slater, P. G.: Flux of Gases across Air-Sea Interface, Nature, 247, 181–184, 1974.

Mackay, D. and Yeun, A. T. K.: Mass-Transfer Coefficient Correlations for Volatilization of Organic Solutes from Water, Environ. Sci. Tech., 17, 211–217, doi:10.1021/Es00110a006, 1983.

McGillis, W. R., Edson, J. B., Ware, J. D., Dacey, J. W. H., Hare, J. E., Fairall, C. W., and Wanninkhof, R.: Carbon dioxide flux techniques performed during GasEx-98, Mar. Chem., 75, 267–280, doi:10.1016/S0304-4203(01)00042-1, 2001.

McGillis, W. R., Edson, J. B., Zappa, C. J., Ware, J. D., McKenna, S. P., Terray, E. A., Hare, J. E., Fairall, C. W., Drennan, W., Donelan, M., DeGrandpre, M. D., Wanninkhof, R., and Feely, R. A.: Air-sea CO_2 exchange in the equatorial Pacific, J. Geophys. Res.-Oceans, 109, C08S02, doi:10.1029/2003jc002256, 2004.

Millet, D. B., Jacob, D. J., Custer, T. G., de Gouw, J. A., Goldstein, A. H., Karl, T., Singh, H. B., Sive, B. C., Talbot, R. W., Warneke, C., and Williams, J.: New constraints on terrestrial and oceanic sources of atmospheric methanol, Atmos. Chem. Phys., 8, 6887–6905, doi:10.5194/acp-8-6887-2008, 2008.

Millet, D. B., Guenther, A., Siegel, D. A., Nelson, N. B., Singh, H. B., de Gouw, J. A., Warneke, C., Williams, J., Eerdekens, G., Sinha, V., Karl, T., Flocke, F., Apel, E., Riemer, D. D., Palmer, P. I., and Barkley, M.: Global atmospheric budget of acetaldehyde: 3-D model analysis and constraints from in-situ and satellite observations, Atmos. Chem. Phys., 10, 3405–3425, doi:10.5194/acp-10-3405-2010, 2010.

Nielsen, R.: Gasaustausch – Entwicklung und Ergebnis eines schnellen Massenbilanzverfahrens zur Messung der Austauschparameter, Dissertation, University of Heidelberg, available at: http://www.ub.uni-heidelberg.de/archiv/5032 (last access: 18 June 2014), 2004.

Nightingale, P. D.: Air-sea gas exchange. Lower Atmosphere Processes, in: Surface Ocean, AGU Books Board, edited by: Le Quéré, C. and Saltzman, E. S., 69–97, 2009.

Nightingale, P. D., Malin, G., Law, C. S., Watson, A. J., Liss, P. S., Liddicoat, M. I., Boutin, J., and Upstill-Goddard, R. C.: In situ evaluation of air-sea gas exchange parameterization using novel conservation and volatile tracers, Global Biogeochem. Cy., 14, 373–387, 2000.

Pozzer, A., Jöckel, P., Sander, R., Williams, J., Ganzeveld, L., and Lelieveld, J.: Technical Note: The MESSy-submodel AIRSEA calculating the air-sea exchange of chemical species, Atmos. Chem. Phys., 6, 5435–5444, doi:10.5194/acp-6-5435-2006, 2006.

Robbins, G. A., Wang, V., and Stuart, D. J.: Using the headspace method to determine Henry's law constants, Anal. Chem., 65, 3113–3118, doi:10.1021/Ac00069a026, 1993.

Rocholz, R.: Spatiotemporal Measurement of Short Wind-Driven Water Waves, Dissertation, University of Heidelberg, available at: http://www.ub.uni-heidelberg.de/archiv/8897 (last access: 18 June 2014), 2008.

Salter, M. E., Upstill-Goddard, R. C., Nightingale, P. D., Archer, S. D., Blomquist, B., Ho, D. T., Huebert, B., Schlosser, P., and Yang, M.: Impact of an artificial surfactant release on air–sea gas fluxes during Deep Ocean Gas Exchange Experiment II, J. Geophys. Res., 116, C11016, doi:10.1029/2011JC007023, 2011.

Saltzman, E.: Introduction to Surface Ocean–Lower Atmosphere Processes, in: Surface Ocean-Lower Atmosphere Processes, Geophysical Research Series, 187, 2009.

Sander, R.: Compilation of Henry's Law Constants for Inorganic and Organic Species of Potential Importance in Environmental Chemistry (Version 3), available at: http://www.henrys-law.org (last access: 18 June 2014), 1999.

Schaffer, D. L. and Daubert, E. T.: Gas-liquid chromatographic determination of solution properties of oxygenated compounds in water, Anal. Chem., 286, 1585–1589, 1969.

Smith, S. D. and Banke, E. G.: Variation of Sea-Surface Drag Coefficient with Wind Speed, Q. J. Roy. Meteorol. Soc., 101, 665–673, 1975.

Snider, J. R. and Dawson, A. G.: Tropospheric light alcohols, carbonyls, and acetonitrile: Concentrations in the southwestern United States and Henry's law data, J. Geophys. Res., 90D, 3797–3805, doi:10.1029/Jd090id02p03797, 1985.

Wanninkhof, R. and McGillis, W. R.: A cubic relationship between air-sea CO_2 exchange and wind speed, Geophys. Res. Lett., 26, 1889–1892, doi:10.1029/1999gl900363, 1999.

Wanninkhof, R., Asher, W., Weppernig, R., Chen, H., Schlosser, P., Langdon, C., and Sambrotto, R.: Gas Transfer Experiment on Georges Bank Using 2 Volatile Deliberate Tracers, J. Geophys. Res.-Oceans, 98, 20237–20248, 1993.

Wanninkhof, R., Asher, W. E., Ho, D. T., Sweeney, C., and McGillis, W. R.: Advances in quantifying air-sea gas exchange and environmental forcing, Annu. Rev. Mar. Sci., 1, 213–244, 2009.

Watson, A. J., Upstill-Goddard, R. C., and Liss, P. S.: Air Sea Gas-Exchange in Rough and Stormy Seas Measured by a Dual-Tracer Technique, Nature, 349, 145–147, 1991.

Weiss, R. F. and Price, B. A.: Nitrous oxide solubility in water and seawater, Mar. Chem., 8, 347–359, doi:10.1016/0304-4203(80)90024-9, 1980.

Williams, J., Holzinger, R., Gros, V., Xu, X., Atlas, E., and Wallace, D. W. R.: Measurements of organic species in air and seawater from the tropical Atlantic, Geophys. Res. Lett., 31, L23S06, doi:10.1029/2004GL020012, 2004.

Wurl, O., Wurl, E., Miller, L., Johnson, K., and Vagle, S.: Formation and global distribution of sea-surface microlayers, Biogeosciences, 8, 121–135, doi:10.5194/bg-8-121-2011, 2011.

Yang, M., Nightingale, P. D., Beale, R., Liss, P. S., Blomquist, B., and Fairall, C.: Atmospheric deposition of methanol over the Atlantic Ocean, P. Natl. Acad. Sci. USA, 110, 20034–20039, doi:10.1073/pnas.1317840110, 2013.

Yaws, C. L.: Handbook of Transport Property Data, Gulf Publishing Company, 1995.

Yaws, C. L. and Pan, X.: Liquid Heat-Capacity for 300 Organics, Chem. Eng., 99, 130–134, 1992.

Zappa, C. J., Asher, W. E., Jessup, A. T., Klinke, J., and Long, S. R.: Microbreaking and the enhancement of air-water transfer velocity, J. Geophys. Res.-Oceans, 109, C08S16, doi:10.1029/2003jc001897, 2004.

6

A wind-driven nonseasonal barotropic fluctuation of the Canadian inland seas

C. G. Piecuch and R. M. Ponte

Atmospheric and Environmental Research, Inc., Lexington, MA 02421, USA

Correspondence to: C. G. Piecuch (cpiecuch@aer.com)

Abstract. A wind-driven, spatially coherent mode of nonseasonal, depth-independent variability in the Canadian inland seas (i.e., the collective of Hudson Bay, James Bay, and Foxe Basin) is identified based on Gravity Recovery and Climate Experiment (GRACE) retrievals, a tide-gauge record, and a barotropic model over 2003–2013. This dominant mode of nonseasonal variability is correlated with the North Atlantic Oscillation and is associated with net flows into and out of the Canadian inland seas; the anomalous inflows and outflows, which are reflected in mean sea level and bottom pressure changes, are driven by wind stress anomalies over Hudson Strait, probably related to wind setup, as well as over the northern North Atlantic Ocean, possibly mediated by various wave mechanisms. The mode is also associated with mass redistribution within the Canadian inland seas, reflecting linear response to local wind stress variations under the combined influences of rotation, gravity, and variable bottom topography. Results exemplify the usefulness of GRACE for studying regional ocean circulation and climate.

1 Introduction

Hudson Bay, James Bay, and Foxe Basin together constitute the Canadian inland seas (CIS; Fig. 1). This set of marginal seas connects to the Labrador Sea and North Atlantic through Hudson Strait and Ungava Bay in the east, and to the Arctic Ocean through Fury and Hecla Strait and the Gulf of Boothia in the north. These seas are shallow, having depths of ~ 90–$150\,\mathrm{m}$, and broad, spanning an area of $\sim 1 \times 10^6\,\mathrm{km}^2$ (MacDonald and Kuyzk, 2011). The mean hydrography in Hudson Bay in summer and autumn is such that a thinner, shallower layer of fresher, warmer water sits atop a thicker, deeper layer of saltier, cooler water; during winter and spring, the surface waters cool, and the mixed layer reaches deeper down in the water column (Prinsenberg, 1986a, 1987). Related climatological features include a seasonal cycle in sea ice, which oscillates between complete ice cover in wintertime and ice-free conditions in summertime (Markham, 1986), as well as volume input due to runoff (Déry et al., 2005, 2011), transport through Fury and Hecla Strait, and flow from Baffin Bay through Hudson Strait along the Baffin Island coast, which is mostly balanced by volume outflow through Hudson Strait along the Québec coast (Straneo and Saucier, 2008a, b).

The CIS play an important role in the ocean general circulation. Numerical simulations suggest that Hudson Strait is one of the most important regions in the world ocean for the dissipation of tidal energy (Egbert and Ray, 2001; Webb, 2014). Barotropic models show how, on synoptic timescales, dynamic response of Hudson Bay to barometric pressure drives flows through Hudson Strait, which generate coastal waves that subsequently affect sea level downstream as they propagate over the continental shelf (Wright et al., 1987; Greatbatch et al., 1996). Direct measurements of the baroclinic boundary current by a moored current array deployed in Hudson Strait reveal that the outflow through Hudson Strait is responsible for a substantial portion of the fresh water supplied to the Labrador Current and the North Atlantic Ocean (Straneo and Saucier, 2008a).

This region is also of interest in the context of changes ongoing in the Arctic system (White et al., 2007). Passive microwave data reveal that concentrations and extents of seasonal sea ice have decreased in Hudson Bay over recent decades, while historical climate data show that surface air temperatures around Hudson Bay have warmed since 1950 (Hochheim and Barber, 2010; Hochheim et al., 2011). Hy-

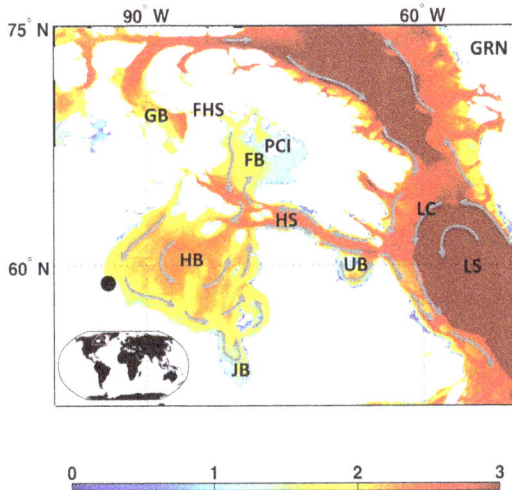

Figure 1. Shading is the logarithm of ocean depth from ETOPO5 5 min gridded elevation data (National Oceanic and Atmospheric Administration 1988) in the Hudson Bay study area. Units are $\log_{10}(m)$. Color shading saturates at a value equivalent to 1000 m. Gray arrows schematically represent the sense of the mean regional surface circulation after Prinsenberg (1986b, c) and Drinkwater (1986). The black circle near 59° N, 94° W is the location of the Churchill tide gauge. Letters are acronyms for major regional features; alphabetically, they are Foxe Basin (FB), Fury and Hecla Strait (FHS), Greenland (GRN), Gulf of Boothia (GB), Hudson Bay (HB), Hudson Strait (HS), James Bay (JB), Labrador Current (LC), Labrador Sea (LS), Prince Charles Island (PCI), and Ungava Bay (UB). Red outlining in the inset displays the study region location relative to the world ocean.

drometric data indicate strong interannual changes in Hudson Bay streamflow along with a marked shift in the seasonality of river discharge (Déry and Wood, 2004; Déry et al., 2011). However, it remains unclear whether the subsurface waters of the CIS have also undergone change.

Despite their relevance to circulation and climate, the CIS have been grossly undersampled: due to their vast expanse, harsh conditions, and remote location, few campaigns have been dedicated to continuously measuring their subsurface waters; even estimates of the bathymetry in this region can show large uncertainties, especially in the more northern reaches of the CIS. Early observational descriptions of circulation patterns and current structures are derived from sparse data (Prinsenberg, 1986b, c; Drinkwater, 1986). More recent data have facilitated more nuanced descriptions, for example, of the spatial structure of the boundary currents and the role of synoptic eddies in transporting fresh water through Hudson Strait (Straneo and Saucier, 2008a; Sutherland et al., 2011) and the mean state and seasonal cycle in the circulation and hydrography in Hudson and James bays (St-Laurent et al., 2012). However, continuous measurements of subsurface properties remain sparse, leaving open basic questions regarding regional ocean behavior on nonseasonal periods longer than a few days.

Concerns over impacts of climate change (Laidler and Gough, 2003) motivate best use of extant data to provide an understanding of anomalous behavior in this region. While a tide gauge situated at Churchill in southwestern Hudson Bay has measured sea level fluctuations since 1940, and the Gravity Recovery and Climate Experiment (GRACE) spacecraft have observed changes in the mass of ocean and ice over the CIS since 2002, only a few studies have made use of these data to understand the nature of variability in the CIS. Based on the Churchill tide gauge and hydrometric data, Gough and Robinson (2000) posit that sea level variations observed at Churchill partly result from local discharge from the Churchill River. Considering GRACE and an atmospheric reanalysis, Piecuch and Ponte (2014) submit that wind setup could effect mass changes in Hudson Bay. However, these hypotheses are based on statistical metrics (correlation coefficients and coefficients of determination), and it remains to test them using a more dynamically rigorous approach.

In this paper, we investigate nonseasonal oceanic behavior in the CIS. We provide an exploration and interpretation of the data from GRACE and the tide-gauge measurements mostly based on a coarse-resolution barotropic model driven by surface wind stress. The remainder of this paper is organized as follows: in Sect. 2, we describe and contrast the observational data; in Sect. 3, we describe the ocean model, comparing it to the available observations as well as output from a higher-resolution ocean/sea-ice model, and then use it to understand the leading mode of nonseasonal behavior of the CIS; in Sect. 4, we summarize and discuss our results.

2 Ocean observations

2.1 Description

2.1.1 Satellite gravimetry

Since their launch in March 2002, the twin GRACE satellites have been monitoring the exchange of water mass between the land and the sea (e.g., Boening et al., 2012). We use monthly ocean bottom pressure estimates derived from Release-05 GRACE time-variable gravity coefficients over the period 2003–2013 to study mass variability in the CIS. The data are taken from the GRACE Tellus server (data version RL05.DSTvDPC1409) and are processed at the University of Texas Center for Space Research (Bettadpur, 2012). Postprocessing by Don P. Chambers (University of South Florida) follows methods described by Chambers and Bonin (2012). Relevant for our purposes, the data are smoothed with a 500 km Gaussian filter, which reduces errors with short wavelengths in the satellite recoveries, but which can also attenuate the magnitudes of the oceanic signals. Relative to Chambers and Bonin (2012), updated estimates are used for degree 2 order 0 coefficients (Cheng et al.,

2011) and glacial isostatic adjustment (A et al., 2013). Global spatial-mean values are subtracted from the ocean mass estimates at each time step. The values are provided on a regular $1° \times 1°$ horizontal grid; at this resolution, Hudson Bay and James Bay are together represented by 75 grid cells, while Foxe Basin is represented by 11 grid cells. Throughout the paper, we quote values in equivalent seawater thickness units.

Given our interest in nonseasonal behavior, we remove a seasonal cycle from all time series, which we compute by averaging together all January entries, February entries, etc., over 2003–2013 into a 12 month time series. To circumvent difficulties of interpreting the gravity data over the ocean in the presence of large rates of glacial isostatic adjustment over Canada and cryospheric mass loss from Greenland (e.g., Tamisiea et al., 2007; Velicogna, 2009; Rignot et al., 2011), linear trends are removed from all time series using least squares.

2.1.2 Tide-gauge data

A tide gauge maintained by the Canadian Hydrographic Service has measured relative sea level at the mouth of the Churchill River in Churchill, Manitoba (Fig. 1), for more than 70 years. Revised local reference monthly data were extracted from the Permanent Service for Mean Sea Level database (Holgate et al., 2013) on 18 August 2014. Data cover 90 % of months between January 1940 and December 2013, with a complete record existing since July 1991. Given the GRACE record, we consider tide-gauge data over 2003–2013.

A set of adjustments is applied to the tide-gauge data. As with the gravimetric estimates, a seasonal cycle and a linear trend are removed. To focus on ocean dynamical signals, we also subtract from the tide-gauge record a global mean sea level time series based on altimetry data (Ablain et al., 2009) as well as the inverted barometer response based on monthly Interim European Centre for Medium-Range Weather Forecasts Reanalysis (ERA-Interim) (Dee et al., 2011) mean sea level pressure fields and Eq. (1) of Ponte (2006). We note that removal of the global mean and inverse barometer signals reduces the detrended monthly variance in the tide-gauge sea level time series over 2003–2013 by 40 %. Given our focus on detrended behavior, we do not make any further corrections for vertical land motion, instead assuming that the relevant geophysical processes (e.g., postglacial rebound) are represented by a linear trend over the analysis period (cf. Santamaría-Gómez et al., 2012).

2.2 Data comparisons

One concern of using GRACE ocean bottom pressure estimates over the CIS is that they might be contaminated by transient terrestrial water storage from surrounding watersheds. Root mean square values of monthly water storage estimated by a land hydrology model can be 5–10 cm equiv-

alent water thickness in parts of the Hudson Bay drainage basin (Landerer and Swenson, 2012). Given the averaging function applied to the gravity data, such land signals could leak into the ocean data (Wahr et al., 1998).

To determine whether the ocean bottom pressure estimates are polluted by leakage of terrestrial water storage, we consider nonseasonal time series of GRACE bottom pressure averaged over the CIS alongside GRACE water storage[1] averaged over the Hudson Bay drainage basin (Fig. 2a).[2] An analogous technique is used by Peralta-Ferriz et al. (2014) to determine whether GRACE data over the Kara and Barents seas are contaminated by land leakage over the respective watersheds that drain into them. The two time series appear visually distinct and their correlation coefficient (0.26) is not statistically significant (Fig. 2a). The two signals do not show meaningful coherence at any frequency (not shown). These results demonstrate that the GRACE data over the CIS are not overwhelmed by leakage of terrestrial water storage at nonseasonal periods. Additional analysis corroborates this conclusion; computing correlation coefficients between the averaged terrestrial water storage curve and bottom pressure time series at individual GRACE ocean grid cells, we find that there are no points within the CIS where the local ocean bottom pressure is significantly correlated with the large-scale land signal on nonseasonal timescales (not shown).

As an additional check on the GRACE ocean data quality, and also to give physical insight, we compare bottom pressure estimates averaged over the CIS to sea level observed at the Churchill tide gauge (Fig. 2b). Notwithstanding the difference in amplitude, which probably partly reflects attenuation of the true ocean bottom pressure signal by the spatial averaging involved in the postprocessing (see Sect. 2.1.1), there is close correspondence between the two curves (Fig. 2b). The overall correlation coefficient between the nonseasonal sea level and bottom pressure (0.58) is statistically significant at the 95 % confidence level. The correspondence between the two time series in Fig. 2b is consistent with our interpretation of the results in Fig. 2a, attesting to the general meaningfulness of the nonseasonal signals in the GRACE gravity data over the ocean. What is more, this result suggests that the nonseasonal behavior at Churchill partly reflects bay-wide, depth-independent (that is, barotropic) variability.

---

[1] GRACE terrestrial water storage estimates were processed at the University of Texas Center for Space Research and postprocessed by Sean Swenson (National Center for Atmospheric Research). The gridded estimates, provided on a $1° \times 1°$ grid, were downloaded from the GRACE Tellus server (data version RS05.DSTvSCS1409) and scaled following Landerer and Swenson (2012).

[2] The Hudson Bay drainage basin has been defined as the union of the Hudson Bay seaboard and Nelson River basins, determined based on watersheds data provided by the Commission for Environmental Cooperation (http://www.cec.org/Page.asp? PageID=924&ContentID=2866).

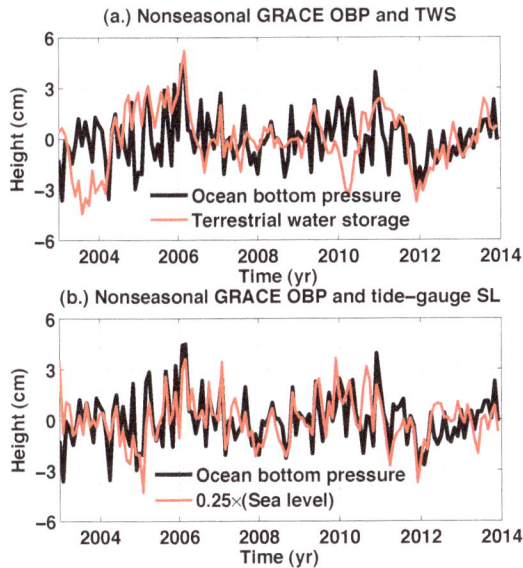

Figure 2. (a) Nonseasonal GRACE Release-05 2003–2013 ocean bottom pressure averaged over the Canadian inland seas (black) and terrestrial water storage averaged over the Hudson Bay drainage basin (red). (b) Nonseasonal GRACE Release-05 2003–2013 ocean bottom pressure averaged over the Canadian inland seas (black) and Churchill tide-gauge sea level (red). Sea level time series is multiplied by 0.25 to fit within axis limits. All quantities shown in equivalent water thickness with units of cm.

3   Interpretation

3.1   Model framework

The main drivers of mass redistribution in a homogeneous ocean are surface loading and wind stress (Hughes, 2008). In the case of synoptic timescales (i.e., periods of a few days), surface loading by barometric pressure can drive a relatively large non-equilibrium sea level response on the continental shelf (standard deviations $\gtrsim 10$ cm) (Greatbatch et al., 1996), related to the fact that propagation speeds of barotropic gravity waves are reduced in shallow regions. Indeed, Ponte and Vinogradov (2007) suggest that the assumption of an inverted-barometer response is not appropriate at periods of about 1 month and shorter for Hudson Bay (their Fig. 6). However, in the case of sub-synoptic timescales (e.g., periods longer than 1 month), non-isostatic adjustment of sea level on the shelf to pressure loading is thought to be comparatively small (standard deviations $\lesssim 1$ cm) (Greatbatch et al., 1996). Thus, we expect that observed sea level variations in Fig. 2 (which are corrected for an inverted barometer response) are driven by wind stress rather than by surface loading.

To assess this expectation, we consider numerical solutions from the Massachusetts Institute of Technology general circulation model (MITgcm; Marshall et al., 1997). The setup solves the primitive equations on a coarse-resolution ($0.5° \times 0.5°$) spherical polar grid having quasi-global spatial coverage, with solid walls imposed at 79° north and south latitude. (At this resolution, the model represents the meridional breadth of Hudson Strait, which varies from ~ 100 to 200 km, depending on longitude (Fig. 1), using between 2 and 4 grid cells.) Boundary conditions are in the form of surface fluxes based on monthly means of instantaneous zonal and meridional turbulent surface wind stresses taken from ERA-Interim (Dee et al., 2011). The reanalysis fields, which are provided on a regular $1.5° \times 1.5°$ grid, are bilinearly interpolated onto the model grid. Bottom topography is based on 5 min gridded elevations and bathymetry for the world (ETOPO5) data (National Oceanic and Atmospheric Administration, 1988); due to sparsity of measurements, it is likely that this bathymetric data set has large uncertainties in the CIS. The bathymetry is averaged within $0.5° \times 0.5°$ bins and then smoothed with a two-dimensional $2° \times 2°$ boxcar function. We use a constant value for density (1029 kg m^{-3}) and a single layer in the vertical. Variable ocean depths are implemented using partial cells (Adcroft et al., 1997). The model uses a linear free surface boundary condition along with a 900 s time step for the momentum equations. We use a vertical eddy viscosity of 5×10^{-4} m^2 s^{-1} and a grid-dependent lateral eddy viscosity varying from about 1×10^4 m^2 s^{-1} at low latitudes to roughly 3×10^3 m^2 s^{-1} at high latitudes. Simulations are run forward in time from rest for 35 years beginning on 1 January 1979 (the temporal range of ERA-Interim). To be consistent with the observations (Fig. 2), we consider monthly averaged model output over 2003–2013 with the seasonal cycle and a linear trend removed.

This framework is admittedly simple; many effects (e.g., mesoscale eddies, sea ice, river runoff, wind stress over the Arctic Ocean) have been precluded. On account of the coarse grid resolution, we do not resolve the topographic gyres and current separations induced by bathymetry in Hudson Bay discussed by Wang et al. (1994) in the context of a finer-resolution model. Due to the lack of ocean stratification, we do not capture the baroclinic boundary current in Hudson and James bays discussed by St-Laurent et al. (2012) among others. Given the lack of an interactive sea-ice model, we do not simulate any role played by sea ice in mediating the transfer of momentum between the atmosphere and the ocean (St-Laurent et al., 2011). However, to the extent that our model agrees with the data of interest, we can conclude that any omitted physics is unimportant in the present context.

3.2   Comparing model and data

Before comparing it to the GRACE ocean data, we smooth the model bottom pressure using the same 500 km spatial filter used in the GRACE postprocessing (Chambers and Bonin, 2012) and then average over the CIS, interpolating onto the GRACE ocean grid. The statistically significant correlation coefficient between the model and data bottom pressure is 0.69, with the model curve explaining 47 % of the data curve's variance (Fig. 3a). The standard deviation from the

a. Bottom pressure over the Canadian Inland Seas

b. Sea level at the Churchill tide gauge site

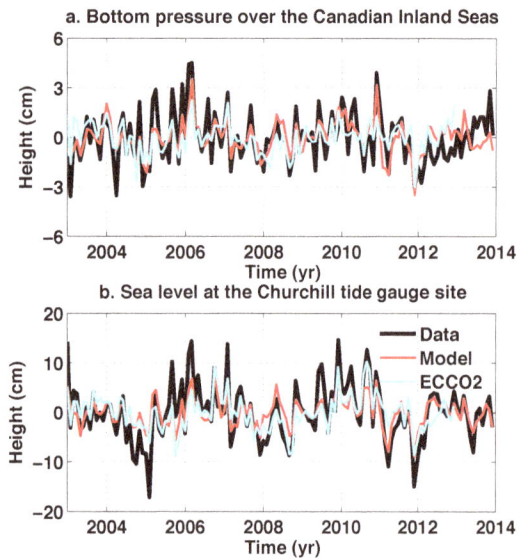

Figure 3. (a) Nonseasonal time series of ocean bottom pressure averaged over the Canadian inland seas from GRACE Release-05 (black), the barotropic model (red), and the ECCO2 solution (cyan). **(b)** Nonseasonal time series of sea level at Churchill measured by a tide gauge (black), the barotropic model (red), and the ECCO2 solution (cyan). Units are cm.

model (1.0 cm) is smaller than the standard deviation from the data (1.5 cm), which could be partly due to residual noise in the data. The gross correspondence between the time series speaks to the realism of the two independent estimates, consistent with our prior assessment of GRACE data quality based on observed water storage and sea level (Fig. 2). It demonstrates that the ocean model suffices for capturing the major features of observed nonseasonal bay-wide fluctuations in ocean mass.

For comparing against the tide-gauge observations, we consider model sea level from the grid cell whose centroid is the closest to the Churchill site (Fig. 3b). The model time series roughly reproduces the gross features of the observed sea level curve. However, the model underestimates the data amplitudes: standard deviations of the observed and simulated signals are 5.7 and 3.0 cm, respectively, possibly reflecting the importance of forcing terms that have been omitted from our model, for example, riverine discharge (Gough and Robinson, 2000), or perhaps indicating that the model underestimates the wind-driven ocean response, for instance, on account of the coarse resolution of the forcing data set. Overall, the correlation coefficient between model and data curves is 0.82, with the model explaining a majority (58 %) of the observed variance.

To gauge the influence of horizontal resolution and missing physics (ocean stratification, sea-ice dynamics, etc.) on the correspondence between the data and our model, we also consider a higher-resolution ocean/sea-ice model. Monthly sea level and bottom pressure from the Estimating the Cir-

culation and Climate of the Ocean Phase-II (ECCO2; Menemenlis et al., 2005) cube92 solution were obtained for 2003–2012. This estimate of the ocean/sea-ice state, generated by the MITgcm coupled with a fully interactive sea-ice model, is defined on a global "cubed sphere" topology, with a nominal horizontal resolution of $0.25° \times 0.25°$ and 50 vertical levels. Surface forcing is essentially based on unadjusted fields from the Japanese 25 year Re-Analysis (JRA-25; Onogi et al. 2007), except for precipitation (JRA-25 adjusted to remove large resultant drifts in salinity and global sea level in the model solution) and runoff (which, for the CIS and the Arctic Ocean, is derived from monthly mean river discharge from the Arctic Runoff Database). Some internal model parameters were previously adjusted to better fit observations. Due to inclusion of ocean stratification, sea level and bottom pressure are not generally equivalent in the ECCO2 solution. Also, in the presence of sea ice, "sea level" is defined as the physical depression of the sea surface plus the sea-ice load in equivalent water thickness units.

Comparable ECCO2 curves for nonseasonal ocean bottom pressure averaged over the CIS and sea level at the Churchill tide-gauge location are overlaid in Fig. 3a and b, respectively. Perhaps surprisingly, for the common period 2003–2012, our simple barotropic model simulation performs as well as (if not better than) the ECCO2 cube92 solution in reproducing the data. While the ECCO2 solution explains 36 and 54 % of the variances in the GRACE and tide-gauge time series, respectively, the barotropic simulation explains 50 and 58 % of the respective observed variances (Fig. 3).

Results in Fig. 3 show that our simple dynamical framework is sufficient to capture the lowest-order monthly nonseasonal behavior observed in the CIS across a range of spatial scales, and imply that the influences of ocean stratification, sea ice, and surface fluxes of mass and buoyancy are higher order. The realism of the model encourages its further exploration to more fully understand the nonseasonal variability in the CIS. In what follows, we focus on sea level, but note that, since sea level and bottom pressure are equivalent quantities in a barotropic ocean, the results also apply to bottom pressure.

3.3 Empirical orthogonal functions

The relationship between observed sea level and bottom pressure changes (Fig. 2b) led us to hypothesize the existence of a bay-wide depth-independent oscillation. To determine more rigorously whether there is in fact such a wind-driven nonseasonal barotropic fluctuation of the CIS, we perform an empirical orthogonal function (EOF) analysis, which boils down to solving for the eigenvalues and eigenvectors of the covariance matrix of a scalar that varies in space and time (von Storch and Zwiers, 1999).

The leading eigenvector of simulated sea level over the CIS shows a single-signed spatial structure (Fig. 4a), but values are larger over the deep interior region and smaller in the

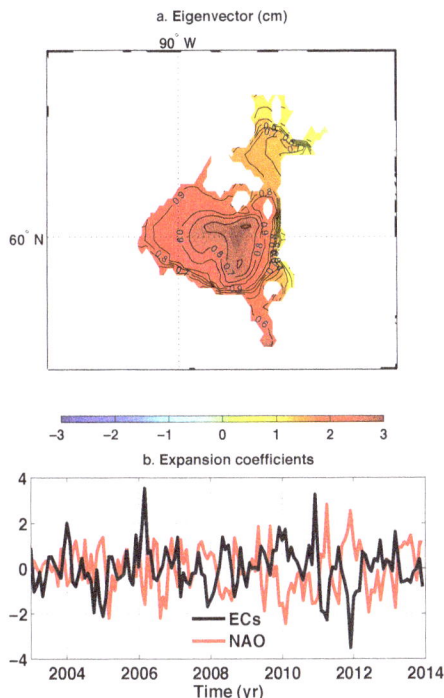

Figure 4. Leading empirical orthogonal function (EOF) of nonseasonal sea level determined from the barotropic model in the Canadian inland seas. (**a**) Shading is the value of the leading eigenvector (cm). Contouring is the local fraction of simulated anomalous sea level variance explained by the leading EOF. (**b**) Nonseasonal time series of the leading expansion coefficients (black) normalized to have unit variance. Also shown is the NAO with a linear trend and a seasonal cycle removed (red). The NAO time series is taken from the National Oceanic and Atmospheric Administration's Earth System Research Laboratory Physical Sciences Division website.

shallow boundary area. The mode could be caused by a combination of local and remote effects, perhaps with remote mechanisms forcing water into and out of the domain and local drivers acting to redistribute mass within the domain. This leading empirical mode explains 69.4 % of the total nonseasonal simulated sea level variance in the CIS (Fig. 4a). Local variance explained is highest (> 90 %) along a "ring" around Hudson Bay separating interior and boundary regions; this ring of strong correlation could reflect rapid Kelvin wave propagation around Hudson Bay, with a possible analogy to the coherent sea level fluctuations along the global continental slope observed by Hughes and Meredith (2006). Explained variance is lowest (< 10 %) in shallow regions of Foxe Basin southeast of Prince Charles Island. Over most of Hudson Bay's shallow boundary (e.g., near Churchill) and deep interior, the mode accounts for about two-thirds of the local sea level variance.

The leading expansion coefficients show variability across all accessible timescales, and a dominant period is not visually obvious (Fig. 4b), while an estimate of the associated power spectral density is slightly red in nature (not shown).

We observe that the expansion-coefficient time series is essentially equivalent to the bay-mean sea level signal: the correlation coefficient between the two curves is ∼ 1 (not shown). There appears to be a relationship between the North Atlantic Oscillation (NAO) and the leading expansion coefficients such that anomalous sea level in the CIS is high when the NAO index is low, and vice versa; the statistically significant correlation coefficient between the two time series (−0.59) confirms this out-of-phase relationship.

Looking further afield, we compute correlation coefficients between the expansion coefficients (Fig. 4b) and nonseasonal sea level time series at each model grid cell over the global ocean (not shown). A statistically significant in-phase relationship is apparent between fluctuations in the CIS, Baffin Bay, and the Mediterranean Sea, while a significant out-of-phase relationship is evident between the expansion coefficients and variations over the midlatitude North Atlantic and along parts of the North Sea. These relationships are similar to those suggested by Piecuch and Ponte (2014), who computed correlations between the leading mode of nonseasonal bottom pressure variability over the midlatitude North Atlantic and anomalous bottom pressure elsewhere based on GRACE data (e.g., their Fig. 3a).

Assuming geostrophy, fluctuations in sea level are proportional to variations in the barotropic stream function, and therefore this mode can be physically interpreted in terms of ocean circulation changes. For example, the domed shape of the eigenvector in Hudson Bay (Fig. 4a) suggests anomalous anticyclonic (cyclonic) circulation when the expansion coefficients are positive (negative). Given the cyclonic sense of Hudson Bay's mean circulation (e.g., St-Laurent et al., 2012), this mode thus corresponds to spin-up and -down of the barotropic component of the mean circulation in Hudson Bay roughly during positive and negative NAO periods, respectively.

Model results in Fig. 4 corroborate our earlier suspicion based on data that there exists a wind-driven barotropic fluctuation of the CIS that explains most of the nonseasonal sea level variance across a range of spatial scales. What is more, this mode of oscillation is correlated with the NAO, implying that some of the anomalous sea level behavior in the CIS is tied to climate variability more broadly over the North Atlantic sector, consistent with suggestions made by Gough and Robinson (2000). It remains to be determined, however, what the important regions of wind forcing are, and what the relevant ocean dynamics are. We turn to these questions in the next section.

3.4 Forcing and dynamics

Seeing as the leading expansion coefficients are correlated with the NAO (Fig. 4b), we now consider the structure of nonseasonal wind stress forcing over the northern North Atlantic sector (Fig. 5). Standard deviations of nonseasonal wind stress are on the order of a few hundredths of a $N\,m^{-2}$.

a. Zonal wind standard deviation (N m⁻²)

b. Meridional wind standard deviation (N m⁻²)

a. Zonal wind correlation

b. Meridional wind correlation

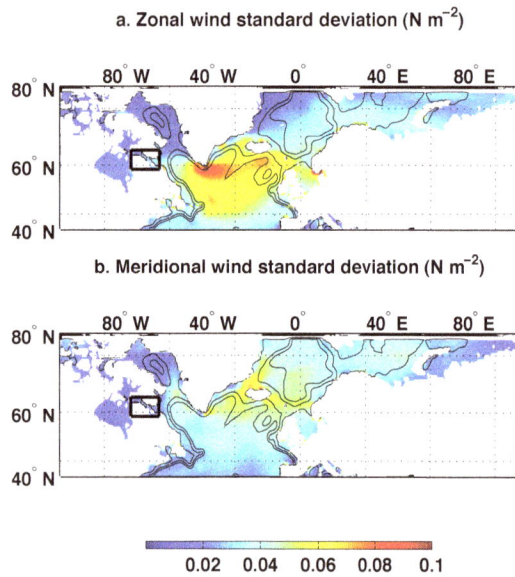

Figure 5. Standard deviations of nonseasonal (**a**) zonal and (**b**) meridional wind stress from ERA-Interim used to force the barotropic model. Units are $N\,m^{-2}$. Thick black box drawn around Hudson Strait is the control volume for the numerical experiment with altered surface wind stress forcing described in Sect. 3.4. Thin black contours are the model's 200, 1000, and 2000 m isobaths, shown for reference.

Figure 6. Correlation coefficients between the expansion coefficients of the leading empirical orthogonal function (Fig. 4b) and nonseasonal (**a**) zonal and (**b**) meridional ERA-Interim wind stress. Only values statistically significant at the 95 % confidence level are shown. Thin black contours are as in Fig. 5.

Noteworthy are strong variations in zonal wind stress near Cape Farewell, Greenland (Moore and Renfrew, 2005).

To suggest relationships between wind stress over the northern North Atlantic and sea level in the CIS, we compute correlations between the expansion coefficients (Fig. 4b) and nonseasonal zonal and meridional wind stress (Fig. 5). Zonal winds over a broad swath extending from Hudson Bay and the Labrador Sea across the northern North Atlantic Ocean to the North, Norwegian and Barents seas are significantly negatively correlated with sea levels over the CIS (Fig. 6a). Sea levels over the CIS are significantly negatively correlated with meridional winds over eastern Hudson Bay and the western Labrador Sea as well as from the northeastern North Atlantic to the Norwegian Sea, while positive correlations are seen off the southwestern coast of Greenland (Fig. 6b). Given the correlation between the expansion coefficients and the NAO (Fig. 4b), these correlation patterns are consistent with wind stress anomalies associated with the NAO; for example, anomalous westerly winds occur during positive phases of the NAO (e.g., Marshall et al., 2001), when sea levels over the CIS are anomalously negative (Fig. 4b).

The strongest correlations between the expansion coefficients and zonal winds occur over Hudson Strait (Fig. 6a), suggesting that the sea level in the CIS might be influenced by wind stress variations over adjacent regions. For example, wind stress along Hudson Strait directed towards the CIS would force flow into the CIS until dynamic balance is established between the along-strait sea level gradient and

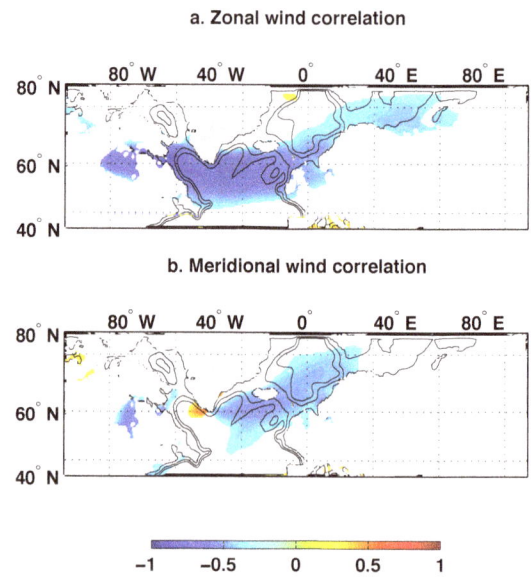

wind stress, i.e., a wind setup (e.g., Csanady, 1981); in the absence of additional boundary forcing, the CIS would undergo barotropic adjustment in response to the mass inflow, likely resulting in a horizontally uniform sea level increase. (An analogous scenario can be entertained for winds over Hudson Strait directed away from the CIS.)

To establish what the influence of wind stress over Hudson Strait is, we perform another numerical simulation based on the model framework described previously (Sect. 3.1) by setting the surface wind stress to zero everywhere except over Hudson Strait (see Fig. 5), all else (e.g., the time period of integration) being equal. This Hudson Strait winds experiment captures some of the variability in the CIS from the baseline experiment examined in Fig. 4. Namely, winds over Hudson Strait effect bay-wide changes in sea level over the CIS, and the correlation between the leading sea level expansion coefficients from the baseline and Hudson Strait winds experiments is 0.58 (Fig. 7), demonstrating that Hudson Strait winds are important to variability in the region.

But there are also disagreements between behaviors generated by the two experiments. Wind stress over Hudson Strait effects sea level changes over the CIS whose amplitudes are horizontally uniform, implying that this experiment lacks important local effects (e.g., winds over the CIS) responsible for generating the spatially varying amplitudes that are manifested in the baseline experiment. What is more, the bay-mean sea level changes from the Hudson Strait winds experiment are smaller than those from the baseline experiment (cf. amplitudes in Figs. 4a and 7a), indicating that wind driv-

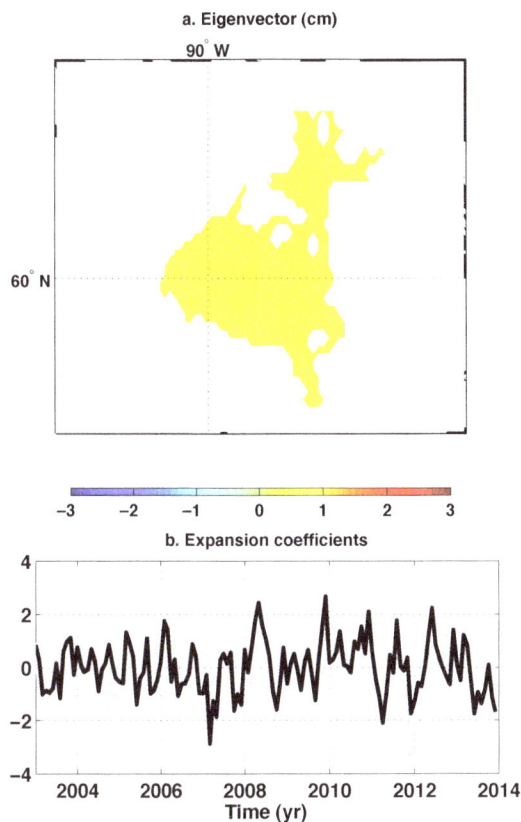

Figure 7. As in Fig. 4 but shown for the experiment driven only by wind stress over Hudson Strait. In (**b**), the nonseasonal time series of the leading expansion coefficients from the baseline experiment is shown in gray for reference.

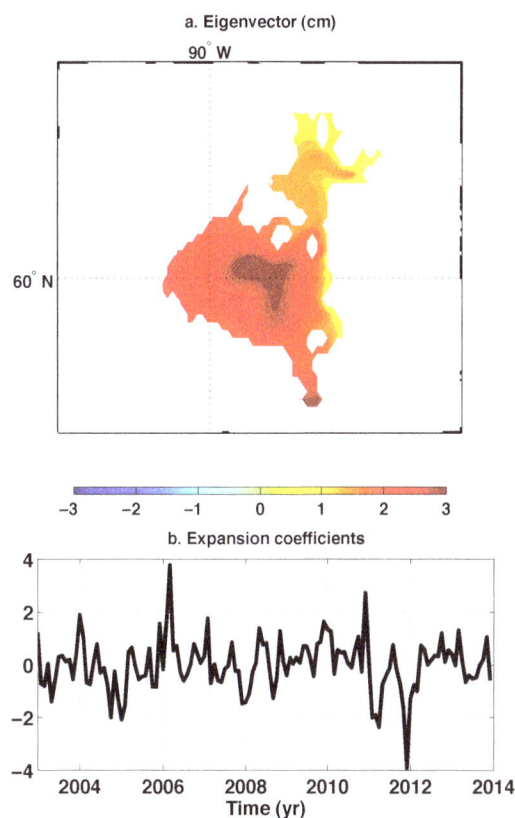

Figure 8. As in Fig. 4 but shown for the experiment driven only by wind stress over regions of statistically significant correlation coefficients shown in Fig. 6. In (**b**), the nonseasonal time series of the leading expansion coefficients from the baseline experiment is shown in gray for reference.

ing in remote regions (e.g., over the North Atlantic Ocean) is also important in forcing water into and out of the CIS.

To more completely account for the behavior from the baseline experiment, another simulation is performed over the same time period by driving the model with wind stress only over regions where the correlation coefficient between wind stress and baseline expansion coefficients is statistically significant (see the colored regions in Fig. 6). This correlated winds experiment generates variability in the CIS that is extremely close to the behavior produced by the baseline simulation (Fig. 8), explaining 95 % of the variance in the bay-mean sea level signal from the latter (Fig. 8b); spatial patterns of the leading modes of variability from the two model experiments are practically identical (Figs. 4a and 8a).

The remote surface forcing is potentially communicated to the CIS by a variety of physical mechanisms that can influence sea level in and around Hudson Strait. We speculate that these could include, for example, coastally trapped propagating waves forced over shallow areas (e.g., around Cape Farewell, Greenland), and planetary and topographic Rossby waves forced by wind curl patterns over the deep ocean.

Finally, to infer what the relevant local dynamics are producing the spatial structure of the variability inside the bay (Fig. 4a), we run a suite of experiments treating the western entrance of Hudson Strait (surrounding Mill, Nottingham, and Salisbury islands) as an open boundary, with flow into and out of the CIS (i.e., bay-mean sea level changes) from the baseline experiment specified a priori. Simulations are identical in all respects other than that we choose to either variously "turn off" one of the terms in the model's momentum equations (e.g., advection, Coriolis, surface forcing, pressure gradient) or change the topography of the CIS.

Turning off winds over the CIS results in a leading mode of sea level variability whose spatial structure is horizontally uniform, similar to the Hudson Strait winds experiment case (Fig. 7a). In contrast, removal of quadratic bottom drag or nonlinear terms from the model dynamics has no perceptible effect, and the variable spatial structure from the baseline experiment is recovered (cf. Fig. 4a). Finally, setting the depth to a constant value (of about 250 m), or turning off either the Coriolis acceleration term or the pressure-gradient contribution, leads to spatial structures for the leading sea level variability very different from those plotted in Fig. 4a. Thus, based on these experiments, we reason that the spatial structure of the leading mode of sea level variability in the CIS reflects linear response to local winds governed by rotation, gravity, and topography.

4 Conclusions

Using satellite gravimetry, a tide gauge, and a barotropic model, we identified a wind-driven nonseasonal barotropic fluctuation of the Canadian inland seas (CIS) that is correlated with the NAO (Figs. 2–6). Anomalous inflows and outflows, which are reflected in spatially averaged changes in sea level and bottom pressure over the CIS, are driven by wind stress over Hudson Strait (Fig. 7), probably through a wind setup, and the northern North Atlantic Ocean (Fig. 8), possibly communicated by means of wave mechanisms. Anomalous mass redistribution within the CIS, which relates to changes in the depth-mean circulation, is governed by the linear ocean response to more local wind stress variations under the joint influences of rotation, gravity, and variable bottom topography. We observe that, while it suggests broad regions of wind forcing over the northern North Atlantic potentially contributing to CIS variability (Fig. 6), our analysis does not unambiguously pinpoint which forcing regions are most relevant – in fact, it could be that forcing over just one small area of the northern North Atlantic is the main driver. An adjoint model – capable of quantifying the sensitivity of a particular modeled quantity at a specific place and time to all model inputs and states at preceding times – could be used to shed more light on which regions of wind forcing most influence the CIS, as is done by Fukumori et al. (2007) to elucidate a near-uniform basin-wide sea level fluctuation of the Mediterranean Sea, but such an analysis is beyond our scope and deferred to future study.

Our findings complement previous modeling work on Hudson Bay (Wang et al., 1994; Saucier and Dionne, 1998; Saucier et al., 2004; St-Laurent, 2011, 2012). Whereas past studies tend to regard Hudson Strait as an open boundary, specifying inflow and outflow at the outset, our forcing experiments highlight wind stress changes over adjacent and remote areas responsible for driving mean sea level changes in the CIS (Figs. 7 and 8), emphasizing the need for accurate estimates of atmospheric variability for modeling the circulation in the CIS. Our ability to reproduce qualitatively the data (Fig. 3) without recourse to ice–ocean interactions accords with St-Laurent et al. (2011), who find that ice plays only a small role in mediating seasonal momentum transfer between air and sea, reflecting the loose, mobile nature of sea ice in Hudson Bay. Similar to Wang et al. (1994), we find that the effects of variable bottom topography, which are ignored in the flat-bottomed conceptual model of Hudson and James bays due to St-Laurent et al. (2012), are an important determinant of circulation changes in the CIS. (We note that St-Laurent et al. (2012) recover some of the effects of topo-graphic steering by assuming that there is a strong current in the boundary region.)

Investigating all periods from monthly data between 1974 and 1994, and based on a correlation analysis, Gough and Robinson (2000) conclude that 43 % of the variance in the Churchill tide-gauge record can be understood in terms of a response to Churchill River discharge. Considering nonseasonal periods from detrended monthly observations over 2003–2013, and using a barotropic model, we submit that at least 58 % of the sea level variance at Churchill is driven by wind stress anomalies (Fig. 3b). The fact that our emphasis (on remote driving and wind stress) differs from that of Gough and Robinson (2000) (on local forcing and river runoff) could reflect the distinct timescales and periods being considered; for example, during the late 1970s and early 1980s, the tide-gauge record is dominated by decadal decline in sea level (Gough and Robinson, 2000, Fig. 3). These considerations, along with the uniqueness of the tide-gauge record, provide ample motivation for more general future works to reconcile the relative roles of wind stress and river runoff, thus painting a more complete portrait of the sea level behavior at Churchill.

One of the challenges of using GRACE data over the ocean in near-coastal regimes is separating oceanic signals from non-oceanic noise (Chambers and Schröter, 2011). In the case of the CIS, this is an especially difficult task, given the large rates of mass loss from the Greenland ice sheet, ongoing postglacial rebound over Canada, and any terrestrial water storage tied to changes in river discharge. Our results (Figs. 2 and 3) suggest that meaningful estimates of nonseasonal ocean bottom pressure behavior can be derived from GRACE retrievals over the CIS.

Acknowledgements. Support came from NASA (GRACE grant NNX12AJ93G) and NSF (grant OCE-0961507). Helpful comments and useful suggestions from the editor and two anonymous reviewers are gratefully acknowledged. GRACE ocean and land data were processed by Don P. Chambers and Sean Swenson, respectively, supported by the NASA MEaSUREs Program, and are available at http://grace.jpl.nasa.gov. Fields from the ECCO2 cube92 solution were downloaded from ftp://ecco2.jpl.nasa.gov/data1/cube/cube92/lat_lon/quart_90S_90N/. We thank D. Menemenlis and H. Zhang for clarifying some details regarding aspects of the ECCO2 solution.

J. M. Huthnance

References

A, G., Wahr, J., and Zhong, S.: Computations of the viscoelastic response of a 3-D compressible Earth to surface loading: an application to Glacial Isostatic Adjustment in Antarctica and Canada, Geophys. J. Int., 192, 557–572, doi:10.1093/gji/ggs030, 2013.

Ablain, M., Cazenave, A., Valladeau, G., and Guinehut, S.: A new assessment of the error budget of global mean sea level rate estimated by satellite altimetry over 1993–2008, Ocean Sci., 5, 193–201, doi:10.5194/os-5-193-2009, 2009.

Adcroft, A., Hill, C., and Marshall, J.: Representation of topography by shaved cells in a height coordinate ocean model, Mon. Weather Rev., 125, 2293–2315, 1997.

Bettadpur, S.: UTCSR Level-2 Processing Standards Document for Level-2 Product Release 0005, GRACE 327-742, CSR Publ. GR-12-xx, Rev 4.0, University of Texas at Austin, Austin, 16 pp., 2012.

Boening, C., Willis, J. K., Landerer, F. W., Nerem, R. S., and Fasullo, J.: The 2011 La Niña: so strong, the oceans fell, Geophys. Res. Lett., 39, L19602, doi:10.1029/2012GL053055, 2012.

Chambers, D. P. and Bonin, J. A.: Evaluation of Release-05 GRACE time-variable gravity coefficients over the ocean, Ocean Sci., 8, 859–868, doi:10.5194/os-8-859-2012, 2012.

Chambers, D. P. and Schröter, J.: Measuring ocean mass variability from satellite gravimetry, J. Geodyn., 52, 333–343, 2011.

Cheng, M., Ries, J. C., and Tapley, B. D.: Variations of the Earth's figure axis from satellite laser ranging and GRACE, J. Geophys. Res., 116, B01409, doi:10.1029/2010JB000850, 2011.

Csanady, G. T.: Circulation in the Coastal Ocean, Part 1, EOS T. Am. Geophys. Un., 62, 9–11, doi:10.1029/EO062i002p00009, 1981.

Dee, D. P., Uppala, S. M., Simmons, A. J., Berrisford, P., Poli, P., Kobayashi, S., Andrae, U., Balmaseda, M. A., Balsamo, G., Bauer, P., Bechtold, P., Beljaars, A. C. M., van de Berg, L., Bidlot, J., Bormann, N., Delsol, C., Dragani, R., Fuentes, M., Geer, A. J., Haimberger, L., Healy, S. B., Hersbach, H., Hólm, E. V., Isaksen, L., Kållberg, P. K., Köhler, M., Matricardi, M., McNally, A. P., Monge-Sanz, B. M., Morcrette, J.-J., Park, B.-K., Peubey, C., de Rosnay, P., Tavolato, C., Thépaut, J.-N., and Vitart, F.: The ERA-Interim reanalysis: configuration and performance of the data assimilation system, Q. J. Roy. Meteorol. Soc., 137, 553–597, 2011.

Déry, S. J. and Wood, E. F.: Teleconnection between the Arctic oscillation and Hudson Bay river discharge, Geophys. Res. Lett., 31, L18205, doi:10.1029/2004GL020729, 2004.

Déry, S. J., Stieglitz, M., McKenna, E. C., and Wood, E. F.: Characteristics and trends of river discharge into Hudson, James, and Ungava Bays, 1964–2000, J. Climate, 18, 2540–2557, 2005.

Déry, S. J., Mlynowski, T. J., Hernández-Henríquez, M. A., and Straneo, F.: Interannual variability and interdecadal trends in Hudson Bay streamflow, J. Mar. Syst., 88, 341–351, 2011.

Drinkwater, K. F.: Physical oceanography of Hudson Strait and Ungava Bay, in: Canadian Inland Seas, Elsevier Oceanogr. Ser., vol. 44, edited by: Martini, I. P., Elsevier, Amsterdam, 237–264, 1986.

Egbert, G. D. and Ray, R. D.: Estimates of M_2 tidal energy dissipation from TOPEX/Poseidon altimeter data, J. Geophys. Res., 106, 22475–22502, doi:10.1029/2000JC000699, 2001.

Fukumori, I., Menemenlis, D., and Lee, T.: A near-uniform basin-wide sea level fluctuation of the Mediterranean Sea, J. Phys. Oceanogr., 37, 338–358, 2007.

Gough, W. A. and Robinson, C. A.: Sea-level variation in Hudson Bay, Canada, from Tide-Gauge data, Arct. Antarct. Alp. Res., 32, 331–335, 2000.

Greatbatch, R. J., Lu, Y., and de Young, B.: Application of a barotropic model to North Atlantic synoptic sea level variability, J. Mar. Res., 54, 451–469, 1996.

Hochheim, K. P. and Barber, D. G.: Atmospheric forcing of sea ice in Hudson Bay during the fall period, 1980–2005, J. Geophys. Res., 115, C05009, doi:10.1029/2009JC005334, 2010.

Hochheim, K. P., Lukovich, J. V., and Barber, D. G.: Atmospheric forcing of sea ice in Hudson Bay during the spring period, 1980–2005, J. Mar. Syst., 88, 476–487, 2011.

Holgate, S. J., Matthews, A., Woodworth, P. L., Rickards, L. J., Tamisiea, M. E., Bradshaw, E., Foden, P. R., Gordon, K. M., Jevrejeva, S., and Pugh, J.: New data systems and products at the permanent service for mean sea level, J. Coast. Res., 29, 493–504, 2013.

Hughes, C. W.: A form of potential vorticity equation for depth-integrated flow with a free surface, J. Phys. Oceanogr., 38, 1131–1136, 2008.

Hughes, C. W. and Meredith, M. P.: Coherent sea-level fluctuations along the global continental slope, Philos. T. Roy. Soc. A, 364, 885–901, doi:10.1098/rsta.2006.1744, 2006.

Laidler, G. J. and Gough, W. A.: Climate variability and climatic change: potential implications for Hudson Bay coastal communities, Polar Geogr., 27, 38–58, 2003.

Landerer, F. W. and Swenson, S. C.: Accuracy of scaled GRACE terrestrial water storage estimates, Water Resour. Res., 48, W04531, doi:10.1029/2011WR011453, 2012.

MacDonald, R. W. and Kuyzk, Z. Z. A.: The Hudson Bay system: A northern inland sea in transition, J. Marine Syst., 88, 337–340, 2011.

Markham, W.: The ice cover, in: Canadian Inland Seas, Elsevier Oceanogr. Ser., vol. 44, edited by: Martini, I. P., Elsevier, Amsterdam, 101–116, 1986.

Marshall, J., Adcroft, A., Hill, C., Perelman, L., and Heisey, C.: A finite-volume, incompressible Navier Stokes model for studies of the ocean on parallel computers, J. Geophys. Res., 102, 5753–5766, doi:10.1029/96JC02775, 1997.

Marshall, J., Johnson, H., and Goodman, J.: A Study of the interaction of the North Atlantic oscillation with ocean circulation, J. Climate, 14, 1399–1421, 2001.

Menemenlis, D., Fukumori, I., and Lee, T.: Using Green's Functions to Calibrate an Ocean General Circulation Model, Mon. Weather Rev., 133, 1224–1240, 2005.

Moore, G. W. K. and Renfrew, I. A.: Tip jets and barrier winds: a QuickSCAT climatology of high wind speed events around Greenland, J. Climate, 18, 3713–3725, 2005.

National Oceanic and Atmospheric Administration: Digital relief of the Surface of the Earth, Data Announcement 88-MGG-02, Natl. Geophys. Data Cent., Boulder, 1988.

Onogi, K., Tsutsui, J., Koide, H., Sakamoto, M., Kobayashi, S., Hatsushika, H., Matsumoto, T., Yamazaki, N., Kamahori, H., Takahashi, K., Kadokura, S., Wada, K., Kato, K., Oyama, R., Ose, T., Mannoji, N., and Taira, R.: The JRA-25 Reanalysis, J. Meteorol. Soc. Jpn., 85, 3, 369–432, 2007.

Peralta-Ferriz, C., Morison, J. H., Wallace, J. M., Bonin, J. A., and Zhang, J.: Arctic Ocean Circulation Patterns Revealed by GRACE, J. Climate, 27, 1445–1468, doi:10.1175/JCLI-D-13-00013.1, 2014.

Piecuch, C. G. and Ponte, R. M.: Nonseasonal mass fluctuations in the midlatitude North Atlantic Ocean, Geophys. Res. Lett., 41, 4261–4269, doi:10.1002/2014GL060248, 2014.

Ponte, R. M.: Low-Frequency Sea Level Variability and the Inverted Barometer Effect, J. Atmos. Ocean. Tech., 23, 619–629, 2006.

Ponte, R. M. and Vinogradov, S. V.: Effects of stratification on the large-scale ocean response to barometric pressure, J. Phys. Oceanogr., 37, 245–258, 2007.

Prinsenberg, S. J.: Salinity and temperature distribution of Hudson Bay and James Bay, in: Canadian Inland Seas, Elsevier Oceanogr. Ser., vol. 44, edited by: Martini, I. P., Elsevier, Amsterdam, 163–186, 1986a.

Prinsenberg, S. J.: The circulation pattern and current structure of Hudson Bay, in: Canadian Inland Seas, Elsevier Oceanogr. Ser., vol. 44, edited by: Martini, I. P., Elsevier, Amsterdam, 187–204, 1986b.

Prinsenberg, S. J.: On the physical oceanography of Foxe Basin, in: Canadian Inland Seas, Elsevier Oceanogr. Ser., vol. 44, edited by: Martini, I. P., Elsevier, Amsterdam, 217–236, 1986c.

Prinsenberg, S. J.: Seasonal current variations observed in Western Hudson Bay, J. Geophys. Res., 92, 10756–10766, doi:10.1029/JC092iC10p10756, 1987.

Rignot, E., Velicogna, I., van den Broeke, M. R., Monaghan, A., and Lenaerts, J. T. M.: Acceleration of the contribution of the Greenland and Antarctic ice sheets to sea level rise, Geophys. Res. Lett., 38, L05503, doi:10.1029/2011GL046583, 2011.

Santamaría-Gómez, A., Gravelle, M., Collilieux, X., Guichard, M., Martín Míguez, B., Tiphaneau, P., and Wöppelmann, G.: Mitigating the effects of vertical land motion in tide gauge records using a state-of-the-art GPS velocity field, Global Planet. Change, 98–99, 6–17, 2012.

Saucier, F. J. and Dionne, J.: A 3-D coupled ice–ocean model applied to Hudson Bay, Canada: the seasonal cycle and time-dependent climate response to atmospheric forcing and runoff, J. Geophys. Res., 103, 27689–27705, doi:10.1029/98JC02066, 1998.

Saucier, F. J., Senneville, S., Prinsenberg, S., Roy, F., Smith, G., Gachon, P., Caya, D., and Laprise, R.: Modelling the sea ice–ocean seasonal cycle in Hudson Bay, Foxe Basin and Hudson Strait, Canada, Clim. Dynam., 23, 303–326, 2004.

St-Laurent, P., Straneo, F., Dumais, J.-F., and Barber, D. G.: What is the fate of the river waters of Hudson Bay?, J. Mar. Syst., 88, 352–361, 2011.

St-Laurent, P., Straneo, F., and Barber, D. G.: A conceptual model of an Arctic sea, J. Geophys. Res., 117, C06010, doi:10.1029/2011JC007652, 2012.

Straneo, F. and Saucier, F.: The outflow from Hudson Strait and its contribution to the Labrador Current, Deep-Sea Res. Pt. I, 55, 926–946, 2008a.

Straneo, F. and Saucier, F.: The Arctic-Subarctic Exchange Through Hudson Strait, in: Arctic-Subarctic Ocean Fluxes: Defining the Role of the Northern Seas in Climate, edited by: Dickson, R. R., Meinke, J., and Rhines, P., Springer, Dordrecht, 249–262, 2008b.

Sutherland, D. A., Straneo, F., Lentz, S. J., and Saint-Laurent, P.: Observations of fresh, anticyclonic eddies in the Hudson Strait outflow, J. Mar. Syst., 88, 375–384, 2011.

Tamisiea, M. E., Mitrovica, J. X., and Davis, J. L.: GRACE gravity data constrain ancient ice geometries and continental dynamics over laurentia, Science, 316, 881–883, 2007.

Velicogna, I.: Increasing rates of ice mass loss from the Greenland and Antarctic ice sheets revealed by GRACE, Geophys. Res. Lett., 36, L19503, doi:10.1029/2009GL040222, 2009.

von Storch, H. and Zwiers, F. W.: Statistical Analysis in Climate Research, Cambridge University Press, Cambridge, UK, 496 pp., 1999.

Wahr, J., Molenaar, M., and Bryan, F.: Time variability of the Earth's gravity field: hydrological and oceanic effects and their possible detection using GRACE, J. Geophys. Res., 103, 30205–30229, doi:10.1029/98JB02844, 1998.

Wang, J., Mysak, L. A., and Ingram, R. G.: A three-dimensional numerical simulation of Hudson Bay summer ocean circulation: topographic gyres, separations, and coastal jets, J. Phys. Oceanogr., 24, 2496–2514, 1994.

Webb, D. J.: On the tides and resonances of Hudson Bay and Hudson Strait, Ocean Sci., 10, 411–426, doi:10.5194/os-10-411-2014, 2014.

White, D., Hinzman, L., Alessa, L., Cassano, J., Chambers, M., Falkner, K., Francis, J., Gutowski, W. J., Holland, M., Holmes, R. M., Huntington, H., Kane, D., Kliskey, A., Lee, C., McClelland, J., Peterson, B., Rupp, T. S., Straneo, F., Steele, M., Woodgate, R., Yang, D., Yoshikawa, K., and Zhang, T.: The arctic freshwater system: changes and impacts, J. Geophys. Res., 112, G04S54, doi:10.1029/2006JG000353, 2007.

Wright, D. G., Greenberg, D. A., and Majaess, F. G.: The influence of bays on adjusted sea level over adjacent shelves with application to the Labrador Shelf, J. Geophys. Res., 92, 14610–14620, doi:10.1029/JC092iC13p14610, 1987.

7

Coastal sea level response to the tropical cyclonic forcing in the northern Indian Ocean

P. Mehra[1], M. Soumya[1], P. Vethamony[1], K. Vijaykumar[1], T. M. Balakrishnan Nair[2], Y. Agarvadekar[1], K. Jyoti[1], K. Sudheesh[1], R. Luis[1], S. Lobo[1], and B. Harmalkar[1]

[1]CSIR-National Institute of Oceanography (NIO), Goa, India
[2]Indian National Centre for Ocean Information Services (INCOIS), Hyderabad, India

Correspondence to: P. Mehra (pmehra@nio.org)

Abstract. The study examines the observed storm-generated sea level variation due to deep depression (event 1: E1) in the Arabian Sea from 26 November to 1 December 2011 and a cyclonic storm "THANE" (event 2: E2) over the Bay of Bengal during 25–31 December 2011. The sea level and surface meteorological measurements collected during these extreme events exhibit strong synoptic disturbances leading to storm surges of up to 43 cm on the west coast and 29 cm on the east coast of India due to E1 and E2. E1 generated sea level oscillations at the measuring stations on the west coast (Ratnagiri, Verem and Karwar) and east coast (Mandapam and Tuticorin) of India with significant energy bands centred at periods of 92, 43 and 23 min. The storm surge is a well-defined peak with a half-amplitude width of 20, 28 and 26 h at Ratnagiri, Verem and Karwar, respectively. However, on the east coast, the sea level oscillations during Thane were similar to those during calm period except for more energy in bands centred at periods of $\sim 100, 42$ and 24 min at Gopalpur, Gangavaram and Kakinada, respectively. The residual sea levels from tide gauge stations in Arabian Sea have been identified as Kelvin-type surges propagating northwards at a speed of $\sim 6.5 \, \mathrm{m \, s^{-1}}$ with a surge peak of almost constant amplitude. Multi-linear regression analysis shows that the local surface meteorological data (daily mean wind and atmospheric pressure) is able to account for ~ 57 and $\sim 69\%$ of daily mean sea level variability along the east and west coasts of India. The remaining part of the variability observed in the sea level may be attributed to local coastal currents and remote forcing.

1 Introduction

Tropical cyclones (TCs) are the most destructive weather systems on the earth, producing intense winds, resulting in high surges, meteotsunamis, torrential rains, severe floods and usually causing damage to property and loss of life. In the northern Indian Ocean, both the Bay of Bengal (BOB) and the Arabian Sea (AS) are potential genesis regions for cyclonic storms. Intense winds associated with TCs, blowing over a large water surface, cause the sea surface to pile up on the coast and leads to sudden inundation and flooding of the vast coastal regions. Also, the heavy rainfall causes flooding of river deltas in combination with tides and surges. A number of general reviews and description of individual cyclones and associated surges in the BOB and the AS have been published previously by several investigators (Murty et al., 1986; Dube et al., 1997; Sundar et al., 1999; Fritz et al., 2010; Joseph et al., 2011). Developments in storm surge prediction in the Bay of Bengal and the Arabian Sea have been highlighted by Dube et al. (2009) and references therein (e.g. Das, 1994; Chittibabu et al., 2000, 2002; Dube et al., 2006; Jain et al., 2007; Rao et al., 2008).

Apart from the studies carried out with a view to assessing the coastal vulnerability, few studies concentrated on the variations in characteristics of different oceanographic parameters in response to tropical cyclones. Joseph et al. (2011) examined the response of the coastal regions of eastern Arabian Sea (AS) and Kavaratti Island lagoon to the tropical cyclonic storm "Phyan", during 9–12 November 2009 until its landfall at the northwest coast of India, based on in situ and satellite-derived measurements. Mehra et al. (2012) reported similarities in the spectral characteristics of sea level oscillations in the Mandovi estuary of Goa in the eastern Arabian Sea due to cyclones (June 2007 and November 2009) and the Sumatra geophysical tsunami (September 2007). Wang et al. (2012) reported the variations in the oceanographic parameters due to the tropical Cyclone Gonu, which passed over a deep autonomous mooring system in the northern Arabian Sea and a shallow cabled mooring system in the Sea of Oman. Near-inertial oscillations at all moorings from thermocline to seafloor were observed to be coincident with the arrival of Gonu. Sub-inertial oscillations with periods of 2–10 days were recorded at the post-storm relaxation stage of Gonu, primarily in the thermocline of the deep array and at the onshore regions of the shallow array. Antony and Unnikrishnan (2013) used hourly tide gauge data at Chennai, Visakhapatnam and Paradip along the east coast of India and at Hiron Point, at the head of Bay of Bengal, to analyse statistically the tide-surge interaction. Recently, Rao et al. (2013) simulated surges and water levels along the east coast of India using an advanced 2-D depth-integrated circulation model (ADCIRC-2DDI).

It is necessary that the problem of storm surge must be seriously addressed by the countries of the various regions through collective efforts and in an integrated manner. In the present study, the objective is to examine the characteristic of the sea level oscillations at different topographic locations in the AS and the BOB due to the meteorological events. Our interest is confined to a few minutes to days and analysis of the spectral features of sea level oscillations in the two basins.

2 Data and methodology

In the present study, we report the response of the sea level to the episodic meteorological events at various coastal and Island locations of India from 1 September 2011 to 31 January 2012. Study encompasses two episodic meteorological events: (i) deep depression in November 2011 (E1) in the AS and (ii) the tropical Cyclone Thane (E2) in the BOB as shown in Fig. 1. Summary of observations is given in Table 1. The radar gauge (RG), which measures sea level, is described in detail by Prabhudesai et al. (2006, 2008) and the evaluation and comparative studies have been reported by Mehra et al. (2013). RG acquires samples over 30 s window at 1 min interval and the average over 5 min is recorded at 5 min interval. The surface meteorological variables are collected by autonomous weather station (NIO-AWS). AWS samples (wind, air temperature, air pressure and relative humidity) data every 10 s over a window of 10 min, averaged and then recorded at every 10 min interval. In the present study, we have used time-series data at 5 (10) min interval from the RG (AWS). Both the systems have been designed and developed in the Marine Instrumentation Division, CSIR-NIO, Goa. Summary of observations from different coastal and Is-

Figure 1. Study location showing the tracks of meteorological events during the year 2011. Note: sea level data at Colombo, Kochi, Karachi, Chabahar, Jask, Masirah, Minocoy and Hanimaadhoo are downloaded from www.gloss-sealevel.org and are shown with red stars. (Time is in Indian standard time (IST).)

land locations of India are provided in Table 1 and the periods covered for different events are as follows:

– Event 1 (E1): 26 November to 1 December 2011, occurrence of deep depression in the Arabian Sea.

– Event 2 (E2): 25–31 December 2011, passage of Thane cyclone in the Bay of Bengal.

The tropical cyclone track data from India Meteorological Department (IMD, www.imd.gov.in), Joint Typhoon Warning Center (JTWC, www.usno.navy.mil/JTWC/) and UNISYS-Unisys Weather (http://weather.unisys.com/hurricane/) are shown Fig. 2. The storm translational speed is calculated using the distance travelled between two consecutive positions and time interval. The average differences in wind speeds as shown in Fig. 2a and d between IMD and JTWC, and IMD and Unisys are -1.1 (-4.2) and -3.7 (-2.8) m s^{-1} during E1 (E2). The sea level pressure reported by IMD and JTWC is similar during E1 (Fig. 2b); however during E2, the minimum sea level pressure differed by ~ -11 mb with a time lag of ~ 3 h (Fig. 2e). The cyclone translation speed estimated using JTWC and Unisys data during E1 varied between 2.5 and 6.4 m s^{-1}, except for two spikes of ~ 9 m s^{-1} observed in Unisys data (Fig. 2c). Similarly, the cyclone translation speed estimated using JTWC and Unisys data during E2 varied between 1.0 and 4.5 m s^{-1}, except for a few spikes of ~ 5–7 m s^{-1} (Fig. 2f). Cyclone translation speed using IMD data is fluctuating as the data are available at every 3 h interval, whereas data from the other two sites is at every 6 h. However, the mean cyclone speed during E1 (E2) from IMD data is 7.8 (2) m s^{-1}.

Sea level data are de-tided using TASK tidal analysis and prediction program (Bell et al., 2000) to obtain sea level residual (SLR). A multi-linear regression model linking sea level and atmospheric parameters has been established. The

Table 1. Summary of observations from different coastal and Island locations of India from 1 September 2011 to 31 January 2012. The CSIR-NIO radar gauge (RG) measures sea level (cm) and the CSIR-NIO autonomous weather station (AWS) provides surface meteorological variables such as winds, atmospheric pressure and air temperature.

Sr No	Measurement Station	Latitude and Longitude		Location type	System	Distance between AWS & RG (m)
		Lat (° N)	Lon (° E)			
1	Gopalpur, Odisha	19.3081	84.9613	Harbour	AWS	255
		19.3069	84.9634		Radar gauge	
2	Gangavaram, Andhra Pradesh	17.6174	83.2322	Harbour	AWS	726
		17.6235	83.2295		Radar gauge	
3	Kakinada, Andhra Pradesh	16.9764	82.2832	Harbour	AWS	2
		16.9764	82.2832		Radar gauge	
4	Mandapam, Tamil Nadu	9.2763	79.1295	Boundary of Palk Strait	AWS	615
		9.2713	79.1321	& Gulf of Mannar	Radar gauge	
5	Tuticorin, Tamil Nadu	8.7500	78.2021	Gulf of Mannar	Radar gauge	–
6	Port Blair, Andaman & Nicobar Islands	11.7099	92.7386	Open ocean	AWS	2984
		11.6884	92.7222		Radar gauge	
7	Karwar, Karnataka	14.8464	74.1317	Open ocean	AWS	5154
		14.8030	74.1144		Radar gauge	
8	Verem, Goa	15.4554	73.8022	Mandovi estuary	AWS	5265
		15.5019	73.8120		Radar gauge	
9	Ratnagiri, Maharashtra	16.8926	73.2758	Cove	AWS	525
		16.8890	73.2853		Radar gauge	

model can be described in general as follows:

$$\eta = B_0 + B_1\tau_U + B_2\tau_V + B_3 A_P + \epsilon, \tag{1}$$

In the above expression, sea level residual (η) is the dependent variable and the independent variables are cross-shore (along-shore) wind stress $\tau_U(\tau_V)$ and atmospheric pressure (A_P). Likewise B_0, B_1, B_2 and B_3 are the coefficients of regression and ϵ is the difference between the measured SLR and estimated SLR using multi-linear regression. The cross-shore (along-shore) wind stress $\tau_U(\tau_V)$ is estimated using cross-shore (U) and along-shore (V) component of winds as follows:

$$\tau_U = \rho_A C_D U \sqrt{U^2 + V^2} \tag{2}$$

$$\tau_V = \rho_A C_D V \sqrt{U^2 + V^2}. \tag{3}$$

$\rho_A = 1.3\,\text{kg m}^{-3}$ is the density of air and $C_D = 1.2 \times 10^{-3}$ is the drag coefficient. The regression is performed using daily mean SLR, τ_U, τ_V and A_P. For each month, coefficients of regression of the daily data are obtained to estimate the SLR, which is then merged to generate the time series of estimated SLR for the duration of September 2011 to January 2012.

3 Observed coastal sea level response to meteorological events

The tracks of the meteorological event under study, which occurred in the AS (the BOB) are shown in Fig. 1. The data

and information about these episodic meteorological events is taken from www.imd.gov.in.

The meteorological event (E1) in AS, developed on 26 November 2011 at 7.5° N, 76.5° E near the southern tip of the Indian sub-continent and moved north-westwards. By 28 November 2011 00:00 UTC, it intensified as a deep depression with maximum sustained surface winds reaching up to 15 m s^{-1} (Fig. 2a) and the minimum estimated central pressure (ECP) $\sim$ 998 mb (Fig. 2b). The average translational speed of this system remained steady to $\sim$ 6.5 m s^{-1}. However, during the minimum ECP, the translation speed also decreased to $\sim$ 2 m s^{-1} on 29 November and increased to $\sim$ 6 m s^{-1} on 30 November (Fig. 2c). The system weakened into a well-marked low pressure area over the west central Arabian Sea.

The cyclonic system named "Thane" initially originated as a depression on 25 December 2011 at 8.5° N, 88.5° E and moved north-westwards (Fig. 1). Thane intensified into a very severe cyclonic storm with maximum sustained surface winds peaked up to $\sim$ 45 m s^{-1} as shown in Fig. 2d and ECP falling to $\sim$ 956 mb (Fig. 2e). The cyclone track turned westwards on 28 December, with an average translational speed of $\sim$ 4 m s^{-1} and then became steady at $\sim$ 3.5 m s^{-1} as shown in Fig. 2f. The translation speed of a storm can exert significant control on the intensity of storms by modulating the strength of the negative effect of the storm-induced sea surface temperature (SST) reduction on the storm intensification (Mei et al., 2012). Thane crossed the Tamil Nadu coast just

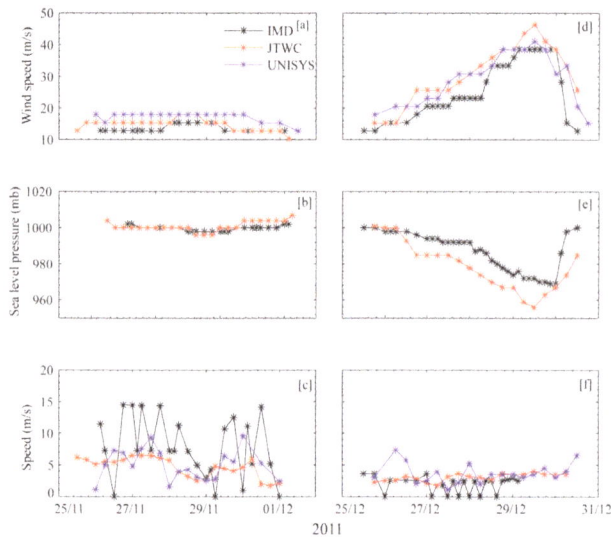

Figure 2. Cyclone parameters during E1 (**a**) wind speed, (**b**) sea level pressure, and (**c**) storm forward translation speed. Similarly, cyclone parameters during E2 (**d**) wind speed, (**e**) sea level pressure, and (**f**) storm forward translation speed. Note: IMD-India Meteorological Department; JTWC-Joint Typhoon Warning Center; UNISYS-Unisys Weather (http://weather.unisys.com/hurricane/).

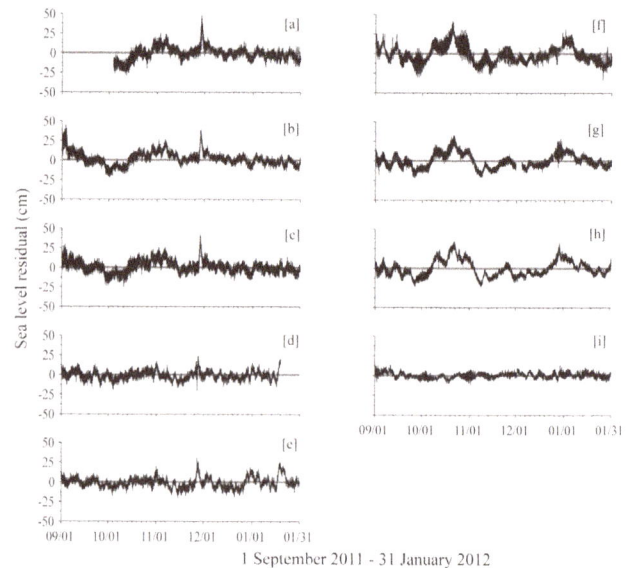

Figure 3. Sea level residual (SLR) at (**a**) Ratnagiri, (**b**) Verem, (**c**) Karwar, (**d**) Tuticorin, (**e**) Mandapam, (**f**) Gopalpur, (**g**) Gangavaram, (**h**) Kakinada and (**i**) Port Blair.

south of Cuddalore between 01:00 and 02:00 UTC of 30 December 2011 and weakened into a well-marked low pressure area over northern Kerala and its neighbourhood.

3.1 Response of sea level to depression in the Arabian Sea (2011)

The sea level residuals (SLR) at Ratnagiri, Verem and Karwar are shown in Fig. 3. The visual observation of SLR indicates that it is normally within ±25 cm at all the three locations. Keila and the subsequent depression from 29 October to 10 November are not able to generate noticeable sea level variations, probably due to large distance of the measurement sites from the cyclonic tracks. For example, the distance of Verem to the trajectory of Keila's ECP is ~ 1554 km. The variance of SLR observed during 29 October to 10 November at Ratnagiri, Verem and Karwar is $\sim 26.1, 21.6$ and 25.8 cm^2, respectively (Fig. 3). However, the deep depression which originated on 26 November 2011 (E1) was in the near proximity to the measurement sites. For example, the distance of Verem from the depression centre on 28 November 2011 was ~ 490 km (Fig. 1). E1 was able to inflict surges at Ratnagiri, Verem and Karwar which peaked up to ~ 43 cm with SLR variance of $\sim 119.4, 95.4$ and 108.2 cm^2, respectively, during E1 (Fig. 3a–c). The storm surge is a well-defined peak with a half- width (see Fandry et al., 1984) of ~ 25 h (Table 2). The local surface meteorological conditions along with SLR are shown in Fig. 4. During E1 (26 November to 1 December 2011), the wind variance was $\sim 1.7, 4.8$ and 0.8 m^2 s^{-2} with wind speeds peaking up to ~ 8.5 m s^{-1} at Ratnagiri

and Verem (Fig. 4a.2–b.2). At Karwar, the wind energy is less compared to the other two sites, still the SLR peaks are of same magnitude (Fig. 4c.1–c.2), indicating the effect of long waves generated by the forcing due to E1 in the open ocean. The wind direction (Fig. 4a.3–c.3) stabilised with respect to the north (Table 2) at Ratnagiri, Verem and Karwar, respectively. The atmospheric pressure anomaly (Fig. 4a.4–c.4) shows a variance of ~ 3.6 mb^2 and falls by ~ 6.0 mb during E1 at the three stations. However, anomalous temperature variations due to E1 were not observed (Fig. 4a.5–c.5), but the range narrowed, which is also the case with relative humidity at the three stations (Fig. 4a.6–c.6).

3.2 Response of sea level to meteorological events on the east coast of India

Response of sea level as storm surges at different sites, to the tropical cyclone Thane, E2, which occurred in the BOB are shown (listed) in Fig. 3 (Table 3). SLR exhibits maximum oscillations (variance) of ~ 27 cm (47.8 and 11.7 cm^2) at Gopalpur (Fig. 5a.1) and Gangavaram (Fig. 5b.1), respectively. At Kakinada, the SLR peaked up to 33 cm, with a variance of 23.3 cm^2 during E2. Minor dip in SLR ~ 14.1, 10.3 and 15.0 cm was also observed at the coastal sites located in the AS (Ratnagiri, Verem and Karwar) due to E2 (Fig. 3a–c). However, at the Island station, Port Blair, the SLR variations are within ±10 cm, and less than at sites north of Thane (Fig. 3i). The SLR variability at Mandapam and Tuticorin was less compared to other sites north of Thane track (Fig. 3d and e), probably due to the following two reasons: (i) the geometrical amplification of the open-ocean waves as they propagate northwards and (ii) wind speeds

Table 2. Meteorological and sea level observations at Ratnagiri, Verem and Karwar during E1 from 26 November to 1 December 2011. Time is in Indian standard time (IST).

Sr No	Variables	Ratnagiri	Verem	Karwar
1	Sea level residual (SLR in cm)	47	39	42
2	SLR rise time from zero-maxima (h)	44.16	39.33	32.58
3	SLR fall time from maxima-zero (h)	48.5	45.25	42.25
4	SLR peak time	29 Nov 2011 03:15	28 Nov 2011 18:00	28 Nov 2011 12:25
5	Maximum wind speed (m s^{-1})	7.4	9.6	4.3
6	Half amplitude surge width duration (h)	20	28	26
7	Wind direction (degrees)	253	112	246
8	Air temperature, reduction in range (°C)	8.3–3.0	13.3–6.8	15.5–8.3
9	Atmospheric pressure fall (mb)	5.8	6.3	5.9
10	Relative humidity range fall (%)	62.0–40.4	65.4–41.3	64.8–33.8

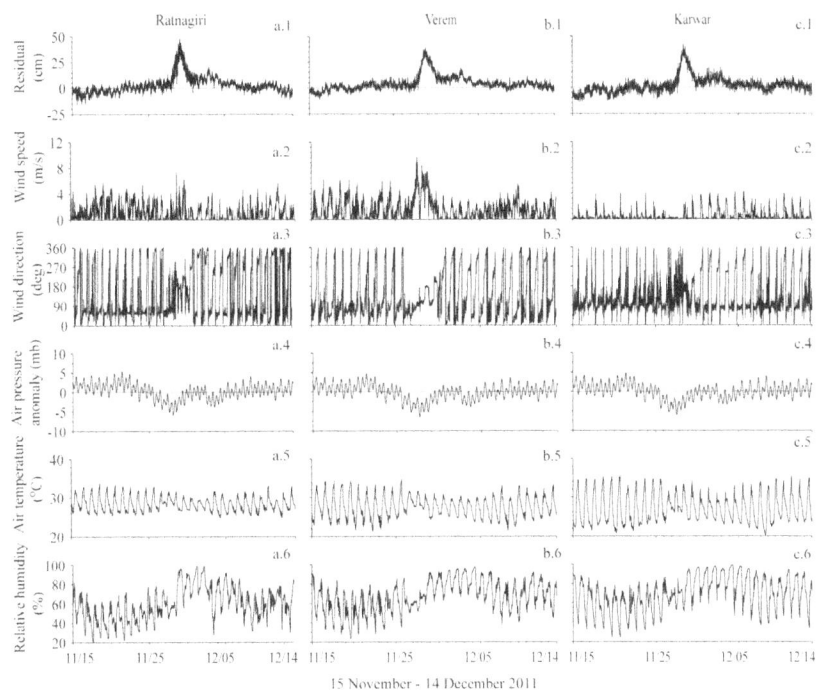

Figure 4. Sea level residual and surface meteorological parameters during the episodic event E1. (**a.1** to **a.6**) SLR, wind speed, wind direction, atmospheric pressure anomaly, air temperature and relative humidity at Ratnagiri, Maharashtra. (**b.1** to **b.6**) same as in (**a**) at Verem, Goa. (**c.1** to **c.6**) same as in (**a**) at Karwar, Karnataka. The atmospheric pressure anomaly is estimated by subtracting the mean atmospheric pressure (1 September 2011 to 31 January 2012) from the measured atmospheric pressure for respective stations.

are less near the central depression point and increases towards the periphery. SLR rise is also seen at Mandapam (Tuticorin) by ∼ 24.3 (23.1) cm even during E1. The local surface meteorological conditions along with SLR are shown in Fig. 5. The large scale extent of E2 is evident in wind and atmospheric pressure measurements at all the three locations and very similar meteorological conditions exist at Gangavaram and Kakinada. At Gopalpur, the winds were weak as compared to the other two southern locations with maximum wind speed reaching up to ∼ 6 m s^{-1}; the direction also fluctuated during E2 and remained southerly after 5 Jan-

uary 2012 and maintained this direction till 10 January 2012 (Fig. 5a.2 and a.3, Table 3). During E2, the wind speed remained high from 26 December 2011 till 4 January 2012. The wind speed peaked up to ∼ 14.0 m s^{-1}, with corresponding wind variance of ∼ 13.7 and 10.3 m^2 s^{-2} at Gangavaram and Kakinada, respectively (Fig. 5b.2 and c.2). The wind direction stabilised and remained north-easterly (Fig. 5b.3 and c.3, Table 3) during E2 at Gangavaram and Kakinada. The atmospheric pressure (Fig. 5a.4–c.4) shows a variance of ∼ 2.7 mb^2 and is devoid of any noticeable fall during E2 at Gopalpur, Gangavaram and Kakinada. Similarly, the anoma-

Table 3. Meteorological and sea level observations at Gopalpur, Gangavaram and Kakinada during E2 from 26–31 December 2011.

Sr No	Variables	Gopalpur	Gangavaram	Kakinada
1	Sea level residual (SLR in cm)	27.4[a]	26.5[a]	32.9
2	SLR rise time from zero-maxima (h)	–	–	123.8
3	SLR fall time from maxima-zero (h)	–	–	233.25
4	Maximum wind speed (m s^{-1})	6.1	15.0	13.3
5	Wind direction (degrees)	184[b]	41.9	60.9
6	Air temperature reduction in range (°C)	10.1–2.6	8.4–3.1	8.5–2.6
7	Relative humidity range reduction (%)	65.7–27.0	57.8–23.8	46–13.6

[a] Maximum of the SLR oscillation at Gopalpur and Gangavaram. [b] The direction fluctuated during E2 and stabilised to ∼ 184° with respect to the north after 5 January 2012 and maintained this direction till 10 January 2012.

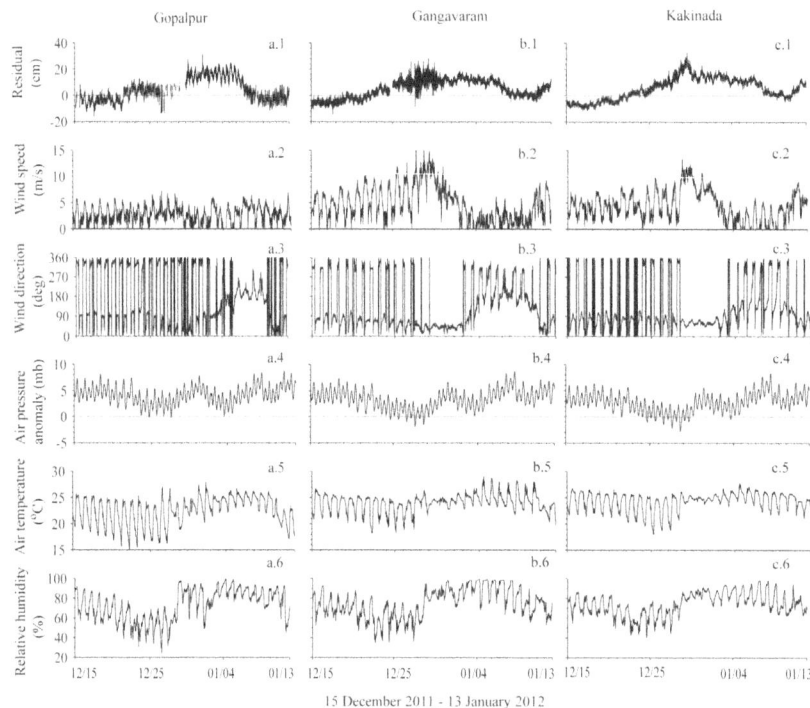

Figure 5. Sea level residual and surface meteorological parameters during the episodic event E2. (**a.1** to **a.6**) SLR, wind speed, wind direction, atmospheric pressure anomaly, air temperature and relative humidity at Gopalpur, Odisha. (**b.1** to **b.6**) same as in (**a**) at Gangavaram, Andhra Pradesh. (**c.1** to **c.6**) same as in (**a**) at Kakinada, Andhra Pradesh. The atmospheric pressure anomaly is estimated by subtracting the mean atmospheric pressure (1 September 2011 to 31 January 2012) from the measured atmospheric pressure for respective stations.

lous variations in temperature due to E2 are not observed; however, the range is narrowed down from ∼ 9.0 to 2.7 °C at the three stations (Fig. 5a.5–c.5). Similarly, a reduction in relative humidity range is also observed at the three stations in the BOB (Fig. 5a.6–c.6, Table 3).

4 Regression model

Multi-linear regression analysis of SLR as the dependent variable with wind stress components (τ_U, τ_V) and atmospheric pressure (A_P) as the independent variable is performed as explained in Sect. 2 (Eq. 1). How well the model

describes the sea level residual is assessed by looking at the percentage of sea level variance explained (Var$_e$) by the model.

$$\mathrm{Var_e} = \left(1 - \frac{\mathrm{variance}(\varepsilon)}{\mathrm{variance\ (measured\ SLR)}}\right) \times 100 \qquad (4)$$

The multi-linear regression performed with 10, 60 min, 6, 12 and 24 h averaged data of Verem and Karwar is shown in the Supplement (Figs. S1–S5) along with the SLR variance (Table S1). The daily average of estimated SLR is comparable with the daily average of measured SLR, and it is able to account for the low frequency variations in the SLR during the

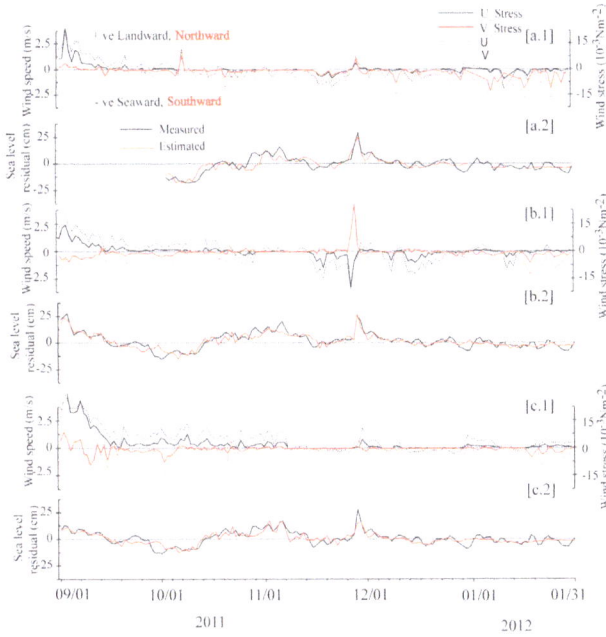

Figure 6. Daily mean wind, wind stress, measured sea level residual and estimated sea level residual from September 2011 to January 2012 at **(a)** Ratnagiri, **(a.1)** daily averaged cross-shore (black) and along-shore (red) winds stress along with respective winds (dotted black or red), **(a.2)** daily mean measured sea level residual (black) and estimated residual (red); **(b)** Verem, **(b.1)** and **(b.2)** same as in **(a)**; **(c)** Karwar, **(c.1)** and **(c.2)** same as in **(a)**. Daily mean estimated SLR is obtained using the multi-linear regression method using daily mean cross-shore (τ_U), along-shore (τ_V) components of winds stress and atmospheric pressure (A_P) as independent variables.

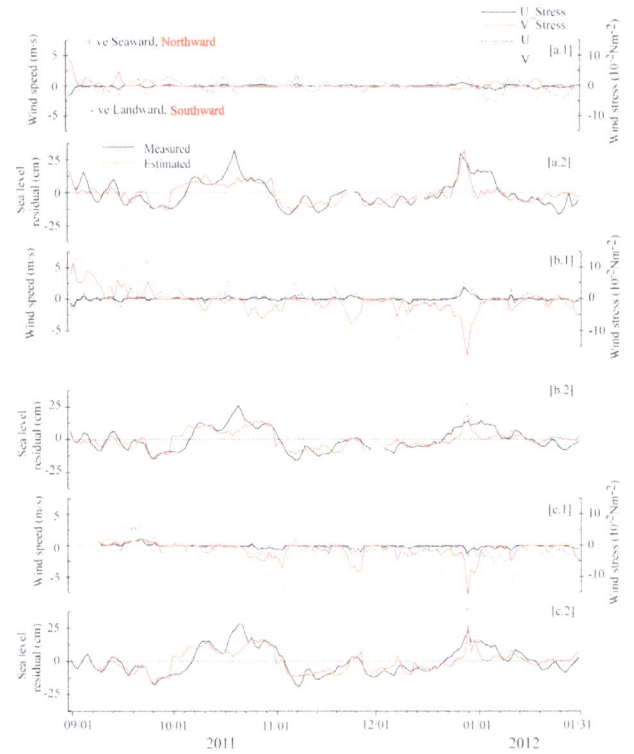

Figure 7. Daily mean wind, wind stress, measured sea level residual and estimated sea level residual from September 2011 to January 2012 at **(a)** Gopalpur, **(a.1)** daily averaged cross-shore (black) and along-shore (red) winds stress along with respective winds (dotted black or red), **(a.2)** daily mean measured sea level residual (black) and estimated residual (red); **(b)** Gangavaram, **(b.1)** and **(b.2)** same as in **(a)**; **(c)** Kakinada, **(c.1)** and **(c.2)** same as in **(a)**. Daily mean estimated SLR is obtained using the multi-linear regression method using daily mean cross-shore (τ_U), along-shore (τ_V) components of winds stress and atmospheric pressure (A_P) as independent variables.

study period of 5 months (September 2011 to January 2012). Therefore, in this section daily averaged data series of SLR and AWS is used to perform multi-linear regression. To begin with, multi-linear regression is performed with cross-shore (τ_U), along-shore (τ_V) components of winds stress and atmospheric pressure (A_P) individually as independent variable to regress the SLR. This would enable us to know, the contribution to the SLR variability by the various surface meteorological variables individually. Then all the three independent variables (τ_U, τ_V, A_P) together are used to regress the daily mean SLR. Results are listed in Table 4 and plotted in Figs. 6 and 7. The cross- and along-shore components are estimated using the local shoreline angles with respect to the north from Google Earth. In AS, the local shoreline angle estimated at Ratnagiri, Verem and Karwar with respect to the north is $\sim -12, -27$ and $-19°$, respectively. Similarly, in the BOB the local shoreline angle with respect to the north is $\sim 52, 45$ and $50°$ at Gopalpur, Gangavaram and Kakinada, respectively. The variations in τ_U, τ_V and A_P results in the SLR variability. The relation between atmospheric pressure and sea level is ~ -1 cm mb^{-1}, inverse barometric effect. The cross-shore wind (U) towards land (sea) will give rise

to increase (decrease) in SLR due to wind stress. Since the study region is in the Northern Hemisphere, the along-shore northward winds with the coast at their right will increase the sea level due to Coriolis force. Therefore, in the AS the positive (negative) U and V will increase (decrease) the SLR (Fig. 6) and in the BOB the positive (negative) U and V will decrease (increase) the SLR (Fig. 7).

At all the three study locations in AS, τ_U, τ_V and A_P individually could explain an average 45 % SLR variability; when τ_U, τ_V, and A_P together are used to regress the daily mean SLR, the total Var$_e$ is ~ 69 % (Table 4). Daily mean $U(\tau_U)$, $V(\tau_V)$ and estimated SLR obtained from independent variables (τ_U, τ_V, A_P) for Ratnagiri is plotted in Fig. 6a. It is observed that the estimated daily mean SLR during E1 is able to peak at up to 96.3 % of the measured daily mean SLR (Fig. 6a.2 and Table 5). The monthly Var$_e$ is ~ 50 % during October and December 2011 (Table 4 and Fig. 8a). However, when E1 occurred in November 2011, τ_U, τ_V, and A_P to-

Table 4. The daily mean sea-level variability explained (Var$_e$) by linear (τ_U, τ_V, A_P individually) and multi-linear (τ_U, τ_V, A_P together) regression during different months from September 2011 to January 2012.

Station	Variable	Total Var$_e$ (%)	Monthly Var$_e$ (%)				
			Sep	Oct	Nov	Dec	Jan
Arabian Sea							
Ratnagiri	τ_U, τ_V, A_P	43.9, 45.0, 59.3					
	τ_U, τ_V, A_P	68.4		52.6	67.8	46.7	3.4
Verem	τ_U, τ_V, A_P	48.9, 39.8, 58.2					
	τ_U, τ_V, A_P	73.9	85.3	72.9	54.5	42.6	7.0
Karwar	τ_U, τ_V, A_P	37.3, 29.5, 45.2					
	τ_U, τ_V, A_P	64.8	64.0	75.7	56.7	38.8	8.3
Bay of Bengal							
Gopalpur	τ_U, τ_V, A_P	43.2, 40.1, 31.2					
	τ_U, τ_V, A_P	53.4	58.8	8.3	44.1	74.7	10.2
Gangavaram	τ_U, τ_V, A_P	42.0, 47.6, 44.2					
	τ_U, τ_V, A_P	56.6	64.4	14.2	37.1	64.3	11.6
Kakinada	τ_U, τ_V, A_P	49.4, 55.6, 44.4					
	τ_U, τ_V, A_P	62.1	77.8	15.4	58.0	68.1	17.7

gether are able to explain the SLR variability up to ∼ 68 %. In January 2012, the Var$_e$ is less than 20 % at all the sites in the AS and the BOB (Fig. 8a and Table 4). At Verem (Fig. 6b), the total Var$_e$ is highest among all the locations in AS, when τ_U, τ_V, and A_P together are used to estimate the daily mean SLR as presented in Fig. 6b.2. During E1, the $-U(V)$ will tend to decrease (increase) the sea level at Verem (Fig. 6b.1), still the estimated daily mean SLR is able to reproduce the comparable response with measured daily mean SLR, with a minor overshoot by ∼ 6.5 % (Table 5). At Karwar also the $U(-V)$ will tend to increase (decrease) the sea level during E1 (Fig. 6c.1); however, the estimated SLR is able to peak only at up to half of the measured daily mean SLR (Fig. 6c.2 and Table 5).

In the BOB, the τ_U, τ_V and A_P individually could explain an average 44 % SLR variability, when τ_U, τ_V, and A_P together are used to regress the daily mean SLR, the total Var$_e$ is ∼ 57 % (Table 4). The monthly Var$_e$ is low < 20 % in October and January for all the three stations in the BOB as shown in Fig. 8b. In the BOB, the daily mean $U(\tau_U)$, $V(\tau_V)$ and estimated SLR is plotted in Fig. 7 for Gopalpur, Gangavaram and Kakinada. As stated earlier, along the east coast of India in the BOB, the positive (negative) U and V will decrease (increase) the sea level. Figure 7a.1 shows the $U(-V)$ at Gopalpur, which are seaward (southward) during E2 favouring the sea-level fall (rise) due to the surface stress. However, by 1 January 2012 both the cross- and along-shore component of wind $-U(V)$ turns landwards (northwards), imposing a sea-level rise (fall). Sea-level appears to be influenced more by alongshore wind, where the estimated SLR follows the forcing (τ_V) of V. Figure 7a.2 is plotted with the estimated and measured daily mean SLR at Gopalpur, dur-

ing E2 the estimated SLR is comparable and is ∼ 9 % more than the measured daily mean SLR (Table 5). It is also observed that the measured SLR remains high (∼ 15 cm) till 5 January 2012, whereas the estimated SLR falls to zero by 31 December 2011. At Gangavaram, the $U(\tau_U)$ $V(\tau_V)$ winds (wind stress) are plotted in Fig. 7b.1, where the daily mean along-shore (V) winds are observed to dominate with a range of ∼ ±10 m s^{-1}. During E2, the measured (estimated) daily mean SLR peaks at up to 13.8 (18.5) cm; the estimate overshoots by ∼ 34 % (Table 5). At Gangavaram, the rise in SLR residual is predominantly due to high along-shore wind ($-V$), as explained by Var$_e$ for December 2011 which is 64 %. Also the measured daily mean SLR remained high from 22 December 2011 to 9 January 2012, whereas the estimated SLR remained high from 25 December 2011 to 2 January 2012 (Fig. 7b.2). Similarly, at Kakinada also the V winds (Fig. 7c.1) appears to dominate with peaks at up to ∼ −10 m s^{-1}. During E2, the measured (estimated) daily mean SLR peaked at up to 22.3 (27.1) cm; the estimate overshoots by 21 % (Table 5). At Kakinada, the rise in sea level residual is predominantly due to strong southward wind (V). The measured SLR started rising above zero on 23 December, reached highest level on 29 December and descended by 9 January 2012, whereas the estimated SLR started ascending on 25 December, reached the highest level on 29 December 2011 and then dropped to zero level by 4 January 2012 (Fig. 7c.2).

5   High frequency response and harbour resonance

Harbour oscillations (coastal seiches) as explained by Rabinovich (2009) are specific type of seiche motion that occur in

Table 5. The peak response of the daily mean sea level residual (SLR) along with the estimated daily mean SLR during E1 & E2.

Station	Event	Measured daily-mean peak SLR (cm)	Estimated daily mean SLR peak (cm)	Difference (%)
Ratnagiri	E1	29.6	28.5	3.7
Verem	E1	25.8	27.4	−6.5
Karwar	E1	27.7	13.6	50.9
Gopalpur	E2	29.7	32.4	−8.9
Gangavaram	E2	13.8	18.5	−34.1
Kakinada	E2	22.3	27.1	−21.5

partially enclosed basins (bays, fjords, inlets and harbours) and are connected through one or more openings to the sea. They are mainly generated by the long waves entering through the open boundary (harbour entrance) from the open sea. An another important property of harbour oscillations is that even small vertical motions (sea level oscillations) may be accompanied by large horizontal motions (harbour currents), resulting in increased risk of damage of moored ships, breaking mooring lines as well as affecting various harbour procedures (Rabinovich, 2009). These waves are similar to a tsunami; however, the catastrophic effects are normally observed in specific bays and inlets. Some specific sites, that have favourable conditions for the resonant generation of extreme ocean waves regularly, have been listed by Monserrat et al. (2006), Rabinovich (2009) and Joseph (2011). Similar phenomena occurs on the southern west coast of India, mainly during pre-southwest monsoon during April or May (Kurian et al., 2009).

In order to understand the harbour oscillations induced by tropical cyclones at various locations, the SLRs are high-pass filtered (time period ≤ 2 h) using a 5th order Butterworth filter (Fig. 9). The amplitude of high frequency SLR (hf-SLR) oscillations in response to E1 at Ratnagiri is $\sim \pm 10$ cm (Fig. 9a), less at Verem and Karwar (Fig. 9b and c). The Karwar station is located in open ocean and therefore does not have the resonance features of a harbour. However, the Verem station is located in Mandovi estuary and Ratnagiri station is located in a cove and may experience resonance with meteorological disturbances. In a similar study at Verem, Mehra et al. (2012) reported the hf-SLR oscillations of $\sim \pm 15$ (10) cm in response to the Cyclone Yemyin (Phyan) which occurred in the BOB (the AS) during 23–25 June 2007 (9–12 November 2009). The hf-SLR oscillations at Tuticorin (Mandapam) are up to 10 (5) cm during E1 (Fig. 9d and e). Mandapam sea level gauge is located on the common boundary of Palk Strait and the Gulf of Mannar, whereas the Tuticorin sea-level gauge is located in the Gulf of Mannar (Fig. 1). The hf-SLR oscillations at the stations located in the BOB are also shown in Fig. 9. The hf-SLR amplitude due to E2 at Tuticorin, Mandapam is not observable, and at Gopalpur a brief amplitude of 10 cm is observed (Fig. 9d, e and f). At Gan-

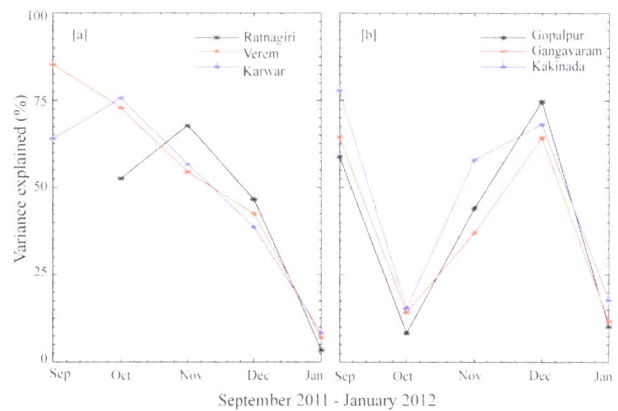

Figure 8. Daily mean sea-level variability explained (Var$_e$ %) by the multi-linear regression during different months from September 2011 to January 2012. **(a)** Var$_e$ (%) at Ratnagiri (black), Verem (red) and Karwar (blue). **(b)** Var$_e$ (%) at Gopalpur (black), Gangavaram (red) and Kakinada (blue).

gavaram (Kakinada) the hf-SLR variations are $\sim \pm 10$ (5) cm as both the gauges are located in the harbour (Fig. 9g and h), and $\sim \pm 4$ cm at Port Blair (Fig. 9i).

Event E1 and background SLR spectra estimated at Karwar, Verem and Ratnagiri are indicated in Fig. 10. The spectrum of SLR data is obtained using "pwelch" function from MATLAB with Hamming window of 256 data points and 50 % overlap. Rabinovich (1997) proposed an approach to separate the influence of source and topography in observed tsunami spectra. The method assumed a linear tide gauge response to external forcing and is based on comparative analysis of tsunami and background spectra. This method will be used to understand the resonant influence of local topography and spectral characteristics of SLR during an event at a particular location. The data duration for estimating the spectrum of the SLR during E1 (background) is from 26 November to 1 December (1 September to 20 November) 2011. Similarly, the data duration during the event E2 (background) is 25–31 December (1 September–10 December) 2011, respectively.

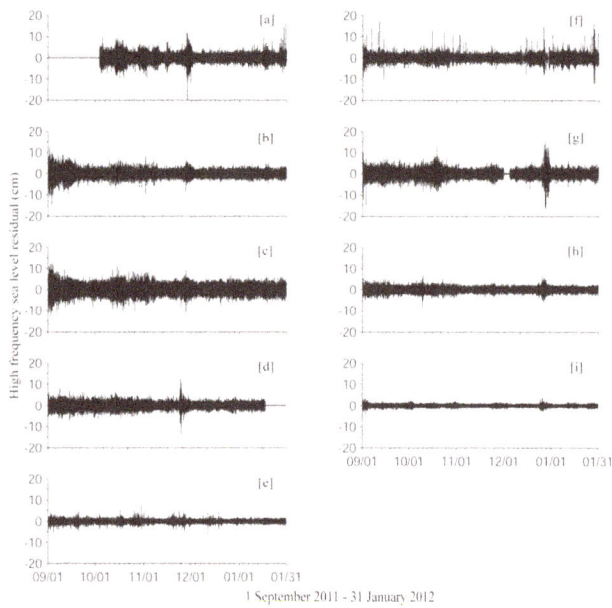

Figure 9. High-pass filtered sea-level residual (hf-SLR) using a 5th order Butterworth filter (time period ≤ 2 h) at (**a**) Ratnagiri, (**b**) Verem, (**c**) Karwar, (**d**) Tuticorin, (**e**) Mandapam, (**f**) Gopalpur, (**g**) Gangavaram, (**h**) Kakinada and (**i**) Port Blair.

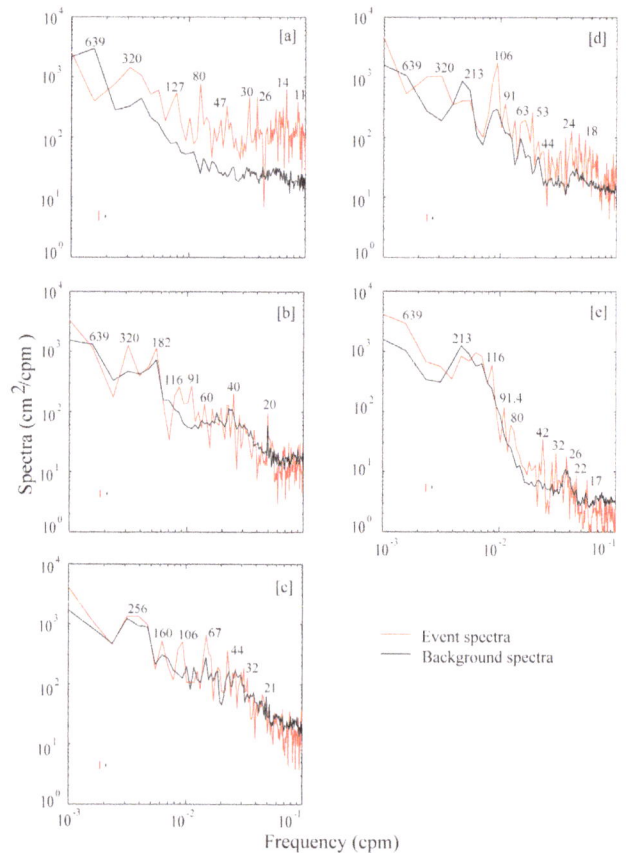

Figure 10. Spectrum of sea level residual (SLR) during E1 at (**a**) Ratnagiri, (**b**) Verem, (**c**) Karwar, (**d**) Tuticorin and (**e**) Mandapam. The data duration for estimating the spectrum of the SLR during E1 (background) is from 26 November to 1 December (1 September to 20 November) 2011. Vertical red (black) line shows the 95 % confidence interval of the event (background) spectrum for the respective stations.

The background spectra of different sites have significant differences at high frequencies as seen in Fig. 10, indicating the influence of local topography. The event spectrum at Ratnagiri is high in energy, well lifted above that of background with major peaks at 127, 80, 47, 30, 26 and 14 min during E1 as shown in Fig. 10a. At Verem the event spectra is intertwined with background spectra with peaks at 182, 91, 40 and 20 min as shown in Fig. 10b. However, the event spectrum at Verem was energetic during the cyclone Yemyin (2007), September Sumatra Tsunami (2007), and the cyclone Phyan (2009), where a distinct peak was observed at ∼43 min (Mehra et al., 2012). Similarly, the event spectrum during E1 (Fig. 10c) at Karwar is similar to the background with some detectable peaks at 106, 67, 44 and 21 min, and further higher frequencies are merged with the background spectra, indicating open-ocean behaviour (lack of harbour resonance). The influence of E1 is also visible at Tuticorin with dominant spectral peaks at 106, 53, 44, 24 and 18 min (Fig. 10d). However, at Mandapam (Fig. 10e) the event spectra are intertwined with background spectra with peaks at 116, 80, 42 and 26 min.

Event E2 and background SLR spectra estimated at Gopalpur, Gangavaram, Kakinada, and Port Blair during E2 are shown in Fig. 11. The event spectrum during E2 (Fig. 11a) at Gopalpur is intertwined with the background spectra with some detectable peaks at 106, 80, 60, 45, 36, 21 and 12 min. The spectral peak at 45 and 21 min are also present in the background signal. The event spectral energy at Gangavaram (Fig. 8b) is higher compared to the back-

ground SLR spectra with peaks at 213, 98, 67, 41, 25 and 18 min; however, the background spectra shows peaks at ∼128, 98 and 17 min. The phenomena of harbour resonance is clearly visible at Gangavaram station (Fig. 5b.1), where it is not the surge but high frequency oscillations triggered by the long waves arriving from the open ocean. At Kakinada (Fig. 11c) the spectra for lower frequencies (time period > 41 min) are similar, but the energy is enhanced for higher-frequency (shorter time-period) oscillations with time periods 41, 37, 25 and 12 min, suggesting resonance occurring in the harbour. At Port Blair, the SLR spectra of both the event and background show similar variability with event peaks at 160, 85, 47, 41, 26 and 17 min (Fig. 11d). The background SLR spectrum has noticeable peaks at 41 and 26 min. Event spectra at shorter time periods are marginally above the background spectra, which indicates the absence of harbour resonance as Port Blair station provides open-ocean conditions.

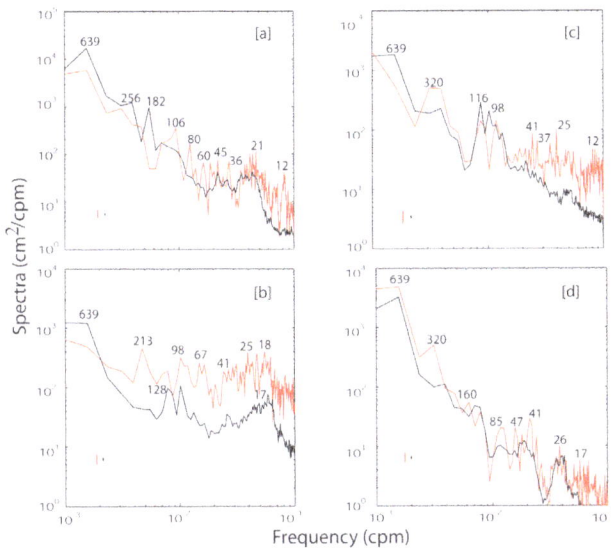

Figure 11. Spectrum of sea level residual (SLR) during E2 at **(a)** Gopalpur, **(b)** Gangavaram, **(c)** Kakinada and **(d)** Port Blair. The data duration for estimating the spectrum of the SLR during E2 (background) is 25–31 December (1 September–10 December) 2011. Vertical red (black) line shows the 95 % confidence interval of the event(background) spectrum for the respective stations.

6   Results and discussion

We summarise the response of sea level of the two events in the AS and the BOB. The SLR rise (fall) in the AS due to E1 reflects the winds as is also seen in the estimated SLR. The estimated SLR peak value at Ratnagiri and Verem is comparable to the measured SLR during E1 but at Karwar it is short by $\sim 50\%$. The $\mathrm{Var_e}$ accounted by local surface meteorological parameters is $\sim 69\%$. $\mathrm{Var_e}$ is small in January at all the locations of the present study (Table 4). In a similar study by Mehra et al. (2010), multi-linear regression analysis (each with a 2-month duration) was also used to resolve the dependence of sea level on various forcing parameters for 2007 and 2008 at Verem, Goa. During the summer monsoon (May–September), the sea level variability attributable to wind was up to 47 and 75 %, respectively, for 2007 and 2008; however, it decreased to $< 20\%$ during the winter monsoon (November–February). A significant part of the variability observed in sea level remains unaccounted for and was attributed to remote forcing. In the BOB, the SLR response to E2 is of a plateau shape with rising peaks and prolonged falls during E2, which the estimated SLR is not able to capture. This persistence of high daily mean SLR state may be attributed to intensity, direction and duration of the event, the distance from the source etc. The distance of Thane (E2) track is ~ 570 km from Gangavaram as shown in Fig. 1. The slope of the continental shelf will also affect the level of surge in a particular area. For example areas with shallow slopes of the continental shelf (as in AS) will allow a greater

storm surge, and areas with deep water just offshore experience large waves, but little storm surge (SLOSH, 2003). At Gangavaram, the daily mean SLR is low (~ 13.8 cm) compared to the other sites on the east coast of India (Table 5), and this could be attributed to distinct harbour oscillations at this location (Fig. 5b.1). Also note that when Thane crossed the Tamil Nadu coast just south of Cuddalore between 01:00 and 02:00 UTC of 30 December 2011, no distinct surge is observed at Mandapam and Tuticorin, even though Mandapam (Tuticorin) is in close proximity ~ 237 (360) km to Thane track. The highest surges usually occur to the right of the storm track (travelling with the storm) at approximately the radius of maximum wind whereas Mandapam and Tuticorin were to the left of the track.

Storm surge is generated partly by the atmospheric pressure variations, but the main contributing factor is wind acting over the shallow water and it is an air-sea interaction problem. The basic mechanism involved in the generation of coastal surges is the influence of a long-shore wind stress, driving an Ekman transport towards the coast, causing the piling-up of water within a Rossby radius of deformation. Only the long-shore wind stress and atmospheric pressure variations associated with a cyclone generate a surge (Thomson, 1970). The surge travels along the coast as a Kelvin wave, away from the finite area of forcing, at a speed $c = \sqrt{(gh)}$, where g is the acceleration due to gravity and h is the depth. Such coastally trapped motions (with the coast to the right (left) in the Northern (Southern) Hemisphere) are called forced Kelvin waves (e.g. LeBlond and Mysak, 1978; Gill, 1982). In the spectrum, the storm surges are centred about 10^{-4} Hz, which corresponds to a period of about 3 h (Platzman, 1971). A few parameter estimates of the E1 and E2 are listed in Table 6. To estimate the surge propagation speed, we have also added a few more locations, where hourly sea level data are available from www.gloss-sealevel.org (marked as red star in Fig. 1). Figure 12 shows the sea level response from Colombo, Sri Lanka to Jask, Iran in the Indian Ocean. Some relevant parameters of the storm surge are listed in Table 7. The average surge propagation speed is estimated to be ~ 6.5 m s^{-1}. The E1 moved northwards with an average along-shore speed of ~ 6.2 m s^{-1}, with the track almost parallel to the west coast of India. The match of the surge propagation speed of 6.5 m s^{-1} with that of E1 alongshore speed is evidence of a forced wave. The residual surge lagged the storm by 3, 4, 6.5 and 8.5 h (Fig. 1 and Table 7) to its nearest proximity at Colombo, Kochi, Karwar and Verem, respectively, with constant peak amplitude of ~ 34.6 cm (Fig. 12). However, at Kochi, the development of secondary peak is clearly visible with time difference of ~ 14 h between the two peaks of ~ 13.5 cm. At Ratnagiri and Karachi the surge peak leads the storm by ~ 1 h with the constant peak amplitude of ~ 33.5 cm.

Similar response as above was observed by Fandry et al. (1984), when Cyclone Glynis moved slowly and almost parallel to the western coast of Australia in February 1970. In

Table 6. Parameter estimates of the events E1 and E2.

Name	Duration	Average eastward velocity U_m (m s^{-1})	Average northward velocity V_m (m s^{-1})	Minimum coastal pressure P_c (mb)	Maximum winds V (m s^{-1})	Maximum stress τ_m (N m^{-2})
DD (E1)	26 Nov–1 Dec 2011	1.1	6.2	998	17	0.9
Thane (E2)	25–31 Dec 2011	−1.1	0.3	969	40.8	5.2

Note: $V = 3.44(1000 - P_c)^{0.644}$ and $\tau_m = 0.000314V^2$ (refer Fandry et al., 1984). P_c is minimum central pressure.

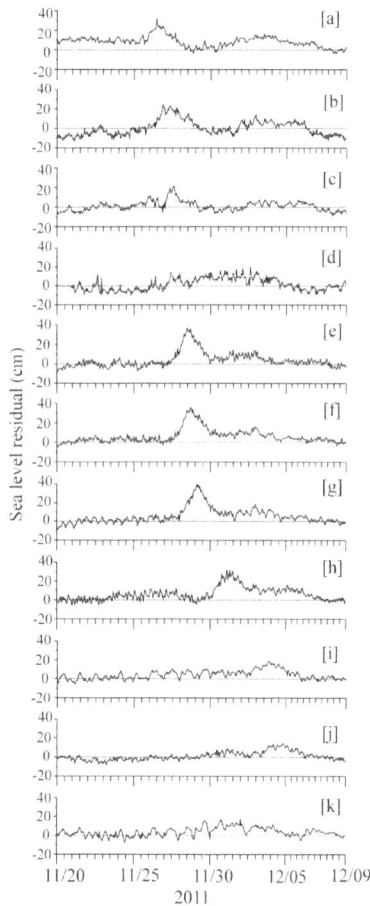

Figure 12. Hourly sea level residual at (**a**) Colombo, (**b**) Mandapam, (**c**) Tuticorin, (**d**) Kochi, (**e**) Karwar, (**f**) Verem, (**g**) Ratnagiri, (**h**) Karachi, (**i**) Chabahar, (**j**) Jask and (**k**) Masirah. Note: 1 – Sea level residual data at Mandapam, Tuticorin, Karwar, Verem and Ratnagiri is hourly averaged. 2 – Sea level data at Colombo, Kochi, Karachi, Chabahar, Jask and Masirah is at hourly interval and downloaded from www.gloss-sealevel.org.

this event, a strong coastal peak travelled down the coast well ahead of the cyclone. In this example $\mu > \frac{U_m f}{C}$ and $V_m < c$ (where f is Coriolis parameter, U_m (V_m) is eastward (northward) velocity of cyclone and μ is decay rate, and theory predicts a coastal peak of constant amplitude moving ahead of the cyclone. In their study, they characterised the sea level response to tropical cyclone as follows:

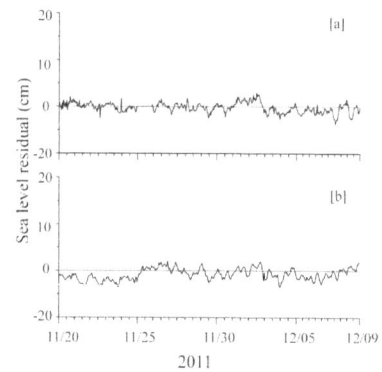

Figure 13. Sea level residual at (**a**) Minicoy and (**b**) Hanimaadhoo. Note: 1 – Sea level data at Minicoy (**b**) Hanimaadhoo is at hourly interval and downloaded from www.gloss-sealevel.org.

a. $\mu < \frac{U_m f}{C}$: coastal peak of increasing in magnitude propagating along the coast with speed V_m. The maximum peak occurs at the edge $y = V_m t$ which is the leading (trailing) edge if $V_m > c (V_m < c)$. Where y is alongshore axis and positive towards north, t is time.

b. $\mu > \frac{U_m f}{C}$: coastal peak of constant magnitude moving behind (ahead of) the cyclone if $V_m > c (V_m < c)$.

In the present study, our observations indicate that the surge peak lagged E1 up to Verem. At Ratnagiri to Karachi the surge peak leads the E1 with an almost constant amplitude (Table 7). Also note that the surge amplitude is almost constant (~ 34 cm) at all the four locations from Karwar to Karachi. The impact of E1 in the BOB at Port Blair (Fig. 3i) and in the northern parts of east coast of India at Kakinada, Gopalpur and Gangavaram (Fig. 3f–h) is not observable. The response of sea level due to E1 at Masirah (Fig. 12k) is also not observable as the location is on the right hand side of the event track. Similarly, the sea level variations at the island locations of Minicoy and Hanimaadhoo are negligible due to E1 (Fig. 13), even though the track of E1 is only ~ 170 and 280 km away, respectively. However, the absence of closed boundary at these Island locations and their locations on the left side of E1 track imply that no surges are predicted. The impact of the E1 is observed only in the AS at the coastal boundary located towards the right side of the event track.

Table 7. Surge propagation parameters during E1 along the west coast of Indian continent.

Location	Peak (cm)	Time of peak (IST)	Path between two locations (km)	Propagation Speed (m s^{-1})
Colombo	31.6	26 Nov 2011 11:30		
Mandapam	24.3	26 Nov 2011 22:00	300	7.9
Tuticorin	21.4	27 Nov 2011 15:00	110	1.8
Kochi (Peak 2)	12.9	27 11 2011 23:30	400	13.1
Kochi (Peak 1)	14.14	27 Nov 2011 09:30		
Karwar	36.9	28 Nov 2011 12:00	577	7.6
Verem	35.5	28 Nov 2011 17:00	90	5.0
Ratnagiri	37.0	29 Nov 2011 03:00	172	4.8
Karachi	31.8	1 Dec 2011 07:30	1094	5.8
Chabahar	18.4	3 Dec 2011 19:30	670	3.1
Jask	13.7	4 Dec 2011 03:30	270	6.7

7   Conclusion

It is being realised increasingly that a near real-time network of sea level and surface meteorological measurements along the coastal and Island locations of India such as ICON (http://inet.nio.org/) established by CSIR-NIO could play an important role in improving the operational (routine) predictions of coastal flooding and enable to understand the fundamental dynamics of these events. Presently, there are only a few mesoscale weather and sea level networks in some coastal segments of the Indian and eastern Atlantic oceans to observe such events. It is also expected that this kind of relatively inexpensive and simple networks, similar to the one in-house developed and established by CSIR-NIO will be affordable to limited-budget institutions in their natural hazard mitigation efforts.

This study attempts to investigate the meteorologically induced surges and water level oscillations along the select locations in response to the passage of the storm (E1) in the Arabian Sea and "Thane (E2)" in Bay of Bengal. The high frequency water level oscillations observed, such as at Gangavaram during the events are found to be due to the result of harbour resonance. Each location has a typical event spectrum related to the local topography. For example, Rabinovich (1997) observed two prominent peaks in Dimitrova Bay. The first peak with a period of 50 min related to the amplification of incoming waves over the shelf of the southern Kuril Islands (i.e to the shelf resonance). The second peak with a period of 18.5 min is caused by the standing oscillation of the bay itself. Some of the locations in the present study showed SLR oscillations around periods of ~ 92, 43 and 23 (100, 42 and 24) min in the AS (the BOB). However, influence of local topography is clearly noticeable at shorter time periods at different stations. During E1, even though the local winds are small in magnitude (4–10 m s^{-1}), the surges are 39–47 cm at Ratnagiri, Verem and Karwar with similar peaks in SLR. In this case, the cyclone track is parallel or at an oblique angle to the coastline and hence the surge occurs on longer stretches of the shoreline. Whereas, during E2 (Thane) the cyclone makes a perpendicular approach to the coast at landfall, and therefore the surges are low at Mandapam and Tuticorin, as these stations are on the left side of the Thane track, as compared to the stations on the right side (Gopalpur, Gangavaram and Kakinada). SLR at Gopalpur, Gangavaram and Kakinada remained high for ~ 9 days even after the land fall of Thane. The residual sea levels from tide gauge stations along the west coast of India and the coast of Pakistan showed surge peak of constant amplitude propagating northwards with a speed of ~ 6.5 m s^{-1} during E1. The propagating surges along the western coast have been identified as forced Kelvin waves with almost constant amplitude at Karwar, Verem, Ratnagiri and Karachi. The multi-linear regression using local daily mean winds (cross- and along-shore) in association with the the atmospheric pressure is able to account for up to 69 % of daily mean sea level residual (SLR) at Ratnagiri, Verem and Karwar during E1. However, in the BOB up to 57 % of daily mean SLR is accounted for at Gopalpur, Gangavaram and Kakinada.

Acknowledgements. The authors acknowledge the support and encouragement provided by S. W. A. Naqvi, director, CSIR-NIO, Goa in carrying out this work. They are grateful to Anil Shirgoankar for his consistent technical support in keeping the systems operational. The authors acknowledge the support of the Finolex Industries Limited, Ratnagiri, Maharashtra; Indian Naval Office at Verem, Goa; Estuary View Resort and Survey of India office, Karwar, Karnataka; CSIR-CECRI, Tuticorin, Tamil Nadu; CMFRI, Mandapam, Tamil Nadu; Kakinada Seaports Ltd., Kakinada, Andhra Pradesh; Ganagvaram Ports Limited, Visakhapatnam, Andhra Pradesh and

Gopalpur Ports Limited, Odisha for providing safe and secured site for sea level and surface meteorological measurements. The authors gratefully acknowledge the financial support (2013–2017) given by the Indian National Centre for the Ocean information Services (INCOIS), Ministry of Earth Sciences, Government of India for maintenance of Autonomous Weather Stations (AWS) under ICON established by CSIR-NIO. Authors thank the two anonymous reviewers for their comments and suggestions. They are also grateful to the editor, John M. Huthnance, for his contributions, which enhanced the quality of the manuscript. This is CSIR-NIO contribution number: 5698.

Edited by: J. M. Huthnance

References

Antony, C. and Unnikrishnan, A. S.: Observed characteristics of tide-surge interaction along the east coast of India and the head of Bay of Bengal, Estuar. Coast. Shelf S., 131, 6–11, doi:10.1016/j.ecss.2013.08.004, 2013.

Bell, C., Vassie, J. M., and Woodworth, P. L.: Tidal Analysis Software Kit 2000 (Task 2000), POL/PSMSL Permanent Service for Mean Sea Level, Proudman Oceanographic Laboratory, UK, 2000.

Chittibabu, P., Dube, S. K., Rao, A. D., Sinha, P. C., and Murty, T. S.: Numerical Simulation of extreme sea levels using location specific high resolution model for Gujarat coast of India, Mar. Geod., 23, 133–142, 2000.

Chittibabu, P., Dube, S. K., Rao, A. D., Sinha, P. C., and Murty, T. S.: Numerical simulation of extreme sea levels for the Tamilnadu (India) and Sri Lanka coasts, Mar. Geod., 25, 235–244, doi:10.1080/01490410290051554, 2002.

Das, P. K.: On the Prediction of storm surges, Sadhana, 19, 583–595, doi:10.1007/BF02835641, 1994.

Dube, S. K., Rao, A. D., Sinha, P. C., Murty, T. S., and Bahulayan, N.: Storm surge in the Bay of Bengal and Arabian Sea: the problem and its prediction, Mausam, 48, 283–304, 1997.

Dube, S. K., Jain, I., and Rao, A. D.: Numerical storm surge prediction model for the North Indian Ocean and the South China Sea, Disaster Dev., 1, 47–63, 2006.

Dube, S. K., Jain, I., Rao, A. D., and Murty, T. S.: Storm surge modelling for the Bay of Bengal and Arabian Sea, Nat. Hazards, 51, 2–27, doi:10.1007/s11069-009-9397-9, 2009.

Fandry, C. B., Leslie, L. M., and Steedman, R. K.: Kelvin-type coastal surges generated by tropical cyclones, J. Phys. Oceanogr., 14, 582–593, 1984.

Fritz, H. M., Blount, C. D., Albusaidi, F. B., and Al-Harthy, A. H. M.: Cyclone Gonu storm surge in Oman, Estuar. Coastal Shelf S., 86, 102–106, 2010.

Gill, A. E.: Atmosphere-Ocean Dynamics, Academic Press, 662 pp., 1982.

Jain, I., Chittibabu, P., Agnihotri, N., Dube, S. K., Sinha, P. C., and Rao, A. D.: Numerical storm surge model for India and Pakistan, Nat. Hazards, 42, 67–73, doi:10.1007/s11069-006-9060-7, 2007.

Joseph, A.: Tsunamis: Detection, monitoring, and early-warning technologies, Elsevier; New York; USA; 436 pp., 2011.

Joseph, A., Desai, R. G. P., Mehra, P., SanilKumar, V., Radhakrishnan, K. V., VijayKumar, K., AshokKumar, K., Agarvadekar, Y., Bhat, U. G., Luis, R., Rivankar, P., and Viegas, B.: Response of west Indian coastal regions and Kavaratti lagoon to the November-2009 tropical cyclone Phyan, Nat. Hazards, 57, 293–312, 2011.

Kurian, N. P., Nirupma, N., Baba, M., and Thomas, K. V.: Coastal flooding due to syntopic scale, meso-scale and remote forcing, Nat. Hazards, 48, 259–273, doi:10.1007/s11069-008-9260-4, 2009.

LeBlond, P. H. and Mysak, L. A.: Waves in the Ocean, Elsevier, 602 pp., 1978.

Mehra, P., Tsimplis, M. N., Prabhudesai, R. G., Joseph, A., Shaw, A. G. P., Somayajulu, Y. K., and Cipollini, P.: Sea level changes induced by local winds on the west coast of India, Ocean Dynam., 60, 819–833, doi:10.1007/s10236-010-0289-z, 2010, 2010.

Mehra, P., Prabhudesai, R. G., Joseph, A., Kumar, V., Agarvadekar, Y., Luis, R., and Viegas, B.: A study of meteorologically and seismically induced water level and water temperature oscillations in an estuary located on the west coast of India (Arabian Sea), Nat. Hazards Earth Syst. Sci., 12, 1607–1620, doi:10.5194/nhess-12-1607-2012, 2012.

Mehra, P., Desai, R. G. P., Joseph, A., VijayKumar, K., Agarvadekar, Y., Luis, R., and Nadaf, L.: Comparison of sea-level measurements between microwave radar and subsurface pressure gauge deployed at select locations along the coast of India, J. Appl. Remote Sens., 7, 16 pp., doi:10.1117/1.JRS.7.073569, 2013.

Mei, W., Pasquero, C., and Primeau, F.: The effect of translation speed upon the intensity of tropical cyclones over the tropical ocean, Geophys. Res. Lett., 39, L07801, doi:10.1029/2011GL050765, 2012.

Monserrat, S., Vilibic, I., and Rabinovich, A. B.: Meteotsunamis: atmospherically induced destructive ocean waves in the tsunami frequency band, Nat. Hazards Earth Syst. Sci., 6, 1035–1051, doi:10.5194/nhess-6-1035-2006, 2006.

Murty, T. S., Flather, R. A., and Henry, R. F.: The storm surge problem in the Bay of Bengal, Prog. Oceanogr., 16, 195–233, 1986.

Prabhudesai, R. G., Joseph, A., Agarvadekar, A., Dabholkar, D., Mehra, P., Gouveia, A., Tengali, S., Vijaykumar, K., and Parab, A.: Development and implementation of cellular-based real-time reporting and Internet accessible coastal sea level gauge – A vital tool for monitoring storm surge and tsunami, Curr. Sci. India, 90, 1413–1418, 2006.

Prabhudesai, R. G, Joseph, A., Mehra, P., Agarvadekar, Y., Tengali, S., and Vijay kumar: Cellular-based and Internet enabled real-time reporting of the tsunami waves at Goa and Kavaratti Island due to Mw 8.4 earthquake in Sumatra on 12th September 2007, Curr. Sci. India, 94, 1151–1157, 2008.

Platzman, G. W.: Ocean tides and related waves, Lectures for the American Mathem-atical Society, 1970, Summer seminars on mathematical problems in the geophysical sciences, Rensselaer Polytechnic Institute, Troy, NY, 94 pp. (also in: Reid, W. H. (Ed.): Mathematical problems in the geophysical sciences, 14, Part 2, 239–291), 1971.

Rabinovich, A. B.: Spectral analysis of tsunami waves: Separation of source and topography effects, J. Geophys. Res., 102, 12663–12676, 1997.

Rabinovich, A. B.: Seiches and Harbor Oscillations, Chapter 9, Handbook of Coastal and Ocean Engineering, edited by: Kim, Y. C., World Scientific Publ., Singapore, 2009.

Rao, A. D., Jain, I., Murthy, M. V. R., Murty, T. S., and Dube, S. K.: Impact of cyclonic wind field on interaction of surge wave computations using Finite- element and Finite-difference models, Nat. Hazards, doi:10.1007/s11069-008-9284-9, 2008.

Rao, A. D., Murty, P. L. N., Jain, I., Kankara, R. S., Dube, S. K., and Murty, T. S.: Simulation of water levels and extent of coastal inundation due to a cyclonic storm along the east coast of India, Nat. Hazards, 66, 1431–1441, doi:10.1007/s11069-012-0193-6, 2013.

SLOSH: Sea, Lake, and Overland Surge from Hurricanes, a computerized model developed by the National Weather Service (NWS), US, to estimate storm surge heights and winds resulting from historical, hypothetical, or predicted hurricanes, SLOSH Display Training, 2003.

Sundar, D., Shankar, D., and Shetye, S. R.: Sea level during storm surges as seen in tide-gauge records along the east coast of India, Curr. Sci. India, 77, 1325–1332, 1999.

Thompson, R. E.: On the generation of Kelvin-type waves by atmospheric disturbances, J. Fluid Mech., 42, 657–670, 1970.

Wang, Z., DiMarco, S. F., Stossel, M. M., Zhang, X., Howard, M. K., and du Vall, K.: Oscillation responses to tropical Cyclone Gonu in northern Arabian Sea from a moored observing system, Deep-Sea Res., 64, 129–145, doi:10.1016/j.dsr.2012.02.005, 2012.

First laboratory study of air–sea gas exchange at hurricane wind speeds

K. E. Krall[1] **and B. Jähne**[1,2]

[1] Institute of Environmental Physics, University of Heidelberg, Im Neuenheimer Feld 229, 69120 Heidelberg, Germany
[2] Heidelberg Collaboratory for Image Processing, University of Heidelberg, Speyerer Straße 6, 69115 Heidelberg, Germany

Correspondence to: K. E. Krall (kerstin.krall@iup.uni-heidelberg.de)

Abstract. In a pilot study conducted in October and November 2011, air–sea gas transfer velocities of the two sparingly soluble trace gases hexafluorobenzene and 1,4-difluorobenzene were measured in the unique high-speed wind-wave tank at Kyoto University, Japan. This air–sea interaction facility is capable of producing hurricane strength wind speeds of up to $u_{10} = 67\,\mathrm{m\,s^{-1}}$. This constitutes the first lab study of gas transfer at such high wind speeds. The measured transfer velocities k_{600} spanned two orders of magnitude, lying between $11\,\mathrm{cm\,h^{-1}}$ and $1180\,\mathrm{cm\,h^{-1}}$ with the latter being the highest ever measured wind-induced gas transfer velocity. The measured gas transfer velocities are in agreement with the only available data set at hurricane wind speeds (McNeil and D'Asaro, 2007). The disproportionately large increase of the transfer velocities found at highest wind speeds indicates a new regime of air–sea gas transfer, which is characterized by strong wave breaking, enhanced turbulence and bubble cloud entrainment.

1 Introduction

Ocean regions, where strong winds usually occur, play an important role in global CO_2 budgets (see Bates et al., 1998). Therefore, a better understanding of gas transfer at high wind speed conditions is essential. Field measurements of air–sea gas exchange velocities under hurricane wind speed conditions are sparse due to the difficulties of sampling under extreme wind conditions. During Hurricane Frances in 2004, McNeil and D'Asaro (2007) measured three transfer velocities of O_2 using unmanned floats at wind speeds larger than $25\,\mathrm{m\,s^{-1}}$, with the highest wind speed being $50.4\,\mathrm{m\,s^{-1}}$.

High wind speeds are associated with the presence of breaking waves. Breaking waves enhance turbulence near the water surface and generate spray and bubble plumes, which increases gas fluxes (see for instance Monahan and Spillane, 1984, and Farmer et al., 1993). Breaking waves enhance gas transfer by several mechanisms: the water surface, across which gas is transferred, is enlarged by waves, and by breaking, waves enhance near-surface turbulence; bubbles and spray provide a limited, mostly short-lived volume of air or water associated with an additional surface area, over which gas transfer can occur (Memery and Merlivat, 1985); and by floating through air and water and bursting through the water surface, bubbles and spray enhance turbulent mixing near the water surface.

Wind-wave tanks provide an alternative to measurements in the field. All the inconveniences and dangers associated with measurements in the field during hurricane wind speed conditions are virtually non-existent in a lab setup. Until now, no gas exchange measurements had been performed in wind-wave tanks at free-stream velocities larger than $20\,\mathrm{m\,s^{-1}}$. The highest gas transfer velocity measured in fresh water is $180\,\mathrm{cm\,h^{-1}}$, at a wind speed close to $20\,\mathrm{m\,s^{-1}}$ (Komori and Shimada, 1995). Higher gas transfer velocities were only measured during the WABEX-93 experiment (Asher et al., 1995) in a freshwater surf pool without wind but with breaking waves. The highest measured gas transfer velocity measured in a laboratory, corrected to a Schmidt number of 600, was $450\,\mathrm{cm\,h^{-1}}$ at a fractional whitecap coverage of 0.067 (Wanninkhof et al., 1995).

In late 2010 the first high-speed wind-wave facility became available at Kyoto University with free-stream wind speeds larger than $40\,\mathrm{m\,s^{-1}}$, opening up new experimental opportunities in the laboratory. It remains an open question, however, whether high wind speed conditions can be adequately simulated in laboratory facilities. This concerns mainly the spatial scale of breaking waves and the deep injection of bubbles with the resulting bubble dissolution flux pathway. Therefore, it makes sense to perform first a pilot study with limited effort to explore the feasibility of such experiments. The results of such a pilot study are reported in this paper.

2   Air–sea gas transfer

The net gas flux j across the air–sea boundary is given as the product of the gas transfer velocity k and the concentration difference as

$$j = k\Delta c = k(c_w - \alpha c_a), \tag{1}$$

with the tracer's air- and water-side concentrations, c_a and c_w, respectively, and the tracer's dimensionless solubility α.

For a sparingly soluble tracer, a dependency of the transfer velocity k on the water-sided friction velocity u_*, a measure for momentum input into the water, is commonly assumed in the form

$$k \propto u_* Sc^{-n}, \tag{2}$$

with the tracer's dimensionless Schmidt number $Sc = \nu/D$, the ratio between the kinematic viscosity of water ν and the tracer's diffusivity in water D. The Schmidt number exponent n is two thirds in the case of a smooth water surface and one half for a rough and wavy surface. More thorough derivations of Eq. (2) can be found in Deacon (1977), Coantic (1986) and Jähne et al. (1989).

Equation (2) can be used to compare the transfer velocities of two tracers, A and B, under the same conditions in the form of Schmidt number scaling,

$$\frac{k_A}{k_B} = \left(\frac{Sc_A}{Sc_B}\right)^{-n}. \tag{3}$$

On the ocean, the gas transfer velocity depends on many different factors such as wind speed, fetch, the presence of surface active material, and atmospheric stability. Wind speed has been identified as the main forcing factor. Many different empirical wind-speed–gas-transfer-velocity parameterizations have been proposed in the last few decades, for instance Liss and Merlivat (1986), Wanninkhof (1992), Nightingale et al. (2000), McGillis et al. (2001), and Wanninkhof et al. (2009). These were all developed in the wind speed region below $15\,\mathrm{m\,s^{-1}}$, where most of them agree reasonably well with each other. Extending these parameterizations to wind speeds observed in a hurricane (see Fig. 1) paints a different picture with large deviations between the different parameterizations. At a wind speed of $50\,\mathrm{m\,s^{-1}}$, the deviations between the highest and the lowest

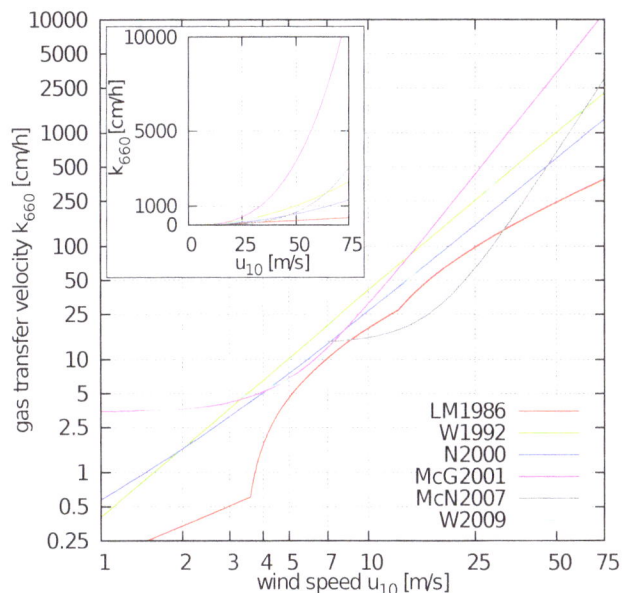

Fig. 1. Some commonly used gas-transfer–wind-speed parameterizations in a double logarithmic plot. The insert shows the same, but with linear axes. LM1986: Liss and Merlivat (1986), W1992: Wanninkhof (1992), N2000: Nightingale et al. (2000), McG2001: McGillis et al. (2001) and W2009: Wanninkhof et al. (2009). The parameterization by McNeil and D'Asaro (2007) (McN2007) is the only one developed for hurricane wind speeds.

predicted transfer velocity is more than one order of magnitude. This highlights the very limited applicability of gas-transfer–wind-speed parameterizations in hurricane conditions. The only parameterization available for hurricane wind speeds by McNeil and D'Asaro (2007), who measured gas transfer velocities during Hurricane Frances, is also shown in Fig. 1.

At high wind speeds, breaking waves generate spray and bubbles. Gas transfer due to single bubbles is well studied experimentally (see for instance Mori et al., 2002, and Vasconcelos et al., 2002) as well as in models (see Memery and Merlivat, 1985). The impact of spray on the gas exchange velocity, however, has not been well studied. In most models of gas exchange at high wind speeds, the effects of breaking waves, spray and bubble clouds are combined into the breaking-wave-mediated transfer velocity, k_b. Then it is assumed that the total gas transfer velocity k can be split up into direct transfer through the surface k_s and the breaking-waves-mediated transfer velocity k_b,

$$k = k_s + k_b, \tag{4}$$

(see Merlivat and Memery, 1983). Examples of parameterizations of k_b can be found in Keeling (1993) and Asher et al. (1996). More complex models are available (see for instance Woolf et al., 2007). All of the models of gas transfer at high wind speeds have in common that the gas exchange of a specific tracer not only depends on the Schmidt number but also on the solubility. Assuming tracers with the same Schmidt number, the transfer velocity due to breaking waves in these empirical models is higher for the tracer with the lower solubility.

3 Method

Classical evasion experiments (see for instance Jähne et al., 1979) were conducted in this study. In an evasion experiment, the decrease in concentration of a tracer, mixed into the water before the start of the experiment, is monitored over time. The simple approach described in Jähne et al. (1979) must be slightly modified and adapted to the Kyoto high-speed wind-wave tank to accommodate for water lost from the system due to spray.

Under the condition of a negligible air-side concentration $\alpha c_a \approx 0$, and small solubility α, as well as the choice of a tracer that is not in the water used to replace the water lost due to spray, the mass balance for a tracer on the water side is found to be

$$V_w \dot{c}_w = -(Ak + \dot{V}_w)c_w. \tag{5}$$

In this equation, the mass of the tracer in the water is expressed using the water-side concentration c_w. A denotes the water surface area, V_w the total water volume, and $\dot{V}_w$ is the rate of water inflow to replace water lost from the flume due to spray being blown out of the tank. The Kyoto high-speed wind-wave tank is an open facility, meaning fresh ambient air is blown over the water surface once and then removed from the system. Choosing a tracer that is not present in ambient air, the condition of a negligible air-side concentration $\alpha c_a \approx 0$, can be met.

Equation (5) can be easily solved,

$$c_w(t) = c_w(0) \cdot \exp\left(-\left(k \cdot \frac{A}{V_w} + \frac{\dot{V}_w}{V_w}\right) \cdot t\right), \tag{6}$$

with $c_w(0)$ being the water-side concentration at time $t = 0$.

The time constant τ of this equation is defined as

$$\frac{1}{\tau} := k \cdot \frac{A}{V_w} + \frac{\dot{V}_w}{V_w}. \tag{7}$$

This time constant τ is acquired from an exponential fit of Eq. (6) to the time series of measured concentrations. The water volume V_w, the water surface area A, as well as the leak rate $\lambda = \dot{V}_w/V_w$ are known or measured during an experiment. The transfer velocity can then be calculated as

$$k = \left(\frac{1}{\tau} - \lambda\right) \cdot \frac{V_w}{A}. \tag{8}$$

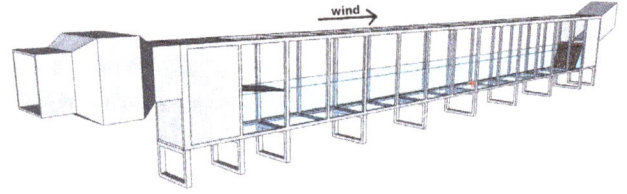

Fig. 2. Schematic view of the flume section of the Kyoto high-speed wind-wave tank. Not shown is the radial fan producing the wind (left side). The red cross marks the approximate sampling position.

4 Experiments

4.1 Tracers

The tracers were chosen such that their diffusivity in water, and thus their Schmidt numbers, were similar, while their solubility differed. Because UV absorption spectroscopy was used to measure tracer concentrations, tracers were chosen which exhibit a high extinction coefficient in the UV range as well as distinctly different spectra . To keep the mass balance described in Sect. 3 simple, the tracers were required to be absent from the ambient air, as well as absent in the tap water. The tracers chosen by these criteria were hexafluorobenzene (HFB) and 1,4-difluorobenzene (DFB). Table 1 lists properties of the tracers as well as carbon dioxide as a reference.

4.2 Experimental setup

4.2.1 The Kyoto high-speed wind-wave tank

The Kyoto high-speed wind-wave tank has a linear flume shape (see Fig. 2). The water flume is 80 cm wide, has a total length of 15.7 m with 12.9 m being exposed to the wind. The total height is 1.6 m, with up to 0.8 m being filled with tap water. The wind is generated by a radial fan. The maximum wind speed that can be reached is $u_{10} = 67.1\ \mathrm{m\,s^{-1}}$. Before the wind enters the air side of the tank, it passes through a honeycomb structure to minimize large eddies. The air is taken from the room surrounding the wind-wave tank and guided out of the building after it has been blown over the water.

There is an external water tank available that holds up to 7 m³ of water, which is connected to the wind-wave flume by two pipes: one pump draws the water out at the downwind end of the flume and into the water tank, and another pump draws the water out of the tank and into the upwind end of the wind-wave flume. For all lower wind speed settings, the amount of water coming out of the lab's water supply lines is sufficient to replace the water lost due to spray. At the highest wind speed setting, the external tank was used as a buffer to keep the water level constant inside the wind-wave tank. Trace gases can be mixed into the water by operating both

Table 1. Molar mass, solubility, diffusivity in water and Schmidt numbers of the tracers hexafluorobenzene (HFB) and 1,4-difluorobenzene (DFB) for a temperature of 20 °C. Also shown is CO_2 for comparison.

Name	M g mol^{-1}	α at 20 °C	D 10^{-5}cm^2 s^{-1}	Sc
HFB	186.1	1.0[a]	0.736[d]	1360[e]
DFB	114.1	3.08[b]	0.815[d]	1225[e]
CO_2	44.01	0.83[c]	1.68[f]	601[f]

[a] calculated from mole fraction solubility from Freire et al. (2005) and vapor pressure from Ambrose et al. (1990). [b] Yaws and Yang (1992). [c] Young (1981). [d] Yaws (1995). [e] calculated from diffusion coefficients taken from Yaws (1995) and water viscosity taken from Kestin et al. (1987). [f] Jähne et al. (1987).

pumps and thus cycling the water between the external tank and the wind-wave flume.

4.2.2 Concentration measurement

Tracer concentrations in the water were monitored using UV absorption spectroscopy. Water was sampled at a fetch of about 6.5 m at a water height of approximately 35 cm with a rate of 7 to 10 L min^{-1}. The approximate sampling location is marked in Fig. 2. Because air bubbles generated by breaking waves would have scattered the light out of the UV spectroscopic measuring cell, it was decided to not spectroscopically analyze the water directly, but to equilibrate the water with a small parcel of air first, and analyze this air. The water extraction and equilibration setup is shown in Fig. 3. A membrane equilibrator called an oxygenator (Jostra Quadrox manufactured by Maquet, Hirrlingen, Germany) was used to equilibrate the water with the air. Because of the large inner surface of the device in relation to its water volume, the time constant for gas equilibration is very fast. Measurements performed by Krall (2013, pp. 56–57) show that the response on a step concentration change of hexafluorobenzene has a time constant between 1.2 and 1.3 min. Because of its higher solubility, the time constant for difluorobenzene should be even faster. Thus the time constant of the gas equilibrator is about five times faster than the shortest e-folding time of hexafluorobenzene gas exchange, which was 6.6 min at the highest wind speed. Therefore, the measured gas transfer velocities are not biased by a limited time constant of the gas equilibrator.

Air is cycled around the closed air loop at a rate of about 150 mL min^{-1}. During the measurements, the valves were set such, that no outside air could enter or leave the air loop. During preparation of the experiment, the valves allowed sampling of ambient air to estimate the background. In addition, the water temperature was monitored.

The gas sampling cell is made of a 1 m-long quartz glass tube with an inner diameter of 3 mm. Light produced by a deuterium lamp enters the tube through a quartz glass lens

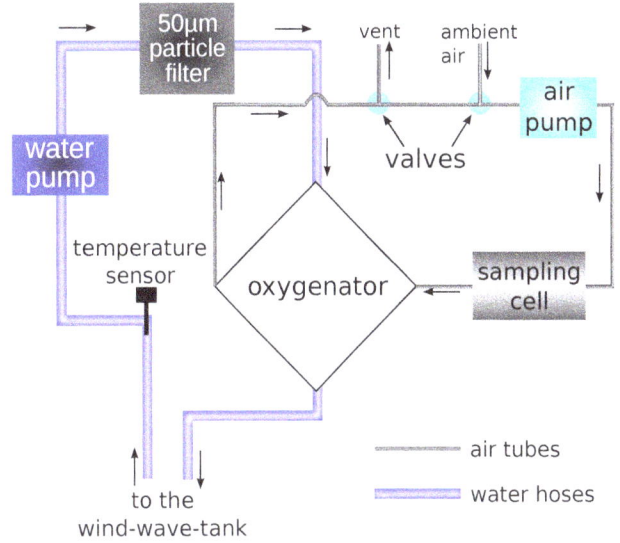

Fig. 3. Gas extraction setup. Water pumped from the wind-wave tank is equilibrated with air using an oxygenator. The air is continuously cycled between the oxygenator and the UV-spectroscopic measuring cell. The valves allow background sampling and are closed during measurements.

with a focal length of 5 cm and a quartz glass window. It leaves the measuring cell through another quartz glass window and lens to be focused on a glass fiber. This glass fiber is connected to a UV spectrometer (Maya2000 Pro by Ocean Optics, Dunedin, USA). This spectrometer can resolve wavelengths from 190.5 to 294.1 nm with a resolution of approximately 0.05 nm. About one spectrum was acquired per second.

During data evaluation, one absorbance value per tracer is calculated from each spectrum in a process described in detail in Krall (2013). Beer's law states that the absorbance A of a tracer is directly proportional to the concentration in the measured air parcel, c_a. According to Henry's law, the air-side concentration is proportional to the water-side concentration c_w; thus $A \propto c_a \propto c_w$. Because only the change in concentration over time is relevant to measure the gas transfer velocity (see Eqs. 6 and 8), no absolute calibration that converts absorbance into the water-side concentrations is needed. Equation (6) can then be converted into the form

$$\frac{c_w(t)}{c_w(0)} = \frac{A(t)}{A(0)} = \exp\left(-\frac{t}{\tau}\right), \tag{9}$$

with the time constant τ that is needed to calculate the gas transfer velocities (see Eq. 8).

4.3 Experimental conditions

A total of 21 experiments at nine different fixed wind speeds were performed. The wind generator's rotational frequency f_{fan} was set and kept constant for each condition. The free-stream wind speed u_{inf}, the air-sided friction velocity u_* as

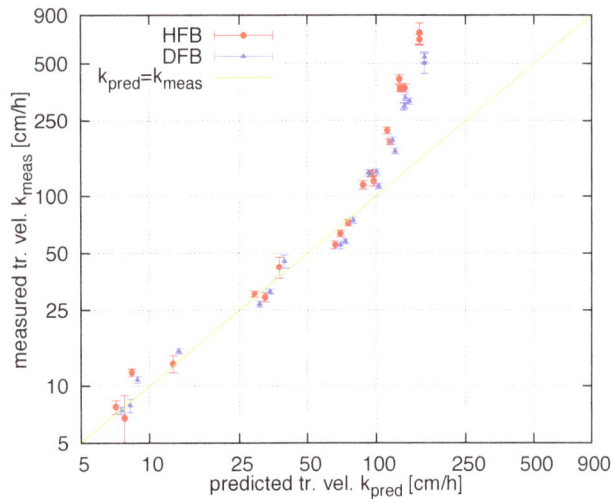

Fig. 4. Measured transfer velocities k_{meas} for HFB and DFB as well as the transfer velocities k_{pred} predicted by Eq. (2) in a double logarithmic plot.

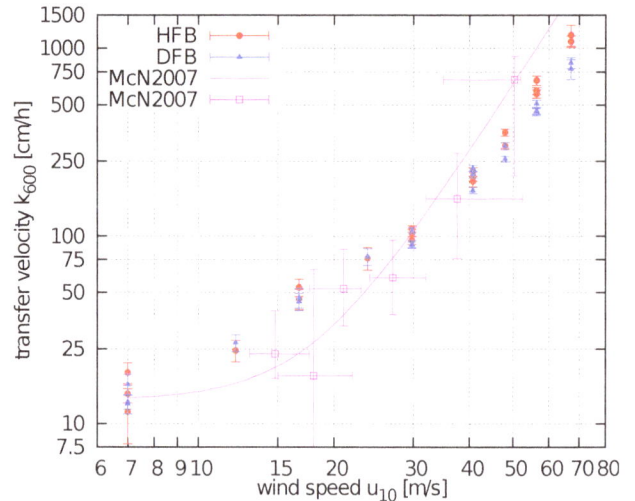

Fig. 5. Measured transfer velocities for HFB and DFB, compared to the data and parameterization by McNeil and D'Asaro (2007), all scaled to a Schmidt number of 600, in a double logarithmic plot.

well as the wind speed at a height of 10 m u_{10}, which is commonly used as a reference, were not measured during the presented campaign, but taken from a table kindly provided by the Japanese colleagues. The wind speed at a height of 10 m u_{10} was extrapolated from the measured friction velocity u_* and free-stream wind speed using a logarithmic wind profile. Water height h_w was measured at the wind inlet before and after each experiment with no wind and no waves. Typically, both water height values differed by no more than 1 %. This ensured that the rate of inflowing water $\dot{V}_w$ was equal to the amount of water lost due to spray as required by the method (see Sect. 3). The conditions used are listed in Table 2. Transfer velocities of both tracers were measured in parallel in each of the experiments, with the exception of one experiment at $f_{fan} = 600$, where only the absorbance time series of DFB could be evaluated.

5   Results

5.1   Comparison with gas exchange model and field measurements

A total of 41 transfer velocities were measured, 21 of which for DFB and 20 for HFB. Figure 4 shows the measured transfer velocities for both tracers versus the transfer velocities predicted by Eq. (2). The momentum transfer resistance parameter β was assumed to be 6.7 (see Krall, 2013). The Schmidt number exponent was chosen to be $n = 0.5$ at medium to high wind speeds. However, to compensate for the smooth water surface visually observed at low fetches during the two lowest wind speeds of 7 m s^{-1} and 12.1 m s^{-1}, the exponent was set to a value of 0.55.

For the low friction velocities of up to $u_* < 6 \, \text{cm s}^{-1}$, which corresponds to a wind speed of $u_{10} < 35 \, \text{m s}^{-1}$, and transfer velocities around 80 cm s^{-1}, the measured transfer velocities agree well with the transfer model's prediction. At higher wind speeds, the measured transfer velocities exceed the ones expected from Eq. (2) by up to around 340 % (HFB) and 220 % (DFB).

Figure 5 shows the transfer velocities, scaled to a Schmidt number of 600 using Schmidt number scaling (Eq. 3) in comparison with the data by McNeil and D'Asaro (2007) acquired on the open ocean, including their proposed parameterization. Within the margin of errors, both data sets agree surprisingly well.

This does not mean that it is possible to transfer these laboratory data directly to the field. The conditions are too different: fresh water was used instead of sea water, the scales of the short-fetch waves in the laboratory are much smaller than at sea, and deep injection of bubbles and the resulting bubble dissolution flux pathway does not occur with a mean water depth of only 0.80 m.

Strictly speaking, scaling to a Schmidt number of 600 is also not correct, because Schmidt number scaling only applies to the transfer across the free-water interface. The different bubble mediated processes scale in a different way, with the tracer's solubility becoming the second key parameter. It can be expected that the applied scaling somewhat underestimates the oxygen–nitrogen-based gas exchange measurements of McNeil and D'Asaro (2007), because these two gases have much lower solubility. Given the limited experimental data from the pilot experiment, it is still the best that can be done. The resulting uncertainty of this approach is probably not larger than the error bars of the field measurements from McNeil and D'Asaro (2007) (Fig. 5).

Table 2. Experimental conditions used at the Kyoto high-speed wind-wave tank. f_{fan} is the frequency of the wind-generating fan, u_* is the friction velocity, u_{inf} denotes the free-stream velocity, and u_{10} is the wind speed at 10 m height. $\dot{V}_{\text{W}}$ is the leak rate. The number of repetitions of each of the conditions is labeled with n. One free-stream velocity u_{inf} was not measured (n.m.).

f_{fan} rpm	u_* cm s^{-1}	u_{inf} m s^{-1}	u_{10} m s^{-1}	$\dot{V}_{\text{W}}$ L min^{-1}	n	notes
100	0.836	4.72	7.0	0	3	
150	1.50	10.36	12.1	0	1	
200	2.34	10.29	16.7	0	2	
250	3.10	n.m.	23.75	0	1	
300	5.19	16.26	29.8	0	3	
400	7.24	22.17	40.7	0	3	
500	8.23	28.47	48.0	3.5	2	
600	9.37	34.75	56.4	14.5	4	only three experiments evaluable for HFB
800	11.5	43.29	67.1	192	2	V_{W} decreased, external tank used

One important conclusion can be drawn nevertheless: if one of the dominant pathways for gas transfer induced by breaking waves were missing in the laboratory experiment, the gas transfer velocities measured there would be significantly smaller than those measured in the field. The experimental data indicate that this is not the case. This encourages further, more detailed wind-wave tunnel experiments, because it is likely that the more important processes are adequately captured in the Kyoto high-speed wind-wave tank. More specifically, this suggests – but still needs to be proven – that the bubble dissolution flux pathway may be not the dominant mechanism at these high wind speeds.

5.2 Enhancement at highest wind speeds

Vlahos and Monahan (2009) and Vlahos et al. (2011) present measured transfer velocities of dimethyl sulfide (DMS), which show a decrease in the gas transfer velocity when bubble clouds are present at high wind speeds. For both tracers used in this study, this decrease was not observed. Up to a wind speed of 35 m s^{-1}, the gas transfer velocity is roughly proportional to $u_{10}^{1.1}$. For higher wind speeds, the proportionality changes to $k \propto u_{10}^3$ for DFB and $k \propto u_{10}^{3.6}$ for HFB (see Fig. 6). This clearly indicates the start of a new regime of air–sea gas exchange starting at around 35 m s^{-1}. In order to compare the theory of Vlahos with our experimental data and to estimate the DMS transfer velocity, it would be necessary to know the total surface of bubbles submerged by breaking waves in relation to the surface area of the facility. These data are not available from this pilot experiment. All that can be said is that difluorobenzene and hexafluorobenzene are flat symmetrical molecules with certainly a much lower surface activity than DMS. Thus it is neither possible to verify nor to dismiss the theory of Vlahos.

At the highest wind speeds, the transfer velocity of HFB increases stronger than the one of DFB, as indicated by the different slopes in Fig. 6. To quantify this, an enhancement

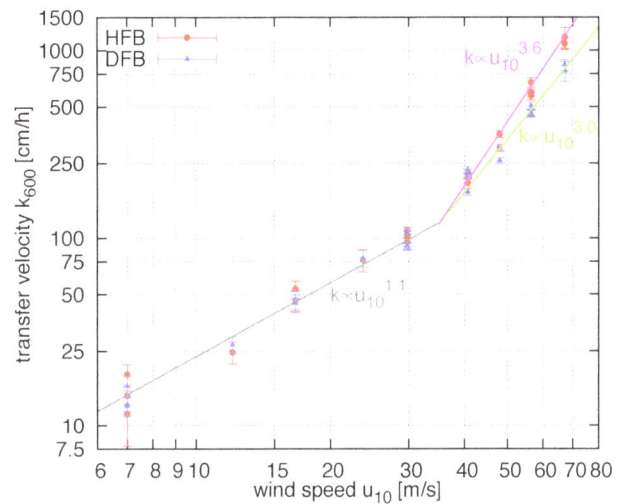

Fig. 6. Measured transfer velocities for HFB and DFB scaled to a Schmidt number of 600. Also shown are lines indicating proportionalities of the transfer velocities k_{600} to u_{10}^x with x depending on the wind speed range and the tracer.

factor E_{f} can be defined by

$$E_{\text{f}} := \frac{k_{600,\text{HFB}} - k_{600,\text{DFB}}}{k_{600,\text{DFB}}} \cdot 100\,\% . \tag{10}$$

Figure 7 shows the enhancement factor E_{f}, averaged on a condition basis. Up to a wind speed of around 40 m s^{-1}, no enhancement is observed. Above 40 m s^{-1}, however, the transfer velocity of HFB is up to 40 % larger than that of DFB with a clear wind speed dependence. This enhancement is expected from bubble models (see Sect. 2), with the less soluble tracer HFB ($\alpha = 1.0$ at 20 °C) having larger transfer velocities than the slightly higher soluble tracer DFB ($\alpha = 3.08$ at 20 °C).

The tracers used in this pilot study span only a small fraction of the Schmidt number – solubility parameter space, (Fig. 8). In particular, gases with low solubilities are missing

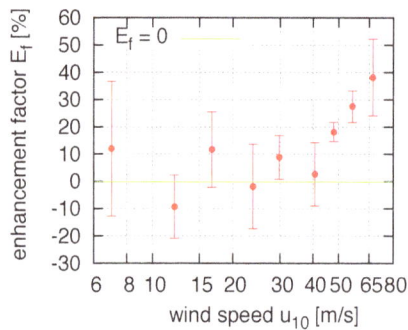

Fig. 7. Mean enhancement of the transfer velocity of HFB over that of DFB, both scaled to a Schmidt number of 600. An E_f of 0 means no enhancement.

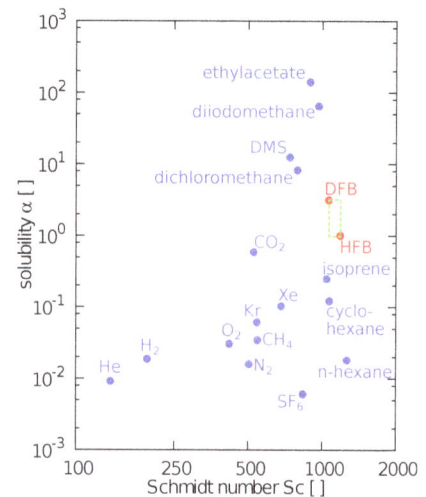

Fig. 8. Double logarithmic Schmidt number–solubility diagram of some environmentally important tracers for a temperature of 25 °C. The tracers used in this study, hexafluorobenzene (HFB) and 1,4-difluorobenzene (DFB), cover a very limited parameter range, marked by the dashed green rectangle.

where bubble-mediated gas transfer can be expected to be even higher. In addition, no bubble density spectra are available. Therefore, a more detailed analysis would make no sense and is omitted in this paper. With such limited data, any model on bubble-mediated gas transfer can be fitted to the data (Krall, 2013) with the result that no conclusive statements are possible. In particular any extrapolation to a gas with lower or higher solubility than the tracers used in this study is highly speculative and very likely incorrect.

6 Conclusions and outlook

The transfer velocities at hurricane strength wind speeds were found to be extremely high. The measured transfer velocities were found to be in agreement with the only other data set of gas transfer at extreme wind speeds (McNeil and D'Asaro, 2007). In wind speeds higher than around 35–40 m s^{-1}, where frequent large-scale wave breaking with bubble entrainment and spray generation occurs, the correlation between gas transfer velocities and wind speed was found to become steeper, indicating a new regime of air–sea gas exchange. The steepness of the relationship between the gas transfer velocity and the wind speed could be linked to the solubility of the tracer. The lower the solubility, the higher the transfer velocities measured.

The results of this pilot study confirm that it is possible to measure realistic air–sea gas exchange velocities in a wind-wave tank at hurricane wind speeds with the method described in this paper. However, due to the aforementioned limitations in solubility and Schmidt number, a physical interpretation as well as physics-based modeling will have to be suspended until detailed measurements of bubble and spray densities and of turbulence have been conducted. For a detailed and robust parameterization, it is also required to perform experiments with many tracers simultaneously, which cover the largest possible range of solubilities and Schmidt numbers. If these requirements are met, it would be highly possible to estimate gas transfer velocities at sea from laboratory measurements.

Acknowledgements. We cordially thank all members of the Environmental Fluids and Thermal Engineering Lab at Kyoto University under the lead of S. Komori for providing excellent working conditions and assistance during preparation and measurements, as well as N. Takagaki and K. Iwano for making wind speed, friction velocity and free-stream velocity data available. We are grateful for the assistance of W. Mischler during all stages of the measurements. Constructive comments of two anonymous reviewers as well as our editor D. Woolf helped to improve this paper. The financial support for this project by Mobility Networks within the Institutional Strategy ZUK49 "Heidelberg: Realizing the Potential of a Comprehensive University" and by the German Federal Ministry of Education and Research (BMBF) joint project "Surface Ocean Processes in the Anthropocene" (SOPRAN, FKZ 03F0611F) is gratefully acknowledged.

Edited by: D. Woolf

References

Ambrose, D., Ewing, M. B., Ghiassee, N. B., and Sanchez Ochoa, J. C.: The ebulliometric method of vapour pressure measurement: vapour pressures of benzene, hexafluorobenzene, and naphthalene, J. Chem. Thermodyn., 22, 589–605, 1990.

Asher, W. E., Higgins, B. J., Karle, L. M., Farley, P. J., Sherwood, C. R., Gardiner, W. W., Wanninkhof, R., Chen, H., Lantry, T., Steckley, M., Monahan, E. C., Wang, Q., and Smith, P. M.: Measurement of gas transfer, whitecap coverage, and brightness temperature in a surf pool: an overview of WABEX-93, in: Air-Water Gas Transfer, Selected Papers, 3rd Intern. Symp. on Air-Water Gas Transfer, edited by: Jähne, B. and Monahan, E., 207–216, AEON, Hanau, 1995.

Asher, W. E., Karle, L. M., Higgins, B. J., Farley, P. J., Monahan, E. C., and Leifer, I. S.: The influence of bubble plumes on air-seawater gas transfer velocities, J. Geophys. Res., 101, 12027–12041, 1996.

Bates, N. R., Knap, A. H., and Michaels, A. F.: Contribution of hurricanes to local and global estimates of air-sea exchange of CO_2, Nature, 395, 58–61, 1998.

Coantic, M.: A model of gas transfer across air–water interfaces with capillary waves, J. Geophys. Res., 91, 3925–3943, 1986.

Deacon, E. L.: Gas transfer to and across an air-water interface, Tellus, 29, 363–374, 1977.

Farmer, D. M., McNeil, C. L., and Johnson, B. D.: Evidence for the importance of bubbles in increasing air-sea gas flux, Nature, 361, 620–623, 1993.

Freire, M. G., Razzouk, A., Mokbel, I., Jose, J., Marrucho, I. M., and Coutinho, J. A. P.: Solubility of hexafluorobenzene in aqueous salt solutions from (280 to 340) K, J. Chem. Eng. Data, 50, 237–242, 2005.

Jähne, B., Münnich, K. O., and Siegenthaler, U.: Measurements of gas-exchange and momentum-transfer in a circular wind-water tunnel, Tellus, 31, 321–329, 1979.

Jähne, B., Heinz, G., and Dietrich, W.: Measurement of the diffusion coefficients of sparingly soluble gases in water, J. Geophys. Res., 92, 10767–10776, 1987.

Jähne, B., Libner, P., Fischer, R., Billen, T., and Plate, E. J.: Investigating the transfer process across the free aqueous boundary layer by the controlled flux method, Tellus, 41B, 177–195, 1989.

Keeling, R. F.: On the role of large bubbles in air-sea gas exchange and supersaturation in the ocean, J. Marine Res., 51, 237–271, 1993.

Kestin, J., Sokolov, M., and Wakeham, W. A.: Viscosity of liquid water in the range −8 °C to 150 °C, J. Phys. Chem. Ref. Data, 7, 941–948, 1978.

Komori, S. and Shimada, T.: Gas transfer across a wind-driven air-water interface and the effects of sea water on CO_2 transfer, in: Air-Water Gas Transfer, Selected Papers, 3rd Intern. Symp. on Air-Water Gas Transfer, edited by: Jähne, B. and Monahan, E., 553–569, AEON, Hanau, 1995.

Krall, K. E.: Laboratory Investigations of Air-Sea Gas Transfer under a Wide Range of Water Surface Conditions, Ph.D. Thesis, available at: http://www.ub.uni-heidelberg.de/archiv/14392 (last access: 14 April 2014), Institut für Umweltphysik, Fakultät für Physik und Astronomie, University of Heidelberg, 2013.

Liss, P. S. and Merlivat, L.: Air-sea gas exchange rates: Introduction and synthesis, in: The role of air-sea exchange in geochemical cycling, edited by: Buat-Menard, P., 113–129, Reidel, Boston, MA, 1986.

McGillis, W. R., Edson, J. B., Hare, J. E., and Fairall, C. W.: Direct covariance air-sea CO_2 fluxes, J. Geophys. Res., 106, 16729–16745, 2001.

McNeil, C. and D'Asaro, E.: Parameterization of air sea gas fluxes at extreme wind speeds, J. Marine Syst., 66, 110–121, 2007.

Memery, L. and Merlivat, L.: Modelling of gas flux through bubbles at the air-water interface, Tellus B, 37, 272–285, 1985.

Merlivat, L. and Memery, L.: Gas exchange across an air-water interface: experimental results and modeling of bubble contribution to transfer, J. Geophys. Res., 88, 707–724, 1983.

Monahan, E. C. and Spillane, M. C.: The role of whitecaps in air-sea gas exchange Gas transfer at water surfaces, in: Gas transfer at water surfaces, edited by: Brutsaert, W. and Jirka, G. H., 495–503, 1984.

Mori, N., Imamura, M., and Yamamoto, R.: An experimental study of bubble mediated gas exchange for a single bubble, in: Gas Transfer at Water Surfaces, edited by: Donelan, M. A., Drennan, W. M., Saltzman, E. S., and Wanninkhof, R., Geophysical Monograph, 127, 311–313, 2002.

Nightingale, P. D., Malin, G., Law, C. S., Watson, A. J., Liss, P. S., Liddicoat, M. I., Boutin, J., and Upstill-Goddard, R. C.: In situ evaluation of air-sea gas exchange parameterization using, novel conservation and volatile tracers, Glob. Biogeochem. Cy., 14, 373–387, 2000.

Vasconcelos, J. M. T., Orvalho, S. P., and Alves, S. S.: Gas-liquid mass transfer to single bubbles: Effect of surface,contamination, AIChE Journal, 48, 1145–1154, 2002.

Vlahos, P. and Monahan, E. C.: A generalized model for the air-sea transfer of dimethyl sulfide at high wind speeds, Geophys. Res. Lett., 36, L21605, doi:10.1029/2009GL040695, 2009.

Vlahos, P., Monahan, E. C., Huebert, B. J., and Edson, J. B.: Wind-dependence of DMS transfer velocity: Comparison of model with recent southern ocean observations, in: Gas Transfer at Water Surfaces 2010, edited by: Komori, S., McGillis, W., and Kurose, R., 313–321, 2011.

Wanninkhof, R.: Relationship between wind speed and gas exchange over the ocean, J. Geophys. Res., 97, 7373–7382, doi:10.1029/92JC00188, 1992.

Wanninkhof, R., Asher, W., and Monahan, E.: The influence of bubbles on air-water gas exchange: results from gas transfer experiments during WABEX-93, in: Air-Water Gas Transfer, Selected Papers, 3rd Intern. Symp. on Air-Water Gas Transfer, edited by: Jähne, B. and Monahan, E., 239–254, AEON, Hanau, 1995.

Wanninkhof, R., Asher, W. E., Ho, D. T., Sweeney, C., and McGillis, W. R.: Advances in quantifying air-sea gas exchange and environmental forcing, Annu. Rev. Mar. Sci., 1, 213–244, 2009.

Woolf, D., Leifer, I., Nightingale, P., Rhee, T., Bowyer, P., Caulliez, G., de Leeuw, G., Larsen, S., Liddicoat, M., Baker, J., and Andreae, M.: Modelling of bubble-mediated gas transfer: Fundamental principles and,a laboratory test, J. Marine Syst., 66, 71–91, 2007.

Yaws, C. L.: Handbook of Transport Property Data, Gulf Publishing Company, 1995.

Yaws, C. L. and Yang, H.-C.: Henry's law constant for compound in water, in: Thermodynamic and Physical Property Data, edited by: Yaws, C. L., Gulf Publishing Company, 181–206, 1992.

Young, C. L. (Ed.): IUPAC Solubility Data Series, Vol. 8, Oxides of Nitrogen, Pergamon Press, Oxford, England, 1981.

Appendix A

Measured transfer velocities

Table A1 summarizes all measured experimental data: friction velocity, wind speed, mean water temperature, as well as both tracer's transfer velocities and Schmidt numbers.

Table A1. Transfer velocities k (not scaled to $Sc = 600$) measured in the Kyoto high-speed wind-wave tank for tracers 1,4-difluorobenzene (DFB) and hexafluorobenzene (HFB) and their respective uncertainties Δk. "n.m." means not measured. The wind speed u_{10} and water-sided friction velocity u_*, kindly provided by the Japanese colleagues, are also given. T denotes the mean water temperature during the measurement. Schmidt numbers Sc of the tracers at this temperature are also shown.

Date (yyyy/mm/dd)	$u_{*,w}$ cm s^{-1}	u_{10} m s^{-1}	k_{HFB} cm h^{-1}	Δk_{HFB} cm h^{-1}	k_{DFB} cm h^{-1}	Δk_{DFB} cm h^{-1}	T $^{\circ}$C	Sc_{HFB}	Sc_{DFB}	n
2011/10/27	9.38	56.4	369.5	14.8	332.1	12.8	17.5	1555	1403	0.5
2011/10/28	7.25	40.7	120.4	7.2	113.5	3.6	17.1	1590	1434	0.5
2011/10/28	9.38	56.4	415.3	23.3	299.2	14.0	17.1	1590	1434	0.5
2011/10/31	2.34	16.7	29.35	1.67	31.40	1.00	18.3	1489	1344	0.5
2011/10/31	8.23	48.0	222.8	9.4	196.9	8.6	17.5	1555	1403	0.5
2011/11/02	5.19	29.8	72.77	2.75	74.86	2.8	19.9	1367	1234	0.5
2011/11/02	8.23	48.0	193.6	6.7	173.0	5.8	18.5	1473	1329	0.5
2011/11/04	0.84	7.0	11.80	0.56	10.84	0.42	19.5	1396	1260	0.55
2011/11/04	9.38	56.4	373.1	19.5	318.1	11.4	19.2	1419	1280	0.5
2011/11/08	5.19	29.8	63.86	2.58	58.13	1.75	17.0	1598	1442	0.5
2011/11/10	11.5	67.1	726.2	92.6	505.3	60.7	17.25	1577	1422	0.5
2011/11/11	9.38	56.4	n.m.	n.m.	299.9	9.7	17.25	1577	1422	0.5
2011/11/14	0.84	7.0	6.77	2.18	7.90	0.65	17.0	1598	1442	0.55
2011/11/15	7.25	40.7	132.4	5.9	134.7	4.8	16.2	1671	1507	0.5
2011/11/16	2.34	16.7	30.5	1.1	26.98	1.0	14.5	1837	1657	0.5
2011/11/16	7.25	40.7	114.9	6.2	133.1	5.3	13.25	1973	1779	0.5
2011/11/17	1.50	12.1	13.1	1.4	15.29	0.48	14.1	1880	1695	0.55
2011/11/17	3.10	23.7	42.3	5.3	45.37	3.71	13.3	1967	1774	0.5
2011/11/18	0.84	7.0	7.75	0.65	7.42	0.34	14.15	1874	1690	0.55
2011/11/19	5.19	29.8	55.6	2.8	55.59	2.8	15.2	1766	1593	0.5
2011/11/19	11.5	67.1	671.7	45.0	543.6	34.2	17.25	1577	1422	0.5

9

Tidal forcing, energetics, and mixing near the Yermak Plateau

I. Fer[1,3]**, M. Müller**[2]**, and A. K. Peterson**[1,3]

[1]Geophysical Institute, University of Bergen, Bergen, Norway
[2]Norwegian Meteorological Institute, Oslo, Norway
[3]Bjerknes Centre for Climate Research, Bergen, Norway

Correspondence to: I. Fer (ilker.fer@uib.no)

Abstract. The Yermak Plateau (YP), located northwest of Svalbard in Fram Strait, is the final passage for the inflow of warm Atlantic Water into the Arctic Ocean. The region is characterized by the largest barotropic tidal velocities in the Arctic Ocean. Internal response to the tidal flow over this topographic feature locally contributes to mixing that removes heat from the Atlantic Water. Here, we investigate the tidal forcing, barotropic-to-baroclinic energy conversion rates, and dissipation rates in the region using observations of oceanic currents, hydrography, and microstructure collected on the southern flanks of the plateau in summer 2007, together with results from a global high-resolution ocean circulation and tide model simulation. The energetics (depth-integrated conversion rates, baroclinic energy fluxes and dissipation rates) show large spatial variability over the plateau and are dominated by the luni-solar diurnal (K_1) and the principal lunar semidiurnal (M_2) constituents. The volume-integrated conversion rate over the region enclosing the topographic feature is approximately 1 GW and accounts for about 50 % of the M_2 and approximately all of the K_1 conversion in a larger domain covering the entire Fram Strait extended to the North Pole. Despite the substantial energy conversion, internal tides are trapped along the topography, implying large local dissipation rates. An approximate local conversion–dissipation balance is found over shallows and also in the deep part of the sloping flanks. The baroclinic energy radiated away from the upper slope is dissipated over the deeper isobaths. From the microstructure observations, we inferred lower and upper bounds on the total dissipation rate of about 0.5 and 1.1 GW, respectively, where about 0.4–0.6 GW can be attributed to the contribution of hot spots of energetic turbulence. The domain-integrated dissipation from the model is close to the upper bound of the observed dissipation, and implies that almost the entire dissipation in the region can be attributed to the dissipation of baroclinic tidal energy.

1 Introduction

The Yermak Plateau (YP, Fig. 1), located northwest of Svalbard, is the main topographic obstacle to the warm Atlantic inflow into the Arctic Ocean. In the vicinity of Svalbard, the marginal ice zone (MIZ), the transition region between open water and dense ice cover, is typically located over the plateau, indicating substantial oceanic heat loss to the melting of sea ice in the region. The YP is identified as a region of enhanced tidal variability for both diurnal and semidiurnal tides (Hunkins, 1986; Padman et al., 1992; D'Asaro and Morison, 1992; Plueddemann, 1992). Strong tidal currents over the sloping flanks of the plateau lead to increased internal wave activity and intense diapycnal mixing of water properties (Padman and Dillon, 1991; Wijesekera et al., 1993; Fer et al., 2010). The mixing in this region is of particular interest because it can contribute to the cooling of the West Spitsbergen Current, which carries warm and saline Atlantic Water through Fram Strait. The heat removed from the Atlantic Water layer can influence the inflowing water mass properties to the Arctic Ocean as well as the regional ice cover north of Svalbard.

The breaking of internal waves is a major source of turbulence-driven diapycnal mixing in the ocean. One generation mechanism of internal waves is the baroclinic response to barotropic tidal flow of a stratified water column over favourable topography, such as continental slopes, isolated ridges, or areas of enhanced seafloor roughness (Garrett and Kunze, 2007). Near sites of internal tide generation, some of the energy dissipates locally through the breaking of high-mode, small-scale waves (e.g. Klymak et al., 2008), whereas the remainder propagates away as low-mode internal tides. The propagation of linear plane internal tides is possible; however, only equatorward of the critical latitude at which the Coriolis parameter, f, equals the tidal wave frequency, ω. The Coriolis parameter changes with the latitude, Ψ, through $f = 2\Omega \sin \Psi$, where $\Omega = 7.292 \times 10^{-5} \, \text{s}^{-1}$ is the Earth's angular rotation. The critical latitudes for the principal lunar semidiurnal tide M_2 and the diurnal tides K_1 and O_1 are near $74°30'$ and $30°$, respectively. Poleward of the critical latitude the wave equation changes form from hyperbolic to elliptic, and no progressive linear plane wave solution is allowed (Vlasenko et al., 2005). The solution is evanescent, exists locally where the barotropic-to-baroclinic energy conversion occurs, and decays exponentially in the vertical and in the cross-slope direction. Baroclinic disturbances in response to tidal flow over topography above the critical latitude are thus topographically trapped near the generation site (a continental slope, a ridge, or a seamount). The energy propagation of topographically trapped waves is possible along the slope, around the topographic feature with negligible radiation away in the cross-slope direction. This is analogous to the sub-inertial baroclinic trapped waves propagating around isolated seamounts (Brink, 1989). Trapped tides dissipate their energy locally, or elsewhere along the topography, leading to substantial vertical mixing (Padman and Dillon, 1991; Tanaka et al., 2010; Johnston and Rudnick, 2014).

In general, the generation of near-inertial and sub-inertial internal tides has not obtained much attention, although there are strong signs that they have the potential to impact decadal climate variations (Tanaka et al., 2012; Müller, 2013). A recent numerical study in the Barents Sea shows internal M_2 tides with bottom-enhanced energy density and dissipation rates, trapped in the vicinity of their generation sites (Kagan and Sofina, 2014). The bottom-trapped tides in the Arctic Ocean are of particular importance because most of the Arctic Basin is located north of critical latitude of the most energetic tidal constituent M_2. Falahat and Nycander (2014) computed an area-integrated energy flux of about 1.1 GW (1 GW $\equiv 10^9$ W) for topography-trapped internal tides (sum of the K_1, O_1, and M_2 constituents, where about 70 % is M_2). Estimates from global numerical models of the conversion rates of barotropic-to-baroclinic tidal energy in the Arctic region are roughly 5 GW (Simmons et al., 2004; Müller, 2013).

At the latitudes of the YP, except from the principal solar semidiurnal tide (S_2, with critical latitude near $85°$), all diurnal frequencies and the principal M_2 semidiurnal frequency are sub-inertial; locally generated internal tides are trapped. The critical latitude also serves as a turning latitude for low-mode internal tides generated elsewhere below

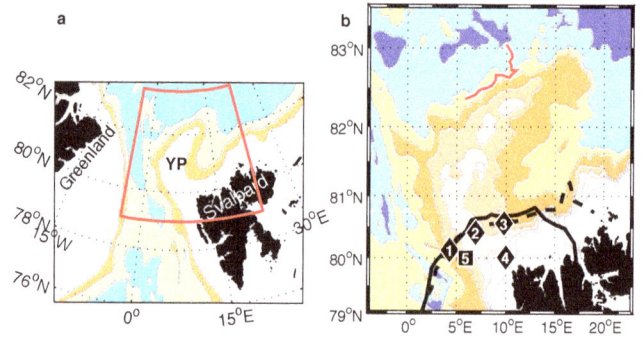

Figure 1. **(a)** Location map of the study area with the Yermak Plateau (YP) enlarged in **(b)**. Isobaths are drawn at 1000 m intervals (Global Relief Model, ETOPO5) in **(a)**, and at 250 m intervals to 1000 m followed by 1000 m intervals (ETOPO1; Amante and Eakins, 2009) in **(b)**. Stations 1–5 are marked. Station 5 is co-located with the short-term mooring (rectangle). The drift of the CEAREX-O camp is shown by the red line on the northern flanks. The ice edge digitized from high-resolution ice charts (http://polarview.org/services/hric.htm) provided by the Norwegian Meteorological Institute are shown for 23 (dashed) and 25 July 2007 (solid).

the critical latitude, and propagating poleward. At the YP, M_2 frequency is only slightly sub-inertial ($f \approx 1.025\omega_{M_2}$), and an anti-cyclonic relative vorticity can shift the effective Coriolis frequency, f_e, relaxing the trapping and the turning latitude. For $f_e < \omega_{M_2}$, progressive solutions are allowed: so-called near-inertial vortex-*trapped* internal waves (Kunze and Toole, 1997). As we show in the following, the YP is a site of substantial barotropic-to-baroclinic energy conversion. It is thus expected that the baroclinic energy extracted from the barotropic tide cannot propagate away and likely dissipates locally, contributing to turbulent mixing where the internal tide is forced. This is supported by observations in the YP region that show localized energetic turbulence and mixing (Padman and Dillon, 1991).

Earlier investigations typically used observations of hydrographic properties and water column velocity sampled from ice camps or by autonomous buoys, see Padman et al. (1992) and Fer et al. (2010) for an overview. Measurements of ocean currents from a drifting buoy showed that energy increased threefold from its level in the Nansen Basin as the buoy passed over the YP (Plueddemann, 1992), dominated by diurnal and near-inertial (semidiurnal) frequencies. Energetic bursts of near-inertial internal wave packets were inferred to propagate upwards, presumably generated by interaction of the barotropic tide with bottom topography. Similarly, D'Asaro and Morison (1992) identified upward propagating near-inertial waves and enhanced eddy diffusivity over the central plateau. Average ocean-to-ice heat fluxes of 22 W m^{-2} were inferred from an automated buoy as it drifted over the northern YP (McPhee et al., 2003). In the pycnocline above the Atlantic Water layer, Padman and Dillon

Table 1. Summary of station position, sampling duration, and vertical coverage; 5-M is the mooring located near station 5. Water depth is the mean ± 1 SD over the station duration. The corresponding stretched depth (in stretched metre; sm) and the covered fraction are listed for a range of 3–500 m for MSS and 12–360 m for the VM-ADCP.

Station	Latitude (N)	Longitude (E)	Duration (h) MSS/ADCP	Water depth (m)	Stretched depth (sm)	Coverage (%)
1	80°7.95′	4°18.10′	24.1/24.3	1253 ± 57	1147	64/52
2	80°25.14′	6°55.24′	23.1/24.7	637 ± 9	715	84/67
3	80°34.00′	9°46.73′	22.7/8.8	1000 ± 86	828	66/50
4	80°0.57′	9°59.25′	24.4/18.8	484 ± 4	567	99/73
5	80°0.10′	5°58.10′	24.2/4.6	879 ± 29	988	70/49
5-M	79°59.78′	5°55.95′	195.5/15.1[a]	889	990	75[b]

[a] Duration of Microcat sampling/duration of Longranger sampling.
[b] Percentage of the stretched depth covered by the moored Microcats.

(1991) observed an upward heat flux of $25\,\mathrm{W\,m^{-2}}$, which decreased to $6\,\mathrm{W\,m^{-2}}$ below the surface mixed layer. These mixing rates and heat fluxes are strong and comparable to those observed over the shelves north of Svalbard and in proximity to the West Spitsbergen Current branches (Fer and Sundfjord, 2007; Sirevaag and Fer, 2009). Here, we hypothesize that spatially varying tides and their baroclinic response over topography are responsible for the observed variability and mixing near the YP.

In this paper we discuss tidal forcing and the role of tides in mixing near the YP. Our study differs from the earlier work in that we report observations from stations, each occupied typically for 1 day, and an 8-day-long time series from a bottom-anchored mooring. The data set is presented earlier in Fer et al. (2010, hereafter F2010), who reported strong variability in the internal wave activity and vertical mixing at the YP. However, the rate of conversion of surface tide energy to baroclinic energy, the spatial distribution of the baroclinic energy fluxes, and the energy dissipation rates are unknown. The limited observational data set is therefore supplemented with results from a global high-resolution ocean circulation and tide model simulation (STORMTIDE, Müller et al., 2012), to obtain barotropic-to-baroclinic energy conversion rates and baroclinic energy fluxes in the region. A thorough evaluation of model–observation comparison and model performance is not intended in this paper. Our approach is to use the (limited) observations to document the substantial tidal variability and levels of turbulence mixing, and use the tidal model results to infer integrated, regional energetics.

2 Methods

2.1 Observational data

The site and measurement details are described fully in F2010; only a brief summary is given here. Observations were made in summer 2007 from the R/V *Håkon Mosby* in southern YP. Sampling was limited by the ice edge in the MIZ, and included five stations and an 8-day-long time series from a mooring (Fig. 1 and Table 1). Each station was occupied for approximately 1 day. Full-depth CTD (conductivity, temperature, depth) profiles were collected using a Sea-Bird Scientific SBE911 plus system. At each station ocean microstructure profiles were taken approximately every 30–60 min (222 casts in total) in the upper 520 m using a loosely tethered free-fall profiler equipped with shear probes (Micro-Structure Sonde, MSS, manufactured by ISW Wassermesstechnik, Germany). Velocity in the water column was measured by a vessel-mounted Teledyne RD Instruments (RDI) Narrowband 150 kHz acoustic Doppler current profiler (VM-ADCP) in the upper 360 m. Additionally, at station 1, six velocity profiles at approximately 4 h interval were collected to the bottom, using eXpendable Current Profilers (XCPs).

The mooring was deployed at a water depth of 889 m, co-located with station 5. The mooring line was instrumented with 19 SBE37 Microcats recording conductivity, temperature and (16 of them) pressure, distributed between 99 and 873 m; a Teledyne RDI 75 kHz Longranger ADCP at approximately 2 m above the bottom looking upward and sampling in 8 m bins; and an Aanderaa (Xylem Inc.) Seaguard current meter at 23 m depth. Sampling interval was 1 min for all moored instruments. Time series of the depth of each instrument is inferred from the pressure record (the depth of the two Microcats without pressure sensors were interpolated from the adjacent sensors). The resulting time-depth structure of temperature and salinity is then gridded onto 20 m regular vertical spacing. This processing accounts for the mooring motion. With the exception of the Longranger that stopped after 15 h because of a leakage, all moored instruments sampled throughout the deployment.

Post-processing of the CTD and current measurements follows common procedures and is briefly summarized in F2010. Post-processing of the microstructure data follows Fer (2006), see also F2010. Profiles of dissipation rate of turbulent kinetic energy (TKE) per unit mass, ε, are obtained from the shear probes to a noise level of $5 \times 10^{-10} \, \mathrm{W \, kg^{-1}}$. The reduced data set collected from the vessel and used in the analysis includes current profiles from the VM-ADCP (hourly temporal and 4 m vertical average) and XCPs (2 m vertical average), temperature, salinity and potential density anomaly, σ_θ, profiles from the ship's CTD and from the MSS (1 m vertical average), and dissipation rate ε from the MSS (1 m vertical average). The data set from the mooring includes hourly averaged and vertically gridded horizontal velocity and σ_θ profiles at 8 and 20 m vertical resolution, respectively.

2.2 Numerical modelling

The numerical model is a global high-resolution ocean circulation and tide model coupled to a thermodynamic sea-ice component (STORMTIDE, Müller et al., 2012). The STORMTIDE model is formulated on a global tripolar grid with an average horizontal resolution of about 10 km, which becomes as small as 5 km in high latitudes. Thus, the mesoscale ocean circulation is implicitly resolved. There are 40 vertical z levels and the time step of the numerical scheme is 600 s, sufficient to simulate the low-mode internal tides. The model physics and numerics are based on the Max Planck Institute Ocean Model (Jungclaus et al., 2006). The astronomical tidal forcing is described by ephemerides and represents the complete lunisolar tidal forcing of second degree. The model uses no internal wave drag, i.e. no additional energy sink for barotropic tides in the deep ocean. The conversion of barotropic-to-baroclinic tidal energy (Sect. 3.1) amounts to 1.14 TW (1 TW $\equiv 10^{12}$ W) in the deep ocean, and is in the range of previous estimates from models, theory, and observations (Egbert and Ray, 2000; Simmons et al., 2004; Nycander, 2005).

A 10-year STORMTIDE simulation is used in the present study, and a detailed description on the set-up of the model simulation including atmospheric forcing, restoring schemes, and physical parametrization can be found in Müller et al. (2014). The hourly output of 32 days of sea level and velocities, starting 1 January, has been used for a harmonic analysis with respect to several tidal constituents (Foreman et al., 2009). Global model products of tidal velocities, energetics, and sea level are available for download at the World Data Centre for Climate (see Müller et al., 2014, for details). The barotropic and low-mode internal tides have been evaluated in Müller et al. (2012). It has been shown that the surface signature of internal tide, evaluated by sea surface satellite products, is reasonably well captured. The barotropic-tide-induced sea level variability is captured to 93 % and a recent barotropic tide model intercomparison shows that STORMTIDE ranks similar to other modern hydrodynamic tide models (Stammer et al., 2014).

3 Methods of analysis

In order to characterize the tidal forcing and energetics at the YP region and at the observation stations, the STORMTIDE and the cruise data, respectively, are used. The methods involved are summarized below. The analysis is based on the baroclinic perturbation fields (e.g. horizontal velocity and pressure) at tidal frequency which are obtained using harmonic analysis of the model and observational baroclinic perturbation fields. The baroclinic perturbations are indicated by a prime; purely sinusoidal baroclinic fluctuations generated using the harmonic analysis results are indicated by a tilde over the corresponding variable.

3.1 Tidal energy conversion and baroclinic energy fluxes from STORMTIDE

The simulated tidal energy flux and conversion of barotropic-to-baroclinic tidal energy are computed from the model by using the derivations presented in Kang and Fringer (2012). The vertically integrated, barotropic-to-baroclinic tidal energy conversion rates are obtained from

$$C = \langle \overline{g\widetilde{\rho}W} \rangle_t, \tag{1}$$

where $\widetilde{\rho}$ is the density perturbation associated with the tidal motion, g the gravitational acceleration, and $\langle \cdot \rangle_t$ and the overbar denote averages in time and integral in the vertical, respectively. Using the model bottom topography and the barotropic tidal currents, the barotropic vertical velocity $W(z)$ at depth z is defined as

$$W(z) = \frac{z}{H} \left(\boldsymbol{u}_{\mathrm{BT}} \cdot \nabla_{\mathrm{H}} H \right), \tag{2}$$

where H is the total water depth, ∇_{H} is the horizontal differential operator, and $\boldsymbol{u}_{\mathrm{BT}}$ the horizontal barotropic tidal current vector.

The vertically integrated, baroclinic tidal energy flux with zonal and meridional components, $\boldsymbol{F} = (F_u, F_v)$, is calculated from

$$\boldsymbol{F} = \langle \overline{\widetilde{p}(z,t)\widetilde{\boldsymbol{u}}(z,t)} \rangle_t, \tag{3}$$

where $\widetilde{p}$ is the perturbation pressure associated with tidal motions and $\widetilde{\boldsymbol{u}}$ is the baroclinic tidal current vector, both obtained using harmonic analysis of the model output. At the latitudes of the YP, $\omega_{M_2} < f$ and the internal wave solution is evanescent (the horizontal wave number is imaginary and the group velocity, $\boldsymbol{c}_{\mathrm{g}}$, is zero). The common approach of relating $\widetilde{p}\widetilde{\boldsymbol{u}}$ to the baroclinic energy flux, $E\boldsymbol{c}_{\mathrm{g}}$ where E is the energy density, should be interpreted with caution (see Sect. 3.3 for discussion). We further derive baroclinic radiation defined as the horizontal divergence of the baroclinic energy flux, $\nabla_{\mathrm{H}} \cdot \boldsymbol{F}$, and define the depth-integrated dissipation, $C - \nabla_{\mathrm{H}} \cdot \boldsymbol{F}$, as the difference between tidal energy conversion and radiation (e.g. Kang and Fringer, 2012).

3.2 Energy and energy flux from observations

The calculation of the baroclinic energetics follows the common methods, see, e.g., Kunze et al. (2002) and Nash et al. (2005), and also accounts for isopycnal heaving by movement of the free surface (Kelly et al., 2010). For a sinusoidal wave, the calculation requires perturbation profiles of velocity and pressure isolated at the corresponding frequency band, sampled over the whole water column, for an integer number of wave periods. We apply harmonic analysis to determine the amplitude and phase of the lunar semidiurnal wave, generate full-period time series, and project them onto flat-bottom vertical modal shapes to obtain full-depth profiles at each station. Section 3.2.1–3.2.5 detail the methods involved in each step. The caveats associated with the flat-bottom assumption and the limited time and vertical span of observations are discussed in Sect. 3.3. At all stations, density profiles are available for approximately two semidiurnal cycles (day-long time series). Comparably long velocity time series are obtained from VM-ADCP at stations 1, 2, and 4 (Table 1). Stations 3 and 5, however, have too short VM-ADCP sampling duration to resolve the semidiurnal currents and are excluded from the analysis. At the mooring location near station 5, approximately full-depth, between 23 and 864 m, and 15.1 h long current profiles recorded by the moored instruments allow for energy flux calculations.

3.2.1 Baroclinic perturbation calculations

Using the observed velocity and density profiles, baroclinic perturbation fields of velocity, u', and buoyancy, b, are calculated similar to the methods described in Kunze et al. (2002) and Nash et al. (2005). The calculation of baroclinic pressure requires cumulative full-depth integrals, and is deferred to Sect. 3.2.5.

Baroclinic perturbation velocity, u', is calculated by removing the depth-averaged velocity profile at each time and then removing the time average at each depth; $u'(z,t) = u(z,t) - \langle u(z,t)\rangle_z - \langle u(z,t)\rangle_t$. $\langle \cdot \rangle_z$ and $\langle \cdot \rangle_t$ indicate averaging over depth and time, respectively. The depth average is an approximation to the barotropic velocity, and is valid because the frictional surface and bottom boundary layers are not covered by the moored instruments and the VM-ADCP. Vertical isopycnal displacement profiles, $\xi(z,t)$, are constructed using displacements of isopycnals from their station time-mean depth. For the mooring data, displacements are calculated relative to 24 h moving average density profiles. The measured buoyancy perturbation is $b = -(g/\rho_0)(\sigma_\theta - \langle \sigma_\theta \rangle_t)$. Equivalent results are obtained using the vertical displacement, $\xi = -bN^{-2}$, where $N(z) = [-(g/\rho_0)(\partial \langle \sigma_\theta \rangle_t / \partial z)]^{1/2}$ is the buoyancy frequency.

3.2.2 Semidiurnal fits

In order to infer the baroclinic semidiurnal energy flux, we isolate the semidiurnal band using harmonic analysis. The semidiurnal frequency, $\omega_{M_2} = 1.405 \times 10^{-4}\,\mathrm{s}^{-1}$, and the inertial frequency at 80° N, $f = 1.44 \times 10^{-4}\,\mathrm{s}^{-1} \sim 1.025\omega_{M_2}$, cannot be distinguished because of short record length. Therefore, semidiurnal must be understood as the near-inertial band when discussing observations. However, in the period spanning the cruise and 4 weeks prior to the cruise, wind forcing in the region was weak (quantified and discussed in Sect. 3.3), and we do not expect significant contribution from wind-induced near-inertial waves.

The semidiurnal fluctuation $\widetilde{x}_{M_2}$ of a perturbation variable x' is estimated using harmonic analysis:

$$\widetilde{x}_{M_2}(z,t) = \langle x' \rangle_t + x_{M_2}(z)\cos\left(\omega_{M_2}t - \phi_{M_2}^x(z)\right)$$

by minimizing the residual $R(z,t)$ in a least-square sense to determine the coefficients in

$$x'(z,t) = A(z)\cos(\omega_{M_2}t) + B(z)\sin(\omega_{M_2}t) + R(z,t),$$

where the amplitude profile is $x_{M_2}(z) = (A^2 + B^2)^{1/2}$ and the phase profile is $\phi_{M_2}^x(z) = \arctan(B/A)$. The harmonic analysis is applied to u', b, and ξ. Semidiurnal fits for the horizontal velocity components explain 30–90 % of the total variance at the stations, whereas the semidiurnal vertical displacement accounts for relatively less (15–65 %) of the observed variance.

When the harmonic analysis is repeated using two constituents ($\omega_{K_1} = 7.292 \times 10^{-5}\,\mathrm{s}^{-1}$ and ω_{M_2}) for the stations with sufficiently long time series, the variance explained by M_2 is within 5 % of the values obtained using the M_2 constituent only (much less than the uncertainty we assigned to harmonic analysis; see Sect. 3.3). At stations 1 and 2, where there are long enough VM-ADCP data to fit both constituents, K_1 accounts for 20–25 % of the horizontal velocity. Diurnal vertical displacement accounts for 20–40 % at stations 1–4 and dominate (57 %) at station 5. The diurnal component is clearly important at the YP. Because of the short station durations, we do not report diurnal baroclinic fluxes from the observations, but discuss them using the STORMTIDE model results.

3.2.3 Vertical modes

Constructed full-period baroclinic semidiurnal time series are used to derive full-depth profiles of $\widetilde{u}$, $\widetilde{\xi}$, and $\widetilde{b}$. At each time, the corresponding vertical profile is projected onto orthogonal vertical modes with vertical structure G_j of each mode j governed by (Phillips, 1977)

$$\frac{\mathrm{d}^2 G_j(z)}{\mathrm{d}z^2} + \left[\frac{N^2(z)}{c_j^2}\right] G_j(z) = 0, \tag{4}$$

where c_j is the eigenspeed for mode j. Vertical velocity and vertical displacement structures of each mode scale with G_j, while the horizontal velocity is proportional to dG_j/dz. Equation (4) is solved numerically for each station, using the station mean $N^2(z)$ and the boundary conditions $G_j(-H) = G_j(0) = 0$, where H is the total depth. Modal amplitudes are then obtained by least-squares fitting (stable solutions are obtained for the first three baroclinic modes). The full-depth profiles are constructed as the sum over modes as, for example, $\widetilde{u}(z,t) = \sum U_j(t)dG_j(z)/dz$, where U_j is the amplitude of the horizontal velocity for mode j. When averaged over each station's duration, the first three modes account for 50–80 % of the semidiurnal baroclinic velocity and 30–65 % of the semidiurnal isopycnal displacement. Vertical modes are orthogonal only over the full depth and over flat bottom. For steep slopes (relative to the wave slope), the horizontal and vertical structures cannot be separated and modal coupling may occur. The associated error in our analysis is discussed in Sect. 3.3.

3.2.4 Wentzel–Kramers–Brillouin scaling

Although the stations are closely spaced, the stratification differs significantly as a result of lateral gradients in proximity to the ice edge (see F2010). In order to account for the varying stratification, we apply Wentzel–Kramers–Brillouin (WKB) scaling (Leaman and Sanford, 1975) using the survey mean stratification of $N_0 = 2.4 \times 10^{-4}\,\mathrm{s}^{-1}$ as the reference buoyancy frequency. The stretched depth for a given station mean $N(z)$ profile is $z_{\mathrm{wkb}} = \int_z^0 N(z')/N_0 dz'$. Horizontal velocity and pressure scale as $(N_0/N(z))^{1/2}$, and vertical velocity and displacement scale as $(N(z)/N_0)^{1/2}$. Table 1 lists the stretched water depth and the percentage of the stretched water column covered by the instruments. Although the sampling was limited to the upper 520 m (MSS) and 360 m (VM-ADCP), 64–99 % (MSS) and 49–73 % (VM-ADCP) of the stretched water column were covered. The portion of the mooring densely equipped with the Microcats covers 75 % of the stretched depth.

3.2.5 Energy and energy flux calculations

Full-depth and full-period baroclinic semidiurnal fields $\widetilde{u}$ and $\widetilde{b}$ are constructed as the sum of the first three baroclinic modes. Baroclinic horizontal kinetic energy (HKE) and available potential energy (APE) in units of $\mathrm{J\,m}^{-3}$ are obtained from

$$\mathrm{HKE}(z) = \frac{\rho_0}{2}\langle \widetilde{u}(z,t)^2 + \widetilde{v}(z,t)^2\rangle_t, \tag{5}$$
$$\mathrm{APE}(z) = \frac{\rho_0}{2}N(z)^{-2}\langle \widetilde{b}(z,t)^2\rangle_t,$$

where $N(z)$ is the station-mean buoyancy frequency profile (calculated using the time-averaged density profile, $\langle \sigma_\theta \rangle_t$). The vertically integrated horizontal baroclinic energy flux,

F, is obtained from Eq. (3). The internal tide baroclinic pressure perturbation, $\widetilde{p}$, in units of Pa, is calculated using hydrostatic assumption and following Kelly et al. (2010) as

$$\widetilde{p}(z,t) = \int_z^0 \rho_0(\widetilde{b} - b^\eta)dz' - c, \tag{6}$$

where ρ_0 is the reference density and c is a constant of integration that ensures $\widetilde{p}$ has zero depth average. The second term in Eq. (6) is buoyancy due to isopycnal heaving by movement of the free surface,

$$b^\eta = -N^2\eta\left(\frac{z+H}{H}\right), \tag{7}$$

where η is the surface displacement (Kelly et al., 2010). We use η from STORMTIDE at the observation station locations. When the pressure perturbation induced by the isopycnal heaving by movement of the free surface is ignored, calculations of Kelly et al. (2010) indicate 10–50 W m^{-1} error in depth-integrated baroclinic energy fluxes, negligible over the energetic Kaena Ridge in Hawaii, but amounting to 10–45 % error over the Oregon continental slope.

3.3 Errors, uncertainty, and caveats

Observations suffer from short sampling duration, and from the uncertainty propagated from modal fits to vertically gappy sampling. Furthermore, in the analysis of the observations there are errors as a result of the study site that is characterized by topography (modal analysis assumes flat bottom) and high latitude (near-inertial waves are in the semidiurnal band). The analysis of the STORMTIDE model data is free from these errors because modal analysis is not employed, and the inertial and tidal frequencies are delineated by the harmonic analysis of sufficiently long time series.

Errors and systematic bias in baroclinic energy flux calculations for a variety of oceanographic sampling schemes are discussed thoroughly in Nash et al. (2005). Using Monte Carlo simulations of synthetic data, including a combination of semidiurnal, near-inertial and internal wave continuum signal, they assess magnitude and parameter dependence of flux estimates made from temporally or vertically imperfect data. For large baroclinic semidiurnal energy fluxes, such as those near a generation site in Hawaiian Ridge, a set of six full-depth profiles spanning 15 h leads to unbiased estimates of semidiurnal depth-integrated energy flux with 10 % error. Even for much weaker fluxes, the error is about 25 %, which is representative of our XCP sampling at station 1 with six casts spanning 20 h. The error further decreases with increasing number of profiles collected in a tidal period. Our shipboard sampling is typically 20–26 profiles (MSS and hourly averaged VM-ADCP) spanning two semidiurnal cycles; hence, errors associated with the harmonic analysis are expected to be small. The analysis of Nash et al. (2006), however, ignores the possible contamination of the semidiurnal band by the inertial waves at high latitudes. Inertial motions in the upper mixed layer are typically excited by energetic wind events and only a small fraction of the energy propagates deeper into the water column as near-inertial waves (see e.g. Alford et al., 2012). Short time series at our stations can be contaminated by near-inertial wave signals generated previous to the cruise at distant poleward locations. During the period 25 June–2 August 2007, spanning the cruise period and 4 weeks prior to the cruise, average ERA-Interim 10 m height wind speed using all grid points within a 100 km radius from station 2 is $5\,\mathrm{m\,s^{-1}}$, with only 4 % stronger than $10\,\mathrm{m\,s^{-1}}$, and none stronger than $12\,\mathrm{m\,s^{-1}}$. We therefore do not expect significant contribution from wind-induced near-inertial waves. Considering the competing effects of weak baroclinic energy fluxes (more error) and more than six profiles per analysis period (less error), and an unquantified but small contamination from near-inertial waves, we assign 20 % error to harmonic analysis.

For the case of vertically gappy sampling, Nash et al. (2005) find that larger gaps can be tolerated near the bottom after WKB scaling (as in the case of our MSS and VM-ADCP sampling), but estimates are sensitive to the data near the surface. Because the energy flux scales as the buoyancy frequency, it is typically surface intensified. For a two-mode fit, when only the top 30 % of the WKB-stretched water column is sampled, Nash et al. (2006) inferred less than 40 % error in F. Using the XCP data (approximately full-depth profiles), we attempt to estimate the error associated with the limited vertical extent of our station data sets. We repeat the XCP-station analysis using the portion of the velocity measurements between 12 and 360 m depth, identical to the VM-ADCP coverage, and the CTD profiles between 3 and 500 m depth, identical to the MSS coverage. Relative to the full-depth analysis, error in the depth-averaged HKE and APE is 7 and 100 %, respectively. The latter is a result of poor modal fits when the lower half of the water column is not sampled. The depth-integrated baroclinic energy flux error is 52 %, comparable to the error reported by Nash et al. (2005).

Our modal decomposition has errors as a result of imperfect depth coverage and sloping bottom. Modes are orthogonal only over the full water depth and over flat bottom, and contamination arises when unresolved variance is projected onto resolved modes, or when the horizontal and vertical structure cannot be separated over steep topography. The error associated with modal analysis can be estimated using the station 4 data. This station has the best vertical coverage and the energetics can be inferred using the observed profiles without modal analysis. The profiles of density cover approximately the entire water column (Table 1), crucial for pressure anomaly calculations. We repeat the station 4 analysis, but without fitting to the modal shapes. The error will thus include both the effect of sloping bottom and the unresolved higher vertical modes. Relative to results without modal fits, error in the depth-averaged HKE and APE is 34

and 36 %, respectively. The depth-integrated baroclinic energy flux error is 48 %.

The errors estimated for each source (harmonic analysis, imperfect vertical sampling, and projection onto flat-bottom normal modes), are not entirely independent, and are not the same for each station. Overall, we assign total errors of 50 % to HKE and 100 % to APE and F.

At the latitudes of the YP diurnal and lunar semidiurnal, internal waves are evanescent with zero group velocity, and our calculations (for both the observations and the model data), using the common approach of $Ec_{\mathrm{g}} = \widetilde{p}\widetilde{u}$, should be interpreted with caution. The water column is not a solid boundary, and forced internal waves in the water column will leak their energy from the generation site to only a limited vertical and lateral extent. Trapped sub-inertial baroclinic motions, however, are also possible and may propagate along the topographic contours (several examples are cited in Sect. 6). Our results thus represent the baroclinic fluxes in a given frequency band, induced by pressure and velocity perturbations associated with tidal response over topography, with possible leaking internal waves and trapped waves.

The STORMTIDE model results are obtained from a global model which has, from a regional perspective, a rather coarse resolution. The model is hydrostatic and thus only permits internal waves with frequencies much less than the buoyancy frequency, which is a reasonable limit for the semidiurnal waves. The horizontal resolution restricts the model to resolve only the low-mode internal tides. The model results are limited to interpretation for linear wave characteristics. The global model thus can not compete with regional models resolving non-linear characteristics of internal wave propagation (e.g. Simmons et al., 2011). However, due to its global characteristics it explicitly avoids the boundary condition issue, which is critical for concurrent simulations of mesoscale ocean circulation and internal tides. The dissipation of internal waves results from numerical parametrization of bottom friction and eddy viscosity and further by numerical dissipation. Previous evaluations against observations from satellite altimetry (Müller et al., 2012), tide gauges (Müller et al., 2014), and a detailed comparison of the generation of internal tides with observations, theory, and inverse models (Müller, 2013) support with confidence that the model simulated internal tides inherit some degree of realism: the internal wave generation and propagation of internal tides are reasonably well represented; near generation sites in the deep ocean, the magnitude of the surface signature of internal tides is reasonably well reproduced; and the magnitudes and the regional characteristics of barotropic-to-baroclinic tidal conversion rates compare well with those in other models, and observational and theoretical studies.

4 Variability at the observation stations

Observations in 2007 were made during neap and transition to spring tides (Fig. 2). The data show strong variability in velocity and vertical isopycnal displacement in both diurnal and semidiurnal bands. Tidal surface elevation inferred at the model grid point nearest to the mooring location shows a tidal range of up to ± 0.5 m, and both its phase and magnitude agree with the record inferred from the moored pressure sensor nearest to the seabed (Fig. 2a), lending further confidence on the model output. The duration of the current record is short; however, the depth-average current agrees well with the tidal flow. A detailed comparison is not attempted since we cannot infer all four tidal constituents (M_2, S_2, K_1, O_1) from the 15.1 h long ADCP record. The time-depth maps of the vertical isopycnal displacement recorded by the moored instruments show up to ± 40 m alternating bands with mixed diurnal and semidiurnal periodicity. On the southern flanks of the plateau, tidal ellipses derived from STORMTIDE show comparably strong semidiurnal (M_2) and diurnal (K_1) currents (not shown). At the observation site, STORMTIDE tidal current amplitudes at station 2 (the most energetic station for all constituents) are (K_1, O_1, M_2, S_2) = (13, 6, 14, 5) cm s^{-1}; i.e. semidiurnal and diurnal bands are of comparable magnitude and typically K_1 and M_2 dominate over O_1 and S_2. Frequency spectra of total velocity at the pycnocline presented in F2010 correspond to about 3 times larger HKE in the diurnal band compared to the semidiurnal band. In agreement with the velocity spectrum, the diurnal band is the most energetic in the vertical isopycnal displacement spectra throughout the water column at the mooring location (F2010). Consistently, depth-integrated baroclinic energy fluxes from STORMTIDE averaged within 10 km of the mooring location (station 5) are dominated by K_1 (Sect. 5; Table 4).

An overview of the semidiurnal amplitudes obtained from the harmonic analysis of the horizontal current and isopycnal displacements at stations 1, 2, and 4 is given in Fig. 3. A detailed model–observation comparison is not attempted; however, the observed semidiurnal amplitudes generally compare well with the STORMTIDE results. Station 2 has baroclinic semidiurnal amplitudes comparable to the barotropic tide, whereas at the other stations the barotropic component is larger. The observed barotropic current amplitude of approximately 5 cm s^{-1} at station 2 is significantly less than the M_2 amplitude from STORMTIDE mentioned above. Overall, the baroclinic semidiurnal current amplitudes are weak; nevertheless, the semidiurnal signal accounts for a significant percentage of the observed profiles (Sect. 3.2.2). The amplitudes are comparable to those from a numerical modelling study of the Barents Sea that show less than 5 cm s^{-1} semidiurnal baroclinic velocity at intermediate depths (Kagan and Sofina, 2014). For comparison, trapped baroclinic diurnal current amplitudes off the southern California Bight (Johnston and Rudnick, 2014) are typically 2–3 times more energetic

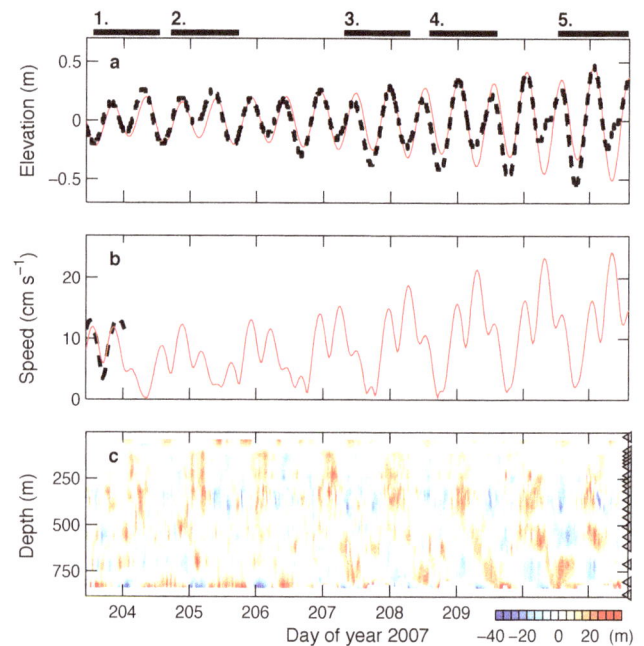

Figure 2. Time series at station 5 of (**a**) tidal surface elevation from STORMTIDE (red; sum of 8 major constituents) and near-bottom pressure anomaly from the deepest moored instrument (dashed), (**b**) tidal current from STORMTIDE (red; sum of M_2, K_1, S_2 and O_1) and depth-averaged current measured by the moored instruments (dashed), and (**c**) vertical isopycnal displacements. The downloaded archived STORMTIDE global data set is limited to these four major constituents for the tidal velocity and phase. The time-average depth of the instruments are shown by triangles on the right. Bars on top show the duration of stations 1–5.

than the semidiurnal current amplitudes at our YP stations. On the northern part of the YP, where the major barotropic-to-baroclinic energy conversion sites are located, larger amplitudes are modelled (Sect. 5). At station 1, baroclinic vertical displacement is typically less than the barotropic component. In the upper 150 m of stations 2 and 4 and between 200 and 300 m at station 4, baroclinic vertical displacement exceeds the barotropic contribution. The phase of the horizontal current increases approximately linearly with depth at station 1, indicating upward energy propagation with a vertical wavelength of at least 500 m. This is supported by the vertical wave number spectra obtained for the bottommost 512 m of the full-depth shear profiles sampled by XCPs suggesting upward energy propagation (F2010). The phase of the vertical displacement, on the contrary, typically decreases with depth. This can be a consequence of the sub-inertial semidiurnal waves trapped along the topography. At stations 2 and 4, phase profiles show 180° jumps suggesting a vertically standing behaviour.

Observed baroclinic energetics inferred from full-depth profiles for the sum of the first three modes are summarized in Table 2. The ratio R of depth-averaged HKE to APE varies

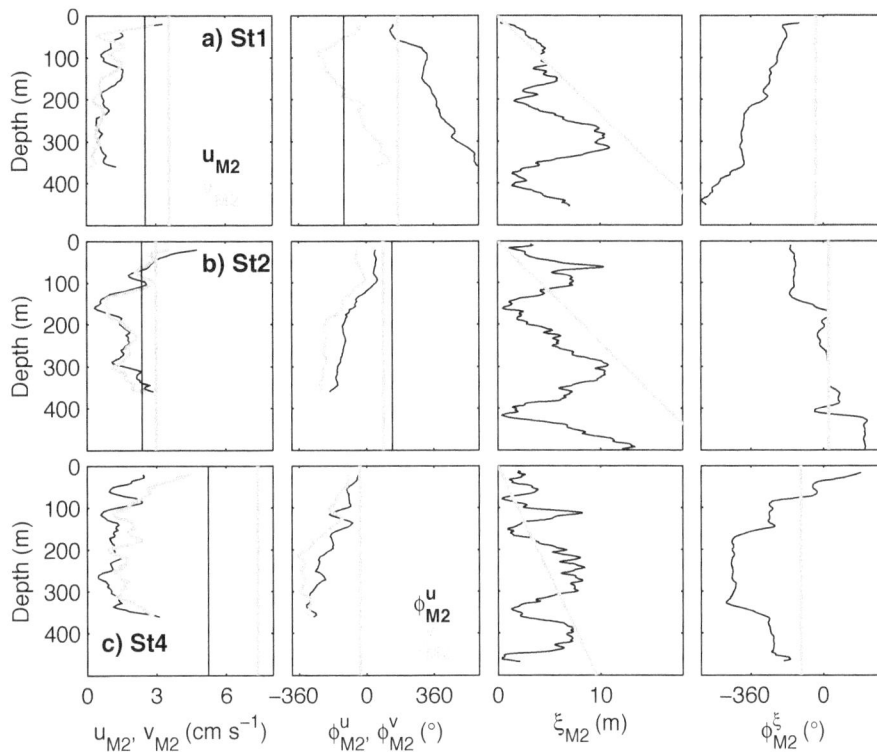

Figure 3. Profiles of amplitude and phase of baroclinic u, v, and ξ obtained from the semidiurnal fits to VM-ADCP and MSS profiles at stations **(a)** 1, **(b)** 2, and **(c)** 4. Also shown are the corresponding barotropic amplitude (vertical lines for u and v, and slanted lines for ξ) and phase (vertical lines), inferred from identical harmonic analysis of the depth-averaged currents and profile of barotropic displacement.

Table 2. Semidiurnal baroclinic energetics at observation stations. Results are shown for the sum of the first three baroclinic modes. An overbar indicates integration over depth; 1-XCP is station 1 with results obtained from the XCP profiles; 5-M is the mooring located near station 5.

Station	$10^3 \times$ HKE $(\mathrm{J\,m^{-3}})$	$10^3 \times$ APE $(\mathrm{J\,m^{-3}})$	R	$\overline{\mathrm{HKE}}$ $(\mathrm{J\,m^{-2}})$	$\overline{\mathrm{APE}}$ $(\mathrm{J\,m^{-2}})$	F_u $(\mathrm{W\,m^{-1}})$	F_v $(\mathrm{W\,m^{-1}})$
1-XCP	211 ± 106	41 ± 41	5 ± 5	205 ± 103	25 ± 25	-29 ± 29	41 ± 41
1	143 ± 72	56 ± 56	3 ± 3	142 ± 71	33 ± 33	-20 ± 20	21 ± 21
2	87 ± 44	22 ± 22	4 ± 4	84 ± 42	12 ± 12	-14 ± 14	8 ± 8
3	–	97 ± 97	–	–	70 ± 70	–	–
4	80 ± 40	15 ± 15	5 ± 5	50 ± 25	8 ± 8	-2 ± 2	-3 ± 3
5	–	39 ± 39	–	–	20 ± 20	–	–
5-M	315 ± 158	47 ± 47	7 ± 7	148 ± 74	12 ± 12	-4 ± 4	15 ± 15

between 3 and 7, but with 100 % error bars. R is equivalent to the shear-strain ratio R_ω discussed in F2010. The values up to 14, when a factor of 2 uncertainty is considered, are typical in the ocean and are consistent with the values inferred in F2010. Estimates from the XCP data at station 1 are larger than those from the MSS/VM-ADCP profiles (but within the error bars), suggesting that the lack of MSS/VM-ADCP sampling in the deepest 50 % of the stretched depth affects the results. The horizontal components of the depth-integrated M_2 baroclinic energy flux are listed in Table 2. At stations 1 and 5, F is directed nearly along the isobaths. The

largest F is observed at station 1 on the western flank. The results are similar when derived from the MSS/VM-ADCP profiles or from the XCP profiles.

The measurements of the dissipation rate will be used in Sect. 6 to characterize the typical volume-integrated dissipation levels in the YP region. A station-mean profile of dissipation rate ε measured by the microstructure profiler is obtained by averaging over all profiles collected at each station (Fig. 4). Depth-integrated dissipation rates are given in Table 3. Station 4 has the largest dissipation rates. Station 4 profile extends to the bottom; observations at other

Table 3. Depth-integrated dissipation rate at observation stations in units of $10^{-3} \times \mathrm{W\,m^{-2}}$. Integrations down to 500 m are obtained from the station mean ε profiles. Integrations from 50 m to the bottom use the assumed minimum and maximum ε profiles described in the text.

Station	$\int_0^{500\,\mathrm{m}} \rho\varepsilon\,\mathrm{d}z$	$\int_{50\,\mathrm{m}}^{500\,\mathrm{m}} \rho\varepsilon\,\mathrm{d}z$	$\int_{50\,\mathrm{m}}^{H} \rho\varepsilon_{\min}\,\mathrm{d}z$	$\int_{50\,\mathrm{m}}^{H} \rho\varepsilon_{\max}\,\mathrm{d}z$
1	3.7	0.4	0.3	1.0
2	2.3	0.6	0.7	1.2
3	0.8	0.5	0.6	1.3
4	1.9	1.3	1.3	1.3
5	0.9	0.3	0.2	0.7

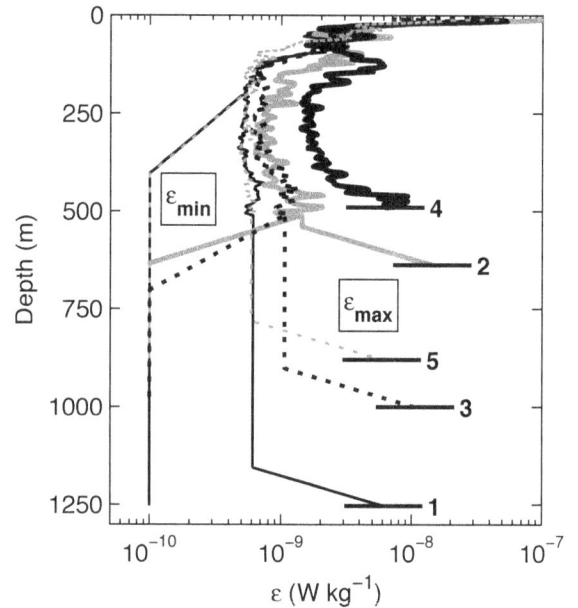

Figure 4. Dissipation rate profiles for stations 1–5. Horizontal lines with the station number indicated mark the mean echo sounder depth for the station. Observed, station-mean profiles are shown in the upper 500 m. Assumed full-depth profiles used in calculating depth-integrated ε are shown for the lower bound $\varepsilon_{\min}$ and the upper bound $\varepsilon_{\max}$.

stations, however, are limited to 500 m depth. Because the turbulence near the surface layer can be dominated by other processes than internal wave-induced mixing, values integrated between 50 and 500 m depth are also listed in Table 3. In order to have an estimate of the full-depth (below 50 m) integrated ε for all stations, we construct a lower bound ($\varepsilon_{\min}$) and an upper bound ($\varepsilon_{\max}$) profile down to the mean echo depth of each station. Profiles of $\varepsilon_{\max}$ are obtained by extending the 450–500 m average value to 100 m height above seabed, and thereafter exponentially increasing this value by a factor of 10 until the bottom is reached (Fig. 4). The near-bottom enhancement for $\varepsilon_{\max}$ is applied to account for bottom boundary turbulence and breaking of internal waves over the slope. The profiles of $\varepsilon_{\min}$ are constructed by extending the observed profiles by an assumed low dissipation of $10^{-10}\,\mathrm{W\,kg^{-1}}$. Observed profiles at stations 1 and 5 reach the noise level of $5 \times 10^{-10}\,\mathrm{W\,kg^{-1}}$ below about 150 m, suggesting that actual values can be lower. We therefore exponentially reduce the 450–500 m average value for stations 2 and 3, and the value at 200 m for stations 1 and 5, to $10^{-10}\,\mathrm{W\,kg^{-1}}$ over a 200 m vertical scale, and extend this value to the seabed (Fig. 4). Depth-integrated dissipation below the surface mixed layer is typically between 0.2 and $1.3 \times 10^{-3}\,\mathrm{W\,m^{-2}}$, with largest values at station 4. Limited by the ice edge, our observations were made only on the southern part of the YP. On the one hand, these stations are not as energetic as the significant conversion sites on the northern flanks. On the other hand, the dissipation rate profiles do cover several regimes (the quiescent station 1, the moderately turbulent station 2, and the turbulent station 4). Using the constructed dissipation profiles described above, we are confident that we obtain lower and upper bounds on the dissipation rate representative of the YP.

5 Energetics from the model

Barotropic tidal fluxes inferred from STORMTIDE are shown in Fig. 5. These can be compared to the K_1 and M_2 tidal fluxes around Svalbard shown in Chen et al. (2009): the limited region shown in Fig. 5 is part of a pattern in which the M_2 energy propagates clockwise around the Svalbard is-

lands, whereas the K_1 tidal energy is mainly trapped along the steep bottom topography. Both the K_1 and M_2 tidal fluxes are elevated along the shelf break north of Spitsbergen and also along the outer flank of the plateau, particularly close to the CEAREX-O camp drift site. These regions of large barotropic fluxes are also associated with high conversion rates and depth-integrated baroclinic fluxes (Fig. 6).

The depth-integrated baroclinic fluxes in the energetic parts of the YP are on the order of 200 W m^{-1} for M_2 and 1000 W m^{-1} for K_1. Near our mooring location, the depth-integrated energy flux is 1 order of magnitude smaller than at many of these energetic sites, and is reduced by another order of magnitude at the other stations.

Barotropic-to-baroclinic energy conversion rate, C, is shown for the semidiurnal and the diurnal tide in Fig. 6. The CEAREX-O site is characterized by moderate M_2 but substantially larger K_1 conversion rates, reaching 0.05 W m^{-2}; the fluxes echo this pattern. The model results are thus consistent with the observations from the CEAREX-O camp where energetic internal waves and enhanced mixing were measured (Padman and Dillon, 1991). Negative conversion rates seen in Fig. 6 are typical of numerical studies on internal tide energetics (Simmons et al., 2004; Kang and Fringer, 2012). A negative conversion occurs when energy is transferred from baroclinic tide to the barotropic tide and indicates interaction between locally and remotely generated baroclinic tides. Relatively large semidiurnal baroclinic fluxes

Figure 5. Barotropic tidal flux obtained from STORMTIDE for the **(a)** M_2 and **(b)** K_1 constituents. Isobaths (grey) are drawn at 500 m intervals from the model topography. The colour is the amplitude of the corresponding flux with colour bar on the right. Every fourth vector is shown with scale indicated. Station positions (open circles), and the drift of CEAREX-O (Padman and Dillon, 1991) (red line over the northern flank) and MIZEX83 (D'Asaro and Morison, 1992) (red line over the plateau) are shown for reference.

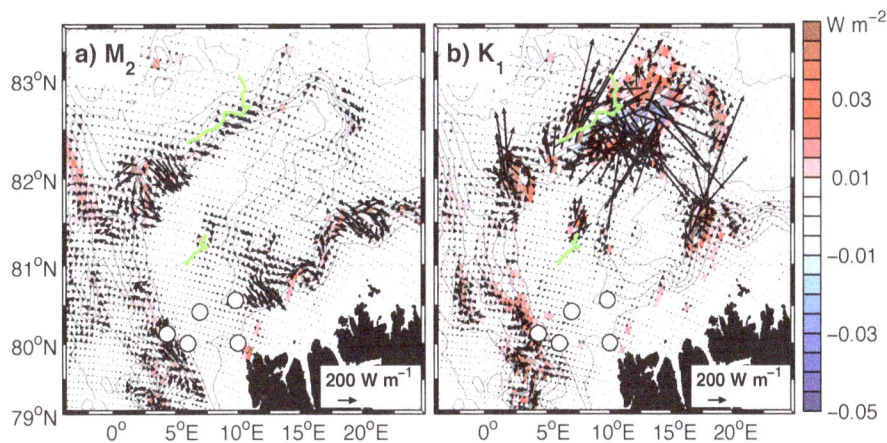

Figure 6. Barotropic-to-baroclinic conversion rate (colour) and the depth-integrated baroclinic energy fluxes (arrows) obtained from STORMTIDE for the **(a)** M_2 and **(b)** K_1 constituents. Isobaths (grey) are drawn at 500 m intervals from the model topography. Station positions (open circles) and the drift of CEAREX-O (Padman and Dillon, 1991) and MIZEX83 (D'Asaro and Morison, 1992) (green) are shown for reference.

can be observed on the western part of the plateau around $82°$ N, and also along the shelf break north of Spitsbergen. Fluxes are typically directed along the topography and with a limited lateral extent consistent with trapped tides.

The site of observational stations is characterized by weak conversion rates and depth-integrated baroclinic fluxes, F, except near stations 1 and 5 on the western slope. From the model, F is calculated as the average over grid points within 10 km in the vicinity of each station. The results, together with 1 standard deviation, are listed in Table 4 for the major diurnal and semidiurnal constituents. The variability at neighbouring grid points is large. The M_2 and K_1 constituents are comparable in magnitude and are greater than the S_2 and O_1 fluxes. The fluxes inferred from the observations (for M_2 only) compare fairly well with the model

results when averaged in the vicinity of the station location (compare Tables 2 and 4); however, the comparison is poor when a single data point from the nearest grid to the station is used. Observed flux vectors are smaller than the size of the station markers in Fig. 6a; hence, they are not shown (a direct comparison on map is presented in Sect. 6).

The conversion rates in the vicinity of the stations are dominated by the M_2 and K_1 constituents (3–10 times larger than S_2 and O_1), with values between 1 and 10×10^{-3} W m^{-2}, and with largest rates near station 1. Model dissipation rates at the station positions are weak, similar for M_2 and K_1, and vary between an order of 10^{-3} W m^{-2} (station 2) and an order of 10^{-2} W m^{-2} (station 1). These can be compared to the observed values of order 10^{-3} W m^{-2} shown in Table 3. The observational stations are limited to the relatively less

Table 4. Depth-integrated baroclinic fluxes, F, in units of Wm^{-1} inferred from STORMTIDE using the grid points within 10 km of the observation stations. The average ± 1 SD over the corresponding grid points are listed for the horizontal components (F_u, F_v) of the four major constituents.

Station	M_2	S_2	K_1	O_1
1	$(-16 \pm 29, 54 \pm 32)$	$(0 \pm 2, -4 \pm 3)$	$(44 \pm 16, 17 \pm 15)$	$(8 \pm 8, 5 \pm 11)$
2	$(-2 \pm 1, -3 \pm 1)$	$(0 \pm 1, 0 \pm 0)$	$(14 \pm 7, 3 \pm 1)$	$(1 \pm 2, 1 \pm 1)$
3	$(5 \pm 6, -22 \pm 11)$	$(1 \pm 2, 0 \pm 3)$	$(2 \pm 2, -2 \pm 3)$	$(0 \pm 1, 0 \pm 2)$
4	$(1 \pm 2, -8 \pm 3)$	$(0 \pm 1, 0 \pm 1)$	$(3 \pm 3, -5 \pm 6)$	$(0 \pm 1, -1 \pm 1)$
5	$(2 \pm 3, 13 \pm 7)$	$(-1 \pm 1, 1 \pm 1)$	$(11 \pm 5, 8 \pm 6)$	$(2 \pm 1, 2 \pm 2)$

energetic southern flanks of the YP. It is therefore crucial to utilize the model data to account for the energetic parts of the plateau. In the following we shall rely on the model results to discuss the energetics in the region and its role for vertical mixing.

The baroclinic radiation and depth-integrated dissipation rates are obtained from the model. A negative radiation is horizontal convergence of the depth-integrated flux and indicates that the location is a sink, whereas a positive radiation indicates that the site is a source for baroclinic energy. Volume-integrated conversion, radiation, and dissipation rates are calculated over the YP region bounded between 5° W and 25° E zonally, 79 and 83°30′ N meridionally, and by the 250 m isobath north of Svalbard. Surface area in this domain with water depth deeper than 250 m is approximately 2×10^{11} m², with a mean water depth of 2300 m. For the M_2 constituent, several outliers in the radiation data (in total 10 grid points, all between the 750 and 2250 m isobaths with radiation values between 0.03 and 0.06 W m^{-2}) are co-located with weak conversion rates of an order of 10^{-3} W m^{-2}, leading to unphysical negative dissipation rates. We removed these points from the calculations. The volume-integrated conversion rate in the region is 322 MW for M_2 and 618 MW for K_1; the radiation is 8 and 1 MW, respectively. This indicates that despite substantial energy conversion into internal tides, only a minute fraction (if any) propagates out of the region, implying substantial local dissipation. The spatial distribution of depth-integrated dissipation rate for the M_2 and K_1 components is shown in Fig. 7. Total dissipation integrated over the domain is 314 and 617 MW for the M_2 and K_1 constituents, respectively.

The conversion is spatially variable, and is particularly concentrated near the 2000 m isobath in the northern flanks of the plateau (Fig. 6). Dissipation is typically elevated where the conversion rates are large and the spatial distribution is characterized by several regions (hot spots) where the dissipation is enhanced (Fig. 7). For quantification of their contribution, we define hot spots as grid-volume-integrated total dissipation exceeding 0.5 MW (corresponding to approximately 10^{-2} W m^{-2} depth-integrated dissipation, strong colours in Fig. 7). The percent of area with hot spots is 7 % for M_2, 11 % for K_1, and 15 % for the

sum of M_2 and K_1. Dissipation of diurnal energy is concentrated on the northern flanks of the plateau, around the CEAREX-O drift site, over the shallow part of the plateau near the MIZEX83 drift, and around 17–18° E north of Svalbard. Dissipation of semidiurnal energy shows a similar pattern, but with relatively less pronounced dissipation near the CEAREX-O drift site. There is a patch of elevated dissipation near station 4 where the highest mixing rates were recorded during our cruise (Fig. 4). These locations of energetic turbulence also coincide with regions of large barotropic tide velocities. The maximum tidal velocity amplitudes during one spring-neap cycle exceed 0.4 m s^{-1} in localized regions over the plateau and approach 1 m s^{-1} over the shelf north of Svalbard east of our station 4. Over the plateau, the largest tidal velocity is near the seamount where D'Asaro and Morison (1992) inferred eddy diffusivities greater than 10^{-4} m² s^{-1} from XCP shear. Close to the CEAREX-O camp, on the northern flanks, the maximum tidal velocity is approximately 25 cm s^{-1}, greater than that at our stations.

The spatial variability is further investigated using the integrated conversion and radiation rates over volumes bounded by increasing isobaths, following Kang and Fringer (2012). The results for the M_2 and K_1 constituents are shown in Fig. 8 using 250 m depth bins. Cumulative rates between chosen isobaths are tabulated in Table 5. The barotropic tide is converted to internal tide at all depths, for both constituents. The semidiurnal conversion shows an increase at isobaths deeper than 1500 m (over the sloping sides of the plateau) and with a peak around the 3000 m isobath (Fig. 8a). The radiation is nil below the 750 m isobath, positive between 750 and 2000 m, and negative for deeper water. All of the semidiurnal energy converted at depths shallower than 750 m, and also the amount generated in deeper water which propagates into shallower depth, are thus dissipated over the shelves. The slope between 750 and 2000 m depth is a generation site for propagating internal tide (positive radiation). The deeper part of the slope is a sink, and dissipates the energy that is locally generated as well as the fraction that is radiated from the upper slope. The radiation for these trapped waves must be interpreted in the context of a decay of energy with an e-folding scale of Rossby radius (Sect. 6).

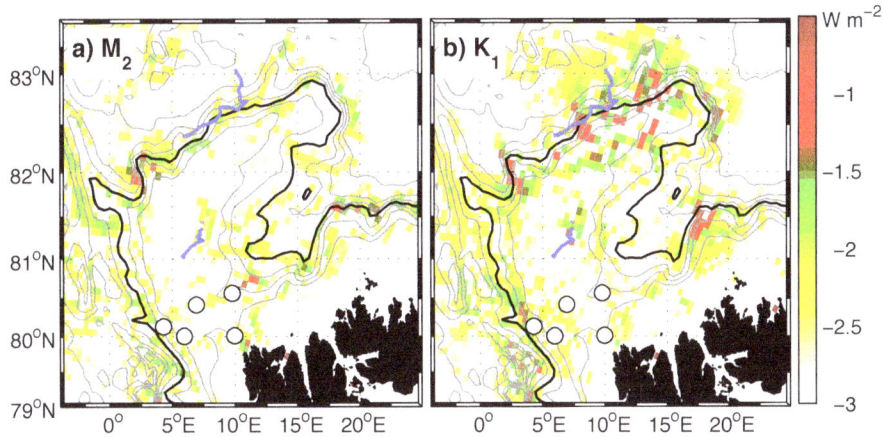

Figure 7. Depth-integrated dissipation rate, presented in base-10 logarithm, for the (**a**) M_2 and (**b**) K_1 constituents obtained from STORMTIDE. Isobaths (grey) drawn at 500 m intervals are from the model topography. The 2000 m isobath is drawn in black for reference. Station positions (open circles) and the drift of CEAREX-O (Padman and Dillon, 1991) and MIZEX83 (D'Asaro and Morison, 1992) are also shown.

Table 5. Volume-integrated conversion and radiation rates for the M_2 and K_1 constituents, bounded by selected isobaths. Shallower than 250 m (not listed), there is approximately 10 MW conversion, for both M_2 and K_1, balanced by local dissipation.

Isobath (m)	M_2 Conversion (MW)	M_2 Radiation (MW)	M_2 Dissipation (MW)	K_1 Conversion (MW)	K_1 Radiation (MW)	K_1 Dissipation (MW)
250–750	19	0	19	42	−4	46
750–2000	85	80	5	175	−23	198
2000–5000	218	−72	290	401	28	373

Volume-integrated conversion of K_1 (618 MW) is approximately twice that of M_2. Similar to the M_2 pattern, there is radiation away from the upper part of the slope, whereas the deeper slope is a sink for the baroclinic K_1 energy flux. The major source and sink regions are relatively more constrained by the isobaths for K_1 (1000–1500 and 1750–2250 m) compared to a broader distributed M_2 (750–2000 and 2250–4000 m). Most of the dissipation occurs deeper than 1500 m. The dissipation curve shows an approximate local conversion–dissipation balance over the shallows and also on the deeper part of the slope, but the dissipation exceeds twice the conversion rates between 1500 and 2000 m depth; all the K_1 energy radiation from the upper slope is dissipated here.

6 Discussion

The model results show substantial barotropic tidal fluxes around the YP and localized regions of large barotropic-to-baroclinic energy conversion rates for both K_1 and M_2 constituents. These conversion sites are shown to be associated with baroclinic energy fluxes. Is it realistic to have baroclinic energy fluxes and radiation for tides above their corresponding critical latitudes? Analogous to barotropic con-

tinental shelf waves, variable bottom topography of ridges, seamounts and plateaus in homogeneous water can support trapped waves (Rhines, 1969; Huthnance, 1974). If, for example, diurnal tidal frequency is close to the natural frequency of one of such free wave modes, the topographically trapped free wave will be resonantly excited by the oscillation of the diurnal tide. Using arbitrary stratification, Brink (1989) showed that sub-inertial baroclinic trapped waves are also supported at isolated seamounts. Wang and Mooers (1976) showed that in a continuously stratified ocean with sloping bottom, topographic Rossby waves are the only form of sub-inertial wave motion (for a negligible coastal wall), and reduce to barotropic shelf waves and to bottom-trapped waves in the limits of small and large stratification, respectively. The energy of the topographically trapped waves propagates along the slope, around the topographic feature with a decay scale of Rossby radius of deformation and negligible radiation in the cross-slope direction. Sub-inertial internal wave energy and energy fluxes have been observed and modelled elsewhere (Allen and Thomson, 1993; Tanaka et al., 2010; Johnston and Rudnick, 2014; Robertson, 2001; Kunze and Toole, 1997). The internal Rossby radius, c_1/f, for the first mode eigenspeed obtained from the modal analysis of our observational data, varies between 3 and 5 km.

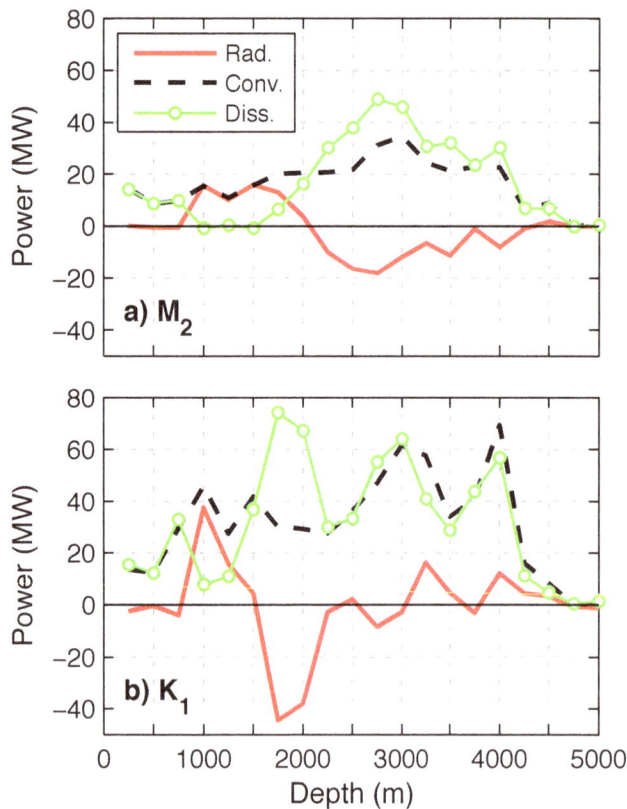

Figure 8. Total baroclinic radiation (red), conversion (dashed black), and dissipation (green, circles) for the **(a)** M_2 and **(b)** K_1 constituents, integrated in 250 m isobath bounded volumes.

The bottom slope between the 1000 and 2000 m isobaths along the northwestern slope, representative of the generation sites, is 0.06 (± 0.03). The cross-slope distance covered between the 1000 and 2000 m isobaths is thus 17 km (± 8). If the trapped wave generated between these isobaths decays with the Rossby radius, a decay to 5 % of the background value occurs between 9 and 15 km. Hence, the substantial inferred radiation near the isobaths where conversion occurs (Fig. 8) is plausible.

The bottom-trapped tides in the Arctic Ocean are of particular importance because most of the Arctic Basin is located north of critical latitude of the most energetic tidal constituent M_2 (Falahat and Nycander, 2014). Using global numerical models, baroclinic tidal energy (including poleward of the critical latitudes) are reported by Niwa and Hibiya (2011) for the major diurnal constituents and by Müller (2013) for the major diurnal and semidiurnal constituents. In an idealized study, Falahat and Nycander (2014) examine the bottom-trapped internal tides for the M_2, K_1, and O_1 tidal constituents over the global ocean. To infer the energy conversion rates from the barotropic tides to bottom-trapped internal tides, they calculate the energy density for linear inviscid waves and assume that the trapped wave energy, for all vertical modes, decays over a 3-day timescale. This ad hoc

timescale is representative of Fieberling Seamount which is a strongly forced and damped system with large dissipation and eddy diffusivity (Kunze and Toole, 1997). In the Arctic Ocean, Falahat and Nycander (2014) inferred an area-integrated energy flux of 1.1 GW for topography-trapped internal tides (sum of the M_2, K_1, and O_1 constituents, where about 70 % is due to M_2). The diurnal component, of about 0.3 GW, is 1 order of magnitude less than the diurnal internal tides in the Arctic Ocean reported in Müller (2013), and hence should be considered as a lower bound.

At high latitudes, diurnal tides with frequencies close to the half of the inertial frequency are a likely source to force resonant-trapped waves (Huthnance, 1974). In the Rockall Bank, the diurnal tidal frequencies are close to the resonance frequency of a natural trapped mode progressing clockwise around the bank, leading to strongly excited diurnal currents consistent with observations (Huthnance, 1974). A similar excitation of diurnal currents was also shown around the Bear Island near the M_2 critical latitude (Huthnance, 1981). Yermak Plateau is an area known to have resonantly enhanced diurnal tides, particularly over the northern flanks (Hunkins, 1986; Chapman, 1989; Padman et al., 1992). Using a barotropic model with an idealized axisymmetric submarine plateau both Hunkins (1986) and Chapman (1989) showed near-resonant diurnal trapped topographic waves propagating around the YP. While Hunkins (1986) looked at the free waves in a frictionless ocean, Chapman (1989) included friction and forcing by rectilinear K_1 tidal currents (i.e. forced and damped trapped waves). The topography of the YP, however, is not axisymmetric (Fig. 1). Padman et al. (1992) showed that dispersion relations derived separately on the northwestern and eastern flanks of the Plateau suggest free diurnal waves on the northwest slope (near the CEAREX-O site) but approximately zero group velocity on the eastern slope. This is inconsistent with the axisymmetric model and with a resonant interaction mechanism related to the path length of a free wave that encircles the entire plateau. They proposed an alternative generation where diurnal energy is due to topographic diurnal waves on the eastern part where the group velocity is near zero, allowing maximum amplification. Although these studies address the barotropic diurnal currents in neutral stratification (for simplicity), they are applicable to baroclinic solutions (Brink, 1989; Wang and Mooers, 1976). For example, Tanaka et al. (2010) uses the baroclinic coastal trapped wave solutions to explain the sub-inertial diurnal baroclinic tidal energy propagating around the Kuril Islands. On the continental slope of the Laptev Sea, Pnyushkov and Polyakov (2012) reported that the baroclinic solutions of the topographically trapped waves show no significant change of the cross-slope structure of tidal current and sea level amplitudes compared with the barotropic experiment.

Sub-inertial internal wave energy and energy fluxes have been observed and modelled elsewhere. Over the Juan de Fuca Ridge, trapped (laterally and vertically) baroclinic subinertial motions were reported (Allen and Thomson, 1993). In a numerical study, Tanaka et al. (2010) show that most of the internal wave energy subtracted from the diurnal barotropic tide is dissipated within the Kuril straits. K_1 tidal frequency is sub-inertial in this area and the tidal energy is fed into topographically trapped waves which propagate along slope around each island with negligible radiation away from the straits. Energy subtracted from the K_1 barotropic tide is approximately 30 GW; most of the energy dissipates locally, only 0.6 GW radiating out from the analysed domain. The local conversion and dissipation balance is similar to what we found near the YP. Johnston and Rudnick (2014) observed topographically trapped diurnal internal waves along the California continental slope and over the Santa Rosa–Cortes Ridge in the southern California Bight. The diurnal (sub-inertial) internal tides are more energetic than the semidiurnal internal tides and are associated with elevated diffusivities near topography. Using current measurements in the upper 200 m in about 2700 m deep water on the continental slope of the Laptev Sea, Pnyushkov and Polyakov (2012) inferred baroclinic tide in the upper 50 m, twice as energetic as the barotropic tidal currents. Numerical solutions of trapped waves over the continental shelf and slope suggest resonance and enhancement of semidiurnal energy consistent with the observations. Poleward of the critical latitudes, near-inertial internal waves have also been observed. For a review of the near-inertial internal tides in the Weddell Sea in Antarctica see Robertson (2001). Over Fieberling Seamount near 32° N Kunze and Toole (1997) reported vortex-trapped near-inertial diurnal internal waves. The dissipation rates are strong enough to dissipate the K_1 motions within 3 days, implying a strongly forced and damped environment. At the YP, M_2 is the strongest semidiurnal component. D'Asaro and Morison (1992) reported that tides here are an attractive source for enhanced near-inertial band energy where sub-inertial M_2 tide generated on the seamount is trapped to the seamount by the barotropic vorticity field.

At the YP region, the trapped diurnal tides are likely generated by resonant forcing of diurnal tides through the processes described in Chapman (1989) and Padman et al. (1992). The trapped semidiurnal tides are possibly generated locally. Internal waves are generated over critical slopes where the ratio $\gamma = \beta/\alpha$ is unity. Here β is the topographic slope and $\alpha = (\omega^2 - f^2)^{1/2}(N^2 - \omega^2)^{-1/2}$ is the characteristic along which linear internal waves with frequency ω propagate (i.e. the horizontal slope of the internal wave ray). Although the critical condition ($\gamma = 1$) is optimal, internal wave generation at supercritical slopes ($\gamma > 1$) is also common. The critical condition also leads to enhanced shear and turbulence (Eriksen, 1985). As the turning latitude where $\omega = f$ is approached; however, the non-linear terms become increasingly important and the parameter γ becomes a crude indicator of the potential for internal wave generation (Vlasenko et al., 2003). Poleward of the turning latitude negative background vorticities can effectively reduce

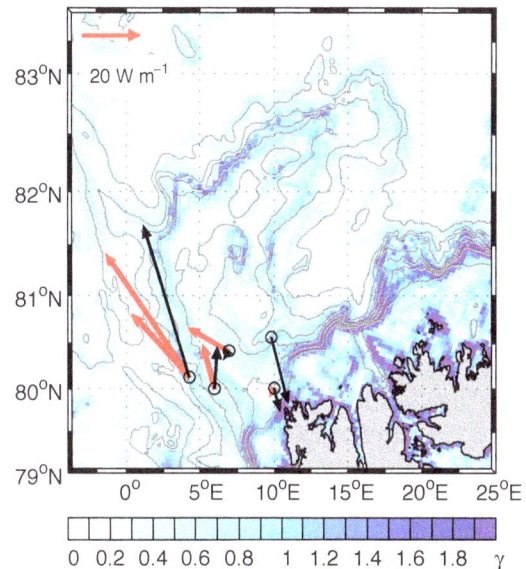

Figure 9. Contours of γ, the ratio of semidiurnal internal wave characteristic to the bottom slope. An effective inertial frequency of $0.95 f$ (i.e. assuming negative, anti-cyclonic, relative vorticity of $0.05 f$) is used in the calculations. Superimposed are the isobaths (250, 500, 750, 1000, 1500, 2000, and 3000 m), station positions (circles), and the depth-integrated baroclinic semidiurnal energy flux (arrows) with scale given on the top left. Red arrows are inferred from the observations, whereas the black arrows are from STORMTIDE (sum of M_2 and S_2) averaged over 10 km range of the stations. At station 1 the two observed flux vectors are obtained from the VM-ADCP and the XCP data.

the inertial frequency and thereby potentially allow for linear sub-inertial internal waves (Kunze and Toole, 1997). We evaluate γ for the M_2 frequency, using an effective inertial frequency of $0.95 f$, and the buoyancy frequency from the survey mean full-depth CTD profiles. Anti-cyclonic loops and negative background vorticity on the order of $10^{-5} \, \mathrm{s}^{-1}$ were previously inferred from floats over YP (Gascard et al., 1995). For the M_2 constituent, the northwestern flanks and the seamount region on the plateau are characterized by values of γ between 0.8 and 1.2 (Fig. 9), which favour the generation of semidiurnal internal tides and suggest elevated shear and mixing, consistent with earlier observations. The amplitude of depth-integrated baroclinic semidiurnal energy flux is about 15–50 W m^{-1} at station 1, and can be compared to the values from STORMTIDE (Fig. 9). The sampling is limited, however, Fig. 9 shows depth-integrated semidiurnal baroclinic energy flux on the southern YP where γ is near unity. The large STORMTIDE baroclinic semidiurnal fluxes apparent on the flanks of the YP and along the slope north of Svalbard (Fig. 6a) are also co-located with near-critical slopes. An entirely different generation mechanism is possible when non-linearity is strong, leading to unsteady lee waves at relatively short horizontal length scales (see, e.g., Vlasenko et al., 2003). The STORMTIDE results, however, do not include the effect of such non-linear internal wave generation.

A crude estimate of the total energy dissipated over the YP can be made from observations as follows. We use the same domain surface area of 2×10^{11} m^2 and the mean water depth of 2300 m inferred from the model. We choose two representative ε profiles: a lower bound ($\varepsilon_{2300\,\text{m-min}}$) and an upper bound ($\varepsilon_{2300\,\text{m-max}}$), each constructed to 2300 m depth. For the lower and upper bounds, we extend the $\varepsilon_{\min}$ and $\varepsilon_{\max}$ profile, respectively, of stations 1 and 2 to 2300 m depth. Station 2, comparably to station 3, is representative of the typical upper level of dissipation rates over the YP, excluding localized, enhanced dissipation observed on the northwestern flanks and the seamount over the plateau. At such locations where the bottom topography is near critical to the semidiurnal waves (Fig. 9), the dissipation profile of station 4 would be more appropriate. Using $\varepsilon_{2300\,\text{m-min}}$ and $\varepsilon_{2300\,\text{m-max}}$, the lower and upper bounds on the total energy dissipated are 100 and 800 MW, respectively. Assuming 10–15 % of the region is characterized by exceptionally enhanced dissipation rates (hot spots, Sect. 5), using 2300 m integrated dissipation from station 4, 400–600 MW can be attributed to localized mixing regions. The bounds on the dissipation rate, including the hot spots, are thus approximately 500 and 1100 MW. This observation-based estimate of the total dissipation includes a contribution from dissipation of internal tides (of diurnal and semidiurnal period) as well as other contributions such as turbulence due to mean shear.

The domain-integrated dissipation from STORMTIDE is 931 MW (sum of K_1 and M_2, where M_2 is 314 MW), in the range inferred from the observations, and suggests that almost the entire dissipation (below 50 m depth) can be explained by the dissipation of baroclinic tidal energy. Using a global domain numerical model with 10 vertical layers, Simmons et al. (2004) reported approximately 1 GW of M_2 conversion rate in the Fram Strait–Yermak Plateau region. Their box for the calculation of this domain covers 76–90° N between 20° W and 15° E. For the same domain, we obtain 0.62 GW for M_2 and 0.63 GW of K_1 from STORMTIDE. According to our model result in the Fram Strait–Yermak Plateau region, about 50 % of the M_2 conversion, and almost 100 % of the K_1 conversion occurs over the YP. Our calculations suggest that 40–80 % of the energy contained in the internal tides in this extended region is dissipated around the YP.

The volume-integrated total dissipation can be converted to an eddy diffusivity using the Osborn model (Osborn, 1980), which can then be related to turbulent vertical heat flux using the mean vertical temperature gradient. We extracted annual average temperature and salinity profiles from the Polar Science Centre Hydrographic Climatology (Steele et al., 2001), from 125 grid points in the domain used in the YP energetics calculations. Domain-averaged annual mean temperature profile below 100 m increases with depth at a rate of 1° C per 100 m down to the core of the Atlantic layer at 250 m depth below which the temperature decreases. If the entire dissipation takes place in this 150 m thick layer between 100 and 250 m (average $N^2 = 1.5 \times 10^{-5}$ s^{-2}), the average upward turbulent heat flux is obtained as 17 W m^{-2}. This value is comparable to the average ocean-to-ice heat flux of 22 W m^{-2} measured by McPhee et al. (2003) in the YP region. The dissipation of baroclinic tidal energy is thus a significant contributor to turbulent mixing and cooling of the Atlantic layer north of Svalbard.

7 Conclusions

Observations made in summer 2007 over the southern part of the Yermak Plateau (YP), together with results from a global high-resolution ocean circulation and tide model simulation (STORMTIDE) are used to investigate the role of tides, topography, and trapped internal tides in turbulent mixing near the YP. The plateau located northwest of Svalbard is of interest because it is the main topographic obstacle for the Atlantic Water carried by the West Spitsbergen Current to the Arctic. Tidal forcing, barotropic-to-baroclinic energy conversion rates, baroclinic energy fluxes, and dissipation rates in the region are discussed. Observational-based analysis suffers from errors as a result of short sampling duration, vertically imperfect sampling, and sloping topography. The STORMTIDE model results are limited to interpretation for linear wave characteristics.

Depth-integrated conversion rates, baroclinic energy fluxes, and dissipation rates show large spatial variability over the YP. The energetics are dominated by the K_1 and M_2 constituents. The volume-integrated conversion rate over the region enclosing the topographic feature is 322 MW for M_2 and 618 MW for K_1. This corresponds to about 50 % of the M_2 and approximately all of the K_1 conversion in a larger domain covering the entire Fram Strait, extended to the North Pole (76–90° N, 20° W–15° E). Despite the large energy conversion, internal tides are trapped with a negligible radiation out of the YP region, implying substantial local dissipation. The suggested enhanced levels of dissipation are supported by past observations showing high dissipation rates and strong mixing over the upper slope on the northern flanks (Hunkins, 1986; Padman and Dillon, 1991) and in the vicinity of a seamount over the plateau (D'Asaro and Morison, 1992). At the YP region, the trapped diurnal tides are likely generated by resonant forcing of diurnal tides through the processes described in Chapman (1989) and Padman et al. (1992). The trapped semidiurnal tides are possibly generated locally over near critical slopes. When a plausible negative background relative vorticity is allowed, we find the bottom topography in these regions critical to the semidiurnal frequency. Similar bottom slopes are also typical over the shelf north of Svalbard, close to one of our station with the largest mixing rates.

An approximate local conversion–dissipation balance is inferred over regions shallower than 1000 m, and also in the deep part of the sloping flanks, deeper than about 2000 m. On average, there is radiation of the baroclinic K_1 and M_2 energy away from the upper part of the slope, which is dissipated over the deeper isobaths. Most of the dissipation occurs in water deeper than 1500 m. Approximately the entire K_1 energy radiated in the region is dissipated between 1000 and 1500 m depth. The dissipation of the radiated M_2 energy is broadly distributed over the 2250–4000 m isobaths.

From observations, we inferred lower and upper bounds on the total dissipation rate of 500 and 1100 MW, of which approximately 400–600 MW can be attributed to the contribution of hot spots. The domain-integrated dissipation from STORMTIDE is in the range inferred from the observations, and suggests that almost all the dissipation in the region can be attributed to the dissipation of baroclinic tidal energy. Using the climatological temperature profiles and stratification averaged above the core of the Atlantic layer, the volume-integrated total dissipation leads to an average upward turbulent heat flux that is comparable to the average ocean-to-ice heat flux measured in the YP region. Although our regional calculations are crude, they underscore that the dissipation of baroclinic tidal energy can be a significant contributor to turbulent mixing and cooling of the Atlantic layer north of Svalbard. The role of tidal forcing in the heat budget of the Arctic Ocean in general merits further studies.

Acknowledgements. This study is funded by the Research Council of Norway through project nos. 178641/S30 and 229786/E10. Authors thank the crew and the participants of the cruise for their efforts during the field work. Helpful comments provided by two anonymous reviewers and the topic editor John M. Huthnance are appreciated.

Edited by: J. M. Huthnance

References

Alford, M. H., Cronin, M. F., and Klymak, J. M.: Annual cycle and depth penetration of wind-generated near-inertial internal waves at Ocean Station Papa in the northeast Pacific, J. Phys. Oceanogr., 42, 889–909, 2012.

Allen, S. E. and Thomson, R. E.: Bottom-trapped subinertial motions over midocean ridges in a stratified rotating fluid, J. Phys. Oceanogr., 23, 566–581, 1993.

Amante, C. and Eakins, B.: ETOPO1 1 Arc-Minute Global Relief Model: Procedures, Data Sources and Analysis. NOAA Technical Memorandum NESDIS NGDC-24. National Geophysical Data Center, NOAA, 2009.

Brink, K. H.: The effect of stratification on seamount-trapped waves, Deep-Sea Res. A., 36, 825–844, 1989.

Chapman, D. C.: Enhanced subinertial diurnal tides over isolated topographic features, Deep-Sea Res., 36, 815–824, 1989.

Chen, C., Gao, G., Qi, J., Proshutinsky, A., Beardsley, R. C. and Kowalik, Z., Lin, H., and Cowles, G.: A new high-resolution unstructured-grid finite-volume Arctic Ocean model (AO-FVCOM): an application for tidal studies, J. Geophys. Res., 114, C08017, doi:10.1029/2008JC004941, 2009.

D'Asaro, E. A. and Morison, J. H.: Internal waves and mixing in the Arctic Ocean, Deep-Sea Res., 39, S459–S484, 1992.

Egbert, G. D. and Ray, R. D.: Significant dissipation of tidal energy in the deep ocean inferred from satellite altimeter data, Nature, 405, 775–778, 2000.

Eriksen, C. C.: Implications of ocean bottom reflection for internal wave spectra and mixing, J. Phys. Oceanogr., 15, 1145–1156, 1985.

Falahat, S. and Nycander, J.: On the generation of bottom-trapped internal tides, J. Phys. Oceanogr., 45, 526–545, doi:10.1175/JPO-D-14-0081.1, 2014.

Fer, I.: Scaling turbulent dissipation in an Arctic fjord, Deep-Sea Res. II, 53, 77–95, 2006.

Fer, I. and Sundfjord, A.: Observations of upper ocean boundary layer dynamics in the marginal ice zone, J. Geophys. Res., 112, C04012, doi:10.1029/2005JC003428, 2007.

Fer, I., Skogseth, R., and Geyer, F.: Internal waves and mixing in the Marginal Ice Zone near the Yermak Plateau, J. Phys. Oceanogr., 40, 1613–1630, 2010.

Foreman, M. G. G., Cherniawsky, J. Y., and Ballantyne, V. A.: Versatile harmonic tidal analysis: improvements and applications, J. Atmos. Ocean. Technol., 26, 806–817, doi:10.1175/2008jtecho615.1, 2009.

Garrett, C. and Kunze, E.: Internal tide generation in the deep ocean, Annu. Rev. Fluid Mech., 39, 57–87, 2007.

Gascard, J. C., Richez, C., and Roaualt, C.: New insights on large-scale oceanography in Fram Strait: the West Spitsbergen Current, in: Arctic oceanography, marginal ice zones and continental shelves, edited by: Smith Jr., W. O. and Grebmeier, J., vol. 49, chap. 5, 131–182, AGU, Washington D.C., USA, 1995.

Hunkins, K.: Anomalous diurnal tidal currents on the Yermak Plateau, J. Mar. Res., 44, 51–69, 1986.

Huthnance, J. M.: On the diurnal tidal currents over Rockall Bank, Deep-Sea Res., 21, 23–35, 1974.

Huthnance, J. M.: Large tidal currents near Bear Island and related tidal energy losses from the North Atlantic, Deep-Sea Res., 28A, 51–70, 1981.

Johnston, T. M. S. and Rudnick, D. L.: Trapped diurnal internal tides, propagating semidiurnal internal tides, and mixing estimates in the California Current System from sustained glider observations, 2006–2012, Deep-Sea Res. II, 112, 61–78, doi:10.1016/j.dsr2.2014.03.009, 2014.

Jungclaus, J. H., Keenlyside, N., Botzet, M., Haak, H., Luo, J. J., Latif, M., Marotzke, J., Mikolajewicz, U., and Roeckner, E.: Ocean circulation and tropical variability in the coupled model ECHAM5/MPI-OM, J. Clim., 19, 3952–3972, doi:10.1175/JCLI3827.1, 2006.

Kagan, B. A. and Sofina, E. V.: Surface and internal semidiurnal tides and tidally induced diapycnal diffusion in the Barents Sea: a numerical study, Cont. Shelf Res., 91, 158–170, doi:10.1016/j.csr.2014.09.010, 2014.

Kang, D. J. and Fringer, O.: Energetics of barotropic and baroclinic tides in the Monterey Bay area, J. Phys. Oceanogr., 42, 272–290, 2012.

Kelly, S. M., Nash, J. D., and Kunze, E.: Internal-tide energy over topography, J. Geophys. Res., 115, C06014, doi:10.1029/2009JC005618, 2010.

Klymak, J. M., Pinkel, R., and Rainville, L.: Direct breaking of the internal tide near topography: Kaena ridge, Hawaii, J. Phys. Oceanogr., 38, 380–399, 2008.

Kunze, E. and Toole, J. M.: Tidally driven vorticity, diurnal shear, and turbulence atop Fieberling Seamount, J. Phys. Oceanogr., 27, 2663–2693, 1997.

Kunze, E., Rosenfeld, L. K., Carter, G. S., and Gregg, M. C.: Internal waves in Monterey Submarine Canyon, J. Phys. Oceanogr., 32, 1890–1913, 2002.

Leaman, K. D. and Sanford, T. B.: Vertical energy propagation of inertial waves: A vector spectral analysis of velocity profiles, J. Geophys. Res., 80, 1975–1978, 1975.

McPhee, M. G., Kikuchi, T., Morison, J. H., and Stanton, T. P.: Ocean-to-ice heat flux at the North Pole environmental observatory, Geophys. Res. Lett., 30, 2274, doi:10.1029/2003GL018580, 2003.

Müller, M.: On the space- and time-dependence of barotropic-to-baroclinic tidal energy conversion, Ocean Model., 72, 242–252, doi:10.1016/j.ocemod.2013.09.007, 2013.

Müller, M., Cherniawsky, J. Y., Foreman, M. G. G., and von Storch, J. S.: Global M_2 internal tide and its seasonal variability from high resolution ocean circulation and tide modeling, Geophys. Res. Lett., 39, L19607, doi:10.1029/2012gl053320, 2012.

Müller, M., Cherniawsky, J., Foreman, M. G., and von Storch, J.-S.: Seasonal variation of the M_2 tide, Ocean Dyn., 64, 159–177, doi:10.1007/s10236-013-0679-0, 2014.

Nash, J. D., Alford, M. H., and Kunze, E.: Estimating internal wave energy fluxes in the ocean, J. Atmos. Ocean. Technol., 22, 1551–1570, 2005.

Nash, J. D., Kunze, E., Lee, C. M., and Sanford, T. B.: Structure of the baroclinic tide generated at Kaena Ridge, Hawaii, J. Phys. Oceanogr., 36, 1123–1135, 2006.

Niwa, Y. and Hibiya, T.: Estimation of baroclinic tide energy available for deep ocean mixing based on three-dimensional global numerical simulations, J. Oceanogr., 67, 493–502, 2011.

Nycander, J.: Generation of internal waves in the deep ocean by tides, J. Geophys. Res., 110, C10028, doi:10.1029/2004jc002487, 2005.

Osborn, T. R.: Estimates of the local rate of vertical diffusion from dissipation measurements, J. Phys. Oceanogr., 10, 83–89, 1980.

Padman, L. and Dillon, T.: Turbulent mixing near the Yermak Plateau during the coordinated Eastern Arctic Experiment, J. Geophys. Res., 96, 4769–4782, 1991.

Padman, L., Plueddemann, A. J., Muench, R. D., and Pinkel, R.: Diurnal tides near the Yermak Plateau, J. Geophys. Res., 97, 12639–12652, 1992.

Phillips, O. M.: The Dynamics of the Upper Ocean, 2nd edn., Cambridge University Press, Cambridge, UK, 1977.

Plueddemann, A. J.: Internal wave observations from the Arctic Environmental Drifting Buoy, J. Geophys. Res., 97, 12619–12638, 1992.

Pnyushkov, A. V. and Polyakov, I. V.: Observations of tidally induced currents over the continental slope of the Laptev Sea, Arctic Ocean, J. Phys. Oceanogr., 42, 78–94, doi:10.1175/JPO-D-11-064.1, 2012.

Rhines, P. B.: Slow oscillations in an ocean of varying depth, 2: Islands and seamounts, J. Fluid Mech., 37, 191–205, 1969.

Robertson, R.: Internal tides and baroclinicity in the southern Weddell Sea 1. Model description, J. Geophys. Res., 106, 27001–27016, 2001.

Simmons, H., Chang, M.-H., Chang, Y.-T., Chao, S.-Y., Fringer, O., Jackson, C., and Ko., D.: Modeling and prediction of internal waves in the South China Sea, Oceanography, 24, 88–99, doi:10.5670/oceanog.2011, 2011.

Simmons, H. L., Hallberg, R. W., and Arbic, B. K.: Internal wave generation in a global baroclinic tide model, Deep-Sea Res. II, 51, 3043–3068, 2004.

Sirevaag, A. and Fer, I.: Early spring oceanic heat fluxes and mixing observed from drift stations north of Svalbard, J. Phys. Oceanogr., 39, 3049–3069, 2009.

Stammer, D., Ray, R. D., Andersen, O. B., Arbic, B. K., Bosch, W., Carrère, L., Cheng, Y., Chinn, D. S., Dushaw, B. D., Egbert, G. D., Erofeeva, S. Y., Fok, H. S., Green, J. A. M., Griffiths, S., King, M. A., Lapin, V., Lemoine, F. G., Luthcke, S. B., Lyard, F., Morison, J., Müller, M., Padman, L., Richman, J. G., Shriver, J. F., Shum, C. K., Taguchi, E., and Yi, Y.: Accuracy assessment of global barotropic ocean tide models, Rev. Geophys., 52, 243–282, doi:10.1002/2014RG000450, 2014.

Steele, M., Morley, R., and Ermold, W.: PHC: A global ocean hydrography with a high-quality Arctic Ocean, J. Clim., 14, 2079–2087, 2001.

Tanaka, Y., Hibiya, T., Niwa, Y., and Iwamae, N.: Numerical study of K1 internal tides in the Kuril straits, J. Geophys. Res., 115, C09016, doi:10.1029/2009JC005903, 2010.

Tanaka, Y., Yasuda, I., Hasumi, H., Tatebe, H., and Osafune, S.: Effects of the 18.6-yr modulation of tidal mixing on the North Pacific bidecadal climate variability in a coupled climate model, J. Clim., 25, 7625–7642, doi:10.1175/Jcli-D-12-00051.1, 2012.

Vlasenko, V., Stashchuk, N., Hutter, K., and Sabinin, K.: Nonlinear internal waves forced by tides near the critical latitude, Deep-Sea Res. I, 50, 317–338, 2003.

Vlasenko, V., Stashchuk, N., and Hutter, K.: Baroclinic tides. Theoretical modeling and observational evidence, Cambridge University Press, 2005.

Wang, D. P. and Mooers, C. N. K.: Coastal-trapped waves in a continuously stratified ocean, J. Phys. Oceanogr., 6, 853–863, 1976.

Wijesekera, H., Padman, L., Dillon, T., Levine, M., Paulson, C., and Pinkel, R.: The application of internal-wave dissipation models to a region of strong mixing, J. Phys. Oceanogr., 23, 269–286, 1993.

10

Ventilation of the Mediterranean Sea constrained by multiple transient tracer measurements

T. Stöven and T. Tanhua

Helmholtz Centre for Ocean Research Kiel, GEOMAR, Kiel, Germany

Correspondence to: T. Stöven (tstoeven@geomar.de)

Abstract. Ventilation is the primary pathway for atmosphere–ocean boundary perturbations, such as temperature anomalies, to be relayed to the ocean interior. It is also a conduit for gas exchange between the interface of atmosphere and ocean. Thus it is a mechanism whereby, for instance, the ocean interior is oxygenated and enriched in anthropogenic carbon. The ventilation of the Mediterranean Sea is fast in comparison to the world ocean and has large temporal variability. Here we present transient tracer data from a field campaign in April 2011 that sampled a unique suite of transient tracers (SF_6, CFC-12, ^3H and ^3He) in all major basins of the Mediterranean. We apply the transit time distribution (TTD) model to the data in order to constrain the mean age, the ratio of the advective / diffusive transport and the number of water masses significant for ventilation.

We found that the eastern part of the eastern Mediterranean can be reasonably described with a one-dimensional inverse Gaussian TTD (IG-TTD), and thus constrained with two independent tracers. The ventilation of the Ionian Sea and the western Mediterranean can only be constrained by a linear combination of IG-TTDs. We approximate the ventilation with a one-dimensional, two inverse Gaussian TTD (2IG-TTD) for these areas and demonstrate a possibility of constraining a 2IG-TTD from the available transient tracer data. The deep water in the Ionian Sea has a mean age between 120 and 160 years and is therefore substantially older than the mean age of the Levantine Basin deep water (60–80 years). These results are in contrast to those expected by the higher transient tracer concentrations in the Ionian Sea deep water. This is partly due to deep water of Adriatic origin having more diffusive properties in transport and formation (i.e., a high ratio of diffusion over advection), compared to the deep water of Aegean Sea origin that still dominates the

deep Levantine Basin deep water after the Eastern Mediterranean Transient (EMT) in the early 1990s. The tracer minimum zone (TMZ) in the intermediate of the Levantine Basin is the oldest water mass with a mean age up to 290 years. We also show that the deep western Mediterranean has contributed approximately 40 % of recently ventilated deep water from the Western Mediterranean Transition (WMT) event of the mid-2000s. The deep water has higher transient tracer concentrations than the mid-depth water, but the mean age is similar with values between 180 and 220 years.

1 Introduction

The Mediterranean Sea is a marginal sea, where the observational record shows significant changes in ventilation (Schneider et al., 2014). The most prominent transient event in the eastern Mediterranean Sea (EMed) is the transfer of the deep water source from the Adriatic Sea to the Aegean Sea and Sea of Crete and vice versa. The observed massive dense water input from the Aegean Sea and Sea of Crete in the early 1990s is known as the Eastern Mediterranean Transient (EMT) event (Roether et al., 1996; Klein et al., 1999; Lascaratos et al., 1999). The extensive deep water formation in the western Mediterranean Sea (WMed) between 2004 and 2006, known as the Western Mediterranean Transition (WMT) event (Schroeder et al., 2008, 2010), is thought to have been triggered by the EMT event (Schroeder et al., 2006). Nevertheless, both events are part of a general circulation pattern which can be observed in the Mediterranean Sea. The surface water in the WMed is supplied by less dense Atlantic water (AW) through the Strait of Gibraltar. The AW flows eastwards at depths < 200 m into the Tyrrhenian Sea and into the EMed via the Strait of Sicily. The salinity of the AW increases along the pathway from 36.5 to > 38 due to net evaporation and is then described as modified Atlantic water (MAW) (Wuest, 1961). The heat loss during winter time in the MAW in the EMed leads to a sufficient increase of density to form the Levantine intermediate water (LIW) at depths between 200 and 600 m (Brasseur et al., 1996; Wuest, 1961). The exact area of the LIW formation process is poorly constrained and possibly variable, but it is expected to be in the eastern part of the EMed near Rhodes (Malanotte-Rizzoli and Hecht, 1988; Lascaratos et al., 1993; Roether et al., 1998). The main volume of the LIW flows back westwards over the shallow sill between Sicily and Tunisia entering the Tyrrhenian Sea along the continental slope of Italy (Wuest, 1961). Parts of the LIW enter the Adriatic Sea via the Strait of Otranto, where it serves as an initial source of the Adriatic Sea overflow water (ASOW). The formation of ASOW in the southern Adriatic pit is based on interactions between the LIW and water masses coming from the northern Adriatic Sea as well as the natural preconditioning factors, for example, wind stress and heat loss (Artegiani et al., 1996a, b). The ASOW flows over the sill of Otranto into the Ionian Sea intruding into the bottom layer and thus representing a source of eastern Mediterranean deep water (EMDW) (Schlitzer et al., 1991; Roether and Schlitzer, 1991). Furthermore, the Ionian Sea is connected with the Levantine Sea via the Cretan Passage, where portions of newly formed EMDW reach the deep water of the Levantine Sea. In 1992–1993, the water-mass conditions in the well-ventilated Aegean Sea and Sea of Crete changed into a more salty and cold state, sufficient enough to initialize the massive dense water input of Cretan deep water (CDW) into the abyssal basins of the EMed (Klein et al., 1999). This EMT event resulted in a disruption of the usual formation pattern of the EMDW. The Adriatic Sea as a major deep water source was thereby replaced by the Aegean Sea and Sea of Crete with the consequence that the bottom layer of the Ionian Sea was now supplied with dense water via the Antikythera Strait and the Levantine Sea via the Kasos Strait. The simultaneous dense water input into both basins, in conjunction with the large amount of the outflow, caused an uplift of the intermediate water layers in the Ionian and Levantine seas. One consequence of the EMT event seemed to be the preconditioning of the WMT event in 2004–2006 by uplifted water masses entering the WMed via the Strait of Sicily. However, the major triggering factor was the heat loss due to the mistral in the Gulf of Lion and the Balearic Sea, which resulted in the extensive deep water formation in the WMed. Although the total magnitude of the WMT event was smaller than the EMT event it was still sufficient to cause a near-complete renewal of the western Mediterranean deep water (WMDW). Recent water-mass analyses indicate, that the EMed is returning to a pre-EMT state with the Adriatic Sea as a major deep water source (Hainbucher et al., 2006; Rubino and Hainbucher, 2007).

The analysis of ventilation processes and their periodicity is an important issue in the understanding of their climate impact. Such analyses are not trivial, however, given the range of methods, views and concepts involved in understanding ventilation processes. For example, transient tracer distributions were used in the EMed by Roether et al. (1996, 2007) and in the WMed by Rhein et al. (1999) to quantify ventilation timescales. The time dependence of the transient tracers were used for first-age estimates based on simple approaches which provided an estimate of an apparent age or tracer age (Roether et al., 1998; Roether and Lupton, 2011). More complex age models, for example, the transit time distribution (TTD) model, account for the influence of mixing processes leading to a more realistic mean age estimate. A TTD related approach of age spectra modeling was carried out by Steinfeldt (2004) for the EMed in 1987, providing one of the first competing age estimates to the tracer age approach. The TTD model was, more recently, used for a time-series analysis of the entire Mediterranean Sea by Schneider et al. (2014). Such concepts of age and timescales in the ocean can also contain commonly used parameters like volume fluxes (changes) per time unit, normally stated as ventilation rate as well as residence time, influence time, tracer age, apparent age and mean age. Some of these parameters are occasionally presented in different contexts and meanings, leading to controversial discussions (Delhez et al., 2013). However, this paper does not include quantitative statements about ventilation rates or residence times. This study was focused on providing methods to constrain the mean age of the Mediterranean Sea within the framework of TTD models, which were then used to describe ventilation in terms of "age" structure and further qualitative characteristics, such as the advective and diffusive behavior of the different water masses. To this end, measurements of the transient tracers dichlorodifluoromethane (CFC-12) and sulfur hexafluoride (SF_6) as well as helium isotopes (^{3}He, ^{4}He) and tritium (^{3}H) were carried out during the *Meteor* expedition M84/3 in 2011 yielding a comprehensive data set of time dependent tracers (Fig. 1). The insights gained in this TTD model application method should also provide possible improvements for continuative estimates of ventilation rates, defined by Primeau and Holzer (2006); Hall et al. (2007) and anthropogenic carbon contents (Tanhua et al., 2008).

2 Materials and method

2.1 Transient tracers

2.1.1 Chronological transient tracers

The uses of chronological transient tracers, such as chlorofluorocarbons (CFCs) and SF_6 to estimate the age of a water mass are based on an increasing tracer concentration in the atmosphere. Concentrations of trace gases in the atmosphere are, for example, measured continuously by the world-wide

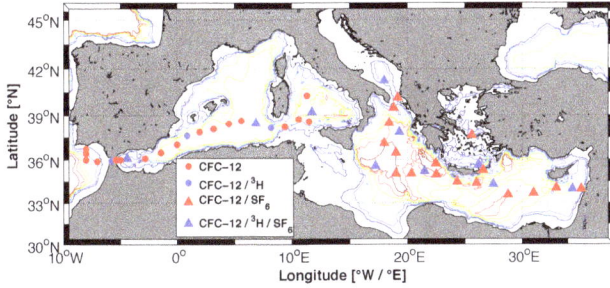

Figure 1. Transient tracer sample stations of the M84/3 cruise from Istanbul to Vigo. Triangles indicate stations including SF_6 measurements and a blue color coding indicates tritium measurements. CFC-12 was measured at all stations on this map. The depth contours are 500 : 500 : 3500.

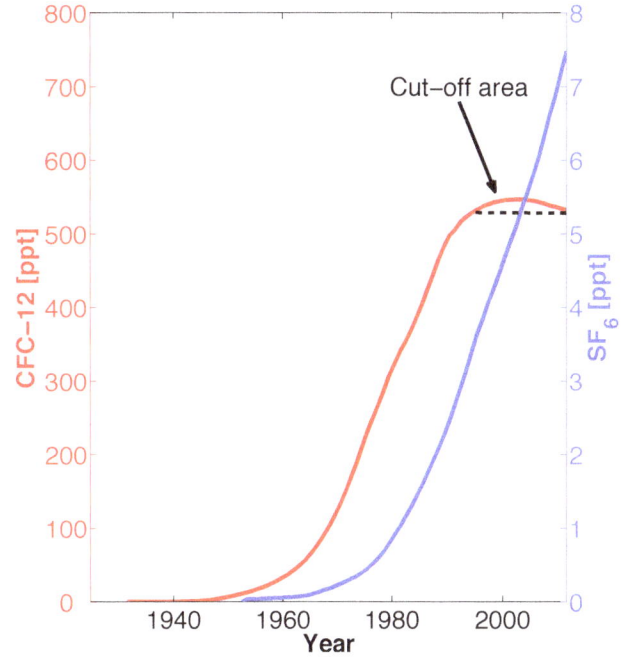

Figure 2. Atmospheric histories of CFC-12 (red) and of SF_6 (blue). The decreasing trend of CFC-12 produces a cut-off area between 1994 and 2011, which means that any CFC-12 concentration above 532 ppt provides only inconclusive information about ventilation.

AGAGE network, so that the emission history and atmospheric variations over time are well known (Walker et al., 2000; Bullister, 2011), which then provide the tracer's input functions. The production of CFCs, and ultimately their emissions, were decreased in the late 1980s and then finally stopped in the 1990s as a result of the Montreal Protocol. This has resulted in a steady decrease of atmospheric CFC-12 concentrations such that in 2011 the concentration was 532 ppt (Bullister, 2011). Because the concentrations of CFC-12 in the atmosphere were increasing prior to the 1990s and then decreased up until 2011, CFC-12 concentrations in seawater that are higher than the 2011 atmospheric concentration could therefore represent one of two dates (i.e., before the 1990s or after the 1990s, Fig. 2). To this end, CFC-12 concentrations are inconclusive for tracer age determination between 1994 and 2011. SF_6 concentrations are still increasing approximately linearly but the emission rate is relatively low so that the concentration in the atmosphere is reportedly below 8 ppt. Some local restrictions are in place for the production and use of SF_6, but an international agreement has yet to be reached, despite its global warming potential of 22 000 (Houghton et al., 1996). Tracers enter the ocean's surface layer via gas exchange and the solubility is a function of temperature, salinity and the physical nature of the molecule. Solubility functions are available for most of the CFCs and SF_6 (Warner and Weiss, 1985; Bullister et al., 2002) and are used to convert the measured gravimetric units (e.g., $pmol\,kg^{-1}$ for CFC-12 and $fmol\,kg^{-1}$ for SF_6), into the partial pressure (ppt) of the tracer. The partial pressure is the preferred choice since it is independent of pressure, salinity and temperature and thus directly comparable within the complete water column and atmosphere.

Chronological tracers are conserved tracers with no significant sources or sinks in the ocean interior. The concentration in the water column depends on the last time the water parcel was in contact with the atmosphere and on the influence of mixing and diffusion.

2.1.2 Radioactive transient tracers

The radioactive tracers such as tritium and its decay product helium-3 (^{3}He) form the second class of transient tracers. Tritium has a natural background concentration of ≈ 0.3 tritium units (TU) in the atmosphere, where 1 TU equals the number of one tritium atom per 10^{18} hydrogen atoms (Ferronsky and Polyakov, 1982). Due to nuclear bomb tests in the late 1950s and 1960s, the tritium concentration increased up to 100 TU in the atmosphere and declined afterwards to a current concentration of 1–1.2 TU in 2011 (Roether et al., 2013). The input of tritium into the ocean surface layer is a function of radioactive decay in the atmosphere, vapor pressure, the variance of location and magnitude of precipitation and fresh water flux by riverine input. Tritium decays to helium-3, known as tritiugenic helium-3 (^{3}He$_{trit}$) which equilibrates with the atmosphere as long as the water parcel remains in the boundary layer of gas exchange. Once the water reaches the oceans interior, radioactive decay serves as time varying sink. However, the total concentration of helium-3 (^{3}He$_{tot}$) in sea water consists of several shares of different sources. The determination of the ^{3}He$_{trit}$ share requires the knowledge of excess helium-3 (^{3}He$_{ex}$), that is, the surface saturation and the terrigenic share (^{3}He$_{terr}$) from the earth crust and mantle (i.e., the sea floor as source of ^{3}He).

The Mediterranean is characterized by higher tritium concentrations than the Atlantic due to continental influences in terms of weather conditions and fresh water input. A commonly used tritium input function (TIF) for the North Atlantic was obtained by Dreisigacker and Roether (1978) and further developed for the EMed by Roether et al. (1992). Based on this data set, another TIF for the EMed was created by R. Steinfeldt (unpublished data) where the data after 1974 was extrapolated by using the decay function of tritium. In the WMed the surface layer is mainly influenced by the inflow of Atlantic water (AW), so that the input function needs to be corrected for the degree of dilution. The difference between the mean surface tritium concentration of the M84/3 cruise and the concentration value of the TIF by R. Steinfeldt (unpublished data) of the same year can be used to determine correction factors for the eastern and western Mediterranean. Under the assumption that the determined offset is constant over years, both factors can be used as an offset correction to create two alternative input functions (Fig. 3) which can be applied to a TTD mixing model (see below). The corrected TIFs have a surface (input) concentration which is 15 % lower in the EMed and 35 % lower in the WMed than suggested in the original input function. Figure 3 shows the recent TIF of the Mediterranean Sea by Roether et al. (2013), which also relies on the data set of Dreisigacker and Roether (1978). This TIF was recalculated for the EMed by using a dilution factor and mean surface tritium concentrations obtained during several cruises between 1974 and 2011. Comparing both recent TIFs of the EMed shows that the shape of both curves is relatively similar. This indicates that both input functions seem to be useful approaches for the EMed despite the different methods used in their estimation. However, by using an interpolated form of the input function of Roether et al. (2013), a higher mean age is yielded compared to the input function we obtained. The main deviation from the decay-based input function is the data point of 1978, following that the interpolated tritium concentrations were significantly elevated between 1975 and 1987, producing differences in mean age. The mean deviation between the different TIFs and the original TIF of the North Atlantic are 86 % and 61 %, respectively (Roether et al., 2013), for the EMed and 43 % for the WMed.

2.2   Tracer age and the transit time distribution

The age of a water parcel can be described in different ways. For chronological transient tracers, the measured concentration of sample c in year t_s (year of sampling) can be set in relation to the same concentration c_0 with the relevant year t_{hist} of the atmospheric history of the tracer (Eq. 1).

$$c(t_s) = c_0(t_{hist}) \tag{1}$$

The difference between the year of sampling t_s and the obtained year t_{hist} defines the tracer age τ (Eq. 2).

$$\tau = t_s - t_{hist} \tag{2}$$

The tracer age of radioactive tracers depends on first order kinetics shown in Eq. (3). The initial concentration c_i, the in

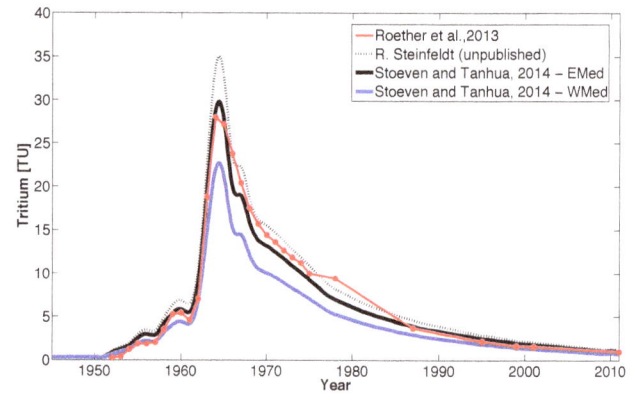

Figure 3. Input functions of tritium. The dotted black curve shows the decay based input function of the Mediterranean Sea by R. Steinfeldt (unpublished data). The black and blue curves describe the off set corrected input functions for the eastern and western Mediterranean Sea. The red curve shows the suggested input function by Roether et al. (2013) of the eastern Mediterranean Sea.

situ concentration c and the decay constant λ are the required parameters to calculate the elapsed time of a tracer in a water parcel.

$$\tau = \frac{1}{\lambda} \cdot \ln\left(\frac{c_i}{c}\right) \tag{3}$$

As mentioned above, tritium has, in addition to the radioactive decay, a relevant input function and thus an unknown part for c_i. Therefore, the share of $^3\mathrm{He_{trit}}$ needs to be determined which replaces the initial concentration of tritium and Eq. (3) can be rewritten as Eq. (4). A generally accepted value for the decay constant of tritium is $\lambda = 0.05576/a$ (Unterweger et al., 1980; Taylor and Roether, 1982).

$$\tau = \frac{1}{\lambda} \cdot \ln\left(1 + \frac{[^3\mathrm{He_{trit}}]}{[^3\mathrm{H}]}\right) \tag{4}$$

The informative value of a tracer age is relatively low because it is based on the assumption of a complete advective behavior neglecting any diffusive mixing process. However, there are also methods such as dilution models (Roether et al., 2013) and the tracer age of CFC-12 and SF_6 with a ≈ 14 year time lag (Tanhua et al., 2013c; Schneider et al., 2014) that allow an estimation of changes in ventilation.

The TTD model is based on the Green's function and was invented to describe atmospheric ventilation processes (Hall and Plumb, 1994). However, the basic idea that a parcel of molecules changes its location under the influence of advection and diffusion can also be applied to ventilation processes of the ocean. Equation (5) is an analytical expression of the Green's function which provides access to use field data within the TTD model (Waugh et al., 2003). It is based on a one-dimensional flow model with constant advective velocity and diffusivity and is therefore known as the one-dimensional inverse Gaussian transit time distribution (IG-TTD).

$$G(t) = \sqrt{\frac{\Gamma^3}{4\pi \Delta^2 t^3}} \cdot \exp\left(\frac{-\Gamma(t-\Gamma)^2}{4\Delta^2 t}\right) \qquad (5)$$

The key variables in this equation are Γ for the mean age and Δ for the width of the distribution. The age spectra t is defined by the initial year t_i of the atmospheric history or the input function of the tracer and the year of sampling t_s. To give a statement on the share of advection and diffusion, the Δ/Γ ratio can be used. A low ratio, such as 0.4–0.8, indicates a high advective part (e.g., extensive deep water formations), whereas a high ratio like $1.2 - 1.8$ indicates a more diffusive character of the water parcel. The definite integral of Eq. (6) contains the link between the measured concentration of a sample $c(t_s)$ and the mean age of the TTD. The parameter r describes the location of the water parcel, t – the time range of the tracer and $e^{-\lambda t}$ – the decay correction for radioactive transient tracers.

$$c(t_s, r) = \int_0^\infty c_0(t_s - t)e^{-\lambda t} \cdot G(t, r)\,dt \qquad (6)$$

A further approach to determine a mean age is the linear combination of two distributions which is shown in Eq. (7). Hereby, α describes the percentage ratio between the two G-functions. Such a two inverse Gaussian TTD (2IG-TTD) can be envisioned for two water masses with different histories (age), but with similar density, that mixes in the ocean interior. This model has been explored by, for instance, Waugh et al. (2002).

$$c(t_s, r) = \int_0^\infty c_0(t_s - t)e^{-\lambda t} \cdot$$
$$[\alpha\, G(\Gamma_1, \Delta_1, t, r) + (1-\alpha)\, G(\Gamma_2, \Delta_2, t, r)]\,dt \qquad (7)$$

The number of possible combinations of distributions and parameters provides a comprehensive concept of age modeling in the ocean. The main complexity consists of finding accurate and reasonable solutions related to the field data. The mean age is then determined by Eq. (8), whereas Γ_1 and Γ_2 are the partial mean age results of each G-function.

$$\Gamma = \alpha \cdot \Gamma_1 + (1-\alpha) \cdot \Gamma_2 \qquad (8)$$

2.3 Practical application of the TTD model

A common procedure described in several published articles (e.g., Schneider et al., 2010; Waugh et al., 2006, 2004) is to apply the IG-TTD with a ratio of $\Delta/\Gamma = 1.0$ to the tracer data to calculate a water-mass mean age. The Δ/Γ ratio has been demonstrated to be close to 1 in large parts of the world ocean (i.e., established as standard ratio). This standard ratio

can be used to analyze transient tracer time series in terms of changes in ventilation, where the rate of age growth yr^{-1} is more in focus than a precise mean age (Huhn et al., 2013). The recently published work by Schneider et al. (2014) is also based on the standard ratio of $\Delta/\Gamma = 1.0$, which allows a comparison to be made between all data sets and thus an analysis of changes during the period of interest. In the case of a time series, it is rarely possible to apply similar constraints to different data sets. The standard ratio should also be used within tracer surveys with only few sample points because local outliers of constrained data points can produce significant flaws in interpolation. For a comprehensive data set, consisting of more than one transient tracer, a constrained TTD model provides an alternative. The determined Δ/Γ ratios provide a first insight into the water-mass structure concerning the advective and dispersive behavior. The further analysis of ventilation processes, rates and recent changes in water masses as well as the estimation of the anthropogenic carbon column inventory is based on the determined mean age and thus dependent on the exact Δ/Γ ratio. However, it is an important principle to identify in which manner a TTD method was applied before comparing different studies.

2.3.1 Constraining the IG-TTD model

There have been several approaches made to constrain a TTD model. For instance, Waugh et al. (2002) uses the lower and upper mean age limit of the transient tracers and plot them against the standard deviation (σ) of the TTD. In the case of an IG-TTD it can be approximated as $\Delta = \sigma/\sqrt{2}$. The area which is spanned by all tracers then constrain the TTD. Other methods are based on property–property plots. For example are tracer concentrations plotted against each other including predicted concentration curves by the TTD for different Δ/Γ ratios, whereas the best fit of a predicted curve to the bunch of data provides a single overall constraint of the TTD (Waugh et al., 2004). A similar method is used by Schneider et al. (2012) where the CFC-12 mean age is plotted vs. the SF_6 mean age for different Δ/Γ ratios. Hereby, a good correlation between the data points and the bisecting line (slope = 1) denotes the ideal ratio. However, each method to constrain a TTD requires a transient tracer couple. The tracers of the couple need to have sufficiently different input functions to constrain the Δ/Γ ratio. Tracer couples with similar atmospheric histories (e.g., CFC-12 and CFC-11) will yield a wide range of possible outcomes and will result in a poorly constrained TTD. Useful couples are CFC-12–SF_6 and tritium–SF_6.

Our approach is based on constraining single data points instead of determining an overall Δ/Γ ratio. Therefore, the first step of data processing includes the calculation of the mean age for Δ/Γ ratios between 0.0 and 1.8 for every data point and tracer, always taking into account the correct input function of the source region. The determined data points of mean age vs. Δ/Γ ratio are used to obtain second-order

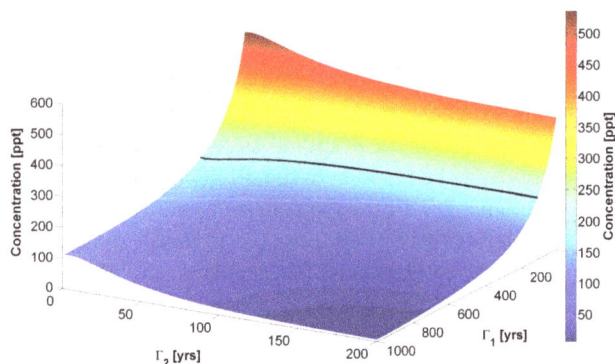

Figure 4. Example mean age matrix of CFC-12 with $^{\Delta_1}/_{\Gamma_1} = 1.4$, $^{\Delta_2}/_{\Gamma_2} = 0.6$ and $\alpha = 80$. The color-coding denotes the concentration of CFC-12 (in ppt, also on the z axis) with a black concentration contour line at 200 ppt; x and y axis denotes the mean age of the 2IG distributions that make up the TTD. The combination of all three-dimensional tracer matrices provides the needed information to constrain a 2IG-TTD, see text.

polynomial regressions. Following this, every sample point of every tracer can be expressed by a mean age function (Eq. 9).

$$\Gamma = a \left(^{\Delta}/_{\Gamma}\right)^2 + b\left(^{\Delta}/_{\Gamma}\right) + c \qquad (9)$$

The intersection between two mean age functions denotes the constrained Δ / Γ ratio and mean age. In some cases where no exact intersection can be found it is useful to determine the local minimum of a combined mean age function in the range of Δ / Γ. A local minimum indicates the point of the smallest difference in mean age, which should be used to a maximum difference of 5 years to ensure also the consideration of a mean analytical error of $\approx 4\%$ (see below). To obtain the mean age of such a minimum function, the average of both mean age values needs to be calculated. However, in some cases it is more meaningful to use one of the tracer's mean age rather than the average mean age. For example, the SF_6 mean age for recently ventilated waters is more significant than the CFC-12 mean age due to the recent non-transient input function of CFC-12. In contrast, the CFC-12 mean age should be used in older water layers where SF_6 concentrations are close to the detection limit (Tanhua et al., 2008).

A further aspect of the IG-TTD model is the validity area of each tracer couple, which defines the possible range of IG-TTD solutions. A rough classification of the specific validity area of a couple can be done by determining the tracer age differences. For example, if the difference of the tracer age between SF_6 and CFC-12 is large (10 years for the sampling year of 2011), it indicates that an IG distribution cannot explain the tracer distribution, and more refined models of the TTD are needed, for instance the linear combination of two IG-TTDs.

2.3.2 Constraining the 2IG-TTD model

Due to the five free parameters $\alpha, \Gamma_1, \Gamma_2, \Delta_1$ and Δ_2, the system of equations is under-determined for any tracer survey with less than five measured transient tracers. Most surveys include two or three transient tracers with sufficiently different atmospheric histories. Here we introduce one way to use an under-determined 2IG-TTD model. Based on oceanographic water-mass analysis one can estimate the composition of the current state of the water masses and roughly the underlying mixing processes. As described earlier, the western and eastern Mediterranean are both affected by an extensive deep water formation with recently ventilated and salty water from the surface and intermediate layers, respectively. The null hypothesis is that an old and more stationary water mass can be described by an IG-TTD which has been intruded by a younger water parcel described by another IG-TTD. Hereby the younger water parcel might be characterized by a more advective behavior with a low ratio (e.g., $\Delta / \Gamma = 0.6$). The ratio of the more stationary water mass is set to $\Delta / \Gamma = 1.4$, describing a typical ratio of a more diffusive/dispersive behavior. By making assumptions about the Δ / Γ ratio of both IG-TTDs one can calculate mean age matrices for different α's with $x = \Gamma_1$, $y = \Gamma_2$ and $z = C_{\text{tracer}}$. The concentration of a measured sample generates different concentration curves for each α in the x–y plane (Fig. 4). The predefined 2IG-TTD is constrained if there is an intersection area of the concentration curves of different tracers describing one mean age (Eq. 8).

2.4 Sampling and measurements

The expedition M84/3 from Istanbul (Turkey) to Vigo (Spain) took place from the 5 to 28 April in 2011 on the German research vessel *FS Meteor* (Tanhua et al., 2013a, b). Figure 1 shows an overview of the sample stations with different symbols denoting which tracers were measured. The transient tracers CFC-12 and SF_6 were sampled at nearly all stations in the EMed, whereas only three stations of SF_6 exist in the WMed. Tritium was sampled at 7 stations in the EMed and 6 stations in the WMed. The sampling depths were chosen to cover the most important water layers in a sufficient resolution. Starting with an minimum sampling depth increment of 25 m in the surface and mixed layer and ending with a maximum increment of 500 m in the deep water layers (see Table 1).

2.4.1 CFC-12 and SF_6

The measurements of CFC-12 and SF_6 were mainly performed on board. The water samples were taken with 250 mL glass syringes or 300 mL glass ampules, under exclusion of atmosphere, from the Niskin bottles. The syringes and ampules were stored in a cooling box filled with water of ≈ 0 °C to prevent outgasing of the tracers. The measurements were

Table 1. Standard sampling depths of the M84/3 cruise in 2011.

Increment [m]	Depth range [m]
25	0–100
50	100–300
100	300–600
200	600–1000
250	1000–2500
500	2500–bottom

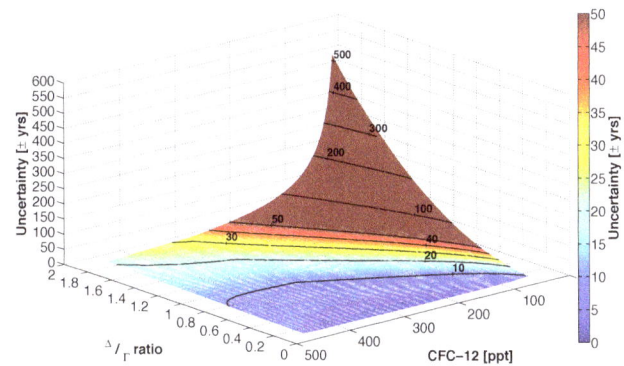

Figure 5. Absolute error of mean age calculations depending on CFC-12 concentrations and $^{\Delta}/_{\Gamma}$ ratios. The color coding is restricted to a maximum of 50 years for an improved error resolution of the main area.

carried out with similar analytical systems as described by Bullister and Wisegarver (2008) and Law et al. (1994). The first measurement system named VS1 consisted of a Shimadzu GC14a gas chromatograph equipped with an electron capture detector (ECD), stainless steel tubing system and Valco valves. An evacuated vacuum sparge tower (VST) was used to transfer the water sample out of the glass ampule into the measurement system. Due to the low pressure in the VST, most of the dissolved gases pass over into the head space during the filling process. The residual was purged out with nitrogen (ECD-quality). The analytes were trapped on a 1/16″ column packed with 70 cm *Heysep D* and then separated with a 1/8″ precolumn, packed with 30 cm *Porasil C* and a 1/8″ main column consisting of 180 cm *Carbograph 1AC* and a 20 cm *Molsieve 5 Å* tail end. The trap was installed in a Dewar filled with a bottom layer of liquid nitrogen. The distance between trap and cooling medium was regulated by a Lab Boy to hold a temperature range between −70 and −60 °C during the purge process. Due to some problems with the VS1 system and a sudden break down of the ECD several samples from key stations have been flame sealed in glass ampules for a later onshore measurement. The sealed ampules were measured during summer 2011 at the IfM-GEOMAR in Kiel with the repaired VS1 instrument and an installed ampule cracker system similar to Vollmer and Weiss (2002).

The second measurement system PT3 consisted of a Shimadzu GC2014 gas chromatograph with a similar basic setup like the VS1 system but with a different column composition, sample chamber and trap system. The 1/8″ precolumn consisted of 60 cm *Porasil C* and 10 cm *Molsieve 5 Å*, the 1/8″ main column of 180 cm *Carbograph 1AC* and 30 cm *Molsieve 5 Å*. Insufficient base line separation prevented a quantitative analysis of SF$_6$ with this column setup. For each measurement, an aliquot of $\approx$ 200 mL was injected into the sample chamber with a sampling syringe and then purged with high purified nitrogen. A pressure regulated ethanol bath was used for keeping the trap cold. The ethanol was cooled by a Julabo cooling finger to a minimum temperature of −68 °C. For the purge and trap process the fill level is raised until the trap dips into the ethanol and is lowered again for the heating process (Bullister and Wisegarver, 2008). The traps of both measurement systems were heated to 90 °C

by an electrical current flow, which was automatically regulated by a proportional–integral–derivative controller (PID). A detailed description of the data set, the sampling, the calibration and measuring procedure including chromatograms and the specific retention times as well as a precise technical overview can be found in the published diploma thesis by Stöven (2011).

2.4.2 Tritium

Water samples for tritium measurements were taken in 1 L plastic bottles and sent to the Institute of Environmental Physics at the University of Bremen where the samples were degassed and stored for several weeks to accumulate ^{3}He$_{\text{trit}}$. The measurements of the tritiugenic helium isotopes were then carried out with a sector field mass spectrometer. Details of the measuring procedure and statistical evaluations can be found in Sültenfuß et al. (2009) and the results are described in Roether et al. (2013).

2.4.3 Uncertainties

The precisions of CFC-12 and SF$_6$ measurements from both instruments can be found in Table 2. The error of calibration routines, that is, standard gas, standard loops, temperature and pressure, is $\approx$ 1 %. The uncertainty of the purge efficiency of CFC-12 is estimated to be 2 % and negligible low for SF$_6$ so that the accuracy of CFC-12 is approximately 3 % and 1 % for SF$_6$. The uncertainty of the atmospheric history is < 1 % for SF$_6$ and CFC-12, whereas for low concentrations of CFC-12 an error of $\leq$ 4 % should be assumed due to the time period prior reliable CFC measurements (Tanhua et al., 2008; Walker et al., 2000). The input functions depend on the degree of saturation during a water-mass formation which is influenced by wind speed, mixed layer depth, convection velocity, pressure and temperature drops as well as the atmospheric emission increase of a tracer, resulting in an approximate 10 % propagation of uncertainty (Haine

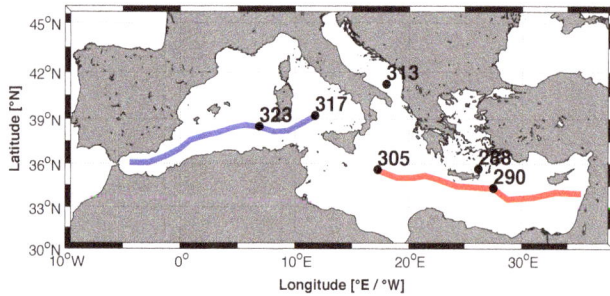

Figure 6. Sections and key stations of the transient tracer analysis. The red line shows the EMed section, the blue line the WMed section and the black dots the key stations.

and Richards, 1995; DeGrandpre et al., 2006; Tanhua et al., 2008). Furthermore, there are some regions were SF$_6$ has been used for release experiments (e.g., 1996 in the Greenland Sea gyre; Watson et al., 1999), which could produce an offset in concentrations. Since 2006 it is recommended to use an alternative tracer for release experiments to avoid such interferences with SF$_6$ of atmospheric origin. However, there was a release experiment using 1.327 mole of SF$_6$ in the Gulf of Lion in 2007 within the Lagrangian transport experiment (LATEX) (Hu et al., 2009). The SF$_6$ was released at shallow depths and it can be assumed that most of the SF$_6$ will be ventilated to the atmosphere, but nonetheless it is a possible error source with an unknown impact on further SF$_6$ surveys in this region. Assuming the worst case scenario of a deep water formation within this SF$_6$ patch, for example, the WMT event with a water renewal volume of $\approx 1.5 \times 10^{14}$ m^3 (Schroeder et al., 2008), the interior concentration of SF$_6$ would be elevated by 0.009 fmol kg^{-1} which is negligible.

The error of tritium measurements is given as ± 3 % and ± 0.02 TU whichever is greater (Roether et al., 2013). The input functions of tritium are in contrast to the atmospheric histories of CFC-12 and SF$_6$ not well documented and have several regional influencing factors as already mentioned above. An uncertainty of up to 15 % might be a realistic estimate of the used input functions.

The uncertainties in mean age is a function of errors in transient tracer concentrations and the Δ / Γ ratio. Figure 5 shows an example for absolute errors in mean age calculations based on CFC-12. The mean age becomes more uncertain for low tracer concentrations and high Δ / Γ ratios. The error functions for SF$_6$ and tritium are similar to the one of CFC-12.

Table 2. Precision of CFC-12 and SF$_6$ measurements.

| System | Precision | |
	SF$_6$	CFC-12
VS1	± 1.4 % / ± 0.05 ppt	± 0.6 % / ± 2 ppt
PT3	–	± 0.3 % / ± 1 ppt
Cracker	± 4.3 % / ± 0.07 ppt	± 1.9 % / ± 5 ppt

3 Results and discussion

3.1 General ventilation pattern

3.1.1 Eastern Mediterranean Sea

The zonal sections of the transient tracer concentrations of the Ionian and Levantine seas show some significant characteristics of their ventilation (Figs. 6, 7). Between 27° E and the coast of Lebanon, a clear tracer minimum zone (TMZ) can be identified by all three tracers which vertically spreads from approximately 700 to 1600 m depths. The core concentration of the TMZ is 106 ppt for CFC-12, 0.3 ppt for SF$_6$ and 0.3 TU for tritium, whereas the lowest values are not visible in the gridded fields shown in Fig. 7. Beneath this TMZ, the tracer concentrations are elevated in the deep water due to the deep water formation in the eastern Mediterranean Sea that led to a high volume input of tracer rich and dense water masses. The bottom concentration of the tracers are ≈ 200 ppt for CFC-12, ≈ 1.1 ppt for SF$_6$ and ≈ 0.6 TU for tritium. In the westerly parts of the section (i.e., the deep Ionian Sea), the tracer concentration is higher in the deep and bottom layer than in the east. This water-mass characteristic belongs to recent intrusions of ASOW coming from the deep water source in the Adriatic Sea. Station 313 in the southern Adriatic pit can be used as representative example for the source region of the ASOW. The concentration profiles show, that the southern Adriatic pit is a well-mixed and ventilated basin with minimum tracer concentrations of CFC-12 > 429 ppt, SF$_6$ > 5 ppt and tritium > 0.9 TU (Fig. 8a). The high concentrations of CFC-12 and tritium throughout the entire water column at station 288 in the Sea of Crete can be related to the time range from the 1990s until the present day (Fig. 8b). The concentration gradient of SF$_6$, however, indicates a recent return to a more layered structure in the Sea of Crete, so that the high concentrations of CFC-12 and tritium in the intermediate and deep water have to be formed before this layering process, probably during the 1990s. This would imply that the EMT source region was a completely mixed basin during the outflow event. This difference in tracer structures is related to the increasing input function of SF$_6$ and the weak input functions of CFCs since the early 1990s (i.e., the onset of the EMT). CFC-12 and tritium concentrations cannot be used to identify the recent change in ventilation of the Sea of Crete. A further

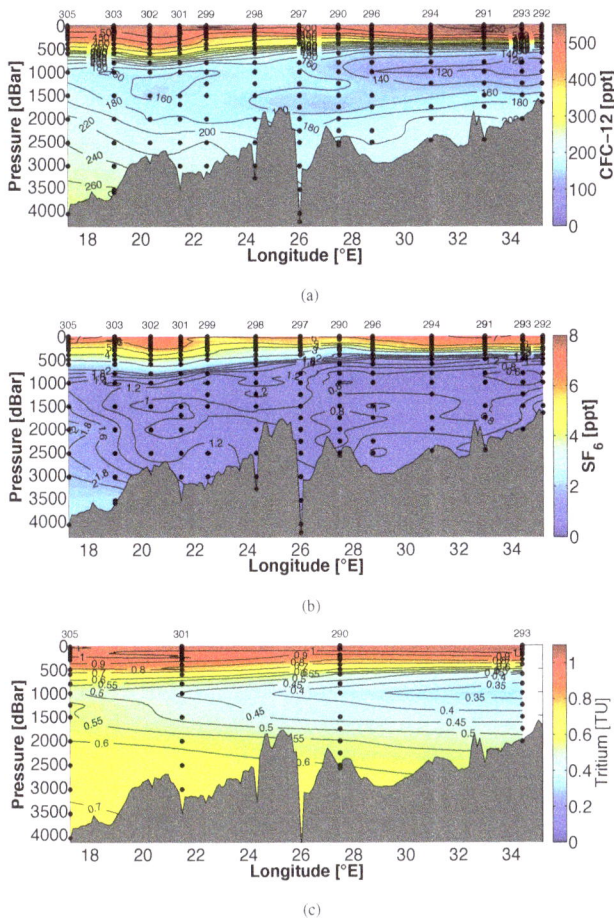

(a)

(b)

(c)

Figure 7. Transient tracer concentrations in the EMed during April 2011 (cruise M84/3), red line in Fig. 6. **(a)** CFC-12 in ppt, **(b)** SF$_6$ in ppt and **(c)** tritium in TU.

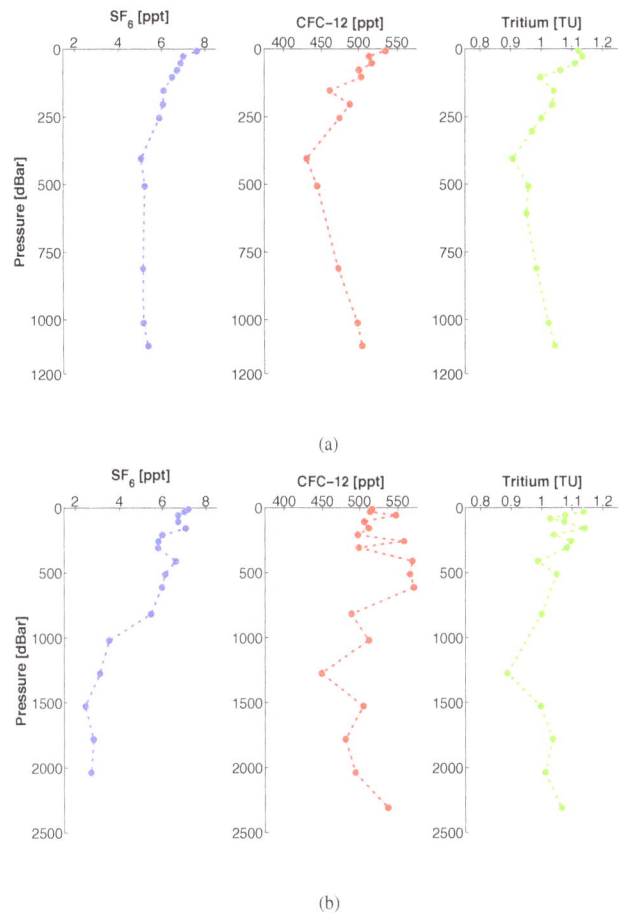

(a)

(b)

Figure 8. Profiles of transient tracers **(a)** in the Adriatic Sea at station 313 and **(b)** in the Sea of Crete at station 288 during April 2011. All three transient tracer show high concentrations throughout the entire water column of the Adriatic Sea. The Sea of Crete profiles show a clear concentration gradient for SF$_6$, whereas the CFC-12 and tritium concentrations scatter around their maximum values.

aspect is the distribution of SF$_6$ in the Ionian and Levantine seas which looks more homogeneous in the intermediate and deep water compared to CFC-12. This is based on the relatively young atmospheric history of SF$_6$, leading to deep water masses which are only less affected by this tracer. Accurate measurements between 1 ppt and the detection limit are needed to have a useful resolution. As mentioned above, CFC-12 has a long atmospheric history with a rapid concentration increase over several decades. Thus, CFC-12 has a large dynamic scale within the measurement range, so that CFC-12 is an important tracer for intermediately old water masses.

3.1.2 Western Mediterranean Sea

The CFC-12 and tritium section from the Tyrrhenian Sea through the western basin into the Alboran Sea show, that the western Mediterranean deep water (WMDW) is completely intruded by recently ventilated water masses coming from the extensive deep water formation during 2004–2006 (Figs. 6, 9) (Schroeder et al., 2008, 2010). The CFC-12

concentrations are > 260 ppt (Fig. 9a) and the tritium concentrations are > 0.5 TU (Fig. 9b) throughout the entire bottom layer of the western basin reaching the channel between Sardinia and Sicily. Figure 10 shows in detail that newly formed WMDW starts to enter the Tyrrhenian Sea along this bottom contour without reaching the interior water masses in 2011. The Tyrrhenian Sea is characterized by CFC-12 concentrations below 180 ppt and tritium concentrations < 0.5 TU suggesting that this basin was not affected by input of the 2004–2006 deep water formation into the deep and bottom water layers by 2011. These relatively low tracer concentrations from the Tyrrhenian Basin rudimentary extends as tongue into the intermediate layer between 700 and 1300 m of the western basin which correlates with the results of Rhein et al. (1999).

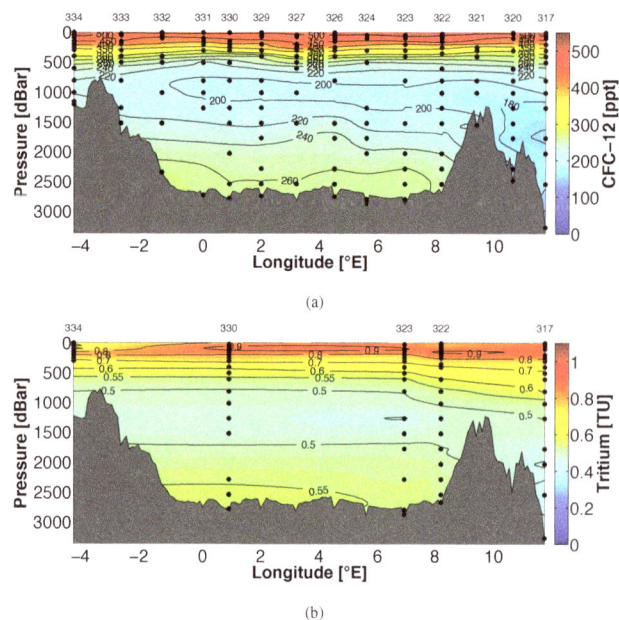

Figure 9. (a) CFC-12 concentration in ppt and (b) Tritium concentration in TU of the western basin and the Tyrrhenian Sea, blue line in Fig. 6.

3.2 Transit time distributions

3.2.1 IG-TTD

The first approach of a mean age is based on the obtained Δ/Γ ratios for the IG-TTD model which also provide further information on water-mass characteristics. The major share of Δ/Γ ratios was determinable for the CFC-12/SF$_6$ tracer couple in a reasonable range for samples of the EMed. Figure 11a shows the section of Δ/Γ ratios as interpolated macrostructure in the EMed based on CFC-12 and SF$_6$ and the IG-TTD model. The sectional interpolation quality is reduced due to few constrainable data points. Figure 11a indicates that the TMZ spreading is more affected by the EMT event than the tracer concentrations suggest. The water below 1200 m in the Levantine Basin has Δ/Γ ratios between 0.4 and 0.6 indicating a high advective behavior of the EMT event. The TMZ has ratios between 1 and 1.3 as expected for a stable water mass where diffusion predominates. The low Δ/Γ ratios of the EMT water masses are also observed in the easterly deep waters of the Ionian Sea, whereas the deep waters further west formed by ASOW have ratios between 1.2 and 1.4. Combining the second tracer couple consisting of SF$_6$ and tritium yields a similar trend of Δ/Γ ratios in the EMed (Fig. 11b). However, there are only four stations with tritium measurements available within the section of the EMed (290, 292, 301, 305) and thus the sectional interpolation is restricted to 34 data points which does not allow a resolution of local phenomena but only provides a rough overview.

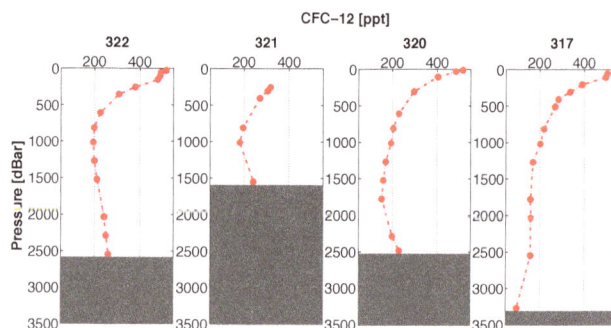

Figure 10. CFC-12 concentration in ppt along the shallow sill between Sardinia and Sicily. The elevated CFC-12 concentration of the bottom layer indicates the overflow of WMDW into the Tyrrhenian Sea. The depth of each station is indicated by the gray patch in each panel.

Comparing the Δ/Γ ratios in the intermediate and deep water layers of the Sea of Crete (Fig. 12a) with the EMT water masses in the Ionian and Levantine seas (Fig. 11a), one can see that the formation of the Cretan Sea overflow water (CSOW) as well as the EMT event itself were based on distinctly advective processes with an expected mean age approaching the tracer age. In contrast, water masses coming from Adriatic deep water (AdDW) seems to be formed by water masses with a more dispersive character belonging to slower formed water layers indicated by significant high Δ/Γ ratios between 1.1 and 1.6 (Fig. 12b). The red dots in Fig. 12b indicate non-constrained data points. It can be supposed that the formation of ASOW is based on slower dispersive processes of different water masses. The westerly bottom water of the Ionian Sea show the same dispersive characteristics as the ASOW (Fig. 11a). This indicates that both states of Δ/Γ ratios were mainly defined by the formation process of ASOW–CSOW source water and only in minor share by mixing processes along the current pathway from source region into the interior of the Levantine and Ionian seas. Therefore, the transient tracer concentrations of both deep water masses are not necessarily simply indicators of their ventilation.

Figure 13 show the determined mean age constrained by either CFC-12 and SF$_6$ or tritium and SF$_6$. The tritium-/SF$_6$-based maximum mean age of the TMZ in the Levantine Sea is 260 years and the deep and bottom water have a mean age between 70 and 80 years. In contrast, the CFC-12/SF$_6$ based maximum mean age of the TMZ is 230 years with a further high mean age in the Ionian Sea between 120 and 180 years. The deep and bottom water from the EMT event is the youngest water with a mean age between 50 and 80 years reaching up to the intermediate layers at station 290 which is near the overflow area of the CSOW. Table 3 shows the mean age related to the results by Schneider et al. (2014) which were obtained by using the standard Δ/Γ ratio of 1.0. The comparison shows clearly the significant influence

(a)

(b)

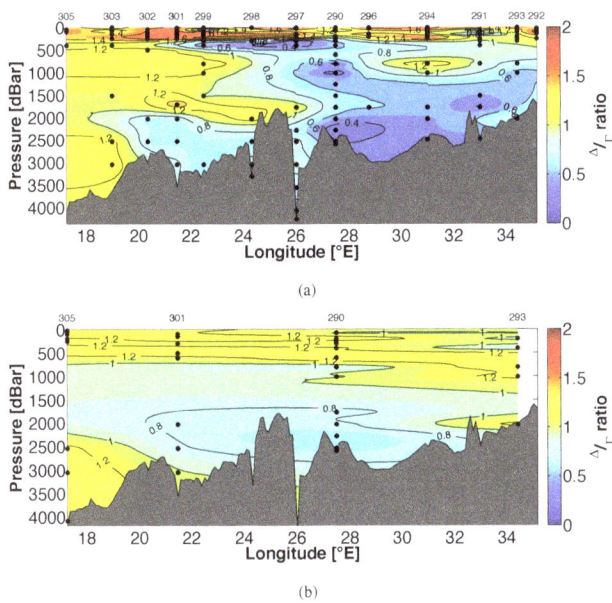

Figure 11. Determined $^\Delta/_\Gamma$ ratios in the EMed based on **(a)** the transient tracer couple CFC-12 and SF_6 and **(b)** tritium and SF_6. The black points indicate the constrained data points of the section.

Figure 12. Determined $^\Delta/_\Gamma$ ratios **(a)** in the Sea of Crete at station 288 and **(b)** in the Adriatic Sea at station 313 obtained by the transient tracer couple CFC-12 and SF_6. The red dots in **(b)** indicate ratios with a mean age difference > 5 years and thus defined as non-constrained within the IG-TTD model.

of the Δ / Γ ratio leading to a mean age difference between both TTD approaches of almost 80 years at 800 m depth in the TMZ. It should be pointed out that the TTD method presented by Schneider et al. (2014) was used in another context (i.e., the analysis of a time series), with the aim of detecting temporal variations of ventilation.

The deep water in the Ionian Basin has a mean age of 100 years which underlines the difference of both deep water formations. Although the Adriatic deep water formation is contemporary and the EMT event replenished large parts of the deep waters in the early 1990s, the deep water layer formed by water masses from the Sea of Crete is much younger than the one from the Adriatic Sea. The mean age of both source regions is quite young with a maximum mean age in the deep water of ≈ 20 years in the Aegean Sea (Fig. 14a), which highly correlates with the timing of the EMT event in the early 1990s, and ≈ 10–17 years in the Adriatic Sea (Fig. 14b). The mean age gradient with depth is much steeper in the Adriatic Sea than in the Sea of Crete which also shows that the ventilation processes of both basins are significantly different. However, the SF_6 mean age indicated by red dots in Fig. 14b at station 313 in the Adriatic Sea fit into the mean age gradient, even though these results are based on minimum functions with a difference of more than 5 years. Such data points are defined as non-constrained, due to the uncertainty by using different transient tracers with the same "approached" constraints (even if there are a number of possible true solutions for one transient tracer). Figure 15 shows the mean age functions of CFC-12 and SF_6 including the analytical error range of 4 %. Figure 15a shows an example

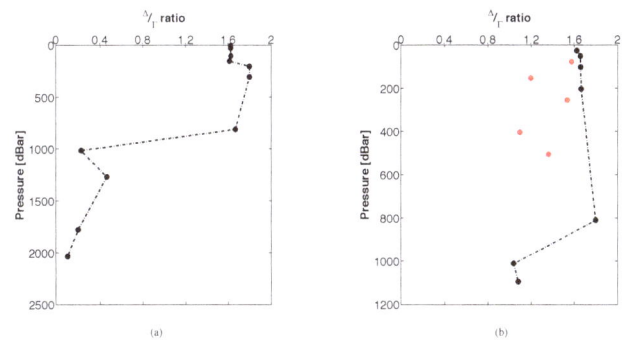

of the CDW and Fig. 15b of the ASDW. Both plots show the differences in water-mass structure, in terms of advective / diffusive ratios, despite uncertainties in the mean age calculations.

Neglecting any systematic errors in sampling and measurements, the TTD mean age is strongly influenced by the input function of a transient tracer (see Sect. 2.4.3). The TIF by Dreisigacker and Roether (1978) and Roether et al. (1992) is a rough estimate of the complex regional impact factors and provides a mean input of tritium into the surface layer of the ocean. Nevertheless, this input function is the best existing approach and the results of the offset corrected input function clearly show that the shape of the curve tends to be correct. The corrected input function of the EMed generates similar but still a slightly higher mean age results compared to the ones obtained by CFC-12 and SF_6 (Fig. 13). This indicates that the corrected input function probably still overestimates the mean input of tritium in the eastern Mediterranean Sea. The IG-TTD provides significant results for the EMed using the tracer couple of CFC-12 and SF_6, whereas SF_6 and tritium should be used with some caution.

The analysis of Δ / Γ ratios in the WMed is restricted to only three stations (317, 323, 334) for which the concentration ratios between the two transient tracer couples were not applicable to the IG-TTD model. The simple approach of water-mass analysis concerning the advective / diffusive ratio and the calculation of a mean age is not possible. Another distribution model is needed to estimate a mean age of non-constrainable data points of both tracer couples in the western Mediterranean Sea.

In this context, it is useful to determine the validity area of a tracer couple, to decide whether the IG or 2IG-TTD model should be applied. Figure 16 shows the validity area of the IG- and 2IG-TTD models by using the tracer age differences of SF_6 and CFC-12 in the EMed. The young and advective EMT water mass can be described by the IG-TTD, whereas the mean age of the intermediate water and parts of the deep

Table 3. Comparison of mean age results between a TTD model with constrained Δ / Γ ratios and a TTD model with standard ratio $\Delta / \Gamma =$ 1.0 (Schneider et al., 2014).

Pressure [dBar]	Constrained ratios		Standard ratio		Difference [yr]
	Mean age [yr]	Δ / Γ ratio	Mean age [yr]	Δ / Γ ratio	
200	27	0.80	29	1.0	2
250	30	0.63	38	1.0	8
300	38	0.71	47	1.0	9
400	47	0.61	69	1.0	22
600	83	0.76	105	1.0	22
800	44	0.21	123	1.0	79
1000	43	0.21	120	1.0	77
1250	57	0.47	105	1.0	48
1500	67	0.63	115	1.0	48
1750	77	0.59	135	1.0	58
2000	48	0.43	99	1.0	51
2250	41	0.19	99	1.0	58
2500	41	0.25	80	1.0	39
2600	52	0.53	94	1.0	42

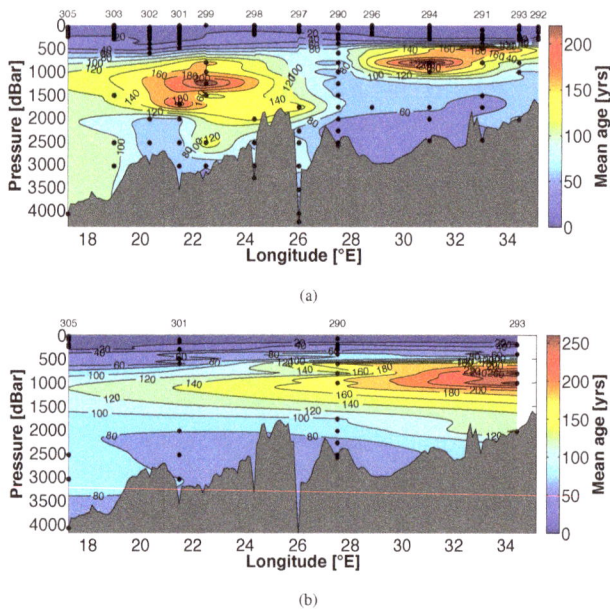

(a)

(b)

Figure 13. Mean age of the EMed based on an IG-TTD constrained by (**a**) the transient tracer couple CFC-12 and SF$_6$ and (**b**) tritium and SF$_6$.

Figure 14. Determined mean age (**a**) in the Sea of Crete at station 288 and (**b**) in the Adriatic Sea at station 313 based on SF$_6$ and an IG-TTD constrained by the transient tracer couple CFC-12 and SF$_6$. The red dots show SF$_6$ mean age results with a difference of more than 5 years compared to the CFC-12 mean age (i.e., non-constrained mean age).

3.2.2 2IG-TTD

The predefined 2IG-TTD was applied to several key stations in the EMed and WMed (290, 305, 317, 323) shown in Fig. 6. As mentioned above we assumed fixed Δ / Γ ratios for both TTDs so that $\Delta_1 / \Gamma_1 = 1.4$ and $\Delta_2 / \Gamma_2 = 0.6$, respectively and that the mean age $\Gamma_2 < \Gamma_1$ under the assumption that Γ_2 describes the younger water parcel. The concentration curve (Fig. 4) of each transient tracer were combined in one matrix to determine the intersections. The weighting factor α was separated in 10 % steps and thus we yielded eleven concentration matrices for each sample point. The determination of the intersections was carried out numerically to obtain a first overview of possible 2IG-TTD results which are shown in Table 4, where the mean age is based on the concentrations

water in the EMed as well as the complete WMed might be better evaluated by a 2IG-TTD. The limiting difference also depends on the tracer concentration to some extend, so that the mixed layer and parts of the pycnocline can also be described by the IG model, although the tracer age difference is larger than 10 years. Development of a clear mathematical definition of validity areas of different tracer couples and distribution models will be part of future work.

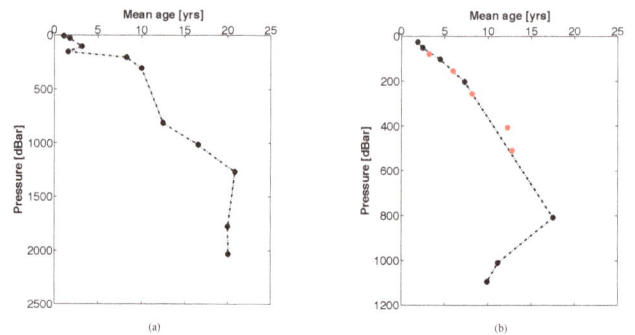

Table 4. Suggested mean age based on a 2IG-TTD at (**a**) station 290, (**b**) station 305, (**c**) station 317 and (**d**) station 323, where n.c. stands for non-constrained data points. The most prominent water layers are labeled according to the OMP of Hainbucher et al. (2013) including modified Atlantic water (MAW), Levantine surface water (LSW), Levantine intermediate water (LIW), eastern Mediterranean deep water (EMDW), western Mediterranean deep water (WMDW), Tyrrhenian deep water (TDW) and Adriatic deep water (AdDW).

(a)	Pressure [dBar]	α [%]	Γ_1 [yr]	Γ_2 [yr]	Mean age [yr]	Water layer
	51	0	0	12	12	LSW
	75	0	0	15	15	transition
	202	0	0	25	25	LIW
	254	0	0	30	30	LIW
	304	0	0	38	38	LIW
	404	0	0	46	46	transition
	607	0	0	60	60	transition
	810	n.c.	n.c.	n.c.	n.c.	transition
	1013	n.c.	n.c.	n.c.	n.c.	EMDW
	1267	n.c.	n.c.	n.c.	n.c.	EMDW
	1522	0	0	67	67	EMDW
	1775	0	0	77	77	EMDW
	2031	n.c.	n.c.	n.c.	n.c.	EMDW
	2286	n.c.	n.c.	n.c.	n.c.	EMDW
	2540	n.c.	n.c.	n.c.	n.c.	EMDW
	2600	0	0	60	60	EMDW

(b)	Pressure [dBar]	α [%]	Γ_1 [yr]	Γ_2 [yr]	Mean age [yr]	Water layer
	26	n.c.	n.c.	n.c.	n.c.	MAW
	52	n.c.	n.c.	n.c.	n.c.	MAW
	77	n.c.	n.c.	n.c.	n.c.	transition
	102	10	59	1	7	transition
	152	40	23	1	10	LIW
	203	50	28	1	15	LIW
	254	60	33	2	20	LIW
	304	50	63	2	33	LIW
	405	60	81	2	50	transition
	506	70	125	2	88	transition
	608	70	207	7	147	transition
	1013	80	320	8	258	AdDW
	1522	80	246	5	197	AdDW
	2032	70	279	4	196	AdDW
	2543	90	138	7	125	AdDW
	3053	80	167	6	134	AdDW
	4087	90	101	4	91	AdDW

of CFC-12, SF$_6$ and to some extent also tritium (see discussion below).

Station 290 in the Levantine Sea can be perfectly described by the IG-TTD which is also indicated by the 2IG-TTD results. As shown in Table 4 the best fits for all samples were obtained for $\alpha = 0$ which is the lower limiting case where the 2IG-TTD turns into an IG-TTD with $\Gamma = \Gamma_2$. Hence, the mean age of the 2IG-TTD is the same as the one from the constrained IG-TTD where $\Delta / \Gamma = 0.6$. The missing data in Table 4 corresponds to the mean age of IG-TTDs with Δ / Γ ratios < 0.6. Figure 17 shows the characteristics of such a concentration curve plot. For $\alpha = 0$, the lines of CFC-12 and SF$_6$ are overlapping, whereas the one of tritium is slightly above, again indicating a higher mean age by this tracer. The

sensitivity of changes with increasing α is of different extend for each tracer. The rate of change is highest for tritium, followed by CFC-12 and lowest for SF$_6$. This results in curve intersections when $\Gamma_2(\text{Tritium}) > \Gamma_2(\text{CFC-12}) > \Gamma_2(\text{SF}_6)$ at $\alpha = 0$.

Such condition can be found at station 305, which is a key station in the Ionian Sea where the IG-TTD is less constrained in the intermediate and deep water. However, CFC-12 and SF$_6$ intersect each other for several values of alpha, so we chose the one with the lowest difference to the tritium intersection (Fig. 18). Table 4 shows the results of the 2IG-TTD for station 305. The intermediate and deep water is characterized by high α values between 80 and 90 %, indicating a stronger influence of more stationary water masses.

Table 4. Continued.

(c)	Pressure [dBar]	α [%]	Γ_1 [yr]	Γ_2 [yr]	Mean age [yr]	Water layer
	51	n.c.	n.c.	n.c.	n.c.	MAW
	102	n.c.	n.c.	n.c.	n.c.	transition
	203	n.c.	n.c.	n.c.	n.c.	LIW
	304	n.c.	n.c.	n.c.	n.c.	LIW
	405	n.c.	n.c.	n.c.	n.c.	LIW
	507	n.c.	n.c.	n.c.	n.c.	LIW
	811	n.c.	n.c.	n.c.	n.c.	transition
	1014	n.c.	n.c.	n.c.	n.c.	TDW
	1267	70	757	3	531	TDW
	1776	80	370	6	297	TDW
	2032	90	230	18	208	TDW
	2542	90	246	4	221	TDW
	3268	n.c.	n.c.	n.c.	n.c.	TDW
(d)	Pressure [dBar]	α [%]	Γ_1 [yr]	Γ_2 [yr]	Mean age [yr]	Water layer
	52	n.c.	n.c.	n.c.	n.c.	MAW
	103	n.c.	n.c.	n.c.	n.c.	transition
	203	50	127	2	64	transition
	304	60	135	2	82	LIW
	406	70	170	3	120	LIW
	812	80	228	8	184	transition
	1523	80	222	4	178	WMDW
	2031	60	411	4	248	WMDW
	2287	60	332	1	200	WMDW
	2542	60	294	3	177	WMDW
	2798	60	364	2	219	WMDW

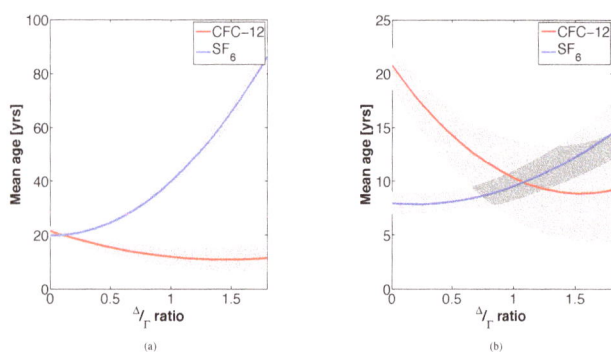

Figure 15. Mean age functions of CFC-12 and SF$_6$ at station 288, (a) in the Sea of Crete at 2030 m depth and (b) at station 313 in the Adriatic Sea at 1094 m depth with an error range of 4 % analytical error.

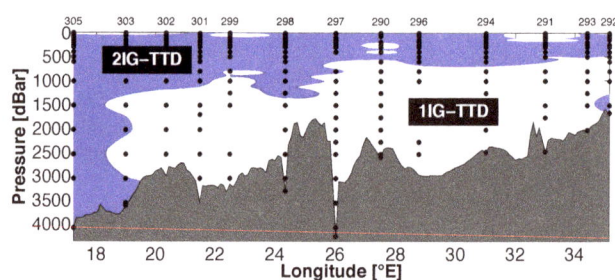

Figure 16. Differences of the tracer age between CFC-12 and SF$_6$. The white shading describes tracer age differences below and the blue shading above 10 years.

Looking at the single mean age results of the distribution, the total mean age is mainly influenced by Γ_1 rather than Γ_2, whereas both single results are not significant for statements about real mixing processes. They are rather part of the predefined model characteristics and provide only tendencies of the water-mass behavior. Whereas the total mean age from the constrained and exact determined TTD model describes the solved equation and thus a significant mean age result of

a water parcel. The highest mean age of **260** years can be found at ≈ 1000 m depth, whereas the mean age decreases to 90 years at the bottom layer. This is in full compliance with the expected younger water masses belonging to the ASOW. Compared to the IG-TTD, which indicates a mean age of 100 years for most of the water column, the 2IG-TTD shows a more differentiated structure with a clear mean age maximum. This case indicates that the 2IG-TTD provides more reasonable results.

The order of the tracer mean age at station 317 in the Tyrrhenian Sea changes from the required standard condition

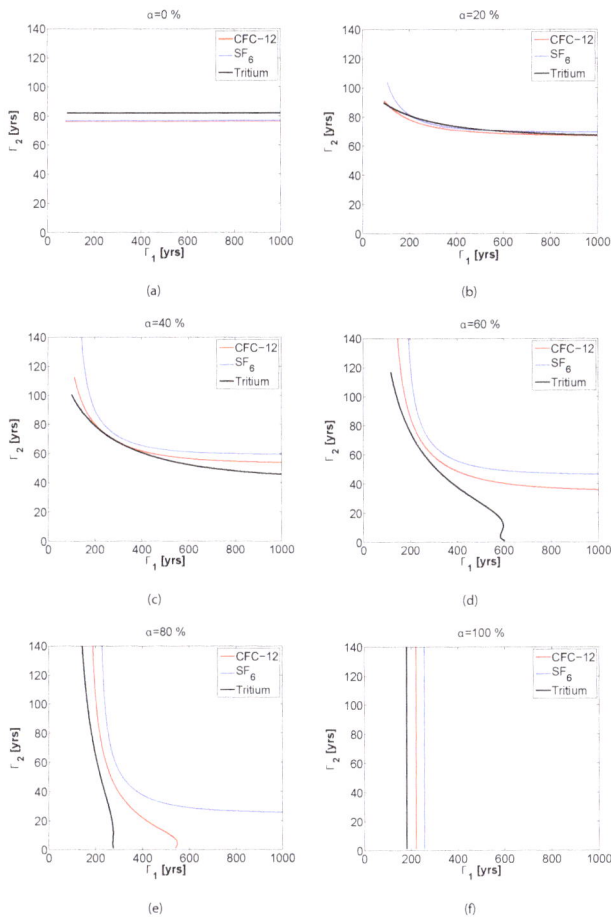

Figure 17. Example of a mean age calculation from three transient tracers (CFC-12, SF_6 and tritium) at station 290 at 1775 m depth. The six panels are for different fractions of the older water mass (α value). The Δ / Γ ratios are set to 0.6 for the younger water mass, and to 1.4 for the older water mass. Panel (**a**) shows the selected result, see text.

Figure 18. Example for a mean age calculation from three transient tracers (CFC-12, SF_6 and tritium) at station 305 at 608 m depth. Panel (**d**) shows a triple intersection of all three transient tracer concentrations ($\alpha = 70\%$) and thus a solution for the mean age at this data point of 147 years.

Station 323 in the Algero-Provençal Basin shows significant characteristics of the newly formed deep and bottom water layer with a constant α of 60 % and a mean age between 170 and 250 years (Table 4, Fig. 19). The lower values of α besides a relatively high mean age describe the extensive intrusion of young water masses into an old deep water layer. The interbasin circulation pattern between the Algero-Provençal Basin and the Tyrrhenian Sea is characterized by high α values denoting less influence of advective water mass input at depth between 1500 and 800 m. However, the mean age is relatively low at this depth range at station 323. Even though the massive inflow of recently ventilated water of 2004–2006 might have notably lowered the mean age, the oldest water masses can be still found in the deep and bottom water. So it can be suggested that these water layers of the western basin were probably less mixed over decadal to centennial timescales. The IG-TTD and tracer age concepts both indicates the oldest waters at intermediate depths (e.g., Rhein et al., 1999; Schneider et al., 2014), whereas our analysis show the highest mean age values in the deep water. This illustrates the power of the 2IG-TTD approach.

3.3 Best mean age approach

The combination of both TTD models allows for an exchange of non-constrainable data points within the IG-TTD by constrained data points of the predefined 2IG-TTD, so that in total 96 % of the data points in the EMed and 81 % in the WMed could be constrained (Table 5). This

into Γ_2(CFC-12) > Γ_2(Tritium) > Γ_2(SF_6) for depths shallower than 1250 m. This change in order is also the limit of the used model, so that only four samples could be determined in the Tyrrhenian Sea (Table 4). The mean age is ≈ 200 years in the deep water and increase up to 531 years at 1250 m. The Tyrrhenian Sea is less affected by intrusion of younger water masses and thus one would expect this high mean age in this basin. There might be several reasons why the major part of the mean age is not determinable in the Tyrrhenian Sea. The values for Γ_1 are increasing with decreasing depth up to 757 years. The used mean age matrices have a size of 1000×200, so that a maximum mean age of 1000 years can be determined for Γ_1. The shape of the curves of CFC-12 and SF_6 show the tendency to have intersections beyond this limit and thus a much higher mean age for Γ_1 than 1000 years. Another possible reason might be the assumed values of Δ / Γ ratios of the 2IG-TTD.

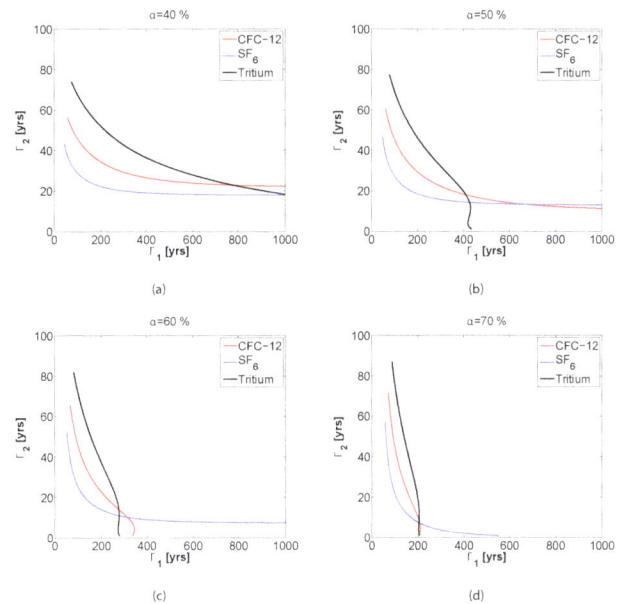

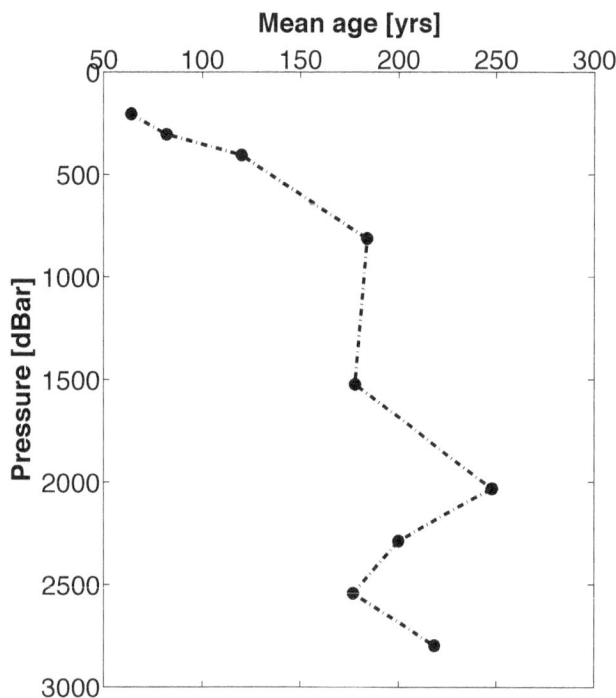

Figure 19. Mean age at station 323 in the WMed based on a 2IG-TTD constrained by the transient tracer couple CFC-12 and SF_6.

Table 5. Number of constrained data points of the IG- and 2IG-TTD using SF_6, CFC-12 and Tritium.

Tracer couple		Data points	IG	2IG
CFC-12/SF_6	Total	412	230	301
	WMed	38	14	23
	EMed	375	216	278
^{3}H/SF_6	Total	206	71	–
	WMed	86	17	–
	EMed	120	54	–

cross-interlocking of both models provides the best estimate of the mean age in the eastern part of the Mediterranean Sea (Fig. 20). The young water masses of the EMT event can be clearly seen in the deep and bottom layers of the Levantine and Ionian Basin with a mean age between 60 and 80 years as well as the newly formed deep water in the westerly parts of the Ionian Basin with a mean age between 120 and 160 years. The mean age maximum layer extends from 600 to 2000 m depth throughout the entire EMed with a mean age between 160 and 290 years. The maximum mean age of 290 years can be found in the TMZ in the Levantine Sea. This high mean age layer is disrupted by a lower mean age of $\approx$ 140 years at the outflow areas of the Sea of Crete (station 299 and 290) and the area near Rhodes (station 293).

4   Conclusions

We have described a comprehensive data set of transient tracer concentrations obtained during the *Meteor* 84/3 cruise in the Mediterranean Sea in April 2011. For the first time measurements of SF_6, CFC-12, tritium and ^{3}He were performed simultaneous on a cruise covering all major basins of the Mediterranean Sea. With this data set, we constrain ventilation characteristics using the transit time distribution (TTD) framework. In particular we constrain the TTD assuming inverse Gaussian (IG) shape of the TTD, either as a one-modal (IG) or two-modal (2IG) distribution. The shape-determining parameters of the IG distribution are the width (Δ) and mean age (Γ), the relative contribution of the two water masses of a 2IG distribution (α), as well as the ratio of Δ / Γ that is indicative of the diffusive to advective transport characteristic of a water mass.

Most regions of the EMed can be described by the IG-TTD model but with widely different Δ / Γ characteristic of the two main deep-water sources, the Cretan deep water (CDW) and the Adriatic Sea overflow water (ASOW), which have developed their Δ / Γ ratio already at their origin region rather than along the flow pathway. The Aegean Sea source water shows a more advective (i.e., a low Δ / Γ ratio) behavior than the Adriatic Sea source water. The majority of the deep water in the eastern Mediterranean of Aegean source originates from the Eastern Mediterranean Transient (EMT) event in the early 1990s. The unusually high deep-water formation rates during this event might explain the advective characteristics of this water mass, which had a mean age between 50 and 80 years. The mean age of the EMT-induced water masses in the Levantine Basin is thus still younger than the deep water in the deep Ionian Sea that has a mean age of approximately 120–160 years using the 2IG-TD model. This is, at first sight, counterintuitive considering the higher CFC-12 concentration in the deep Ionian Sea due to recent contribution from the deep water source in the Adriatic Sea. All tracers show a distinct minimum at around 1000 m depth throughout the Med, although the origin of this tracer minimum zone (TMZ) is slightly different for the eastern and western basins. For the EMed horizontal gradients in the Δ / Γ ratio lead to a mean age distribution that not directly correlated to the tracer concentrations, with lower a mean age southeast of Crete, in the outflow region of water from the Sea of Crete, and a higher age in the Levantine Basin and Ionian Sea.

Although most of the EMed ventilation can be described by an IG-TTD model, there are areas where this model of mixing cannot explain the observed tracer distributions. We found the IG-TTD approximation to be invalid for water samples with a tracer age difference exceeding $\approx$ 10 years (for the SF_6-CFC-12 couple sampled in 2011). Thus we use a 2IG model to constrain the TTD in the western part of the Ionian Sea and most of the WMed. The Western Mediterranean Transition (WMT) event with enhanced deep-water

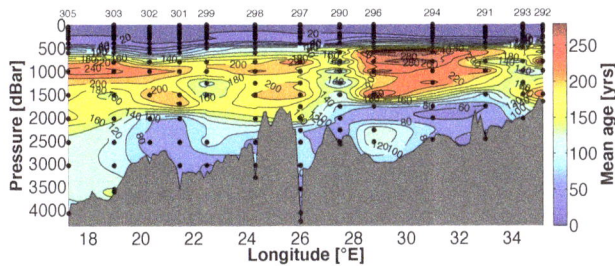

Figure 20. Best estimate of the mean age of the EMed based on the combination of an IG-TTD and a predefined 2IG-TTD constrained by the transient tracer couple CFC-12 and SF_6.

formation during the winters of 2004–2006 in the western basin can be characterized by the 2IG-TTD model, where the recently ventilated deep-water with advective properties mixes with more stationary water of more diffusive character. The contemporaneous deep water in WMed has approximately 40 % contribution of recently ventilated waters from the WMT, leading to a mean age of about 200 years, which is similar or higher than the mean age higher up in the water column that have lower CFC-12 and SF_6 concentrations. This highlights the need to constrain the TTD in order to understand the "age" of a water mass based on transient tracer concentrations. The recently ventilated deep water was observed as elevated concentrations of CFC-12 in excess of 200 ppt along the bottom in the Sardinia Channel towards, but not yet within, the Tyrrhenian Sea. The horizontally relatively homogeneous CFC-12 concentration of the western basin suggest that the TTD structure determined at station 323, SW of Sardinia, can be extrapolated to large parts of the WMed.

The results presented here point to the need to consider alternatives to the commonly applied IG-TTD model, particularly in regions with variable ventilation or where more water masses mix with each other. For observational constraints of more complex TTDs, a suite of transient tracers needs to be measured and interpreted. For such regions in the ocean, we show that a predefined 2IG-TTD provides a useful tool.

Acknowledgements. The authors want to thank the captain and crew on the research vessel *Meteor* for the excellent cooperation during the campaign. The *Meteor* cruise M84/3 and the transient tracer measurements were supported by a grant from the Deutsche Forschungsgemeinschaft Senatskommission für Ozenographie (DFG), and from a grant from the DFG; TA 317/3-1. Furthermore we would like to thank Wolfgang Roether for his support, help and inspiring discussions on particular matters of helium and tritium. Our special thanks goes to Damian Grundle and Martha Gledhill for linguistic improvements of the manuscript.

The service charges for this open access publication have been covered by a Research Centre of the Helmholtz Association.

Edited by: S. Sparnocchia

References

Artegiani, A., Bregant, D., Paschini, E., Pinardi, N., Raicich, F., and Russo, A.: The Adriatic Sea General Circulation. Part I: Air-Sea Interactions and Water Mass Structure, J. Phys. Oceanogr., 27, 1492–1514, 1996a.

Artegiani, A., Bregant, D., Paschini, E., Pinardi, N., Raicich, F., and Russo, A.: The Adriatic Sea General Circulation. Part II: Baroclinic Circulation Structure, J. Phys. Oceanogr., 27, 1515–1532, 1996b.

Brasseur, P., Beckers, J., Brankart, J., and Schoenauen, R.: Seasonal temperature and salinity fields in the Mediterranean Sea: Climatological analyses of a historical data set, Deep Sea Res.-Pt. I, 43, 159–192, 1996.

Bullister, J. L.: Atmospheric CFC-11, CFC-12, CFC-113, CCl_4 and SF_6 Histories (1910–2011), Carbon Dioxide Information Analysis Center, available at: http://cdiac.ornl.gov/oceans/new_atmCFC.html (last access: 1 June 2014), 2011.

Bullister, J. L. and Wisegarver, D.: The shipboard analysis of trace levels of sulfur hexafluoride, chlorofluorocarbon-11 and chlorofluorocarbon-12 in seawater, Deep Sea Res.-Pt. I, 55, 1063–1074, 2008.

Bullister, J., Wisegarver, D., and Menzia, F.: The solubility of sulfur hexafluoride in water and seawater, Deep Sea Res.-Pt. I, 49, 175–187, 2002.

DeGrandpre, M. D., Koertzinger, A., Send, U., Wallace, D. W. R., and Bellerby, R. G. J.: Uptake and sequestration of atmospheric CO_2 in the Labrador Sea deep convection region, Geophys. Res. Lett., 33, doi:10.1029/2006GL026881, 2006.

Delhez, E., de Brye, B., de Brauwere, A., and Deleersnijder, E.: Residence time vs influence time, J. Marine Syst., 132, 185–195, doi:10.1016/j.jmarsys.2013.12.005, 2013.

Dreisigacker, E. and Roether, W.: Tritium and ^{90}Sr in North Atlantic surface water, Earth Planet. Sci. Lett., 38, 301–312, 1978.

Ferronsky, V. and Polyakov, V.: Environmental isotopes in the hydrosphere, John Wiley & Sons, Chichester, New York, 1982.

Hainbucher, D., Rubino, A., and Klein, B.: Water mass characteristics in the deep layers of the western Ionian Basin observed during May 2003, Geophys. Res. Lett., 33, L05608, doi:10.1029/2005GL025318, 2006.

Hainbucher, D., Rubino, A., Cardin, V., Tanhua, T., Schroeder, K., and Bensi, M.: Hydrographic situation during cruise M84/3 and P414 (spring 2011) in the Mediterranean Sea, Ocean Sci. Discuss., 10, 2399–2432, doi:10.5194/osd-10-2399-2013, 2013.

Haine, T. W. N. and Richards, K. J.: The influence of the seasonal mixed layer on oceanic uptake of CFCs, J. Geophys. Res.-Oceans, 100, 10727–10744, doi:10.1029/95JC00629, 1995.

Hall, T. and Plumb, R.: Age as a diagnostic of stratospheric transport, J. Geophys. Res., 99, 1059–1070, 1994.

Hall, T., Haine, T., Waugh, D., Holzer, M., Terenzi, F., and LeBel, D.: Ventilation rates Estimated from Tracers in the Presence of Mixing, J. Phys. Oceanogr., 37, 2599–2611, doi:10.1175/2006JPO3471.1, 2007.

Houghton, J., Meira Filho, L., Callander, B., Harris, N., Kattenberg, A., and Maskell, K.: Climate Change 1995, The Science of Climate Change, Cambridge University Press, 1996.

Hu, Z., Doglioli, A., Petrenko, A., Marsaleix, P., and Dekeyser, I.: Numerical simulations of eddies in the Gulf of Lion, Ocean Model., 28, 203–208, doi:10.1016/j.ocemod.2009.02.004, 2009.

Huhn, O., Rhein, M., Hoppema, M., and van Heuven, S.: Decline of deep and bottom water ventilation and slowing down of anthropogenic carbon storage in the Weddell Sea, 1984-2011, Deep Sea Res.-Pt. I, 76, 66–84, doi:10.1016/j.dsr.2013.01.005, 2013.

Klein, B., Roether, W., Manca, B., Bregant, D., Beitzel, V., Kovacevic, V., and Luchetta, A.: The large deep water transient in the Eastern Mediterranean, Deep Sea Res.-Pt. I, 46, 371–414, doi:10.1016/S0967-0637(98)00075-2, 1999.

Lascaratos, A., Williams, R., and Tragou, E.: A Mixed-Layer Study of the Formation of Levantine Intermediate Water, J. Geophys. Res., 98, 14739–14749, 1993.

Lascaratos, A., Roether, W., Nittis, K., and Klein, B.: Recent changes in deep water formation and spreading in the eastern Mediterranean Sea: a review, Prog. Oceanogr., 44, 5–36, doi:10.1016/S0079-6611(99)00019-1, 1999.

Law, C., Watson, A., and Liddicoat, M.: Automated vacuum analysis of sulphur hexafluoride in seawater: derivation of the atmospheric trend (1970–1993) and potential as a transient tracer, Mar. Chem., 48, 57–69, 1994.

Malanotte-Rizzoli, P. and Hecht, A.: Large scale properties of the eastern Mediterranean: a review, Oceanol. Acta, 11, 323–335, 1988.

Primeau, F. and Holzer, M.: The Ocean's Memory of the Atmosphere: Residence-Time and Ventilation-Rate Distribution of Water Masses, J. Phys. Oceanogr., 36, 1439–1456, doi:10.1175/JPO2919.1, 2006.

Rhein, M., Send, U., Klein, B., and Krahmann, G.: Interbasin deep water exchange in the western Mediterranean, J. Geophys. Res.-Oceans, 104, 23495–23508, doi:10.1029/1999JC900162, 1999.

Roether, W. and Lupton, J. E.: Tracers confirm downward mixing of Tyrrhenian Sea upper waters associated with the Eastern Mediterranean Transient, Ocean Sci., 7, 91–99, doi:10.5194/os-7-91-2011, 2011.

Roether, W. and Schlitzer, R.: Eastern Mediterranean deep water renewal on the basis of chlorofluoromethane and tritium data, Dynam. Atmos. Oceans, 15, 333–354, 1991.

Roether, W., Schlosser, P., Kuntz, R., and Weiss, W.: Transient-tracer studies of the thermohaline circulation of the Mediterranean, Reports in Meteorology and Oceanography, 41, 291–317, 1992.

Roether, W., Manca, B., Klein, B., Bregant, D., Georgopoulos, D., Beitzel, V., Kovacevic, V., and Luchetta, A.: Recent Changes in Eastern Mediterranean Deep Waters, Science, 271, 333–335, 1996.

Roether, W., Klein, B., Beitzel, V., and Manca, B.: Property distributions and transient-tracer ages in Levantine Intermediate Water in the Eastern Mediterranean, J. Marine Syst., 18, 71–87, 1998.

Roether, W., Klein, B., Manca, B. B., Theocharis, A., and Kioroglou, S.: Transient Eastern Mediterranean deep waters in response to the massive dense-water output of the Aegean Sea in the 1990s, Prog. Oceanogr., 74, 540–571, doi:10.1016/j.pocean.2007.03.001, 2007.

Roether, W., Jean-Baptiste, P., Fourrè, E., and Sültenfuß, J.: The transient distribution of nuclear weapon-generated tritium and its decay product ^{3}He in the Mediterranean Sea, 1952–2011, and their oceanographic potential, Ocean Sci., 9, 837–854, doi:10.5194/os-9-837-2013, 2013.

Rubino, A. and Hainbucher, D.: A large abrupt change in the abyssal water masses of the eastern Mediterranean, Geophys. Res. Lett., 34, L23607, doi:10.1029/2007GL031737, 2007.

Schlitzer, R., Roether, W., Oster, H., Junghans, H., Hausmann, M., Johannsen, H., and Michelato, A.: Chlorofluoromethane and oxygen in the Eastern Mediterranean, Deep Sea Res.-Pt. A, 38, 1531–1551, 1991.

Schneider, A., Tanhua, T., Koertzinger, A., and Wallace, D.: High anthropogenic carbon content in the eastern Mediterranean, J. Geophys. Res., 115, C12050, doi:10.1029/2010JC006171, 2010.

Schneider, A., Tanhua, T., Koertzinger, A., and Wallace, D.: An evaluation of tracer fields and anthropogenic carbon in the equatorial and the tropical North Atlantic, Deep Sea Res.-Pt. I, 67, 85–97, doi:10.1016/j.dsr.2012.05.007, 2012.

Schneider, A., Tanhua, T., Roether, W., and Steinfeldt, R.: Changes in ventilation of the Mediterranean Sea during the past 25 year, Ocean Sci., 10, 1–16, doi:10.5194/os-10-1-2014, 2014

Schroeder, K., Gasparini, G., Tangherlini, M., and Astraldi, M.: Deep and intermediate water in the western Mediterranean under the influence of the Eastern Mediterranean Transient, Geophys. Res. Lett., 33, L21607, doi:10.1029/2006GL027121, 2006.

Schroeder, K., Ribotti, A., Borghini, M., Sorgente, R., Perilli, A., and Gasparini, G.: An extensive western Mediterranean deep water renewal between 2004 and 2006, Geophys. Res. Lett., 35, L18605, doi:10.1029/2008GL035146, 2008.

Schroeder, K., Josey, S., Herrmann, M., Grignon, L., Gasparini, G., and Bryden, H.: Abrupt warming and salting of the Western Mediterranean Deep Water after 2005: Atmospheric forcings and lateral advection, J. Geophys. Res., 115, C08029, doi:10.1029/2009JC005749, 2010.

Steinfeldt, R.: Ages and age spectra of Eastern Mediterranean Deep Water, J. Marine Syst., 48, 67–81, 2004.

Stöven, T.: Ventilation processes of the Mediterranean Sea based on CFC-12 and SF$_6$ measurements, GEOMAR OceanRep, available at: http://oceanrep.geomar.de/id/eprint/13936 (last access: 1 June 2014), diploma thesis, Christian-Albrechts-Universität zu Kiel, 2011.

Sültenfuß, J., Roether, W., and Rhein, M.: The Bremen mass spectrometric facility for the measurement of helium isotopes, neon, and tritium in water, Isot. Environ. Health Stud., 45, 1–13, 2009.

Tanhua, T., Waugh, D., and Wallace, D.: Use of SF$_6$ to estimate anthropogenic CO$_2$ in the upper ocean, J. Geophys. Res., 113, 2156–2202, doi:10.1029/2007JC004416, 2008.

Tanhua, T., Hainbucher, D., Cardin, V., Álvarez, M., Civitarese, G., McNichol, A. P., and Key, R. M.: Repeat hydrography in the Mediterranean Sea, data from the Meteor cruise 84/3 in 2011, Earth Syst. Sci. Data, 5, 289–294, doi:10.5194/essd-5-289-2013, 2013a.

Tanhua, T., Hainbucher, D., Schroeder, K., Cardin, V., Álvarez, M., and Civitarese, G.: The Mediterranean Sea system: a review and an introduction to the special issue, Ocean Sci., 9, 789–803, doi:10.5194/os-9-789-2013, 2013b.

Tanhua, T., Waugh, D., and Bullister, J.: Estimating changes in ocean ventilation from the early 1990s CFC-12 and late SF_6 measurements, Geophys. Res. Lett., 40, 927–932, doi:10.1002/grl.50251, 2013c.

Taylor, C. and Roether, W.: A uniform scale to report low-level tritium measurements in water, Int. J. Appl. Radiat. Is., 33, 377–382, 1982.

Unterweger, M., Coursey, B., Schima, F., and Mann, W.: Preparation and calibration of the 1987 National Bureau of Standards tritiated-water standards, Int. J. Appl. Radiat. Is., 31, 611–614, 1980.

Vollmer, M. and Weiss, R.: Simultaneous determination of sulfur hexafluoride and three chlorofluorocarbons in water and air, Mar. Chem., 78, 137–148, 2002.

Walker, S. J., Weiss, R. F., and Salameh, P. K.: Reconstructed histories of the annual mean atmospheric mole fractions for the halocarbons CFC-11 CFC-12, CFC-113, and carbon tetrachloride, J. Geophys. Res.-Oceans, 105, 14285–14296, doi:10.1029/1999JC900273, 2000.

Warner, M. and Weiss, R.: Solubilities of chlorofluorocarbons 11 and 12 in water and seawater, Deep Sea Res.-Pt. A, 32, 1485–1497, 1985.

Watson, A., Messias, M.-J., Fogelqvist, E., Van Scoy, K., Johannessen, T., Oliver, K., Stevens, D., Rey, F., Tanhua, T., Olsson, K., Carse, F., Simonsen, K., Ledwell, J., Jansen, E., Cooper, D., Kruepke, J., and Guilyardi, E.: Mixing and convection in the Greenland Sea from a tracer-release experiment, Nature, 401, 902–904, doi:10.1038/44807, 1999.

Waugh, D. W., Vollmer, M. K., Weiss, R. F., Haine, T. W. N., and Hall, T. M.: Transit time distributions in Lake Issyk-Kul, Geophys. Res. Lett., 29, 84.1–84.4, doi:10.1029/2002GL016201, 2002.

Waugh, D., Hall, T., and Haine, T.: Relationships among tracer ages, J. Geophys. Res., 108, 3138, doi:10.1029/2002JC001325, 2003.

Waugh, D., Haine, T. W., and Hall, T. M.: Transport times and anthropogenic carbon in the subpolar North Atlantic Ocean, Deep Sea Res.-Pt. I, 51, 1475–1491, 2004.

Waugh, D., Hall, T., McNeil, B., Key, R., and Matear, R.: Anthropogenic CO_2 in the oceans estimated using transit time distributions, Tellus B, 58, 376–389, 2006.

Wuest, G.: On the vertical circulation of the Mediterranean Sea, J. Geophys. Res., 66, 3261–3271, 1961.

On the tides and resonances of Hudson Bay and Hudson Strait

D. J. Webb

National Oceanography Centre, Southampton SO14 3ZH, UK

Correspondence to: D. J. Webb (djw@soton.ac.uk)

Abstract. The resonances of Hudson Bay, Foxe Basin and Hudson Strait are investigated using a linear shallow water numerical model. The region is of particular interest because it is the most important region of the world ocean for dissipating tidal energy.

The model shows that the semi-diurnal tides of the region are dominated by four nearby overlapping resonances. It shows that these not only affect Ungava Bay, a region of extreme tidal range, but they also extend far into Foxe Basin and Hudson Bay and appear to be affected by the geometry of those regions. The results also indicate that it is the four resonances acting together which make the region such an important area for dissipating tidal energy.

1 Introduction

In his study of tidal dissipation on the world ocean, Miller (1966) estimated that Hudson Bay and the Labrador Sea dissipated 140 GW of M2 tidal energy. This made it the fifth most important region of tidal dissipation, the first four being the Bering Sea, the Sea of Okhotsk, the Northwest Australian Shelf and the European Shelf.

More recent studies have completely changed this picture. Le Provost and Rougier (1997), using a numerical model, found that the M2 tide dissipated 313 GW in the Hudson Bay region. This made it their most important region for tidal dissipation.

Egbert and Ray (2001) assimilated satellite altimeter data into an ocean model and again found the Hudson Bay region to be the most important, the M2 tide dissipating 261 GW. Next in importance was the European Shelf (~ 208 GW), followed by the Northwest Australian Shelf (~ 158 GW), the Yellow Sea (~ 149 GW) and the Patagonian Shelf (~ 112 GW).

The importance of the Hudson Bay complex is also emphasised if one plots the M2 energy flux vectors for the North Atlantic. This has been done in Fig. 1, making use of the satellite-derived tidal fields of Egbert and Erofeeva (2002). In the eastern North Atlantic the figure shows a northward flux of tidal energy associated with a propagating Kelvin wave. Part of the energy is lost to the European Shelf but a large amount continues north. It then turns westwards and passes south of Greenland before converging on Hudson Strait.

If the fluxes of Fig. 1 are integrated along lines between 44° W, 42° N and the coasts of Spain and Greenland, the results show that the M2 tide fluxes 490 GW northwards into the northeast Atlantic and that 220 GW passes south of Greenland towards Hudson Strait. At the entrance to Hudson Strait the flux of energy is 250 GW – the increase being due to a small northward flow of tidal energy on the western side of the Atlantic.

Thus not only is the Hudson Bay complex the major tidal dissipation region of the global ocean, it is also so effective that no energy-transporting Kelvin wave continues southwards along the coast of Labrador and Newfoundland. This is in marked contrast to the behaviour on the eastern side of the ocean where a large fraction of the energy flux continues past the resonant European continental shelves. Even though the entrance to Hudson Strait is only 70 km wide, it transmits much more energy than the Celtic Sea, at the entrance to the English Channel and Irish Sea, which is over 400 km wide.

Continental shelf regions with large amounts of tidal dissipation are usually associated with resonances of the shelf. Examples are the Bristol Channel (Fong and Heaps, 1978; Webb, 2013a), the Patagonian Shelf (Huthnance, 1980) and the Northwest Australian Shelf. Such regions are usually associated with high tides, so one possible reason why Hudson Bay was overlooked is that it is only recently that Leaf

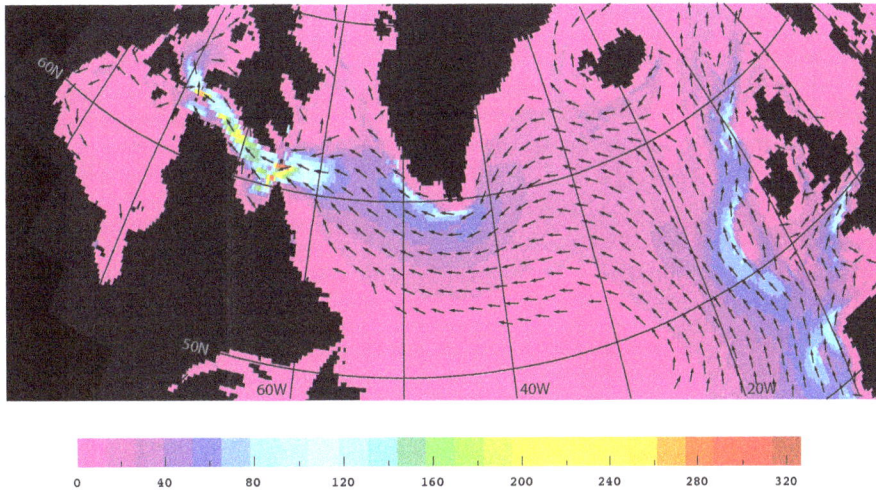

Figure 1. Energy flux vectors for the M2 tide in the north Atlantic, based on from data from OTIS2 (Egbert and Erofeeva, 2002). Energy flux in units of $MW\,m^{-1}$.

Bay, part of Ungava Bay[1] near the entrance to Hudson Strait, has been reported as having the world's second highest tidal range (O'Reilly et al., 2005).

The high tides of Ungava Bay were studied by Arbic et al. (2007) using a time-dependent numerical model. This showed that the tides of the region were affected by a quarter-wavelength resonance between the coast and the deep ocean. The resonance mode also showed maxima elsewhere in Foxe Basin and Hudson Bay which, following the recent study of the English Channel and Irish Sea (Webb, 2013a), might indicate the presence of additional resonances affecting the tides.

The English Channel and Irish Sea study used a time-independent model. This has the advantage that it can be run for complex values of angular velocity and so allows a detailed study of the resonant structure of a region.

The study uses a method of analysis in regular use by physicists (e.g. Morse and Feshbach, 1953; Mathews and Walker, 1965; Courant and Hilbert, 2008; Riley et al., 1998) but not widely used by physical oceanographers. In this, the response to forcing of a linear (or approximately linear) system is shown to depend on the resonance properties of the system. Use is also made of the fact that, for such systems, the response to periodic forcing is described by an analytic function, the response function, whose poles correspond to the resonant angular velocities or eigenvalues of the system. Once the resonance eigenvalues and eigenfunctions are known, the response of the system to any kind of forcing can be calculated.

Although the tides are affected by nonlinearities, the amplitude of the nonlinear tidal constituents are small over most of the ocean. As a result, the assumption of linearity is a good first approximation for any study of the large-scale behaviour of the tides.

If the system is frictionless, the poles of the response function lie on the real angular velocity axis and the eigenfunctions corresponding to the different resonances are orthogonal. If friction is present, as it is within the ocean, the poles lie off the real axis and the imaginary component of angular velocity then equals the inverse decay time of the resonance. With friction the eigenfunctions are also not orthogonal. This complicates the analysis, but it is still tractable making use of the adjoint set of equations and eigenvalues.

A simple example of the approach, using a 1-D tidal model, is given in Webb (2011). Further details and examples, solving Laplace's tidal equations in more realistic regions of ocean, are given in other papers from the present series (Webb, 2012, 2013a, b, 2014). Results from three much earlier papers on the tides (Webb, 1973, 1976, 1982) are also relevant.

In the present paper, a model similar to that used for the English Channel study is used to study the Hudson Bay region. The study aims to investigate the resonances and also the impact of three unusual properties of the region.

The first of these concerns Hudson Strait. With central depths of over 300 m, the strait is much deeper than is normal for continental shelf regions. The extra depth means that tidal wavelengths are longer than normal and frictional effects are smaller. As a result there is a potential for quarter and three-quarter wavelength resonances extending far into Hudson Bay in the south and into Foxe Basin in the north.

A second concerns the complex pattern of bays and channels in Foxe Basin. These all have the potential for supporting resonances and introducing extra complexity into the system.

---

[1]Leaf Bay is adjacent to Hopes Advanced Bay, indicated by the letter U in Fig. 2. Other locations discussed in the text are also shown in this figure.

Finally, the open region of Hudson Bay in the south is large enough to support a circulating wave. This may act like a resonator but a damped circulating wave might also have the properties of a damped infinite channel.

Section 2 of the paper gives the details of the model used for the study. Section 3 then reports on the results obtained using real values of angular velocity and Sect. 4 extends this to complex values of angular velocity. Section 5 is concerned with the main resonances affecting the semi-diurnal tides and Sect. 6 investigates how these combine to generate the observed response within the tidal band. Finally the discussion section reviews the main results of this study and considers their implications.

2   The numerical model

The model used to study the Hudson Bay, Hudson Strait and Foxe Basin region is based on that described by Webb (2013a). It solves the linear form of Laplace's tidal equations at a single angular velocity using finite difference equations based on an Arakawa C-grid distribution of model variables.

The model covers the region bounded by the latitude and longitude lines at 94.25° W, 57.5° W, 51.125° N and 70.375° N, with a resolution of 0.125° in the east–west direction and 0.25° in the north–south direction. The other free parameters are the linear coefficient of bottom friction and the minimum cell depth. These were set to $0.2 \, \mathrm{cm \, s^{-1}}$ and 2.5 m, as in Webb (2013a).[2]

Coastlines and cell depths are based on the GEBCO coastline depth data sets (IOC et al., 2003). The GEBCO depth data is at a higher resolution (1/60°) than that used for the model grid, so depths were calculated such that the volume of each model grid cell is the same as the corresponding GEBCO region. Away from coastlines this is equivalent to them having the same average depth.

As discussed in Webb (2013a), coastlines are specified to pass through velocity points, such that the normal velocity at the coast can be specified to be zero. Open boundaries are specified to follow lines of sea surface height points.

In the extreme northwest, the model closes off the narrow Fury and Hecla Strait which connects Foxe Basin with the Gulf of Boothia. The east of the strait is blocked by Elder and Ormonde islands, with the largest of the narrows being only 2 km wide. The total cross-section is so small that little tidal energy can flow into or out of Foxe Basin via this route.

The open boundary includes part of the Labrador Sea with depths extending down to 3000 m. In this region of the model the northern and southern limits are at 57.125° N and

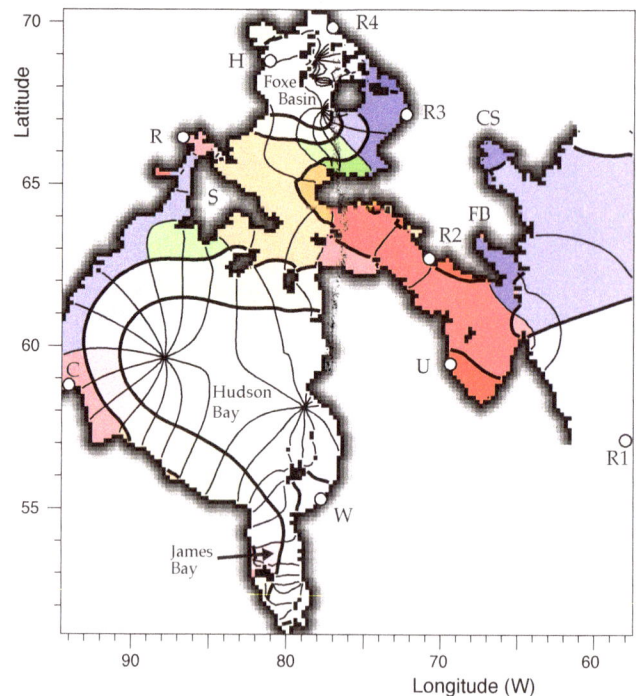

Figure 2. Model solution for the M2 tide. Thick lines are contours of amplitude at 0.5 m, 1 m, 2 m and 3 m. Thin lines are contours of phase, at intervals of 30 degrees, relative to the equilibrium tide at Greenwich. Colours denote phase quadrant (red, 0°–90°; orange, 90°–180°; green, 180°–290°; blue, 270°–360°) with the more intense colours denoting higher amplitudes. The tide gauge stations are C, Churchill; H, Hall Beach; R, Repulse Bay; W, Great Whale River; U, Hopes Advance Bay (Ungava Bay). Locations R1 to R4, S (Southampton Island), FB (Frobisher Bay) and CS (Cumberland Sound) are referred to elsewhere in the paper.

66.875° N. As a result the open boundary in the east includes almost all of the energy inflow region shown in Fig. 1.

For the open boundary, Webb (2013a) used Dirichlet boundary conditions in which the tidal height on the boundary is fixed by observations. The tangential velocity is set to the value one row in from the open boundary.

The present version of the model uses the same open boundary condition for the validation stage where the deep ocean tide is known. Then in the remainder of the study it is changed to allow radiation of energy back into the deep ocean. The radiation scheme is based on that of Flather (1976) which is often used for time-dependent models. Details are given in Appendix B1.

3   The M2 tide

The model was validated using a simulation of the M2 tide. For this the model was forced at the open boundary with tidal amplitudes and phases taken from a global assimilation of satellite altimeter data (Egbert and Erofeeva, 2002). The

---

[2]The model allows the linear coefficient of bottom friction to be a function of position. As discussed by Hunter (1975) this allows the mean effect of a realistic non-linear bottom friction term to be included. However, this requires additional runs of a fully non-linear model and has not be used for the study reported here.

model result is shown in Fig. 2 and the values at representative stations compared with tidal observations (IHB, 1954) in Table 1.

The agreement is good at Hopes Advance Bay, in the west of Ungava Bay near Leaf Basin. The phase in the deep ocean is around 300° so the coast here is approximately 70° out of phase. Given that some of the highest tides in the world are found in Leaf Basin, this phase difference between the coast and the deep ocean is less than the 90° expected from a pure quarter-wavelength resonance.

Agreement is also good at Churchill in the far west of Hudson Bay and, after the tidal wave has propagated around the south of Hudson Bay, the reduction in amplitude is represented reasonably well at Great Whale River in the southeast. However, the model tide arrives there early. This may indicate that the model depths are too deep or it may be because the model is not correctly representing the nearby amphidrome.

Unfortunately, in the north of the model region, analyses are available from only three stations and these are based on only a month's data. Two of these (Repulse Bay and Hall Beach) are included in the table. The third, Rowley Island, lies close to Hall Beach.

At both Repulse Bay and Hall Beach the model amplitude is not unreasonable but the phase is almost 180° out of agreement. Two series of tests were carried out to see if the difference could be understood.

In the first, the model was run with different values of the friction coefficient. This was found to affect the amplitude at inland stations but have only a small effect on phase. Thus halving the friction coefficient increased the amplitude at Churchill by over 50 % but only changed the phase by 11°. At Repulse Bay the increase was over 150 % and the phase change 12°.

The second set of tests arose from the observation that model phases close to the observed phase at Repulse Bay were found in the southwest of Foxe Basin. There is also a minimum amplitude in the channel joining the two regions and an area of rapid phase change, implying the presence of an amphidrome or nearby virtual amphidrome.

As the channel may have been too small in the model, tests were carried out where this was made wider and had its shallows removed. The changes affected the amplitude at Repulse Bay but had very little effect on the phases or the position of the amphidrome. A 180° phase change at Repulse Bay only appears possible if the amphidrome is moved to the channel running south towards Hudson Bay.

The phase agreement at Churchill indicates that amplitudes and phases are good in Hudson Bay, so the amphidrome can only be moved if the depths in the south of Foxe Basin are increased to raise the speed of the tidal wave through the region. Similarly, the phase error at Hall Beach, which is also in a region of low amplitudes, may be explained if the position of the model amphidrome is incorrect due to errors in model depths.

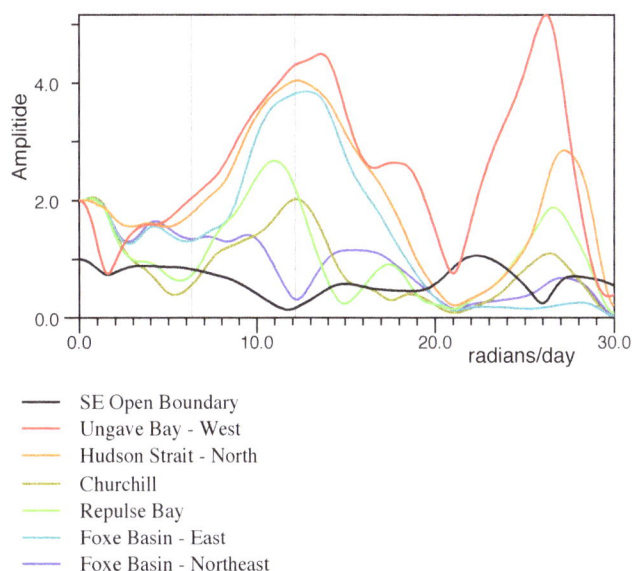

Figure 3. In colour: amplitude of the response function plotted as a function of angular velocity for real values of angular velocity. Colours: red, Ungava Bay (Hopes Advance Bay); orange, location R2 (see Fig. 1), north side of Hudson Strait; brown: Churchill; green, Repulse Bay; light blue, R3, East Foxe Basin; dark blue, R4, North Foxe Basin. In black: amplitude of outgoing (radiated) wave at boundary point R1.

Good depth data from Foxe Basin is limited, as it is covered in ice for much of the year. Freely floating sea-ice should not affect the tidal wave. Sea-ice fixed to the shore should result in increased turbulence and damping but is not expected to have a significant effect on wavelength.

This is because, as discussed in Webb (2011), with the friction coefficient and depths used in the model, friction has a significant effect on amplitudes and decay times but only a small effect on wavelength. Similarly, the use of a linear friction coefficient, instead of a fully nonlinear one, should affect decay times and amplitudes but have little effect on wavelength.

The suspicion therefore remains that the model phase errors result from a problem with the depths. No more can be done at this stage but the phase errors need to be kept in mind in the remainder of this analysis.

4   The response at real values of angular velocity

For the remainder of the study, the model was forced at the open boundary with an incoming wave of unit amplitude. Figure 3 shows the amplitude of the resulting model response at representative points within the region plus the amplitude of the outgoing wave at one point of the open boundary.

The chosen locations include Ungava Bay, Repulse Bay and model points R2 and R3. The latter two are on the north side of Hudson Strait and the east side of Foxe Basin where

Table 1. Model M2 tidal amplitude (m) and phases (degrees relative to the equilibrium tide at Greenwich) compared with tide gauge analyses (IHB, 1954).

		Model		Tide gauge	
		Amp. (m)	Phase°	Amp. (m)	Phase°
Hopes Advance Bay	69.6° W 59.4° N	3.16	13	3.88	10
Churchill	94.2° W 58.8° N	1.46	13	1.52	24
Great Whale River	77.8° W 55.3° N	0.54	345	0.63	17
Repulse Bay	86.5° W 66.4° N	1.63	18	1.88	192
Hall Beach	81.2° W 68.8° N	0.36	186	0.22	25

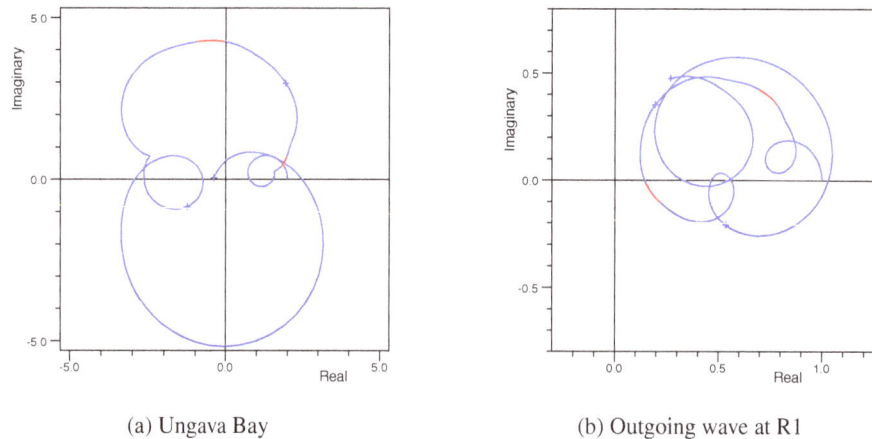

(a) Ungava Bay   (b) Outgoing wave at R1

Figure 4. Real and imaginary components of (**a**) the response function on the west side of Ungava Bay and (**b**) the outgoing wave at point R1 on the open boundary, plotted for real values of angular velocity between zero and 30 rad day^{-1}. Crosses at intervals of 10 rad day^{-1}. The red sections correspond to the diurnal and semi-diurnal tides. At zero rad day^{-1} the response function at Ungava Bay has the value $(2 + i0)$ and the outgoing wave has the value $(1 + i0)$.

the model semi-diurnal tides are highest. A station in north Foxe Basin is also included because the response there appears to be very different to that of east Foxe Basin. In the south, the figure includes Churchill, on the western shore of Hudson Bay, but stations further south are omitted as the responses there are similar to Churchill but with lower amplitudes.

Finally, the figure includes the outgoing wave at point R1 on the open boundary (see Fig. 2). This is chosen because it is in deep water at the foot of the continental slope and is in a position where it should give an indication of the behaviour of Kelvin waves travelling south out of the model region. It also lies near the positions where the resonances, discussed later, show the maximum energy losses through the open boundary.

At zero angular velocity both the ingoing and outgoing waves at the open boundary have unit amplitude, so the amplitude everywhere equals two. As angular velocity increases, the amplitudes initially tend to decrease but they then increase to maxima near the semi-diurnal tidal band. There is then a general decrease to minima around

21 rad day^{-1}[3] after which there are further maxima between 25 and 30 rad day^{-1}.

Within this large-scale behaviour there are many individual maxima and changes in curvature, which, on the basis of previous work, are likely to be due to individual resonances or groups of resonances. Sometimes the maxima and changes in curvature occur at the same angular velocity, indicating that such regions are coupled and affected by the same resonance. However, at other angular velocities the same regions may have very different response to the forcing.

A noticeable feature of the outgoing wave at R1 is that it has a minimum where the resonances are largest near 12 rad day^{-1} and a maximum near 22 rad day^{-1} where the large-scale response is least. It then has another minimum near 26 rad day^{-1} where Ungava Bay has a maximum.

An alternative view of the system's response is obtained by plotting the real and imaginary components of the response function, as in Fig. 4. As shown in Appendix A, the response function $\boldsymbol{R}(x, \omega)$ at position x and angular velocity ω can be

---

[3]The paper uses units of radians per day (rad day^{-1}). Thus the semi-diurnal tides with periods around 12 h have angular velocities around 4π rad day^{-1}.

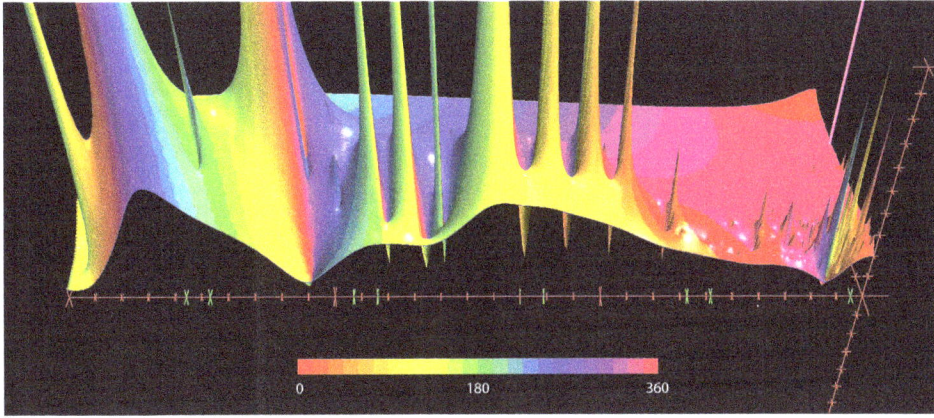

Figure 5. Response function amplitude on the west side of Ungava Bay plotted as a function of complex angular velocity. The colours denote complex phase, in degrees, as denoted on the scale below the main figure. Values at real values of angular velocity are plotted in blue. The origin (0,0) is on the right with the positive real axis (in red) running from right to left and the negative imaginary axis running into the figure. Both are marked by red crosses every 1 rad day^{-1}. The vertical axis is marked similarly at unit intervals. On the real axis green crosses indicate the limits of the tidal bands near 2π (diurnal), 4π (semi-diurnal), 6π and 8π rad day^{-1}.

expressed as a sum over the resonance contributions,

$$R(x, \omega) = \sum_j \boldsymbol{\psi}_j(x) r_j / (\omega - \omega_j), \tag{1}$$

where ω_j is the angular velocity of the jth resonance, $\boldsymbol{\psi}_j(x)$ describes the spatial structure of the resonance and r_j depends on how the system is forced.

At a fixed location when the resonances are well separated, the response function $\boldsymbol{R}(\omega)$ near each resonances has the form

$$\boldsymbol{R}(\omega) = A_j / (\omega - \omega_j) + B(\omega), \tag{2}$$

where $B(\omega)$ represents the smooth background due to of distant resonances.

When this function, without the background term, is plotted as in Fig. 4, its locus generates a simple circle. This starts at the origin when ω is minus infinity. It then moves in an anticlockwise direction, reaching maximum amplitude when ω equals ω_j, and returning to the origin as ω approaches infinity.

When the background is added, isolated resonances generate single loops on the smooth background. Where resonances are overlapping, more complicated structures may be formed but the contribution of each resonance still has the same simple underlying form.

Applying these ideas to Fig. 4, at Ungava Bay the response function is seen to contain four main loop structures. These consist of two small loops generating the amplitude minima near 2 and 21 rad day^{-1} and two larger ones which generate the maxima near 13 and 26 rad day^{-1}.

The outgoing wave at R1 also shows large and small loops at roughly the same angular velocities, but this time the large loops reduce the amplitude of the radiated wave. Near 12 rad day^{-1} the amplitude drops to near 0.1, and as

the radiated power depends on the square of the amplitude it implies that the radiated power is close to 1 % of the incident power.

5 The response at complex values of angular velocity

The previous figures illustrate the type of information that can be obtained with models that can investigate real values of angular velocity. However, an advantage of the present model is that it can also obtain solutions for complex values of angular velocity and thus explore the full resonant structure of the system.

Figures 5–7 show the response function at four stations plotted on the complex plane. In these figures the vertical coordinate indicates the response function amplitude and the colour represents phase, zero phase being the phase of the incoming wave on the open boundary.

The functions can be interpreted using Eq. (1). The poles in the complex plane correspond to the resonances, the coordinate of the poles giving the real and imaginary components of each eigenvalue ω_j. The position of the resonances is the same in each figure but their strengths differ due to changes in the eigenvector $\boldsymbol{\psi}_j(x)$ between locations.

The figures show only a quarter of the complex angular velocity plane, the quarter where the real component of angular velocity is zero or positive and the imaginary component is zero or negative. There is no need to show more because, as discussed in Appendix A, causality means that there can be no resonances with positive imaginary components and symmetry requires the function to be mirrored about the imaginary axis. Thus each resonance with angular velocity ω_j and residue r_j has a twin with angular velocity $-\omega_j^*$ and residue $-r_j^*$.

(a) Churchill

(b) R3, east side of Foxe Basin

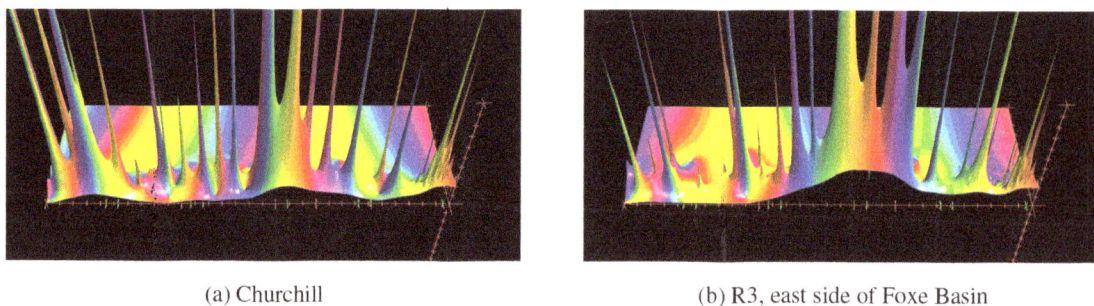

Figure 6. The response function amplitude plotted as a function of complex angular velocity (**a**) at Churchill and (**b**) location R3 on the eastern side of Foxe Basin. Colours and axes as in Fig. 5.

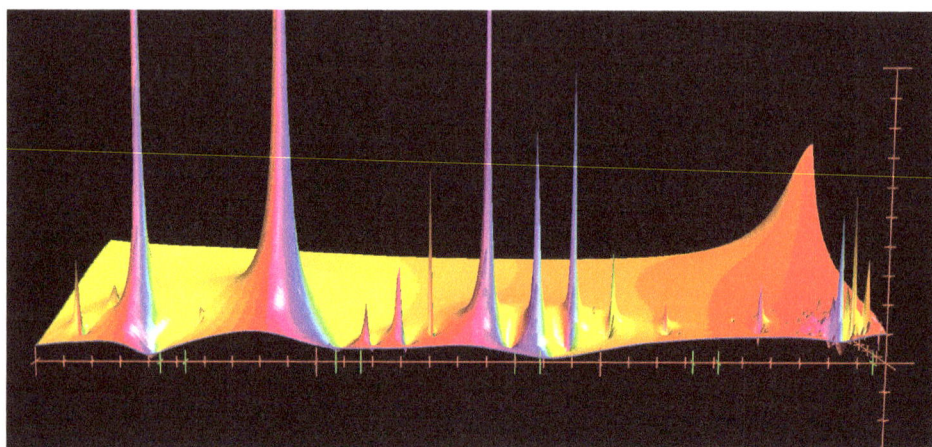

Figure 7. Response function amplitude of the outgoing wave at point R1 on the open boundary. Colours and scales as in Fig. 5.

In Fig. 5, Ungava Bay is plotted because of the interest in its extreme tides and because the region appears to support a classic quarter-wavelength resonance. Churchill and location R3 (Fig. 6) are chosen to represent Hudson Bay and Foxe Basin. Location R1 (Fig. 7) lies beyond the outer edge of the continental slope and is chosen to capture any Kelvin wave progressing southward out of the model region.

In the limit of zero angular velocity, both the incoming and outgoing wave on the open boundary have unit amplitude. As a result, as shown in Figs. 5 and 6, the amplitude at the origin is two. Figure 7 shows only the outgoing wave on the boundary, so its amplitude at the origin is one.

As angular velocity increases along the real axis, the phase tends to increase. This arises because the ratio of the phase to the angular velocity is a measure of the time taken for the wave to propagate to each location from the forcing region. Off the real axis the behaviour is more complicated due to the phase increase of 2π radians close around each pole which arises from Eq. (2).

Starting from near the origin, there is initially a dense group of weak resonances which extends to near 6 rad day^{-1}. As in Webb (2013a) these are the continental shelf and Rossby wave modes. There is then a series of gravity wave

modes extending to higher angular velocities. The angular velocities of these modes, calculated using the method outlined in Appendix B2, are given in Table 2.

Near the real axis, the colours show that the phase at Ungava Bay increases slowly, reaching 90° in the region of the semi-diurnal tides. In contrast, there is a more rapid phase change at both Churchill and at location R3, the colours indicating that near the semi-diurnal band these phases are approximately 450°, i.e. one and a quarter wavelengths different from the forcing.

The response function for Ungava Bay shows that the large-amplitude region near the semi-diurnal tidal band (12 rad day^{-1}) is associated with four main resonances. These are resonances D to G of Table 2. The figures show that the same resonances are also responsible for large amplitudes in Hudson Bay and Foxe Basin, the strength of individual resonances there often being larger than in Ungava Bay.

In some ways this result is surprising. Ungava Bay is known to have high tides, and given the local bathymetry and distance from the shelf edge and the phase of the response function, the high semi-diurnal response might be expected to be due to a single quarter-wavelength resonance. Hudson

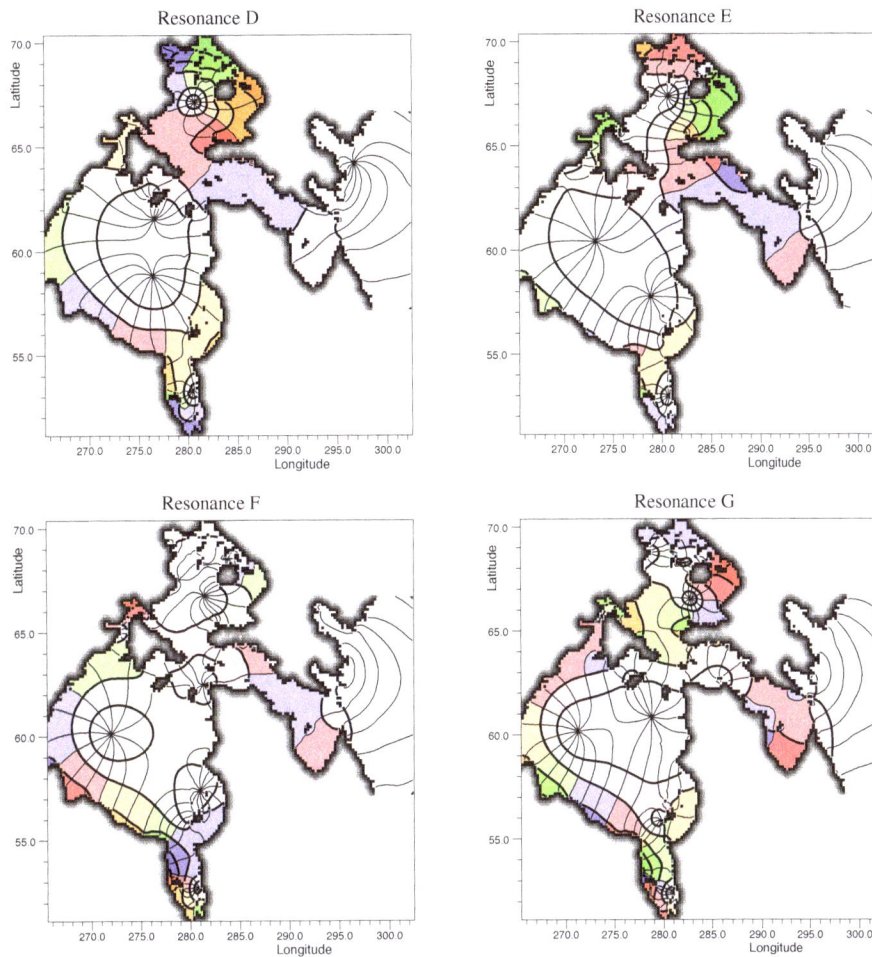

Figure 8. The amplitude and phase of resonances D to G of Table 2. Amplitudes normalised to one and phases are relative to the west side of Ungava Bay. Thick lines are contours of amplitude at 0.1, 0.2, 0.4, 0.6 and 0.8. Thin lines are contours of phase at intervals of 30 degrees. Colours as in Fig. 1, with zero phase between the red and blue areas.

Bay and Foxe Basin are much further from the shelf edge and on the same basis they must involve at least three-quarter wavelength resonances and possibly one and a quarter wavelength resonances.

At higher angular velocities the Ungava Bay response function shows two further strong resonances. The first, resonance Q near 21 rad day^{-1}, has little impact on the response at real values of angular velocity. The second, resonance U near 27 rad day^{-1}, has a much greater impact on the response in Ungava Bay. It is also responsible for an increased amplitude at Churchill but has essentially no impact at position R3. Further investigation of their structure showed that resonance Q is primarily a quarter-wavelength resonance of Cumberland Sound and U a similar resonance of Frobisher Bay.

5.1 Radiation at the open boundary

Figure 7 indicates that the first of the large loops in Fig. 4b is associated with resonances D to G and the second large loop

with resonance U. The first of the small loops is associated with the shelf and Rossby wave resonances. The cause of the second loop is not so clear but it is likely to be Q together with K and L.

In the previous study of the English Channel and Irish Sea (Webb, 2013a), the high semi-diurnal tides in the Bristol Channel were found to result from two resonances which both had slightly higher angular velocities. It was assumed that these were also responsible for the large amount of tidal energy that was dissipated in the region.

The present result indicates that the strong absorption of tidal energy by the Hudson Bay system is a result of four resonances which straddle the tidal band. The strong absorption thus may result from straddling the tidal band or it may result from the fact that four resonances are involved, each of which can provide an independent contribution to resonant absorption of tidal energy.

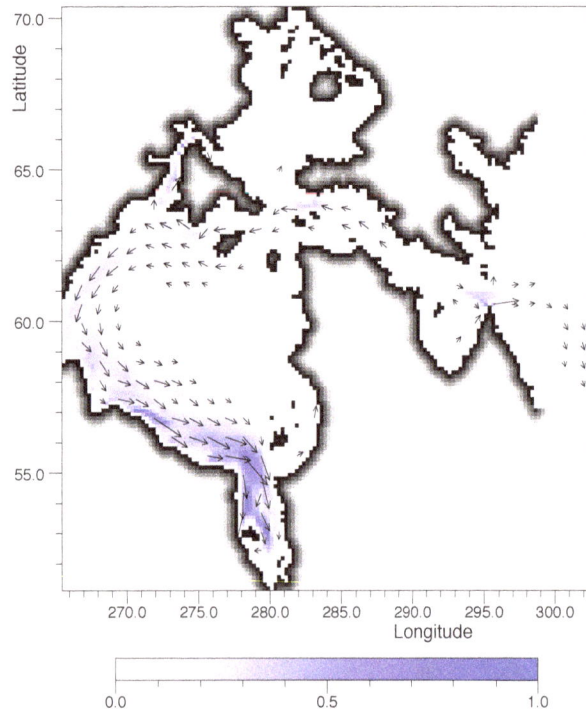

Figure 9. Energy flux vectors for resonance F. The flux is normalised so that its maximum value is one.

Table 2. Real and imaginary components of angular velocity (in radians per day) for the gravity wave resonances.

	Angular velocity			Angular velocity	
	Real	Imag.		Real	Imag.
A	4.1435	−1.1716	N	19.9437	−2.5796
B	5.1052	−1.6125	O	20.2071	−1.5208
C	7.5167	−1.2326	P	20.9670	−1.8545
D	9.4433	−1.5199	Q	21.4514	−2.1738
E	10.8327	−1.7144	R	23.2567	−1.6131
F	12.1613	−1.5320	S	24.4043	−1.3558
G	13.9215	−1.6782	T	26.1338	−1.6607
H	15.6169	−3.9629	U	26.5379	−1.2928
I	16.0036	−1.5008	V	27.7020	−2.0412
J	16.5089	−4.4841	W	28.7601	−1.1003
K	17.1372	−1.4164	X	28.9116	−1.7231
L	18.2980	−1.2522	Y	29.0517	−4.0404
M	19.7361	−1.5471			

6   Structure of the main semi-diurnal resonances

The spatial structures of the four largest resonances near the semi-diurnal tidal band were calculated using the method outlined in Appendix B2. The results are shown in Fig. 8. The solutions have been normalised so that the maximum amplitude is one and the phase is zero in the west of Ungava Bay. For two of the modes, the maxima are on the eastern side of Foxe Basin. The other two have maxima on the western side of James Bay.

The energy flux vectors for resonance F are shown in Fig. 9. This and similar plots (not included) for the other modes show that, in each case, the flux is away from the deep regions of Hudson Strait, both eastwards towards the open boundary and westwards towards the high-amplitude regions in Foxe Basin or Hudson Bay.

None of the modes can be characterised by a single unique feature, such as a standing wave in a limited region of shelf. Instead they appear to involve the coupling of a series of such simple or underlying modes.

The first of these is the quarter-wavelength resonance involving Ungava Bay. The justification for this is the fact that all four modes show an approximate 90 degree phase lag between the west side of Ungava Bay and the shelf edge or open boundary – although depending on the precise point chosen the value can vary between 75 and 130 degrees.

A second is a three-quarter wavelength mode between the shelf edge or open boundary and the east side of Foxe Basin.

As seen best in resonance E where there is a maximum in Hudson Strait and a node to the north of Southampton Island, the locations lying roughly one-quarter and one-half of a wavelength from the open boundary.

A third is a half-wavelength resonance trapped between the northwest and southeast coasts of Foxe Basin. This is best seen in resonance D, but as the model had poor agreement with the measured tide in the NW of Foxe Basin, this possibility should be treated with caution.

Finally, Hudson Bay itself appears to support its own underlying mode, consisting of a single wavelength that circles the bay in an anticlockwise direction. This is best seen in resonance F.

The results imply that although resonances involving the shelf edge are important, the grouping of resonances around the semi-diurnal tides is partly due to standing waves within the interior of the Hudson Bay system.

7   Resonant contributions to the semi-diurnal tides

As discussed in Webb (2012), the response function near a tidal band can be split into the contribution of nearby resonances and a smooth background due to distant resonances. Thus,

$$R(x, \omega) = \sum_j A_j(x)/(\omega - \omega_j) + S(x, \omega) + B(x, \omega), \qquad (3)$$

where $R(x, \omega)$ is the response function at position x and angular velocity ω. The sum j is over key nearby resonances and $A_j(x)$ is the residue at w_j, the resonance angular velocity. Methods for calculating the residue are described in Appendix B3. For this part of the analysis, the symmetry term $S(x, \omega)$, due to the mirror images of the resonances in the summation, is simple to calculate and so is split off from the background.

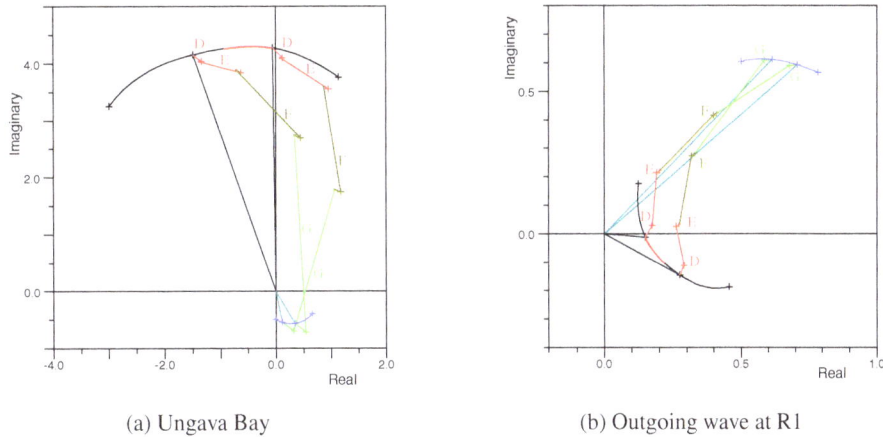

(a) Ungava Bay (b) Outgoing wave at R1

Figure 10. In black: real and imaginary components of the response function (**a**) on the west side of Ungava Bay and (**b**) the outgoing wave at point R1 on the open boundary, plotted for real values of angular velocity between 11 and 14 rad day^{-1}. Coloured vectors indicate the contributions of the resonances D–G of Table 2 and their conjugates. The blue line is the residual background. The red section of the main curve corresponds to the semi-diurnal tides.

In Fig. 10, Eq. (3) has been used to determine how the four resonances of Fig. 8 contribute to the high semi-diurnal tides of Ungava Bay and the low outgoing wave at position R1 on the open boundary.

It shows that in Ungava Bay the high tides are primarily due to resonances F and G. They both have large amplitudes and, as their phases are similar, they reinforce each other. Resonance E makes some contribution, but D is insignificant, its amplitude being smaller than the background.

Resonances F and G are also the dominant ones at position R1. At 12 rad day^{-1} they both have the effect of reducing the amplitude of the outgoing wave by about a third. At the same point resonances D and E are contributing to a reduction of the outgoing wave, but by 13 rad day^{-1} this is no longer the case.

The net effect of the resonances at 12 rad day^{-1} is to reduce the response function amplitude from 0.91 to 0.15. Assuming that the power is proportional to the amplitude squared, this means that the four resonances are absorbing over 97 % of the incident tidal energy. Resonances F and G together reduce the energy by approximately 90 % and although the contributions of resonances D and E are much smaller they are responsible for absorbing over 70 % of the remaining energy.

8 Conclusions

The study has shown that the semi-diurnal tides of the Hudson Bay region are dominated by four resonances. These straddle the semi-diurnal tidal band and contribute both to high tidal amplitudes within the region and to very low amplitudes in the tidal wave radiated away from the region.

The previous study, made using a time-dependent model (Arbic et al., 2007), was only able to identify one resonance affecting the semi-diurnal tides. The new result therefore emphasises the usefulness of the present approach.

The study does not explain why the Hudson Bay system is such a good absorber of tidal energy and more effective than the English Channel and Irish Sea but it does give hints that need to be followed up.

The first is the primary result that the Hudson Bay system has four significant resonances close to and straddling the semi-diurnal tidal band. In the case of the Bristol Channel and Gulf of St Malo there are only two resonances and they lie to one side of the tidal band.

The study also used one of the points on the open boundary as an analogue of the reflected wave for the case of a uniform amplitude incident wave at all points on the open boundary. Although the uniform amplitude incident wave is a special case, it is a plausible first approximation to the way the M2 tide forces the region.

The results show that each of the four main resonances acted to reduce the amplitude of the reflected wave in the semi-diurnal tidal band. At 4π rad day^{-1}, two of the resonances together absorbed ~ 90 % of the incident energy and the other two, although weaker, absorbed ~ 70 % of the remainder. The small reflection coefficients imply that all four resonances have impedances which are well matched to that of the deep ocean.[4]

The study shows that the closeness of the resonances is partly due to the complex topography of the region. As well as the "classical" 1/4 wavelength wave between Ungava Bay and the shelf edge, the deep Hudson Strait also allows the development of 3/4 wavelength, and possibly 5/4 wavelength,

_____

[4]Impedance matching is critical in the design of waveguides and in other systems involving propagating waves.

resonances between the shelf edge and features far to the west. There, both Foxe Basin and Hudson Bay are of the right size to support standing waves of near-tidal period.

The depth of Hudson Strait also means that energy dissipation is reduced compared with a normal shelf. The mean depth (~ 300 m) is approximately four times that of a normal shelf (~ 80 m), so the frictional effect per wavelength in the strait should be halved. This probably helps in matching the Foxe Basin and Hudson Bay components of the resonances to the deep ocean.

The effectiveness of a continental shelf region in absorbing tidal energy is also likely to depend on both the length of continental shelf involved and the angle at which the tidal wave approaches the shelf. These features have not been studied here but as it passes the English Channel and Irish Sea, the semi-diurnal tidal wave in the deep ocean runs roughly parallel to the shelf edge. In contrast, as it approaches Hudson Strait the wave approaches roughly at right angles to the shelf edge.

To conclude, the present study has given new insights into the properties of the Hudson Bay region and the complex interactions that are involved. The study has shown that there is still much to be learnt about the physics of the region but the results presented here should provide a useful basis for further work.

Appendix A: Solution in terms of resonances

Laplace's tidal equations with a linear friction term are

$$\rho \partial u/\partial t + \rho f \times u + \rho g \nabla \eta + (\kappa/h)u = \rho g \nabla \eta_0, \quad (A1)$$
$$\partial \eta/\partial t + \nabla \cdot (hu) = 0,$$

where u is horizontal velocity, η tidal height, ρ density, f the Coriolis parameter, g gravity, h water depth, κ the linear coefficient of bottom friction and η_0 the equilibrium tide forcing the ocean.

If $u(x,t)$ and $\eta(x,t)$ at location x and time t are represented by the vector Ψ,

$$\Psi(x,t) = \begin{pmatrix} u(x,t) \\ \eta(x,t) \end{pmatrix}, \quad (A2)$$

and if

$$\Psi_0(x,t) = \begin{pmatrix} 0 \\ \eta_0(x,t) \end{pmatrix}, \quad (A3)$$

then the tidal equations can be written in the form

$$\partial \Psi(x,t)/\partial t + \mathcal{L}\Psi(x,t) = \mathcal{L}\Psi_0(x,t) \quad (A4)$$

where $\mathcal{L}$ is a matrix operator discussed further in Webb (2014).

Consider the response when the system is forced at angular velocity ω. The solution will then have the form

$$\Psi(x,t) = (\Psi(x,\omega)\exp(-i\omega t) + c.c., \quad (A5)$$
$$= 2Re(\Psi(x,\omega)\exp(-i\omega t)), \quad (A6)$$

where $Re()$ represents the real part and $c.c.$ represents complex conjugate. Substituting in Eq. (A4),

$$(\mathcal{L} - i\omega\mathbf{1})\Psi(x,\omega) = \mathcal{L}\Psi_0(x,\omega), \quad (A7)$$

where $\Psi_0(x,\omega)$ and $\Psi(x,\omega)$ are the forcing and ocean response at angular velocity ω and $\mathbf{1}$ is the unit matrix.

Equations of this form can be solved using the eigenfunctions of the operator $\mathcal{L}$ and its adjoint $\widetilde{\mathcal{L}}$. The basic method is described in Morse and Feshbach (1953). Webb (2014) discusses its application to Laplace's tidal equations and shows how a suitable definition of the inner or dot product generates a physically meaningful adjoint.

Dropping the x coordinate, if ω_j and ψ_j are the eigenvalues and eigenfunctions of the equation

$$(\mathcal{L} - i\omega_j\mathbf{1})\psi_j = 0 \quad (A8)$$

and if λ_k and ϕ_k are the eigenvalues and eigenfunctions of the adjoint operator

$$(\widetilde{\mathcal{L}} - i\lambda_k\mathbf{1})\phi_k = 0, \quad (A9)$$

then (Morse and Feshbach, 1953; Webb, 2014)

$$\int dx \, \phi_j^*(x) \cdot \psi_k(x)(\omega_j + \lambda_k) = 0. \quad (A10)$$

Thus either λ_k equals $-\omega_j^*$ or the integral is zero, so the eigenfunctions are orthogonal.

Normalise $\psi_k(x)$ so that the above integral with $\phi_k(x)$ equals one. Then expanding $\Psi(x,\omega)$ and the equilibrium tide $\Psi_0(x,\omega)$ in terms of the eigenfunctions

$$\Psi(x,\omega) = \sum_j a_j(\omega)\psi_j(x), \quad (A11)$$

$$\Psi_0(x,\omega) = \sum_j \psi_j(x) \int dx'\phi_j^*(x') \cdot \Psi_0(x',\omega). \quad (A12)$$

Substituting in Eq. (A7),

$$(\mathcal{L} - i\omega) \sum_j a_j(\omega)\psi_j(x) =$$
$$\mathcal{L}\sum_j \psi_j(x) \int dx'\phi_j^*(x') \cdot \Psi_0(x',\omega),$$
$$\sum_j a_j(w)(i\omega_j - i\omega)\psi_j(x) =$$
$$\sum_j \psi_j(x)i\omega_j \int dx'\phi_j^*(x') \cdot \Psi_0(x',\omega). \quad (A13)$$

Multiplying by $\phi_k^*(x)$ and integrating over x,

$$a_k(\omega) = (-\omega_k/((\omega - \omega_k)) \int dx' \, \phi_k^*(x') \cdot \Psi_0(x',w). \quad (A14)$$

Thus the solution to Eq. (A4) is

$$\Psi(x,\omega) =$$
$$\sum_j \psi_j(x) \frac{-w_j}{w - w_j} \int dx' \, \phi_j^*(x') \cdot \Psi_0(x',\omega). \quad (A15)$$

In the type of problem investigated in this paper, the forcing term $\Psi_0(x,t)$ can be separated into a spatial term $\Psi_0(x)$ and a time-dependent term $\Psi_0(t)$, with Fourier transform $\Psi_0(\omega)$. Then,

$$\Psi(x,\omega) = R(x,\omega)\Psi_0(\omega), \quad (A16)$$

$R(x,\omega)$ is the (vector) response function

$$R(x,w) = \sum_j \psi_j(x) \frac{r_j}{w - w_j} \quad (A17)$$

$$r_j = -\omega_j \int dx' \, \phi_j^*(x') \cdot \Psi_0(x'). \quad (A18)$$

The response functions plotted in Figs. 5 and 6 are the sea surface elevation component of R.

A1 Forcing at an open boundary

A similar result is obtained when forcing is due by a tidal wave entering through an open boundary. In this case the eigenfunctions are first obtained with η equal to zero on the open boundary and the equilibrium tide term in Eq. (A1) is replaced by a function which is zero within the region studied but has a step change along a line just inside the open boundary,

$$\eta_0(x) = -\Theta(n)\eta_b(s),\qquad(A19)$$

where n and s are the coordinates normal and tangential to each point on the open boundary and $\eta_b(s)$ is the forced wave at position s. The function Θ is zero inside the boundary and equals one on and outside the open boundary.

To confirm that Eq. (A19) has the right form, integrate Eq. (A1) between a point a distance ϵ within the region and a point on the boundary. As the distance ϵ tends to zero, all terms tend to zero except for

$$\int_{-\epsilon}^{0} dn\ \rho g(\partial/\partial n)\eta = -\int_{-\epsilon}^{0} dn\ \rho g(\partial/\partial n)\Theta(n)\eta_b(s).\qquad(A20)$$

But η is zero on the boundary and the integral of the gradient of a step function is one, so after integrating and rearranging, the value of sea level just inside the boundary equals

$$\eta = \eta_b(s),\qquad(A21)$$

as required.

The derivation of Eq. (A15) follows as before, the main change occurring in Eq. (A13) which now reads as

$$(\mathcal{L} - i\omega\mathbf{1})\sum_j a_j(\omega)\boldsymbol{\psi}_j(x) = \begin{pmatrix} -g\nabla(\Theta(n)\eta_b(s)) \\ 0 \end{pmatrix}\quad(A22)$$

When multiplied by $\boldsymbol{\phi}_k^*(x)$ and integrated over x, the right-hand side becomes an integral around the open boundary s,

$$-g\oint ds\ \phi_k^{n*}(x(s))\cdot(\partial/\partial n)(\Theta(n)\eta_b(s)),\qquad(A23)$$

where ϕ_k^n is the velocity component of eigenvector $\boldsymbol{\phi}_k$ which is normal to the boundary. Integrating by parts, this becomes

$$g\oint ds\ \partial/\partial n(\phi_k^{n*}(x(s)))\cdot\eta_b(s).\qquad(A24)$$

Thus the integral of Eq. (A15) is replaced by an integral around the open boundary.

A2   Symmetry

Equation (A5) corresponds to a Fourier transform representation when only a single angular velocity is present. For the general case where the system contains a full range of angular velocities, the Fourier transform between the time and angular velocity representation of $\boldsymbol{\Psi}(x)$ is

$$\boldsymbol{\Psi}(x,t) = \int_{-\infty}^{\infty} d\omega\ \exp(-i\omega t)\ \boldsymbol{\Psi}(x,\omega),\qquad(A25)$$

$$\boldsymbol{\Psi}(x,\omega) = \frac{1}{2\pi}\int_{-\infty}^{\infty} dt\ \exp(i\omega t)\ \boldsymbol{\Psi}(x,t).\qquad(A26)$$

$\boldsymbol{\Psi}(x,t)$ is real so

$$\boldsymbol{\Psi}(x,-\omega^*)^* = \frac{1}{2\pi}\int_{-\infty}^{\infty} dt\ \exp(i\omega t)\ \boldsymbol{\Psi}(x,t),\qquad(A27)$$

and

$$\boldsymbol{\Psi}(x,-\omega^*)^* = \boldsymbol{\Psi}(x,\omega).\qquad(A28)$$

$\boldsymbol{\Psi}_0(x,t)$ is also real, so

$$\boldsymbol{\Psi}_0(x,-\omega^*)^* = \boldsymbol{\Psi}_0(x,\omega).\qquad(A29)$$

Similarly, for the response function $\boldsymbol{R}(x,\omega)$ relating $\boldsymbol{\Psi}(x,\omega)$ and $\boldsymbol{\Psi}_0(x,\omega)$,

$$\boldsymbol{R}(x,-\omega^*)^* = \boldsymbol{R}(x,\omega).\qquad(A30)$$

In Eqs. (A15) and (A17), this means that if there is a pole at ω_j with residue $r_j(x)$, there must also be one at $-\omega_j^*$ with residue $-r_j(x)^*$.

A3   Causality

Substituting the response function equation (Eq. A16) into Eq. (A25),

$$\boldsymbol{\Psi}(x,t) = \int_{-\infty}^{\infty} d\omega\ \exp(-i\omega t)\ \boldsymbol{\Psi}(x,\omega)\qquad(A31)$$

$$= \int_{-\infty}^{\infty} d\omega\ \exp(-i\omega t)\ \boldsymbol{R}(x,\omega)\Psi_0(\omega),\qquad(A32)$$

$$= \int_{-\infty}^{\infty} d\omega\ \exp(-i\omega t)\ \boldsymbol{R}(x,\omega)\frac{1}{2\pi}$$
$$\int_{-\infty}^{\infty} dt'\ \exp(i\omega t')\Psi_0(t').\qquad(A33)$$

Let the forcing $\Psi_0(t)$ be an impulse at time zero only. Such an impulse can be represented by the product $\Psi_0\delta(t)$. Ψ_0 is a constant and the delta function $\delta(t)$ has the property that for any function $F(t)$,

$$\int dt\ \delta(t)F(t) = F(0).\qquad(A34)$$

Then, using Eq. (A17),

$$
\Psi(x,t) = \int_{-\infty}^{\infty} d\omega \ \exp(-i\omega t) \sum_j \psi_j(x) \frac{r_j}{w-w_j} \frac{1}{2\pi}
$$

$$
\int_{-\infty}^{\infty} dt' \ \exp(i\omega t')\delta(t')\Psi_0 \qquad (A35)
$$

$$
= \frac{1}{2\pi} \int_{-\infty}^{\infty} d\omega \ \exp(-i\omega t))
$$

$$
\sum_j \psi_j(x) \frac{r_j}{w-w_j} \Psi_0. \qquad (A36)
$$

The integrand is an analytic function of ω, so the integral can be completed using the method of contour integration. If t is positive, $\exp(-i\omega t)$ tends to zero as ω tends to minus infinity, so the contour can be completed in a clockwise direction around the negative imaginary half-plane. Thus,

$$
\Psi(x,t) = -i \sum_j \exp(-i\omega_j t) \ \psi_j(x) r_j \Psi_0, \qquad (A37)
$$

where the sum is over the poles in the negative imaginary half plane. Each resonance oscillates independently and dies away at its own natural rate.

If t is negative, the contour can be completed around the positive imaginary half-plane. If there are any poles there, the result is non-zero, so the system is responding before any forcing is applied. This is impossible for physically realistic systems as it breaks causality. As a result there can be no poles in the positive imaginary half-plane, i.e. no resonances with positive imaginary components of angular velocity.

Appendix B: Mathematical and numerical details

The model used for the present study represents Laplace's tidal equations as a set of finite difference equations on an Arakawa C-grid, as described in Webb (2013a). The model assumes a time dependence of the form $\exp(-i\omega t)$, where t is time and ω the angular velocity of the ocean wave. If the model variables are represented by a vector y, then the finite difference equations can be written as a matrix equation,

$$
(L - i\omega 1)y = z, \qquad (B1)
$$

where 1 is the unit matrix. The term $(-i\omega 1)$ results from the time-dependent terms in Laplace's tidal equations and the matrix L contains the contributions from all the other terms in the set of finite difference equations. The vector z represents the forcing. If the variables are numbered in a systematic manner, L becomes a band matrix and the equations can be solved using efficient band matrix algorithms.

B1 The open boundary condition

In previous versions of the model, sea surface height (SSH) points on the open boundary were treated explicitly as part of the model vector y. However, an investigation of the properties of the adjoint system (Webb, 2014) showed that it was better to treat the open boundary condition implicitly, that is as an additional term acting on the normal velocities at points adjacent to the open boundary.

Webb (2014) also showed that it is better to set the tangential velocities at the open boundary to zero, so this was done in the present model. The change ensures that the finite difference Coriolis terms conserve energy and also ensures that the resulting matrix equation has the expected adjoint symmetry.

The previous model also used Dirichlet boundary conditions in which SSH on the open boundary is fixed. To allow radiation of energy through the open boundary, a scheme has been developed based on the one proposed by Flather (1976) which is often used for time-dependent models.

Let ζ represent sea surface height and u the velocity normal to an open boundary placed at the origin of coordinate x. In a plane wave propagating in the positive direction in a region of constant depth h, the sea surface height and velocity are related by

$$
u = (c_0/h)\zeta, \qquad (B2)
$$

where the wave speed c_0 equals $(gh)^{1/2}$.

The new boundary condition assumes that the solution in the neighbourhood of a boundary point can be expressed as the sum of two such waves each propagating in opposite directions normal to the boundary,

$$
\zeta = A\exp(ikx - i\omega t) + B\exp(-ikx - i\omega t), \qquad (B3)
$$

$$
u = A(c_0/h)\exp(ikx - i\omega t) - B(c_0/h)\exp(-ikx - i\omega t).
$$

If A represents the unknown outgoing wave and B the known incoming wave, then eliminating A, the open boundary condition becomes

$$
\zeta - (c_0/h)u = 2B. \qquad (B4)
$$

As the open boundary value of ζ is not part of the model vector, this equation is used to replace the term involving the open boundary ζ in the equation for the normal velocity point closest to the boundary.

Webb (2013b) proposed a similar scheme but one which allowed for the differing position of the SSH and normal velocity points. Unfortunately, the resulting open boundary condition is a function of ω and as a result the matrix L also becomes a function of ω. Tests were carried out with both boundary conditions, to see the effect of the change on the calculated eigenvalues and eigenfunctions. The effect was small, the differences in the calculated resonance eigenvalues being less than 0.01 rad day^{-1}.

B2   Calculation of eigenvalues and eigenvectors

Initial estimates of the eigenvalues ω_j were obtained from the data sets used to generate Figs. 5 and 6 by fitting the four points around each maximum of the response function, for a fixed x_k, to the expansion

$$R(x_k, \omega) = R_j(x_k)/(\omega - \omega_j) + B(x_k) + C(x_k)\omega. \quad \text{(B5)}$$

Accurate values of the eigenvector and eigenvalue were then obtained by inverse iteration, i.e. by solving the set of equations

$$(L - i\omega_j' 1)\psi_j^{[n]} = \psi_j^{[n-1]}/N_j^{[n-1]}, \quad \text{(B6)}$$

where ω_j' is the initial estimate of ω_j, $\psi_j^{[n]}$ is the solution following the nth iteration and $N_j^{[N]}$ is a normalising constant equal to the maximum element of vector $\psi_j^{[n]}$.

The sequence converged to the order of the machine rounding error after less than 10 iterations. Then if ψ_j is the converged eigenvector and N_j the converged normalisation constant,

$$(L - i\omega_j 1)\psi_j = 0. \quad \text{(B7)}$$
$$(L - i\omega_j' 1)\psi_j = \psi_j/N_j.$$

Subtracting the equations, the true eigenvalue ω is given by

$$\omega_j = \omega_j' - i/N_j. \quad \text{(B8)}$$

The results were checked by obtaining the corresponding eigenvectors ϕ_j of the Hermitian adjoint matrix equation with eigenvalue $-w_j^*$. These were then normalised so that the dot product $(\psi_j^* \cdot \phi_j)$ equalled one. Under these conditions the dot product $(\psi_j^* \cdot \phi_k)$ should be zero when $j \neq k$. This was found to be correct to within the machine rounding error.

B3   The residue

In Sect. 7 of the paper, Eq. (3) requires the SSH residue $A_j(x_m)$ of the eigenvector at position x_m. Let y be the solution of the matrix equation B1,

$$y = \sum_k a_k \psi_k. \quad \text{(B9)}$$

Then using the matrix eigenvectors and eigenvalues defined in Appendix B2,

$$(L - i\omega 1) \sum_k a_k \psi_k = z, \quad \text{(B10)}$$

$$\sum_k a_k(i\omega_k - i\omega)\psi_k = z. \quad \text{(B11)}$$

Taking the dot product with ϕ_j and rearranging,

$$a_j = \frac{i}{(\omega - \omega_j)}(\phi_j^* \cdot z). \quad \text{(B12)}$$

Thus the residue at x_m is

$$A_j(x_m) = \psi_{j,m} i(\phi_j^* \cdot z). \quad \text{(B13)}$$

The residue can also be obtained from the inverse iteration sequence (Eq. B6) without solving for the adjoint eigenvalue. If the iterations are initialised with $\psi_j^{[0]}$ equal to z and $N_j^{[0]}$ equal to 1, then after n iterations

$$\psi_j^{[n]} = \psi_j^{[n-1]} i(\phi_j^* \cdot z) \prod_{k=1}^{n} (N_j^{[k]}/N_j^{[n]}) + \epsilon, \quad \text{(B14)}$$

where ϵ is the contribution from other resonances. Once the solution has converged then, to within the machine rounding error, ϵ is zero and $\psi_j^{[n]}$ equals $\psi_j^{[n-1]}$, so

$$i(\phi_j^* \cdot z) = \prod_{k=1}^{n} (N_j^{[n]}/N_j^{[k]}), \quad \text{(B15)}$$

$$A_j(x_m) = \psi_{j,m} \prod_{k=1}^{n} (N_j^{[n]}/N_j^{[k]}). \quad \text{(B16)}$$

Acknowledgements. I wish to acknowledge the support of the Marine Systems Modelling group of the United Kingdom's National Oceanography Centre. The group provided essential material support for this research and also funded publication of the results. Thanks also to the reviewers for their detailed comments.

Edited by: N. Wells

References

Arbic, B. K., St-Laurent, P., Sutherland, G., and Garrett, C.: On the resonance and influence of the tides in Ungava Bay and Hudson Strait, Geophys. Res. Lett, 34, L17606, doi:10.1029/2007GL030845, 2007.

Courant, R. and Hilbert, D.: Methods of Mathematical Physics, Volume 1, John Wiley & Sons, 2008.

Egbert, G. D. and Erofeeva, S. Y.: Efficient Inverse Modeling of Barotropic Ocean Tides, J. Atmos. Oceanic Technol., 1919, 183–204, doi:10.1175/1520-0426(2002)019<0183:EIMOBO>2.0.CO;2, 2002.

Egbert, G. D. and Ray, R.: Estimates of M_2 tidal energy dissipation from TOPEX/Poseidon altimeter data, J. Geophys. Res., 106, 22475–22502, 2001.

Flather, R. A.: A Tidal Model of the North-west European Continental Shelf, Mémoires Société Royale des Sciences de Liége, 10, 141–164, 1976.

Fong, S. and Heaps, N.: Note on the quarter-wave tidal resonance in the Bristol Channnel, Institute of Oceanographic Sciences, Report No., 63, 15 pp., 1978.

Hunter, J. R.: A Note on Quadratic Friction in the Presence of Tides, Est. Coast. Mar. Sci., 3, 473–475, 1975.

Huthnance, J. M.: On shelf-sea resonance with application to Brazilian M3 tides, Deep Sea Res., 27A, 347–366, 1980.

IHB: Tides, List of Harmonic Constants, Special Publication No. 26, International Hydrographic Bureau, Monaco, 1954.

IOC, IHO, and BODC: Centenary Edition of the GEBCO Digital Atlas, published on CD-ROM on behalf of the Intergovernmental Oceanographic Commission and the International Hydrographic Organization as part of the General Bathymetric Chart of the Oceans, British Oceanographic Data Centre, Liverpool, UK, 2003.

Le Provost, C. and Rougier, F.: Energetics of the barotropic ocean tides: An estimate of bottom friction dissipation from a hydrodynamic model, Progr. Oceanogr., 40, 37–52, 1997.

Mathews, J. and Walker, R. L.: Mathematical Methods of Physics, W. A. Benjamin, Inc., 1965.

Miller, G. R.: The flux of tidal energy out of the deep ocean, J. Geophys. Res., 71, 2485–2489, 1966.

Morse, P. M. and Feshbach, H.: Methods of Theoretical Physics: Volume 1, McGraw-Hill, 1953.

O'Reilly, C. T., Solvason, R., and Solomon, C.: Where are the World's Largest Tides, in: BIO Annual Report: 2004 in Review, edited by: Ryan, J., 44–46, Biotechnol. Ind. Org., Washington, DC, 2005.

Riley, K. F., Hobson, M. P., and Bence, S. J.: Mathematical methods for Physics and Engineeering, Cambridge University Press, 1998.

Webb, D. J.: Green's Function and Tidal Prediction, Rev. Geophys. Space Phys., 12, 103–116, 1973.

Webb, D. J.: A Model of Continental Shelf Resonances, Deep-Sea Res., 23, 1–15, 1976.

Webb, D. J.: Tides and Tidal Energy, Contemporary Physics, 23, 419–442, 1982.

Webb, D. J.: Notes on a 1-D Model of Continental Shelf Resonances, Research and Consultancy Report 85, National Oceanography Centre, Southampton, available at: http://eprints.soton.ac.uk/171197 (last access: 19 May 2014), 2011.

Webb, D. J.: On the shelf resonances of the Gulf of Carpentaria and the Arafura Sea, Ocean Sci., 8, 733–750, doi:10.5194/os-8-733-2012, 2012.

Webb, D. J.: On the shelf resonances of the English Channel and Irish Sea, Ocean Sci., 9, 731–744, doi:10.5194/os-9-731-2013, 2013a.

Webb, D. J.: On the Impact of a Radiational Open Boundary Condition on Continental Shelf Resonances, National Oceanography Centre, Internal Document 06, National Oceanography Centre, Southampton, available at: http://eprints.soton.ac.uk/349401 (last access: 19 May 2014), 2013b.

Webb, D. J.: On the adjoint of Laplace's Tidal Equations, National Oceanography Centre, Internal Document 07, National Oceanography Centre, Southampton, available at: http://eprints.soton.ac.uk/361041 (last access: 19 May 2014), 2014.

12

An automated gas exchange tank for determining gas transfer velocities in natural seawater samples

K. Schneider-Zapp[1]**, M. E. Salter**[2]**, and R. C. Upstill-Goddard**[1]

[1]Ocean Research Group, School of Marine Science and Technology, Newcastle University, Newcastle upon Tyne, UK
[2]Department of Applied Environmental Science, Stockholm University, Stockholm, Sweden

Correspondence to: K. Schneider-Zapp (klaus.schneider-zapp@ncl.ac.uk)

Abstract. In order to advance understanding of the role of seawater surfactants in the air–sea exchange of climatically active trace gases via suppression of the gas transfer velocity (k_w), we constructed a fully automated, closed air–water gas exchange tank and coupled analytical system. The system allows water-side turbulence in the tank to be precisely controlled with an electronically operated baffle. Two coupled gas chromatographs and an integral equilibrator, connected to the tank in a continuous gas-tight system, allow temporal changes in the partial pressures of SF_6, CH_4 and N_2O to be measured simultaneously in the tank water and headspace at multiple turbulence settings, during a typical experimental run of 3.25 h. PC software developed by the authors controls all operations and data acquisition, enabling the optimisation of experimental conditions with high reproducibility. The use of three gases allows three independent estimates of k_w for each turbulence setting; these values are subsequently normalised to a constant Schmidt number for direct comparison. The normalised k_w estimates show close agreement. Repeated experiments with Milli-Q water demonstrate a typical measurement accuracy of 4 % for k_w. Experiments with natural seawater show that the system clearly resolves the effects on k_w of spatial and temporal trends in natural surfactant activity. The system is an effective tool with which to probe the relationships between k_w, surfactant activity and biogeochemical indices of primary productivity, and should assist in providing valuable new insights into the air–sea gas exchange process.

1 Introduction

Air–sea gas exchange is a critical global process, providing the fundamental link between reactive trace gas production and consumption in the oceans and global atmospheric processes. For example, the oceans are the largest single sink for tropospheric carbon dioxide (CO_2) (Khatiwala et al., 2009), contribute around one-third of tropospheric nitrous oxide (N_2O) (IPCC, 2007) and make significant contributions to the global biogeochemical budgets of several other climate-active gases including methane (CH_4), carbon monoxide (CO), dimethyl sulfide (CH_3SCH_3) and some other sulfur gases, and a range of halocarbons and hydrocarbons (Upstill-Goddard, 2011). Understanding the physical and biogeochemical controls of air–sea gas exchange is therefore necessary for establishing biogeochemical models for predicting regional- and global-scale trace gas fluxes and feedbacks.

For a sparingly soluble gas, which applies to almost all gases of global biogeochemical interest, the flux F (in $mol\,cm^{-2}\,h^{-1}$) across the air–sea interface can be considered as a diffusion-limited process in a typically $20\,\mu m$ to $200\,\mu m$ thick "diffusive sub-layer" on the water side of the interface (Jähne, 2009). It is written as the product of the driving force, i.e. its concentration difference between air C_a and sea water C_w (in $mol\,cm^{-3}$), and the air–sea gas transfer velocity k_w (in $cm\,h^{-1}$):

$$F = k_w(\alpha C_a - C_w), \tag{1}$$

where α is the Ostwald solubility coefficient. Similar equations apply to the exchange of heat and momentum. The concentration difference term ($\alpha C_a - C_w$) can be directly measured quite routinely but k_w cannot. Moreover, the magnitude of k_w varies with the degree of near-surface turbulence; increasing turbulence reduces the depth of the diffusive sub-layer, resulting in an increase in k_w. Indirect approaches are therefore required to estimate k_w in situ and evaluate its variability in response to environmental forcing functions generating turbulence. One often-used method for in situ measurement is the so-called "dual tracer technique", which measures the relative rates of evasion to air of two purposefully released, inert volatile tracers: sulfur hexafluoride (SF_6) and helium-3 (^3He). Temporal changes in the ratio of their seawater concentrations, typically measured over 24 to 48 h time intervals, are used to derive k_w estimates which are then scaled to corresponding values for CO_2 and other reactive trace gases of interest, using diffusivity-based relationships (Wanninkhof et al., 1993, 1997; Watson et al., 1991; Nightingale et al., 2000). The dual tracer technique is, however, time-consuming, logistically complex and expensive, and it provides little, if any, information on the spatio-temporal variability of k_w. Moreover, when the dual tracer k_w data are plotted against wind speed as the primary driver of turbulence, they show a high degree of scatter. Indeed, uncertainty over k_w variability presents one of the greatest challenges to quantifying the net global air–sea exchange of CO_2 (Takahashi et al., 2009). While a recent analysis shows that around one half of this uncertainty may be ascribed to experimental and measurement errors inherent in the dual tracer technique (Asher, 2009), this still leaves significant k_w variability unaccounted for. This remaining variability reflects other environmental forcing functions in addition to wind speed, including wind fetch, atmospheric stability, sea state, wave breaking, white capping and bubble bursting, sea surface temperature, rain and the presence of surfactants and other organics (Upstill-Goddard, 2006, 2011).

Surfactants are well known to greatly suppress k_w and consequently the rate of air–sea gas exchange, mostly by modifying sea surface hydrodynamics and hence turbulent energy transfer, but also by forming a monolayer physical barrier (McKenna and McGillis, 2004). Natural surfactants are ubiquitous in seawater, being primarily phytoplankton exudates such as polysaccharides, proteins and lipids, and their degradation products (Gašparović, 2012; Žutić et al., 1981), with additional sources via terrestrial inputs of humic and fulvic acids to coastal waters. Spatio-temporal distributions are therefore highly variable. Surfactants tend to be enriched in the diffusive sub-layer relative to underlying water up to high wind speeds (Wurl et al., 2011) but their precise effects on k_w are not well characterised, studies with natural seawaters being comparatively scarce. Most data have been derived using wind flumes and/or open exchange tanks in the laboratory (e.g. Goldman et al., 1988; Frew, 1997; Bock et al., 1999); this in part reflects the specialised nature of the analyses and a need to simplify experiments by using single, "model surfactants" in controlled conditions. While deliberate releases of man-made surfactant in tandem with the dual tracer technique have yielded some information on potential surfactant

effects in situ (Salter et al., 2011), practical considerations currently preclude isolating the effects on k_w of natural surfactant levels during field experiments. This makes recourse to laboratory-based experiments inescapable.

In order to make further progress in this regard, we have devised a laboratory procedure that enables us to directly evaluate the contrast in k_w between natural seawaters of varying surfactant content under controlled and reproducible conditions of turbulence. To simplify the system, we generate water-side turbulence, allowing us to concentrate on the important aspects of comparative k_w measurements at constant turbulence and avoiding unnecessary complication of the system. This simplifies process-understanding and also overcomes difficulties associated with simulating wind-induced turbulence in a laboratory. Even though the absolute k_w values may not be strictly comparable to in situ conditions, this simplification is wholly adequate to achieve our goal of comparative k_w measurements at constant turbulence, and is an important step in understanding more completely the relationships between surfactants, turbulence and k_w. Once this is established, more elaborate experiments in wind–wave facilities can provide valuable supplementary information. By measuring wave spectra in the tank, a record of the (temporal) wave field is available for each measurement, facilitating further comparison of experimental conditions between runs.

Our system uses a sealed laboratory gas exchange tank with integral water equilibrator. The partial pressures of experimental gases (SF_6, CH_4 and N_2O) in the water and air phases of the system are measured using gas chromatography. Turbulence in the tank can be routinely selected and modified such that inter-sample differences in k_w exclusively reflect differences in sample surfactant content, something that cannot currently be achieved through field experiments.

2 System design

2.1 Prior considerations

We previously used a sealed gas exchange tank for examining the microbial controls of air–sea gas transfer (Upstill-Goddard et al., 2003). This featured semi-automated analysis of the tank headspace gases (SF_6 and CH_4) but the water phase sampling required manual operation and large sample volumes, necessitating large and cumulative volume corrections, and turbulence control was comparatively rudimentary. Analytical uncertainty was therefore high but experimental reproducibility and accuracy were not rigorously assessed. Overall system reliability was also rather variable. To adequately resolve inter-sample differences in k_w due to variability in surfactant content demands a redesigned system with vastly improved performance and flexibility. An important aspect of the system described here is its full automation. PC control software developed by the authors controls all operations and data acquisition. This is a key aspect of the system because it guarantees that all experiments are run under identical conditions, free of operator-induced variability. It also facilitates the optimisation of experimental run times, which is critical for surfactant-focused work because it is essential to preserve sample integrity when several samples have been collected along lateral seawater transects, for example. Natural surfactants can degrade rapidly and significantly on storage. Such changes are variable and can be difficult to predict due to differences in organic composition between water samples (Schneider-Zapp et al., 2013). Consequently, when sample storage is unavoidable it should be minimised as far as is practically possible.

2.2 Selection of experimental gases

The use of SF_6, CH_4 and N_2O in the gas exchange experiments is based on our prior experience with them, in particular their ease of analysis, but most importantly because of their relevance in the context of air–sea gas exchange and global biogeochemistry. SF_6 is inert and is essentially all man-made, although there is a small geological source (Harnisch and Eisenhauer, 1998), and it is routinely used for estimating k_w in situ (Wanninkhof et al., 1993, 1997; Watson et al., 1991; Nightingale et al., 2000). CH_4 and N_2O both have significant marine sources and sinks, are infrared active and have long atmospheric lifetimes (IPCC, 2007). N_2O is also involved in stratospheric O_3 regulation via NO_x generation (Nevison and Holland, 1997) and CH_4 participates in stratospheric water formation and in the photochemical regulation of tropospheric OH and O_3 (Crutzen and Zimmermann, 1991). Both gases are currently increasing in the troposphere, but at variable rates that are not well understood (Dlugokencky et al., 1998, 2001; Khalil, 1993; Prinn et al., 1990; Rigby et al., 2008). Quantifying the constraints on their air–sea exchange rates is therefore critical.

2.3 System overview

A schematic of the gas exchange system is shown in Fig. 1. Its principal components are: (i) a sealed acrylic gas exchange tank that can be approximately half filled with seawater (Fig. 2 details its major features); (ii) an equilibration system used in preparing tank water subsamples for analysis; (iii) two gas chromatographs (GCs) identically configured for the analysis of SF_6, CH_4 and N_2O, in the tank headspace and in air that has been equilibrated with a tank water subsample. These components form a continuous, sealed circuit that can be decoupled and reconnected as required via the operation of solenoids. Details of the components are given in Sect. 3. The gas equilibration system is shown in Fig. 3 and the GC configuration in Fig. 4.

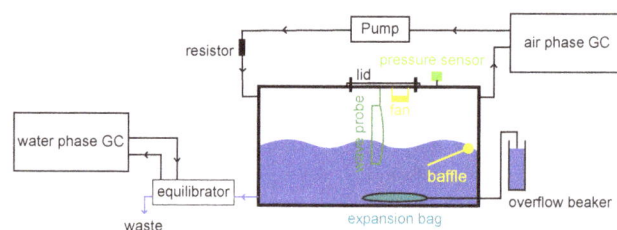

Figure 1. Schematic of the gas exchange experiment. The seawater sample (93 L) is contained in a gas-tight tank (0.73 m × 0.48 m × 0.48 m internally). Water-side turbulence is created with a baffle driven by a stepper motor. An automatic equilibration circuit (details see Fig. 3) regularly takes water samples and equilibrates the water with an "equilibrator gas" of known SF_6, CH_4 and N_2O composition, before measuring resulting gas partial pressures in a gas chromatograph (GC). Air phase gas partial pressures are constantly measured with a second GC. Wave spectra at a single point are acquired with a capacitance wave probe. Pressure and temperature in the tank and the equilibrator as well as total ambient pressure are continuously monitored.

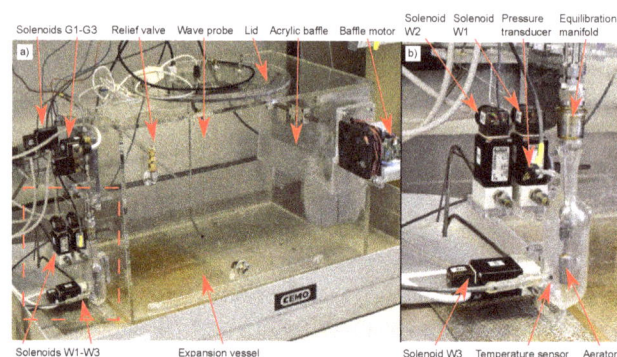

Figure 2. Annotated photograph of the gas exchange apparatus: **(a)** the gas exchange tank. Note that the overflow vessel usually connected to the expansion vessel is omitted for clarity; **(b)** the equilibration vessel shown by the dashed red box in image **(a)**.

3 System components

3.1 Gas exchange tank

The basic structure of the gas exchange tank (Fig. 2) was custom-built (Bay Plastics Ltd, UK) for our k_w-surfactant work. Subsequent modifications, principally the installation of mechanical and electronic components (see below), were carried out in-house. The tank has an internal base area of 0.73 m × 0.48 m, is 0.48 m in height internally and is constructed from 12 mm acrylic. Stainless steel bulkhead connectors are used for all tank connections. The incorporation of a headspace pressure relief valve precludes system overpressurisation and subsequent damage to the tank structure in the event of system malfunction. The tank is filled with sample gravimetrically to a notional volume of (93.0 ± 0.1) L, leaving a corresponding notional 74.1 L headspace $(1.11 \pm$

0.01 L is accounted for by the baffle, expansion bag and holders). This method of filling was selected because it is important to reproduce the sample volume precisely. Small differences in the fill level can have a large effect on the degree of tank water turbulence, which is selected and controlled using an internal acrylic baffle mounted across the full width of the tank on a transverse shaft. The gravimetric procedure overcomes this problem. A stepper motor (PD2-116-60-SE: Trinamic, Germany) is located outside the tank and connected to the baffle shaft via a gas-tight bearing and Viton® seal. The motor is operated via a serial RS232 link and allows precise control of the forward and reverse motion of the baffle. The tank air-phase (headspace) is continuously mixed using a low throughput fan (Sanyo Denki 9S1212F4011: RS components, UK) that does not create any detectable water turbulence. The fan is mounted on the inside of a removable circular tank lid, along with a wave height gauge (Sect. 3.3). The lid facilitates access for maintenance, for internal tank cleaning and for filling with sample and is sealed using a double Viton® O-ring. Viton® is compatible with gaseous hydrocarbons and its use also precludes SF_6 memory effects that may be encountered with some other seal materials (Upstill-Goddard et al., 2003).

During operation (Sect. 4.2) aliquots of the tank water are automatically transferred to the equilibration vessel (Sect. 3.2). In order to prevent a progressive decrease in tank internal pressure due to this procedure, an expandable plastic bag (Supel Inert Film 10 L gas sampling bag, Sigma-Aldrich, UK) inside the tank is connected to a small external water reservoir containing artificial seawater (ASW) of salinity ≥ 45 via a water-tight bulkhead fitting. The density contrast between the ASW and the tank seawater (maximum salinity ≈ 35) ensures a negative buoyancy that prevents the expandable bag from rising from the bottom of the tank. A secondary, larger ASW reservoir is used to maintain the water level inside the small reservoir at the same level as in the exchange tank using a peristaltic pump; an overflow returns any excess back into the larger reservoir, thereby keeping the water level in the smaller reservoir constant.

Pressure in the tank headspace (P_h) is continuously monitored using a transducer (40PC001B: range ± 66.7 hPa; accuracy 0.5 hPa: Honeywell, USA). The analogue output voltage is converted to a digital signal using a USB-6008 (12 bit) ADC (National Instruments, USA). Temperature in the tank water phase is recorded on an autonomous mini data logger (Minilog 8, Vemco, Canada; accuracy 0.2 °C) that is retrieved for data download at the end of each experiment.

3.2   Equilibration vessel

The analysis of dissolved gases by gas chromatography necessitates either a pre-extraction or equilibration step, followed by the measurement of gas partial pressures in the resulting gas phase and corrections for air and water volumes and gas solubilities (Upstill-Goddard et al., 1996). Extraction

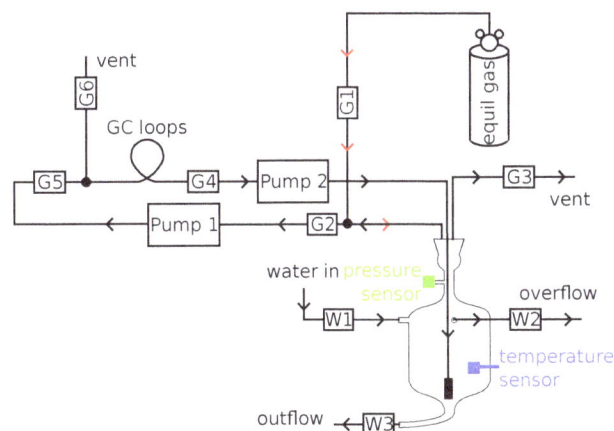

Figure 3. Schematic of the equilibration system. For circuit operation see text.

techniques often involve pre-concentration procedures which can be complicated and the overall extraction efficiency can vary significantly. By contrast, automated gas equilibration has been shown to be highly reproducible (Upstill-Goddard et al., 1996). We therefore incorporated a water–air equilibration vessel as an integral component of the gas exchange tank apparatus. The equilibration vessel has a total internal volume of 183 cm³ and has two principal components: a glass vessel equilibrator and a removable stainless steel equilibration manifold (Fig. 3). The design derives from a system we constructed for the high-precision analysis of dissolved gases at sea (Upstill-Goddard et al., 1996). The equilibration manifold comprises three lengths of stainless steel tubing silver-soldered through a tapered stainless steel plug machined to seat precisely in the neck of the glass vessel to give a gas-tight seal. Two tubes are cut flush to the base of the plug and a third is connected to a stainless steel aerator frit near the bottom of the glass vessel. The frit is a standard chromatography solvent filter (Thames Restek, UK). The equilibration vessel has three water inlet/outlets (all 4 mm i.d.), each connected via a short length of flexible Tygon® tubing to a solenoid (Burkert 0124 2/2 way for aggressive media, Burkert, Germany; W1–W3 in Fig. 3). A digital temperature sensor (DS18B20+; resolution 0.1°; accuracy < 0.5°: Maxim, USA) is housed in a side arm. A second side arm in the equilibrator neck houses a pressure transducer identical to that used for monitoring tank headspace pressures (Sect. 3.1). A cylinder containing compressed air of known SF_6, N_2O and CH_4 composition (the "equilibrator gas") is connected via solenoid G1. The gas is circulated through G2, the GC sample loops and back through the equilibrator frit. Solenoid G3 allows venting of the equilibrator during its filling with seawater via W1. Solenoids G1–G3 are Burkert 6013A 2/2 way (Burkert, Germany).

Prior to equilibration the equilibration vessel and all associated GC sample tubing is flushed with the equilibrator gas via solenoid G1. Next the vessel is completely filled

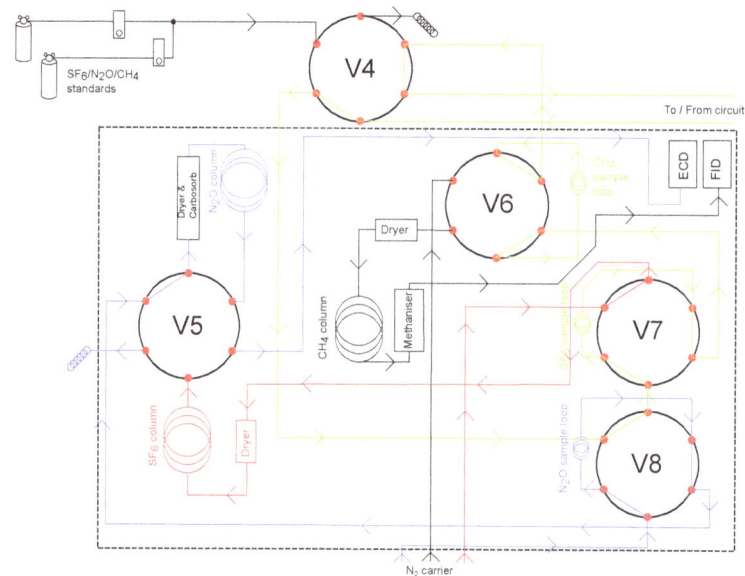

Figure 4. Schematic of the GC system used to measure SF_6, N_2O, and CH_4, here in N_2O inject mode. Each sample loop can be injected separately using the corresponding valve (V3, V4, and V5, respectively). For the ECD, the N_2O and the SF_6 column can be selected with valve V2. For detailed description see text.

with tank water via solenoid W1, all air being displaced via solenoid G3. A headspace of equilibrator gas is then introduced via G1, displacing sample via the overflow and solenoid W2. The fill/displacement cycle is then repeated to ensure the removal of all traces of previous sample and equilibrator gas. The procedure facilitates a reproducible headspace to water volume ratio (Sect. 4.4.2) which is required for accurately correcting for solubility-driven phase partitioning during equilibration (Upstill-Goddard et al., 1996). All solenoids are then closed and G2, G4 and G5 are opened. Two gas sampling pumps (NMP015.1.2KNL: KNF Neuberger AG, Switzerland) circulate the equilibrated sample gas around the closed circuit, through the GC sample loops and back through the equilibrating water sample via the aerator frit inside the equilibrator for 4.35 min. Equilibration-time curves (Fig. 5) show that all three gases are fully equilibrated within 3 min. Two pumps are necessary to equalise pressure gradients and thus maintain the internal equilibrator pressure at ambient. Pumping rates are regulated via 8 bit pulse width modulation of the 12 V supply. The speed of pump 1 is kept constant and that of pump 2 is regulated in response to the equilibrator internal pressure.

Following equilibration the pumps are switched off, all solenoids are closed and G6 is opened for 20 s to allow the GC sample loops to reach ambient atmospheric pressure before injection onto the GC carrier gas lines. This avoids pressure effects that might otherwise interfere with the detector responses.

Each equilibration step removes 549 mL of water from the gas exchange tank. This is around 0.59 % of the initial tank water volume and the total cumulative volume removed

during a typical experiment is therefore less than 4 % of the initial volume. This is accounted for in subsequent data processing, the required corrections being smaller than for our experiments with an earlier gas exchange tank (Upstill-Goddard et al., 2003).

Determining equilibration volumes

The relative volumes of water to headspace V_a/V_w involved in the equilibration step must be accurately known in order to facilitate corrections for solubility-driven phase partitioning (Upstill-Goddard et al., 1996). V_w can be determined gravimetrically by repeatedly generating headspace in the equilibrator. By contrast, system configuration precludes directly measuring V_a. To overcome this we directly estimated V_a/V_w by equilibration, similar to Upstill-Goddard et al. (1996). The gas exchange tank was filled with 93 L Milli-Q water (resistivity typically $18.2\,M\Omega\,cm^{-3}$: Millipore Corporation, USA) enriched with SF_6, N_2O and CH_4, sealed and equilibrated by operating the baffle until the gas partial pressures in both the equilibrator headspace and in the tank headspace remained constant for > 12 h. For this measurement ultrahigh purity (UHP) N_2 (> 99.999 % N_2, no detectable SF_6, N_2O or CH_4) was used as equilibrator gas, i.e. $C_0 = 0$. Concentrations in the tank headspace and equilibrator were then determined multiple times and averaged. These values were used together with the appropriate Ostwald solubilities of SF_6 (Bullister et al., 2002), CH_4 (Wiesenburg and Guinasso, 1979) and N_2O (Weiss and Price, 1980) at the temperature and pressure of equilibration and the tank headspace and

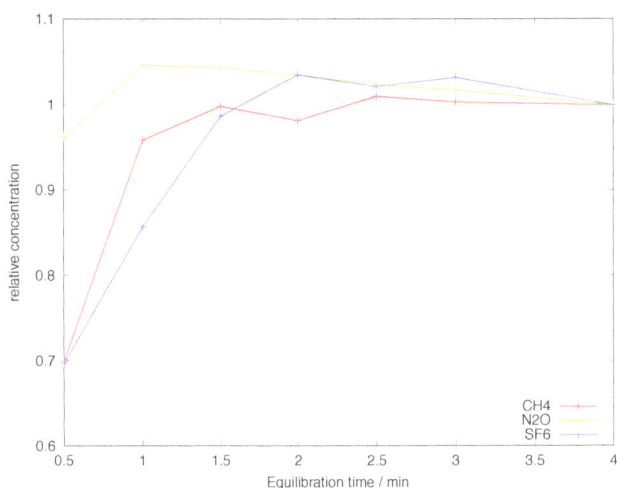

Figure 5. Relative gas concentration vs. equilibration time for SF_6, N_2O and CH_4.

water volumes, to calculate V_a/V_w according to Eq. (17). For the system as currently configured $V_a/V_w = 0.79 \pm 0.02$.

3.3 Wave height gauge

A capacitance-type high-precision wave height gauge (AWP-24; 30 cm double strand sensing wire, Akamina Technologies, Canada) is used. Analogue output voltage is digitised at 400 Hz (USB-6008 ADC, National Instruments, USA). The output voltage of the device is linearly proportional to the water level.

The probe is routinely calibrated to determine the relationship between water depth and output voltage by filling the gas exchange tank with sample water and progressively immersing the probe step-wise into the water. This is done by mounting the probe on a rod with precisely machined holes at 1 cm intervals. For each step the rod is bolted through one of the holes to a sturdy mount secured to the tank. After waiting for the water surface to settle, the output voltage is averaged over 10 s. A line is fitted to the immersion depth–voltage relation.

3.4 Ancillary measurements

Absolute pressure (P_0) and temperature (T) in the laboratory are measured using a digital sensor (Sensortec BMP085; pressure range 300–1100 hPa; absolute pressure accuracy 1 hPa; absolute temperature accuracy 0.5 °C: Bosch, Germany).

3.5 Gas chromatography

The need to determine the partial pressures of SF_6, CH_4 and N_2O in both the air and water phases during an experiment precludes using a single GC; this would necessitate long sampling intervals and/or a long experimental duration,

with consequent loss of experimental resolution. Therefore, two identically configured GCs were used (both HP 5890), one for analysing tank headspace ("air-phase GC") and one for analysing equilibrator air following water sample equilibration ("water-phase GC"). The analysis is identical in each GC, being isothermal (60 °C) and based on methods developed in our laboratory (Upstill-Goddard et al., 1990, 1996, 2003). A schematic is shown in Fig. 4. A series of motor-driven stainless steel chromatography valves, V1–V5 in Fig. 4 (Valco: Vici AG, Switzerland), allow the selective switching of tank headspace, equilibrator headspace and calibration standards onto the separating columns (one each for SF_6, N_2O and CH_4 in each GC) and detectors, via fixed volume sample loops (internal volume: SF_6 10 mL, CH_4 1 mL, N_2O 1.5 mL). Chromatographic separation of SF_6 is on Molecular Sieve 5A columns (4 m × 1.75 mm i.d.), whereas N_2O and CH_4 are both separated on 80–100 mesh Porapak Q columns (CH_4, 4 m × 1.75 mm i.d.; N_2O, 5 m × 1.75 mm i.d.). The GC carrier gas is UHP N_2. Flow rates are typically around 25 cm^3 min^{-1} for CH_4 and N_2O, and 50 cm^3 min^{-1} for SF_6. Water vapour produced during sample equilibration is removed using Mg(ClO4)2 and CO_2 is removed using NaOH (Upstill-Goddard et al., 1996). Detection of CH_4 uses a flame ionisation detector (FID) at 300 °C whereas detection of N_2O and SF_6 uses an Electron Capture Detector (ECD) with a ^{63}Ni source at 350 °C.

The GC responses are integrated automatically using proprietary GC software (Clarity: DataApex, Prague, Czech Republic). Method calibration uses a series of mixed calibration standards prepared by pressure dilution in UHP N_2 (Upstill-Goddard et al., 1990, 1996). Analytical precisions are typically ± 1 % CH_4, ± 0.8 % N_2O and ± 1 % SF_6. All three gases are analysed in less than 8 min.

3.6 Limits of detection

Minimum detectable levels of SF_6, N_2O and CH_4 have been determined by estimating the detector responses corresponding to a signal-to-baseline-noise ratio of 2 and dividing by the detector peak width (peak area/peak height) in seconds (Upstill-Goddard et al., 1996). Minimum detectable levels are 0.5 pptv SF_6, 0.2 ppbv N_2O and 10 ppbv CH_4. However, in practice the partial pressures of all three gases in the equilibrator gas combined with solubility considerations (Bullister et al., 2002; Wiesenburg and Guinasso, 1979; Weiss and Price, 1980) preclude operating the detectors close to these limits.

4 Experimental procedure

Prior to use, all ancillary equipment is thoroughly cleaned with ethanol and rinsed with Milli-Q water. All metal and glassware is subsequently baked at 450 °C overnight.

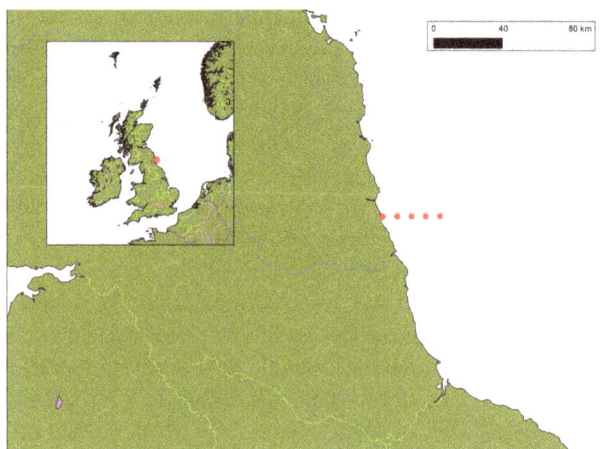

Figure 6. The sampling stations (red circles) on the transects in the coastal North Sea off northeast England.

4.1 Field sampling

Large-volume water samples (≈ 100 L) for the gas exchange experiments were collected during two coastal North Sea transects of R/V *Princess Royal*, on 4 October 2012 and 13 February 2013. On both, samples were collected from five fixed locations approximately equally spaced up to 20 km off-shore of the UK Northumberland coast (Fig. 6). The samples were drawn with an on-board sampling pump and stored on-deck in "aged" polyethylene seawater carboys (i.e. all leachable components removed using concentrated HCl solution). Additional samples for the measurement of surfactant activity were collected from the surface microlayer (SML) using a Garrett (1965) screen (for further details of our Garrett screen samplers, see Schneider-Zapp et al., 2013) and decanted into sterile polypropylene sampling tubes; for samples from the underlying water (ULW), water was drawn using a stainless steel bucket, then sterile polypropylene tubes were opened below the water surface of the bucket, filled, and closed using nitrile gloves to avoid sampling SML. All these samples were kept refrigerated in the dark at 4 °C to minimise degradation during pre-analysis storage (Schneider-Zapp et al., 2013). Salinity and temperature were measured using a hand-held probe. Meteorological data were acquired via an on-board weather station.

4.2 Gas exchange experiments

The inside of the gas exchange tank is repeatedly cleaned and filled/rinsed with Milli-Q water until surfactant activity (SA) in the tank water surface microlayer (SML) is analytically identical to that of fresh Milli-Q (within between 0.02 eq. mg L^{-1} T-X-100 and max. 0.04 eq. mg L^{-1} T-X-100). The SML is routinely collected with a small Garrett (1965) screen and SA is measured by hanging mercury drop AC voltammetry (see Sect. 4.3).

Following cleaning and thorough rinsing of the inside of the tank with sample, the tank is filled with 93 L of sample (measured gravimetrically), which is added directly from sampling carboys using a peristaltic pump. During this procedure 1.1 L of sample is decanted directly from one of the carboys into a sealable glass bottle and enriched with SF$_6$, N$_2$O and CH$_4$ as follows: 80 cm^3 of 10 ppmv SF$_6$ in UHP N$_2$, prepared by pressure dilution (Upstill-Goddard et al., 1990, 1996), 20 cm^3 of $\geq$ 99.99 % research grade CH$_4$ and 1 cm^3 of 99.998 % research grade N$_2$O are injected into the vessel and equilibrated with the sample water for 20 min using a small pump and aerator. This gas-enriched subsample is added to the tank at the end of the filling procedure to give the final, notional 93 L volume and the tank is then sealed.

To estimate the sample volume, the carboys (including that of the gas enriched subsample) are weighed before and after filling. The weight difference quantifies the sample mass, which is converted to volume using sample density derived from its salinity and temperature. These are measured in the residual sample in the carboys directly after filling, using a pre-calibrated hand-held probe, in order to preclude any possible contamination of the tank water sample. Continuous weighing during the filling process allows the filling procedure to be terminated when the correct nominal volume is reached.

The enriched subsample creates a disequilibrium in gas partial pressures between the tank water and tank headspace, so as to drive measurable water-to-air exchange of the measured gases during the experiment. Note that for a subset of our tests, the N$_2$O chromatography was not yet fully operational. To be consistent, we decided to only show data for CH$_4$ and SF$_6$, where measurements are available for all data sets.

A configuration file defining all experimental settings is read by the control program which executes the experiment. The system is optimised to run a gas exchange experiment in just under 3.25 h. We routinely use three sequential fixed levels of turbulence, each of 64.5 min duration; corresponding baffle frequencies are 0.6, 0.7, and 0.75 Hz. GC analysis of the tank headspace and equilibrated water phases is every 10.75 min, enabling six measurements of each phase for each selected level of turbulence during an experiment. Temperatures and pressures are logged every 0.1 min. Immediately prior to and following each experiment, calibration gas standards are repeatedly measured on each GC for data calibration and to allow an estimate of detector drift. This is usually less than 2 % over the course of a typical experiment and is corrected for by applying a time-dependent linear fit to the detector responses.

Gas concentration uncertainties are estimated using the standard deviation of the standard measurements. For all other measured quantities, specified instrument accuracies are used. In the calculations, Gaussian error propagation (Tayler, 1996, chapter 3; for details of the calculations see the Appendix) is used to propagate uncertainties. This over-estimates the uncertainty, because not all uncertainties are independent, but covariance matrices are not known. For example, the GC errors include the detector drift of up to 2 % as estimated from the standard measurements. This effect cannot be separated from the statistical error or corrected for, as the drift is irregular.

Estimates of k_w are obtained from Eq. (13) using weighted linear regression (Sect. 4.4.1). The uncertainty is estimated from the weighted fit. Convention is to scale all measured values of k_w to Schmidt numbers of either 660 or 600, being the values for CO_2 in freshwater and salinity 35 seawater respectively, at 20 °C. Therefore, k_w estimates are scaled to Schmidt number 660 using Eq. (19) (Sect. 4.4.3). Schmidt numbers are obtained from Wanninkhof (1992). From the Milli-Q data, a Schmidt number exponent of $n = 1/2$ was determined using Eq. (20), corresponding to a wavy surface.

The wave frequency energy spectrum (Phillips, 1980) is calculated from the sampled wave height using the method of Welch (1967) (see Harris, 1978, for more information) with a Hann window of length 131 072. For the measurement frequency of 400 Hz, used here, the Nyquist frequency is 200 Hz $\approx$ 1257 rad s^{-1}; however, noise begins to affect the measurements at frequencies lower than this. We plotted the results up to 500 rad s^{-1} $\approx$ 80 Hz, since we consider these data to be unaffected by noise.

4.3 Surfactant measurement

Surfactant activity (SA) is measured using AC voltammetry (Ćosović and Vojvodić, 1982) (Metrohm 797 VA Computrace, Metrohm, Switzerland) with a hanging mercury drop, a silver/silver chloride reference electrode and a platinum wire auxiliary electrode. Samples are brought to salinity 35 prior to measurement by adding surfactant-free NaCl solution. For each measurement, a new mercury drop is created and the first few drops discarded. Surfactants accumulate on the drop at $V = -0.6$ V for 15 or 60 s with stirring (1000 rpm). Alternating voltage scans of 10 mV at 75 Hz produce a current which is measured. Instrument calibration uses the non-ionic soluble surfactant Triton T-X-100. Each response is corrected for the added NaCl solution and expressed as an equivalent T-X-100 concentration.

4.4 Theory

4.4.1 Tank gas exchange

For a sealed gas exchange tank containing seawater and air and without gas sources or sinks, Eq. (1) can be used to derive a mass balance:

$$\begin{pmatrix} \frac{\partial C_a}{\partial t} V_a \\ \frac{\partial C_w}{\partial t} V_w \end{pmatrix} = k_w \begin{pmatrix} C_w - \alpha C_a \\ \alpha C_a - C_w \end{pmatrix} A. \tag{2}$$

The solubility α, volumes V_a and V_w, and surface area A are assumed to be constant. In reality, α depends on temperature. In practice changes in experimental temperature

are of the order of 0.5 °C. For such a change in temperature at 20 °C, the change in α is < 1.6 % for SF$_6$, < 1.4 % for N$_2$O and < 1 % for CH$_4$ according to published parameterisations (Bullister et al., 2002; Wiesenburg and Guinasso, 1979; Weiss and Price, 1980).

Let the height of phase i (i.e. air a or water w) be h_i, i.e. $h_i := V_i/A$, and let

$$D := C_w - \alpha C_a. \tag{3}$$

Re-arranging Eq. (3), taking the derivative and using the chain rule results in

$$\frac{\partial C_a}{\partial t} = \frac{1}{\alpha}\left[\frac{\partial C_w}{\partial t} - \frac{\partial D}{\partial t}\right] = \frac{k_w}{h_a}D, \tag{4}$$

where Eq. (2) (top) has been used in the last equality. Substituting $\frac{\partial C_w}{\partial t}$ into Eq. (2) (bottom), a differential equation in D is obtained:

$$\frac{\partial D}{\partial t} + k_w \underbrace{\left[\frac{1}{h_w} + \frac{\alpha}{h_a}\right]}_{=:\beta} = 0. \tag{5}$$

Solving this gives

$$D = D_0 \exp(-k_w \beta t), \tag{6}$$

where $D_0 := D(t = 0)$.

Due to conservation of mass, the total amount of gas N in the system must remain constant. Hence,

$$N = C_a V_a + C_w V_w = \text{const.} \tag{7}$$

This relation serves as a routine check of the experimental results and system integrity; a change in the value of N during the experiment implies system leaks and/or defective chromatography.

At each equilibration step the tank water volume V_w decreases (here assumed instantaneously) by volume V_s, such that $V_{w>} = V_{w<} - V_s$, where $V_{w<}$ and $V_{w>}$ are the tank water volumes before and after drawing the sample, respectively. The effect is to change the value of β so that the differential equation no longer has constant coefficients. However, within each interval $[t_n, t_{n+1}]$ between two measurements at times t_n and t_{n+1}, the volumes are constant and Eq. (6) can be used. At each sampling step the values of the coefficients change instantaneously; the variables at the end of the previous interval become the initial conditions for the next interval. If $V_{e,n} = nV_s$ is the total water volume already extracted at t_n, using the abbreviations $V_{w,n} := V_w|_{t=t_n}$, $D_n := D(t = t_n)$ and

$$\beta_n := \beta|_{t=t_n} = \frac{A}{V_{w,n-1} - V_s} + \frac{A\alpha}{V_a}$$
$$= \frac{A}{V_{w,0} - nV_s} + \frac{A\alpha}{V_a} = \frac{1}{h_{w,0} - nh_s} + \frac{\alpha}{h_a} \tag{8}$$

derives

$$D_n = D_{n-1} \exp(-k_{\text{w}} \beta_n (t_n - t_{n-1})). \tag{9}$$

The first water sample is drawn at t_0, the experiment starts running with the reduced water volume $V_{\text{w}} - V_{\text{s}}$ and thus the system response is

$$D_n = D_0 \exp\left(-k_{\text{w}} \sum_{j=1}^{n} \beta_j (t_j - t_{j-1})\right). \tag{10}$$

Note that for $n = 0$, the sum is zero and the equation is identical to the original Eq. (6). The new solution Eq. (10) is not in the form $\exp{-k_{\text{w}} \beta t}$ but has a sum of different t_j in its exponential. It can be solved for k_{w}; however, it diverges for $n = 0$. This is overcome by conversion to the form $\exp{-k_{\text{w}} \beta t}$ as

$$D_n = D_0 \exp\left(-k_{\text{w}} \underbrace{\left[\sum_{j=1}^{n} \beta_j \frac{t_j - t_{j-1}}{t_n - t_0}\right]}_{=:B_n} (t_n - t_0)\right) \tag{11}$$

with

$$B_n := \sum_{j=1}^{n} \beta_j \frac{t_j - t_{j-1}}{t_n - t_0}$$
$$= \frac{A}{t_n - t_0} \sum_{j=1}^{n} \frac{t_j - t_{j-1}}{V_{\text{w},0} - jV_{\text{s}}} + \frac{A\alpha}{V_{\text{a}}}. \tag{12}$$

Note that $B_0 = \beta_0$ (Eqs. 11 and 6). For $V_{\text{s}} = 0$, we obtain $B_n = \beta_0$ and the solution reduces to Eq. (6) with $t_0 = 0$. With $V_{\text{s}} > 0$, the value of B_n increases (the denominator in each summand is decreased) and consequently D decreases progressively more rapidly with increasing experimental run time during which further water is extracted from the tank. Consequently, some fraction of the decrease in D is due to volume extraction. Without any correction for this, k_{w} is overestimated. The factor $\frac{t_j - t_{j-1}}{t_n - t_0}$ is applied as a weight factor for any given water volume during the experiment.

The solution can be expressed in logarithmic form to derive a linear fit obtaining k_{w} as

$$\chi_n := \frac{1}{B_n} \ln \frac{D_0}{D_n} = k_{\text{w}}(t_n - t_0). \tag{13}$$

The mass balance Eq. (7) also has to be adjusted to account for the water loss on sampling:

$$N_n = V_{\text{w},n} C_{\text{w},n} + \sum_{j=1}^{n} C_{\text{w},j-1} V_{\text{s}} + V_{\text{a}} C_{\text{a},n}. \tag{14}$$

4.4.2 Water sample equilibration

We can consider a water sample of volume V_{w} in the equilibrator with an initial dissolved gas concentration C_{w} at in situ pressure P_1 and temperature T_1. The number of moles of gas in the water is $N_1 = C_{\text{w}} V_{\text{w}}$.

The water subsample then equilibrates with a head space of volume V_{a} and initial gas concentration C_0. The total number of moles of gas in the equilibrator is then $N = N_1 + N_0 = V_{\text{w}} C_{\text{w}} + V_{\text{a}} C_0$. During equilibration the gas partitions according to $\alpha C_{\text{a}}' = C_{\text{w}}'$, where α is the Ostwald solubility coefficient. N remains constant (conservation of mass), thus

$$V_{\text{a}} C_{\text{a}}' + V_{\text{w}} C_{\text{w}}' = V_{\text{a}} C_{\text{a}}' + V_{\text{w}} \alpha C_{\text{a}}' = V_{\text{a}} C_0 + V_{\text{w}} C_{\text{w}}. \tag{15}$$

Solving for C_{w} results in

$$C_{\text{w}} = \frac{V_{\text{a}}}{V_{\text{w}}} (C_{\text{a}}' - C_0) + \alpha C_{\text{a}}', \tag{16}$$

which can be used to back-calculate the gas concentration in the water sample using C_{a}'.

For evaluating Eq. (16), the water–headspace volume ratio $V_{\text{a}}/V_{\text{w}}$ is required. It is determined by a measurement with known C_{w} and C_{a}' so that Eq. (15) is then solved for $V_{\text{a}}/V_{\text{w}}$:

$$\frac{V_{\text{a}}}{V_{\text{w}}} = \frac{C_{\text{w}} - \alpha C_{\text{a}}'}{C_{\text{a}}' - C_0}. \tag{17}$$

4.4.3 Schmidt number scaling

The value of k_{w} for any gas is a function of its Schmidt number Sc, which is defined as the ratio of the viscosity of water to the corresponding gas diffusivity at the requisite temperature, i.e. $Sc = \nu/D$. Theory predicts the scaling

$$k_{\text{w}} = \frac{u_*}{R} Sc^{-n} \tag{18}$$

where u_* is the friction velocity, R is the resistance for momentum transfer and the exponent n is equal to $2/3$ for a smooth surface and $1/2$ for a rough surface with a smooth transition (Richter and Jähne, 2010). This relation allows the interconversion of k_{w} for any given gas to k_{w} for any other specified gas. Given two gases 1 and 2 with transfer velocities $k_{\text{w}1}$ and $k_{\text{w}2}$ and Schmidt numbers Sc_1 and Sc_2, respectively, one obtains

$$k_{\text{w}1} = \left(\frac{Sc_1}{Sc_2}\right)^{-n} k_{\text{w}2}. \tag{19}$$

Simultaneous measurements of two gases with different Schmidt numbers can be used to calculate the exponent:

$$n = \frac{\ln \frac{k_{\text{w}1}}{k_{\text{w}2}}}{\ln \frac{Sc_2}{Sc_1}} = \frac{\ln \frac{k_{\text{w}1}}{k_{\text{w}2}}}{\ln \frac{D_1}{D_2}}. \tag{20}$$

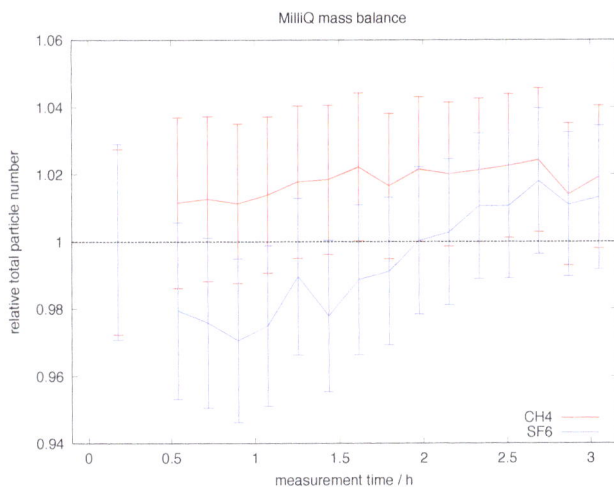

Figure 7. Total amount of gas in the tank according to Eq. (14), normalised to the first measurement, vs. time of a typical gas exchange experiment with Milli-Q water. It shows that no gas is measurably lost or acquired during the experiment.

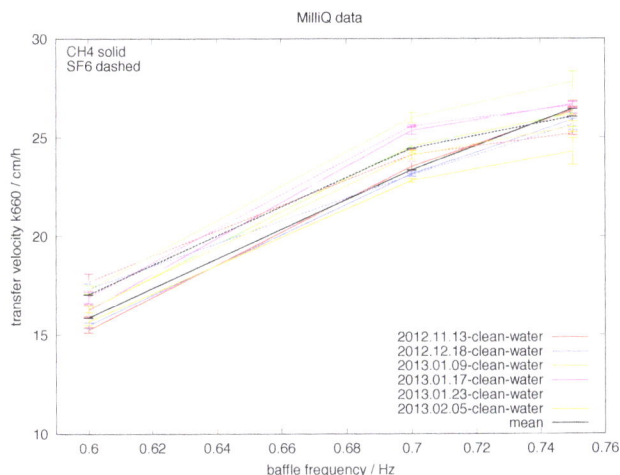

Figure 8. Estimated transfer velocities of Milli-Q water scaled to Schmidt number 660 at three different baffle settings for 6 different experiments, for both CH_4 and SF_6.

5 Results and discussion

5.1 Evaluation procedure

To test the evaluation procedure, synthetic data were used. After choosing a nominal value of k_w, an initial condition $C_{w,0}$, $C_{a,0}$, and constants T and S, the true system response was calculated using Eq. (10), with volumes and dimensions from the experimental setup. Gaussian noise with mean 0 and variance of 2 % was added to the true gas concentrations to model the measurement process. The data were then put into the evaluation procedure. The original transfer velocity k_w was always within the uncertainty of the estimated transfer velocity.

Table 1. Estimated values of k_{660} derived from CH_4 and SF_6 for six different Milli-Q experiments.

Baffle frequency (Hz)	k_{660} (cm h^{-1})	
	CH_4	SF_6
0.6	15.5 ± 0.6	16.7 ± 0.6
0.7	22.8 ± 0.9	24.0 ± 1.0
0.75	25.9 ± 0.6	25.5 ± 0.9

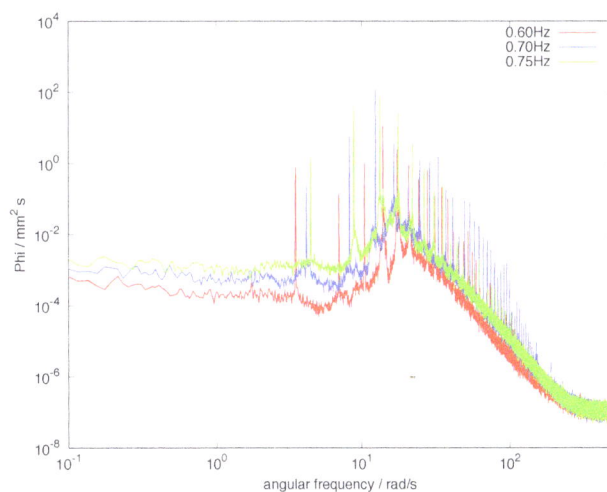

Figure 9. Wave spectra of a typical gas exchange experiment with Milli-Q water. The first peak for each boundary condition corresponds to the respective baffle speed, peaks at higher frequencies to harmonics.

5.2 Milli-Q experiments

A number of experiments with Milli-Q water were conducted to validate the experiment and to test the reproducibility. Figure 7 shows the mass balance, i.e. the total amount of gas within the tank, for a typical exchange experiment. Deviations are within the error, proving that the setup is gas-tight, i.e. no gas is lost or acquired during the run.

Estimated values of k_{660} (k_w scaled to a Schmidt number of 660) derived from CH_4 and SF_6 for six different Milli-Q experiments are shown in Fig. 8. Weighted means and standard deviations of these data are summarised in Table 1. For individual experiments the two independent k_{660} estimates show very close agreement; any small discrepancies most likely include uncertainties in the Schmidt number values and the solubility parameterisations. Thus, even in the worst case changes of $2 \, \text{cm} \, \text{h}^{-1}$ are significant with 95 % probability; for the baffle speed of 0.6 Hz the significance level is $1.2 \, \text{cm} \, \text{h}^{-1}$ with 95 % probability. The weighted standard deviation is 4 % for all baffle speeds and both gases.

Wave spectra for a selected experiment are shown in Fig. 9. As expected, the wave energy is higher for higher baffle frequencies. The first peak for each boundary condition corresponds to the respective baffle speed, showing that the baffle

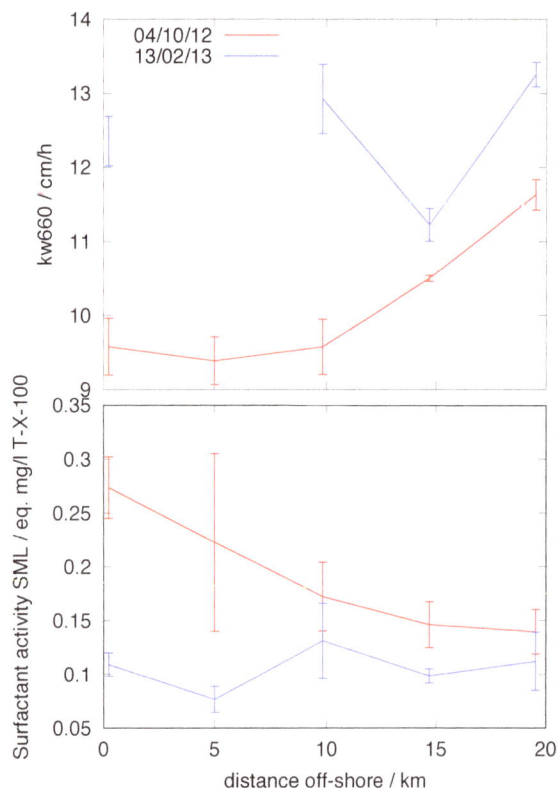

Figure 10. Top: estimated transfer velocities of seawater samples scaled to Schmidt number 660 for two transects in the coastal North Sea off northeast England. Bottom: measured surfactant activity (SA) of the surface microlayer (SML) on transects in the coastal North Sea off northeast England. Results are expressed as equivalent concentration of the calibration surfactant Triton T-X-100.

has a reproducible and stable frequency. Further peaks at higher frequencies are the harmonics caused by reflection and refraction inside the tank.

5.3 Seawater experiments

Estimated values of k_w from the coastal North Sea transects, derived from CH_4 at a baffle speed of 0.6 Hz, are shown in Fig. 10 (top). For the winter transect (13 February 2013) k_w was between (12.4 ± 0.3) cm h^{-1} (near-shore) and (13.2 ± 0.2) cm h^{-1} (off-shore), while corresponding autumn (4 October 2012) values were between (9.4 ± 0.3) cm h^{-1} (near-shore) and (11.6 ± 0.2) cm h^{-1} (off-shore). Comparing these transfer velocities with the surfactant activity (SA) of the SML samples (Fig. 10 bottom) clearly shows a correlation. The spatial and temporal differences in k_w are a function of SA. The spatial gradients in k_w are consistent with a decreasing influence of terrestrially derived surfactants in river outflow with distance offshore. Higher k_w suppression by surfactant during the autumn reflects higher SA arising from primary productivity, whereas lower winter suppression presumably reflects lower SA arising from surfactant degradation processes.

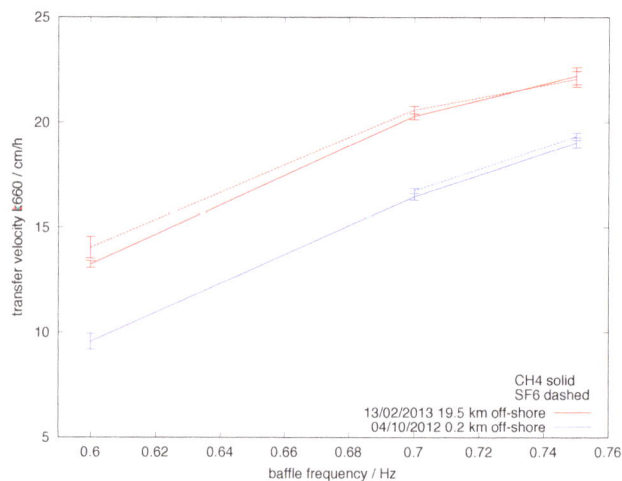

Figure 11. Estimated transfer velocities k_w vs. baffle frequency for the most landwards station of the autumn transect (low k_w) and the most off-shore station from the winter transect (high k_w) for CH_4 and SF_6. Note that for the 4 October 2012 transect, the datum for SF_6 at $F = 0.6$ Hz is excluded due to chromatography problems for this particular point.

For the most landwards station of the autumn transect (low k_w) and the most off-shore station from the winter transect (high k_w), k_w vs. baffle frequency is shown for CH_4 and SF_6 in Fig. 11. The agreement between the two gases is acceptable, the discrepancies being largely attributable to uncertainties in the Schmidt number parameterisations, with additional small contributions arising from GC detector drift, which is somewhat larger for SF_6 than for CH_4. Nevertheless, the observed trends are clearly significant within the analytical error. Our experimental procedures are evidently well suited to examine the relative natural variability of k_w between seawater samples containing varying levels of surfactant.

6 Conclusions

We have developed a laboratory gas exchange tank and associated analytical methodology that enables fully automated, routine determination of the gas transfer velocities of SF_6, CH_4 and N_2O in natural seawaters under strictly controlled conditions of turbulence. Repeated experiments with Milli-Q water demonstrated a typical measurement accuracy of 4 % for k_w. Experiments with natural seawater samples collected on two North Sea coastal transects showed a clear influence of surfactant activity on the strong spatial and temporal gradients in k_w that we observed. During ongoing and planned work, both in the coastal North Sea and in the open ocean, we aim to establish clear relationships between k_w, surfactant activity and biogeochemical indices of primary productivity. In so doing we hope to better understand the spatio-temporal variability of k_w and thereby, to contribute valuable new insights into the air–sea gas exchange process.

Appendix A: Error propagation

Uncertainties of measured quantities were calculated using Gaussian error propagation. For a quantity y which is a function of N statistically independent variables x_i with Gaussian-distributed uncertainties σ_{x_i}, $i \in 1, \ldots, N$, i.e. $y = y(x_1, \ldots, x_N)$, the uncertainty of y is

$$\sigma_y^2 = \sum_{i=1}^{N} \left(\frac{\partial y}{\partial x_i} \right)^2 \sigma_{x_i}^2 . \tag{A1}$$

Application to Eqs. (12) and (13) leads to

$$
\begin{aligned}
\sigma_{B_n}^2 &= \left[\frac{A}{t_n - t_0} \sum_{j=1}^{n} \frac{t_j - t_{j-1}}{\left(V_{w,0} - j V_s\right)^2} \right]^2 \sigma_{V_{w,0}}^2 \\
&+ \left[\frac{A}{t_n - t_0} \sum_{j=1}^{n} \frac{j(t_j - t_{j-1})}{\left(V_{w,0} - j V_s\right)^2} \right]^2 \sigma_{V_s}^2 \\
&+ \left(\frac{\alpha A}{V_a^2} \right)^2 \sigma_{V_a}^2 + \left(\frac{A}{V_a} \right)^2 \sigma_\alpha^2 \tag{A2}
\end{aligned}
$$

and

$$\sigma_{\chi_n}^2 = \frac{1}{B_n^2} \left(\frac{1}{D^2} \sigma_D^2 + \frac{1}{D_0^2} \sigma_{D_0}^2 \right) + \left(\frac{\ln\left(\frac{D_0}{D} \right)}{B_n^2} \right)^2 \sigma_{B_n}^2 . \tag{A3}$$

Acknowledgements. We wish to thank our colleagues in the workshop of the School of Marine Science and Technology at Newcastle for modifications to the basic tank structure and for installing the electronic components, and in the glassblowing workshop in the School of Chemistry at Newcastle for manufacturing the equilibration vessel. We acknowledge and appreciate funding provided by the German Research Foundation in support of K. Schneider-Zapp (DFG research fellowship) and we thank the UK Natural Environment Research Council (NERC) for awarding a NERC small grant NE/IO15299/1 to R. Upstill-Goddard.

Edited by: M. Hoppema

References

Asher, W. E.: The effects of experimental uncertainty in parameterizing air-sea gas exchange using tracer experiment data, Atmos. Chem. Phys., 9, 131–139, doi:10.5194/acp-9-131-2009, 2009.

Bock, E. J., Hara, T., Frew, N. M., and McGillis, W. R.: Relationship between air–sea gas transfer and short wind waves, J. Geophys. Res., 104, 25821–25831, doi:10.1029/1999JC900200, 1999.

Bullister, J. L., Wisegarver, D. P., and Menzia, F. A.: The solubility of sulfur hexafluoride in water and seawater, Deep-Sea Res. Pt. I, 49, 175–187, doi:10.1016/S0967-0637(01)00051-6, 2002.

Ćosović, B. and Vojvodić, V.: The application of ac polarography to the determination of surface-active substances in seawater, Limnol. Oceanogr., 27, 361–369, 1982.

Crutzen, P. J. and Zimmermann, P. H.: The changing photochemistry of the troposphere, Tellus B, 43, 136–151, doi:10.1034/j.1600-0889.1991.t01-1-00012.x, 1991.

Dlugokencky, E. J., Masarie, K. A., Lang, P. M., and Tans, P. P.: Continuing decline in the growth rate of the atmospheric methane burden, Nature, 393, 447–450, 1998.

Dlugokencky, E. J., Walter, B. P., Masarie, K. A., Lang, P. M., and Kasischke, E. S.: Measurements of an anomalous global methane increase during 1998, Geophys. Res. Lett., 28, 499–502, doi:10.1029/2000GL012119, 2001.

Frew, N. M.: The role of organic films in air–sea exchange, in: The Sea Surface and Global Change, edited by: Liss, P. S. and Duce, R. A., Cambridge University Press, 121–172, 1997.

Garrett, W. D.: Collection of slick-forming materials from the sea surface, Limnol. Oceanogr., 10, 602–605, 1965.

Gašparović, B.: Decreased production of surface-active organic substances as a consequence of the oligotrophication in the northern Adriatic Sea, Estuar. Coast. Shelf Sc., 115, 33–39, doi:10.1016/j.ecss.2012.02.004, 2012.

Goldman, J. C., Dennett, M. R., and Frew, N. M.: Surfactant effects on air–sea gas exchange under turbulent conditions, Deep-Sea Res., 35, 1953–1970, doi:10.1016/0198-0149(88)90119-7, 1988.

Harnisch, J. and Eisenhauer, A.: Natural CF_4 and SF_6 on Earth, Geophys. Res. Lett., 25, 2401–2404, doi:10.1029/98GL01779, 1998.

Harris, F. J.: On the use of windows for harmonic analysis with the discrete Fourier Transform, in: Proceedings of the IEEE, Vol. 66-1, 1978.

IPCC: Climate Change 2007 – The Physical Science Basis: Working Group I Contribution to the Fourth Assessment Report of the IPCC, Cambridge University Press, Cambridge, UK and New York, NY, USA, 2007.

Jähne, B.: Air–sea gas exchange, in: Encyclopedia Ocean Sciences, edited by: Steele, J. H., Turekian, K. K., and Thorpe, S. A., 3434–3444, Elsevier, doi:10.1016/B978-012374473-9.00642-1, 2009.

Khalil, M.: Atmospheric Methane: Sources, Sinks and Role in Global Change, Springer, New York, 1993.

Khatiwala, S., Primeau, F., and Hall, T.: Reconstruction of the history of anthropogenic CO_2 concentrations in the ocean, Nature, 462, 346–349, doi:10.1038/nature08526, 2009.

McKenna, S. P. and McGillis, W. R.: The role of free-surface turbulence and surfactants in air–water gas transfer, Int. J. Heat Mass Tran., 47, 539–553, 2004.

Nevison, C. D. and Holland, E.: A reexamination of the impact of anthropogenically fixed nitrogen on Atmospheric N_2O and the stratospheric O_3 layer, J. Geophys. Res., 102, 25519–25536, doi:10.1029/97JD02391, 1997.

Nightingale, P., Malin, G., Law, C. S., Watson, A. J., Liss, P. S., Liddicoat, M. I., Boutin, J., and Upstill-Goddard, R. C.: In situ evaluation of air–sea gas exchange parameterizations using novel conservative and volatile tracers, Glob. Biogeochem. Sci., 14, 373–387, doi:10.1029/1999GB900091, 2000.

Phillips, O. M.: The Dynamics of the Upper Ocean, Cambridge University Press, New York, 1980.

Prinn, R., Cunnold, D., Rasmussen, R., Simmonds, P., Alyea, F., Crawford, A., Fraser, P., and Rosen, R.: Atmospheric emissions and trends of nitrous oxide deduced from 10 years of ALE-GAGE data, J. Geophys. Res.-Atmos., 95, 18369–18385, doi:10.1029/JD095iD11p18369, 1990.

Richter, K. and Jähne, B.: A laboratory study of the Schmidt Number dependency of air–water gas transfer, in: Proc. 6th Int. Symp. Gas Transfer at Water Surfaces, 2010.

Rigby, M., Prinn, R. G., Fraser, P. J., Simmonds, P. G., Langenfelds, R. L., Huang, J., Cunnold, D. M., Steele, L. P., Krummel, P. B., Weiss, R. F., O'Doherty, S., Salameh, P. K., Wang, H. J., Harth, C. M., Mühle, J., and Porter, L. W.: Renewed growth of atmospheric methane, Geophys. Res. Lett., 35, L22805, doi:10.1029/2008GL036037, 2008.

Salter, M. E., Upstill-Goddard, R. C., Nightingale, P. D., Archer, S. D., Blomquist, B., Ho, D. T., Huebert, B., Schlosser, P., and Yang, M.: Impact of an artificial surfactant release on air–sea gas fluxes during Deep Ocean Gas Exchange Experiment II, J. Geophys. Res., 116, C11016, doi:10.1029/2011JC007023, 2011.

Schneider-Zapp, K., Salter, M. E., Mann, P. J., and Upstill-Goddard, R. C.: Technical Note: Comparison of storage strategies of sea surface microlayer samples, Biogeosciences, 10, 4927–4936, doi:10.5194/bg-10-4927-2013, 2013.

Takahashi, T., Sutherland, S. C., Wanninkhof, R., Sweeney, C., Feely, R. A., Chipman, D. W., Hales, B., Friederich, G., Chavez, F., Sabine, C., A., W., Bakker, D. C. E., Schuster, U., N., M., Yoshikawa-Inoue, H., Ishii, M., Midorikawa, T., Nojiri, Y., Körtzinger, A., Steinhoff, T., Hoppema, M., Olafsson, J., Arnarson, T. S., Tilbrook, B., Johannessen, T., Olsen, A., Bellerby, R., Wong, C. S., Delille, B., Bates, N. R., and de Baar, H. J. W.: Climatological mean and decadal change in surface ocean pCO_2, and net sea–air CO_2 flux over the global oceans, Deep-Sea Res. Pt. II, 56, 554–577, doi:10.1016/j.dsr2.2008.12.009, 2009.

Tayler, J. R.: An introduction to error analysis: the study of uncertainties in physical measurements. 2nd edition, University Science Books, Sausalito, 1996.

Upstill-Goddard, R. C.: Air–sea gas exchange in the coastal zone, Estuar. Coast. Shelf S., 70, 388–404, doi:10.1016/j.ecss.2006.05.043, 2006.

Upstill-Goddard, R. C.: The production of trace gases in the estuarine and coastal environment, in: Treatise on Estuarine and Coastal Science, edited by: Wolanski, E. and McLusky, D., 271–309, Academic Press, London, 2011.

Upstill-Goddard, R. C., Watson, A. J., Liss, P. S., and Liddicoat, M. I.: Gas transfer velocities in lakes measured with SF_6, Tellus B, 42, 364–377, doi:10.1034/j.1600-0889.1990.t01-3-00006.x, 1990.

Upstill-Goddard, R. C., Rees, A. P., and Owens, N. J. P.: Simultaneous high-precision measurements of methane and nitrous oxide in water and seawater by single phase equilibration gas chromatography, Deep-Sea Res. Pt. I, 43, 1669–1682, doi:10.1016/S0967-0637(96)00074-X, 1996.

Upstill-Goddard, R. C., Frost, T., Henry, G. R., Franklin, M., Murrell, J. C., and Owens, N. J. P.: Bacterioneuston control of air–water methane exchange determined with a laboratory gas exchange tank, Global Biogeochem. Cy., 17, 19.1–19.15, doi:10.1029/2003GB002043, 2003.

Wanninkhof, R.: Relationship between wind speed and gas exchange over the ocean, J. Geophys. Res., 97, 7373–7382, doi:10.1029/92JC00188, 1992.

Wanninkhof, R., Asher, W., Weppernig, R., Chen, H., Schlosser, P., Langdon, C., and Sambrotto, R.: Gas transfer experiment on Georges Bank using two volatile deliberate tracers, J. Geophys. Res., 98, 20237–20248, 1993.

Wanninkhof, R., Hitchcock, G., Wiseman, W. J., Vargo, G., Ortner, P. B., Asher, W., Ho, D. T., Schlosser, P., Dickson, M. L., Masserini, R., Fanning, K., and Zhang, J. Z.: Gas exchange, dispersion and biological productivity on the West Florida Shelf: results from a Lagrangian tracer study, Geophys. Res. Lett., 24, 1767–1770, 1997.

Watson, A. J., Upstill-Goddard, R. C., and Liss, P. S.: Air–sea gas exchange in rough and stormy seas measured by a dual tracer technique, Nature, 349, 145–147, 1991.

Weiss, R. F. and Price, B. A.: Nitrous oxide solubility in water and seawater, Mar. Chem., 8, 347–359, doi:10.1016/0304-4203(80)90024-9, 1980.

Welch, P. D.: The use of fast Fourier transform for the estimation of power spectra: a method based on time averaging over short, modified periodograms, IEEE Transactions on Audio and Electroacoustics, 15, 70–73, doi:10.1109/TAU.1967.1161901, 1967.

Wiesenburg, D. A. and Guinasso, N. L.: Equilibrium solubilities of methane, carbon monoxide, and hydrogen in water and sea water, J. Chem. Eng. Data, 24, 356–360, doi:10.1021/je60083a006, 1979.

Wurl, O., Wurl, E., Miller, L., Johnson, K., and Vagle, S.: Formation and global distribution of sea-surface microlayers, Biogeosciences, 8, 121–135, doi:10.5194/bg-8-121-2011, 2011.

Zutić, V. B., Ćosović, E., Marčenko, E., and Bihari, N.: Surfactant production by marine phytoplankton, Mar. Chem., 10, 505–520, doi:10.1016/0304-4203(81)90004-9, 1981.

13

Upper ocean response to the passage of two sequential typhoons

D. B. Baranowski[1], **P. J. Flatau**[2], **S. Chen**[3], **and P. G. Black**[4]

[1]Institute of Geophysics, Faculty of Physics, University of Warsaw, Warsaw, Poland
[2]Scripps Institution of Oceanography, University of California San Diego, San Diego, CA, USA
[3]Naval Research Laboratory, Monterey, CA, USA
[4]Science Applications International Corporation, Monterey, CA, USA

Correspondence to: D. B. Baranowski (dabar@igf.fuw.edu.pl)

Abstract. The atmospheric wind stress forcing and the oceanic response are examined for the period between 15 September 2008 and 6 October 2008, during which two typhoons – Hagupit and Jangmi – passed through the same region of the western Pacific at Saffir–Simpson intensity categories one and three, respectively. A three-dimensional oceanic mixed layer model is compared against the remote sensing observations as well as high-repetition Argo float data. Numerical model simulations suggested that magnitude of the cooling caused by the second typhoon, Jangmi, would have been significantly larger if the ocean had not already been influenced by the first typhoon, Hagupit. It is estimated that the temperature anomaly behind Jangmi would have been about 0.4 °C larger in both cold wake and left side of the track. The numerical simulations suggest that the magnitude and position of Jangmi's cold wake depends on the precursor state of the ocean as well as lag between typhoons. Based on sensitivity experiments we show that temperature anomaly difference between "single typhoon" and "two typhoons" as well as magnitude of the cooling strongly depends on the distance between them. The amount of kinetic energy and coupling with inertial oscillations are important factors for determining magnitude of the temperature anomaly behind moving typhoons. This paper indicates that studies of ocean–atmosphere tropical cyclone interaction will benefit from denser, high-repetition Argo float measurements.

1 Introduction

In recent years understanding the predictability of tropical cyclone formation, intensification, and structure change has been a subject of intense interest. On the experimental front, THe Observing system Research and Predictability EXperiment (THORPEX) has concentrated on improving the skill of 1–14-day forecasts of high-impact weather. Throughout the THORPEX program, several regional campaigns have been undertaken to address these events. One such multi-national program was conducted during August–September 2008 over the western North Pacific (WPAC) as the summer component of THORPEX Pacific Asian Regional Campaign in collaboration with the Office of Naval Research (ONR)-sponsored Tropical Cyclone Structure 2008 experiment, referred to as TPARC/TCS08. The focus of TPARC/TCS08 has been on various aspects of typhoon activity, including formation, intensification, structure change, motion, and extratropical transition (Elsberry and Harr, 2008). Another recent joint program – The Impacts of Typhoons on the Ocean in the Pacific and Tropical Cyclone Structure 2010 (ITOP/TCS10) – took place during September–October 2010 also in WPAC (D'Asaro et al., 2011). It too was a multi-national field campaign that aimed to study the ocean response to typhoons in the western Pacific Ocean with scientific objectives related to the formation and dissipation of typhoon cold wakes, the magnitude of air–sea fluxes for winds greater than 30 m s^{-1}, the influence of ocean eddies on typhoons, the surface wave field under typhoons, and typhoon genesis in relation to environmental factors. These projects built upon the earlier air-sea interaction observations obtained in the Atlantic during the Coupled Boundary Layer Air-Sea Transfer (CBLAST) experiment conducted in the western Atlantic (Black et al., 2007).

One feature of particular interest related to the air–sea interaction associated with tropical cyclone passage is the formation and dissipation of cold wakes (D'Asaro et al., 2007; Price et al., 2008; Fisher, 1958). It is known that cold wakes can persist for several weeks after the passage of a typhoon (Price et al., 2008) during which time they evolve due to a number of processes such as solar heating, lateral mesoscale stirring, lateral mixing by baroclinic instability and vertical mixing, which act to determine the rate and character of the wake dissipation (Price, 1981; Price et al., 1986). The cold ocean wake is also expected to modify the atmospheric boundary layer (Wu et al., 2007; Cione and Uhlhorn, 2003; Ramage, 1974) and the biology and chemistry of the upper ocean (Babin et al., 2004; Lin et al., 2003). The wake development itself depends on complex non-linear atmospheric forcing related to the tropical cyclone size, strength and transitional speed as well as background ocean conditions like the upper ocean stratification and depth of the thermocline (Price et al., 1994; Hong et al., 2007; Schade and Emanuel, 1999). Based on observations, data analysis (Uhlhorn and Shay, 2011), and modeling (Uhlhorn and Shay, 2013), it has been shown that preexisting upper ocean kinetic energy structure plays also an important role in the ocean response to the tropical cyclone forcing.

Even though some progress towards understanding the oceanic response to cyclone passage has been recently documented, direct measurements of the oceanic state for such events are limited (D'Asaro et al., 2007, 2011; Mrvaljevic et al., 2013). As the tropical cyclone development and intensification is sensitive to the sea surface temperature (SST) (Wu et al., 2007; Emanuel et al., 2004), it follows that cold wakes would likely affect the development and intensification of subsequent tropical cyclones that happen to follow a similar track to their predecessor within a short period of time. Interactions and interlinks between two tropical cyclones occurring in vicinity of each other have been recognized as an important forecasting issue (Brand, 1971; Falkovich et al., 1995), but more detailed studies have become available only in recent years (Wada and Usui, 2007). These studies indicate that upper ocean thermal and energy structure disturbed by the preceding cyclone might influence the evolution of the following cyclone. Therefore, the proper assessment of the upper ocean heat content, which represents upper ocean thermal structure, is important for tropical cyclone forecasting systems, including forecasts of initiation, evolution, trajectory and foremost intensification. The interaction of two cyclones closely related in time and space thus provides a natural laboratory for studying ocean response and cold wake persistence.

To our knowledge, the question of how the cold wake of the trailing typhoon might change as a result of the disturbed oceanic state generated by its predecessor has not been previously addressed. Such an interaction might be important for proper assessment and forecast of ocean stratification, SST and evolution of atmospheric convection. In this article, we address this question by concentrating on a system of two typhoons that followed nearly identical tracks within a time span of 7 days. One of the unique aspects of the study is the analysis of in situ daily ocean profile observations, which were provided by an Argo float deployed by the Japan Agency for Marine-Earth Science and TEChnology (JAMSTEC) as part of the PALAU2008 field experiment (in collaboration with TPARC/TCS08). As Argo floats typically resurface once every 10 days, this daily-profiling Argo float thus provides a unique opportunity to study the ocean temperature and salinity changes associated with the passage of two consecutive typhoons.

We begin with climatology of the region in Sect. 2 in an effort to define the basic state of the ocean that existed prior to the passage of the typhoon couplet of interest in this study. This is followed by description of the Argo data as well as the ocean model and wind fields used to conduct numerical simulations of the cold wake dynamics. The in situ observations of mixed layer (ML) evolution as obtained by the daily JAMSTEC Argo float are presented in Sect. 3, while Sect. 4 presents modeling results, which include the comparison model runs and numerical experiments. A summary of the primary results and discussion can be found in Sect. 5.

2 Oceanic state and description of methods

2.1 Basic oceanic state

WPAC has the largest number and frequency of tropical cyclones (TCs) of all the ocean basins (Frank and Young, 2007). During the typhoon season in this basin, which usually lasts from mid-July until mid-November, there are typically around 30 tropical cyclones (TCs), half of which achieve typhoon strength (Lander, 1994). Our analysis of the typhoon tracks in WPAC during 2000 to 2010 shows there were a total of 288 TCs, of which 33 % had recurring tracks (defined as those TCs that are separated temporally by no more than a 2-week period and exhibit storms tracks that lie no more than 100 km apart (Fig. 1)). Of the TCs reaching the typhoon stage, 17 % (or 32 typhoons) were found to have recurring tracks. While the use of more stringent search criteria of 50 km reduces the number of repeating storms to about 10 % (28 typhoons), it is clear that, on average, there was more than one pair of typhoons exhibiting repeating tracks for this basin in each season examined.

For the recurring typhoons there is generally only a short period of time for the ocean to fully recover to the initial undisturbed state prior to the passage of the subsequent storm. One thus expects in such cases that the oceanic mixed layer (ML) encountered by the trailing storm is perturbed and may have an impact on the subsequent typhoon development. In a effort to establish the undisturbed conditions for our subsequent analyses, the mean temperature, salinity and density profiles from the Argo-based climatology

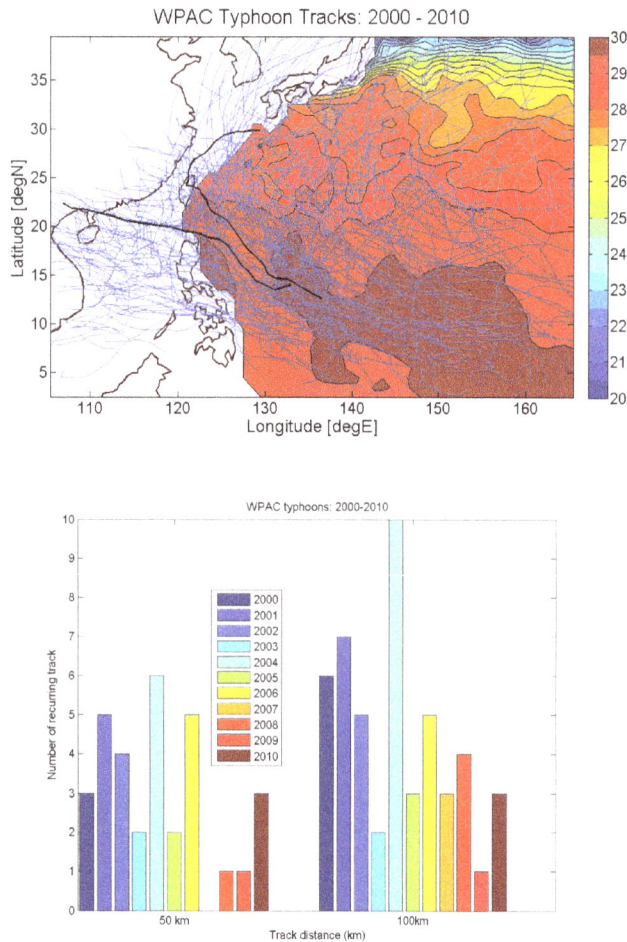

Figure 1. Climatology of 288 western Pacific typhoon tracks during year 2000 to 2010 and SST climatology for September (top panel) and the number of recurring tracks for each year (bottom panel). Two criteria of track radius, 50 and 100 km, were used to compute the statistics. On the top panel tracks of typhoons Hagupit and Jangmi are black solid lines.

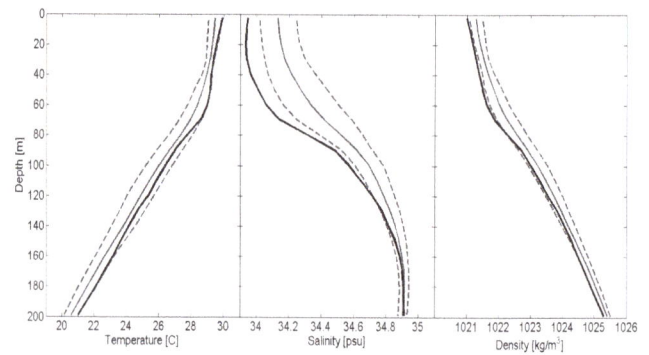

Figure 2. Monthly climatology of temperature, salinity and density from Argo-based climatology (Roemmich and Gilson, 2009) for the 1 degree × 1 degree grid centered at 17.5° N, 126.5° E. Solid grey line represents the mean value, and dashed grey lines represent deviation from the mean (standard deviation). Solid black line shows the September 2008 value.

26 September (see Figs. 3 and 4, respectively). The passage of the Hagupit–Jangmi typhoon pair was itself preceded in this area by developing typhoon Sinlaku in early September and typhoon Nuri in mid-August. While precipitation from the Hagupit–Jangmi typhoon pair may have itself contributed to a fresher than usual ML, we suspect that September 2008 was less salty than normal due to the passage through the same area by these two previous typhoons. Sinlaku developed into a tropical storm on 8–9 September passing through the middle of the region only 12 days prior to Hagupit. Typhoon Nuri, meanwhile, passed through the northern portion of the region as a CAT 1 intensity storm on 19 August. The Naval Research Laboratory (NRL) microwave satellite imagery (http://www.nrlmry.navy.mil/tc-bin/tc_home2.cgi) for Sinlaku showed that the eye of this storm was just developing and was associated with weaker rainbands that passed through the analysis area. The Nuri imagery indicated that the heaviest precipitation region associated with this typhoon was centered over the southern portion of our area of interest. Therefore, the winds over this region would have been light during both of these events, but the rainfall would likely have been extensive and heavy, probably resulting in the less saline conditions for September 2008 but with very little ML modification due to wind forcing.

2.2 Argo data

Of the large number of Argo floats present in the western Pacific region in September 2008, only one happened to be fortuitously located right between the two sequentially occurring typhoons Hagupit and Jangmi. This instrument, float ID 5901579, was deployed by JAMSTEC as part of the PALAU2008 field program and was set to conduct a vertical profile of the ocean from a depth of 500 m to the surface everyday. This float was initially released from R/V *Mirai* and was intended to measure air–sea interactions with

(Roemmich and Gilson, 2009) were computed for the 1 degree × 1 degree box region centered at 17.5° N, 126.5° E that corresponds to the mean location for our area of interest (Fig. 2). This data set is based on Argo profiles for years 2004–2012 and shows the overall time-mean profiles along with their standard deviation for the month of September (given by the grey curves in the plot). When these longer term climatological profiles are compared with the conditions during September 2008 associated with the development of Hagupit and Jangmi, we find that the ocean state accompanying these storms was warmer and less salty by approximately one and two standard deviations, respectively. Climatological SST for September based on the same data set is shown in Fig. 1; SST in the area of interest is horizontally uniform. Typhoon Hagupit passed through the analysis region on 21 September as a Saffir–Simpson category (CAT) 1 typhoon followed by the CAT 3 typhoon Jangmi on

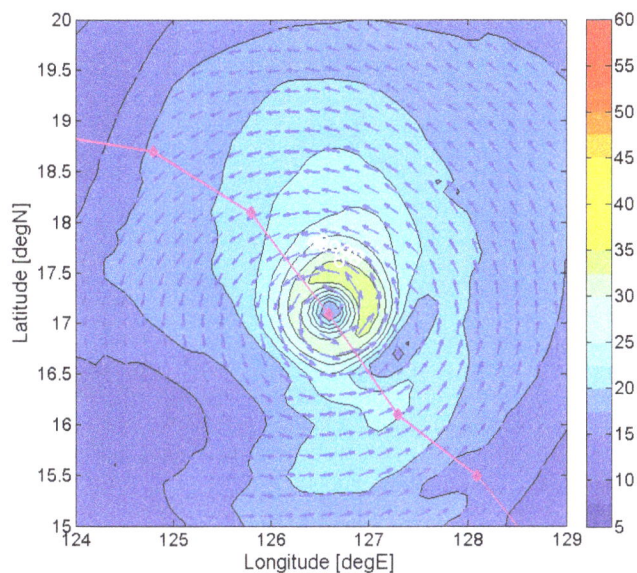

Figure 3. The wind speed [m s^{-1}] and wind direction (blue arrows) at 06Z on 21 September 2008. The best track of typhoon Hagupit is plotted as the magenta line with the diamonds denoting the 12 h storm position. The white circles indicate the daily Argo float locations between 18 September and 29 September 2008.

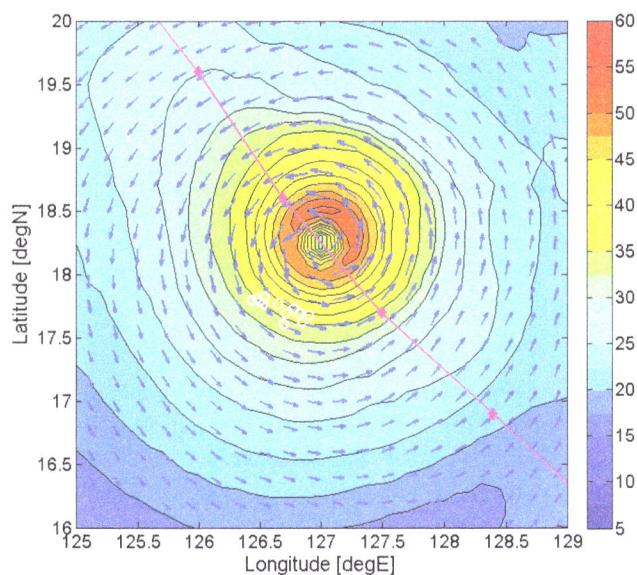

Figure 4. The wind speed [m s^{-1}] and wind direction (blue arrows) on 26 September 2008. The best track position of typhoon Jangmi is plotted as the magenta line with diamonds denoting the 12 position of the storm center. The white circles indicate the daily Argo float locations between 18 September and 29 September 2008.

the particular objective of improving understanding of intra-seasonal variations in the western Pacific. For that reason it was set up to take one profile everyday with the vertical resolution of about 5 m from the ocean surface down to 200 m; below 200 m the float's resolution is about 10 m. The data

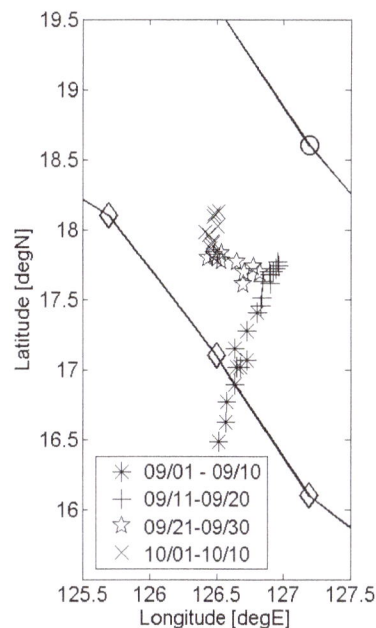

Figure 5. Position of the Argo float ID 5901579 between 1 September and 10 October with respect to the typhoon Hagupit (diamonds) and Jangmi (circles) best tracks. The various symbols represent different group time periods as denoted on the plot.

obtained thus provide a unique opportunity to study daily changes in the upper ocean during the sequential passage of these two active typhoons. Only data flagged "good" are used; therefore some data points are skipped. Presence of this particular instrument at that time and location was highly fortunate. Both typhoons Hagupit and Jangmi were intensively studied as part of TPARC/TCS08 field campaign, but none of the instruments deployed as part of targeted measurements were in the position to measure response to repeated typhoon passages.

Figure 5 shows that the float was located near 17.5° N, 126.5° E during the month of September 2008 and exhibited overall displacements less than 100 km during the 20-day period during which Hagupit and Jangmi passed through the area. From Figs. 3 and 4 we see that this float was positioned about 40 km to the right of the track of typhoon Hagupit and approximately 100 km to the left of the track of typhoon Jangmi. It was thus ideally situated for use in subsequent analyses of the cold wake as well as input for use in the idealized model studies that we describe next.

2.3 The three-dimensional Price–Weller–Pinkel (3DPWP) ocean model

A three-dimensional, primitive-equation, hydrostatic numerical ocean model (3DPWP) (Price et al., 1994, 1986) is used for numerical simulations of the cold wakes examined in this study. The model runs were performed with a 5 km horizontal resolution and a variable vertical grid resolution that was set

Table 1. Description of the numerical experiments.

Experiment name	Initial ocean condition	Number of typhoons	Number of model simulations
HH	Hagupit	1	1
JJ	Jangmi	1	1
HJ	Hagupit	1	1
HHJ	Hagupit	2	5
HHH	Hagupit	2	10

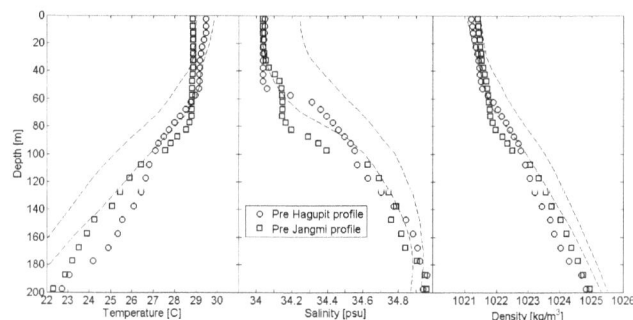

Figure 6. The initial temperature [°C], salinity [psu], and density [kg m^{-3}] profiles used in the 3DPWP simulations HH and HJ (circles) and JJ (squares) described in the text. The dashed black lines show the mean climatology profiles with added and subtracted standard deviations for September 2008.

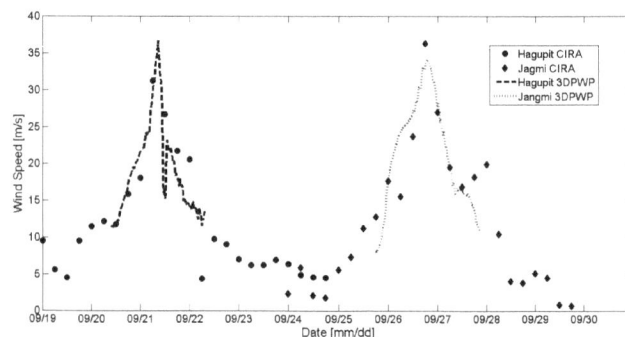

Figure 7. Wind speed as a function of time at the Argo float location (17.5° N, 126.5° E) based on observations and model results. The dots and diamonds represent CIRA data for typhoons Hagupit and Jangmi every 6 h. The dashed and dotted lines are wind speeds used in the model forcing (based on only one initial velocity field, but advected with the center of TC movement) for typhoons Hagupit and Jangmi, respectively.

to 5 m between the surface and 100 m, and a constant 10 m resolution below a depth of 100 m. The total domain depth was set to 2000 m. The model was initialized with the wind field obtained from Cooperative Institute for Research in Atmosphere (CIRA) as the surface forcing (Figs. 3 and 4) and employed the drag coefficient defined by Powell et al. (2003).

The model initial ocean conditions were assumed to be horizontally homogenous and correspond to the undisturbed Argo profile shown in Fig. 6. Five model experiments were performed – two control simulations and three sensitivity experiments (Table 1). For the "control" simulation, typhoon Hagupit (hereafter referred to as HH) was initialized based on the pre-Hagupit ocean conditions. The typhoon Jangmi control case (hereafter referred to as JJ) was initialized from the pre-Jangmi oceanic conditions. These simulations were performed to establish the model's three-dimensional ocean response from these two TC forcings. They are compared with in situ observations from Argo float and the sensitivity experiments. The sensitivity "experiment 1" (HJ) is the same as the Jangmi control except the initial ocean is from the pre-Hagupit conditions. Sensitivity "experiment 2" simulates two sequential typhoons, and it was initialized with the pre-Hagupit conditions (HHJ). The two model typhoon trajectories are separated by 140 km, which corresponds to the observed track distance between these two typhoons (see Fig. 5). In the final set of experiments, we investigate the ocean's response to two identical Hagupit-like forcings following the same track but separated by 55–195 h in time. Pre-Hagupit conditions are used as initial conditions in this experiment (HHH).

For the "pre-Hagupit" conditions, the Argo profile for 19 September is used, and for the "pre-Jangmi" conditions the profile from 25 September is used (Fig. 6). The translational velocity of the typhoon is set to 6 m s^{-1} in all numerical experiments based on the best tracks data from Japan Meteorological Agency (JMA) and Joint Typhoon Warning Center (JTWC). These assumptions reproduced a good agreement between the model and 6 hourly CIRA wind speeds at the Argo float location (Fig. 7).

For the control simulations HH and JJ, we compare the mixed layer temperature (MLT), temperature tendencies, mixed layer depth (MLD), and stratification in the thermocline. We exercised some care when doing such direct model

and measurements comparisons. For example, it is observed (Price, 1981; Firing et al., 1997; Sanford et al., 2011) that after a typhoon's passage there are inertial oscillations in the ocean that cause vertical oscillations of the thermocline and therefore change the MLD. The Argo profiles were collected once a day in our case; thus it is not possible to distinguish the phase of the inertial oscillation (and its MLD modifications). Nevertheless, they provide a sanity check for the model-simulated MLT, temperature tendencies, and thermocline stratification.

2.4 Wind stress

The typhoon wind fields used to initialize the model for Hagupit and Jangmi were obtained from the CIRA Multiplatform Satellite Surface Wind Analysis Product (images available at http://rammb.cira.colostate.edu/products/tc_realtime/). These data are available at 6 h intervals and have a

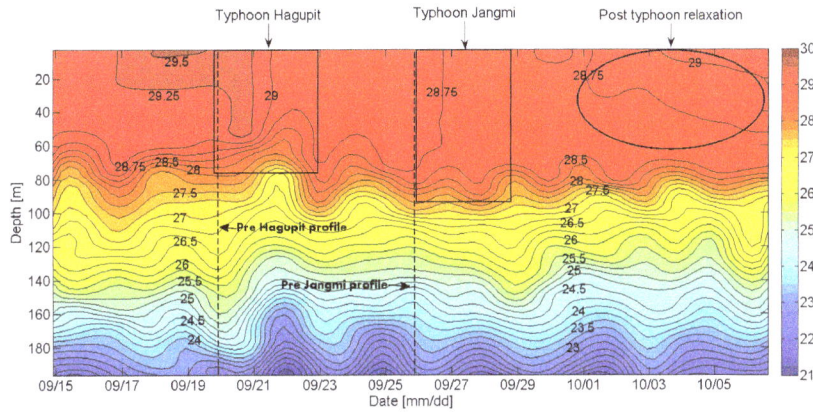

Figure 8. Temperature [°C] observed by the Argo float ID 5901579. Cooling due to typhoons Hagupit and Jangmi is visible between 20 and 27 September.

0.1 degree ×0.1 degree spatial resolution. The wind field from 21 September 06Z analysis was selected for the Hagupit model runs (Fig. 3), while the 26 September 18Z analysis (Fig. 4) was chosen for the Jangmi model runs. Both choices are based on the closest distance of each storm from the position of the Argo float. For Hagupit, the Argo float was located at a distance approximately 1.5 times the radius of maximum winds (R_{max}) to the right of the TC where maximum wind speeds were on the order of $30\,\mathrm{m\,s^{-1}}$, whereas for the stronger Jangmi the float was approximately $2.5 \times R_{max}$ to the left of the TC where the maximum wind speeds were on the order of $35\,\mathrm{m\,s^{-1}}$ (Figs. 3 and 4).

Figure 7 shows the time series of wind speed corresponding to the location of the daily Argo float ascent to the surface. The first wind maximum evident on 21 September corresponds to the passage of typhoon Hagupit, while the second wind maximum, on 26 September, corresponds to the passage of typhoon Jangmi. The deviation between wind speed in the model and the CIRA data for the same location is due to evolution of typhoon wind field. One specific wind field (Figs. 3 and 4) was used for each individual typhoon, while the CIRA-based wind time series show wind speeds at float location obtained from sequential (6 h interval) wind fields. Such an assumption is only valid when changes in overall wind field of the typhoon are small for the 24 h period.

3 Ocean evolution during the typhoon passage

In this section we focus on the period between 15 September and 6 October 2008 when the float's position (Fig. 5) was between the tracks of the two typhoons. At the location of the float, a decrease of 0.67 °C SST is visible after Hagupit's passage on 21 September (Fig. 8). This cooling is associated with approximately 20 m deepening of the ML from ~ 60 m to ~ 80 m that occurred as Hagupit passed through the region. While the SST cooling due to the passage of typhoon Jangmi was about 0.22 °C, no substantial changes in

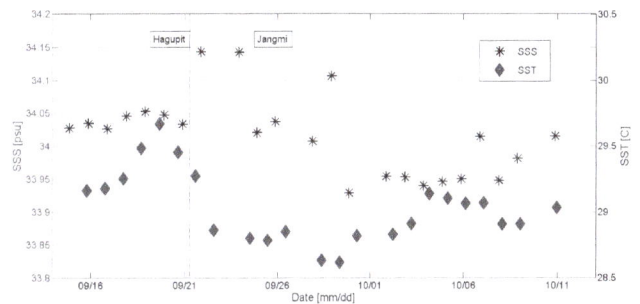

Figure 9. Argo float surface salinity (SSS) [psu] (left axis) and sea surface temperature [°C] (right axis) at 4 m depth.

the mixed layer depth (MLD) was detected by the Argo float. The minimum SST of 28.6 °C is observed on 28 September and is followed by warming of the ML due to the solar heating at the surface and its gradual downward propagation due to mixing (recovery process).

The temperature and salinity at a depth of 4 m is plotted as a function of time in Fig. 9 for the time period between 20 September and 10 October. The position of the float during this period did not vary substantially indicating small currents at the times when the float resurfaced. This suggests that the background large-scale horizontal advection not associated with inertial movement forced by typhoon had a limited role in the observed behavior. The drop in temperature beginning on 19 September is associated with a large variation in the salinity values (of magnitude 0.2 psu) observed between 19 September and 1 October. We speculate that the salinity variation was caused by the mixing of more saline water from beneath the thermocline modulated by the freshwater input from typhoon precipitation. It was suggested (Jacob and Koblinsky, 2007; Price, 2009) that freshening of the ML acts to increase the stability and reduce the subsequent entrainment of the cold water from thermocline. At the float location the wind speed was comparable for both typhoons

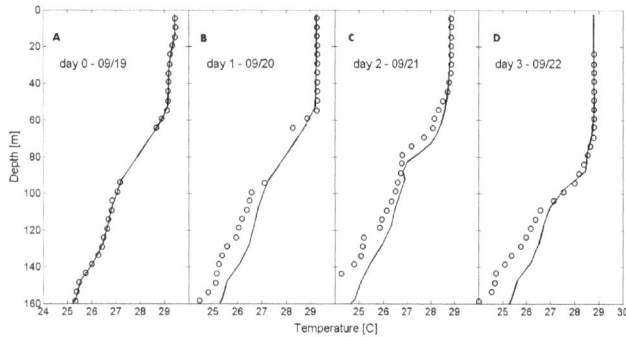

Figure 10. Comparison between Argo profile temperature observations (circles) and the 3DPWP model with the typhoon Hagupit setup (black solid line). Four plots are presented (**A–D**) showing a comparison between 19 September (**A**) and 22 September (**D**).

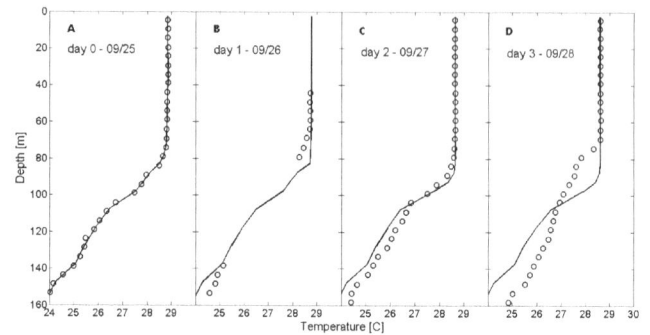

Figure 11. A comparison between the Argo profile temperature observations (circles) and the 3DPWP model with typhoon Jangmi setup (black solid line). Four plots are presented (**A–D**) showing a comparison between 25 September (**A**) and 28 September (**D**).

(Fig. 5), but the mixed layer temperature and salinity response to typhoon Jangmi (26–27 September) was weaker, most probably due to the float's different relative position with respect to the typhoon track.

The wind stress coupling with inertial oscillations may also enhance the mixing on the right side of the typhoon track (Price, 1981), possibly helping to explain the results shown here. Observations taken during the passage of Jangmi, on the other hand, were limited to the left side of the typhoon track where mixing is not supported by coupling with inertial currents curl.

4 Modeling

4.1 Model – in situ comparison

In this section we present the control run results for HH and JJ simulations, and we compare the modeled ocean response with the available Argo float observations. For comparison with in situ measurements, we use the results at the float location, which were, respectively, 40 km to the right of the track of typhoon Hagupit and 100 km to the left of the track of Jangmi. Figure 10 shows such profile comparisons for typhoon Hagupit. Overall, one can observe a very good correspondence between the observed and modeled cooling during the early typhoon approach and passage of typhoon Hagupit for the four sequential Argo profiles plotted in Fig. 10. The maximum modeled cooling of 0.8 °C matches the observations quite well, for example, as do the temperature tendencies within the mixed layer. The day-3 Argo profile in Fig. 10 is missing the upper 20 m of data, and therefore it is not possible to compare it with modeled response, though modeled temperature below 20 m and depth of the thermocline are represented well when compared to observations. It shows that model well reproduces cooling processes in this case. It is also apparent that temperature gradients within the upper thermocline are also well captured

by the model. Substantial deviation between model and observations is evident far below the ML, and these may result from internal wave forcing from the model lateral boundaries. During typhoon Hagupit passage, in situ observations were made within area of maximum modeled temperature anomaly. This comparison implies that 3DPWP model performs well within the area of the cold wake.

Figure 11 shows a comparison of the observed and modeled vertical profiles of temperature for typhoon Jangmi between 25 and 28 September at a time when the float was approximately 100 km to the left of the storm track, in the opposite region from the right side maximum cooling predicted by the model. As with Hagupit, it can be seen that the mixed layer temperature predicted by the model agrees well with the observed conditions. Although part of the day-1 in situ profile is missing above 40 m, one expects that it is well mixed and similar to the prior profile under strong wind conditions evident in Fig. 5. The model also does well in predicting the depth of the mixed layer particularly during days 1 and 2. Note that the observed cooling comes to an end on 27 September after the SSTs reach a minimum temperature of 28.6 °C. The model performs well in this regard with no change in temperature detected between days 2 and 3. The model on day 3 shows a relatively deeper MLD compared with the observations but that might be due to thermocline depth modified by the internal waves or other physical processes that were not represented in the model. Note that modeled profile has not changed from the previous day, when thermocline structure of observed profiles is clearly different. Since the typhoon at this time already passed by the float location, it is likely that the observed change is due to factors other than interaction with typhoon Jangmi. Model estimates of the MLD and temperature gradient within the thermocline are quite good with the exception of day 3.

Although typhoon Jangmi case has stronger winds, the modeled cold wake is weaker than for Hagupit. One possible explanation may lie in the role of the initial ocean static stability and MLD in contributing to the overall wake

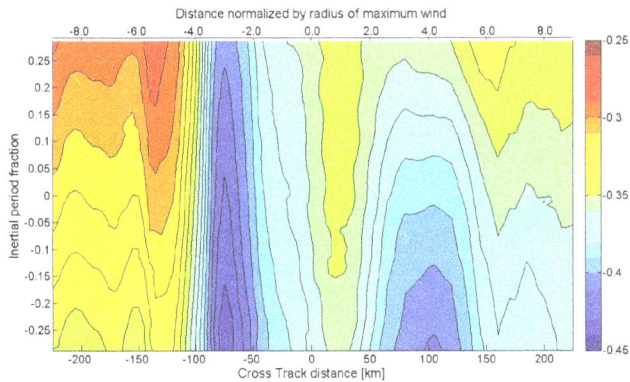

Figure 12. Double-TC surface temperature anomaly (see Eq. 1). Here, the distance perpendicular to the best track is plotted on the lower x axis with negative values representing the distance to the left of the track and positive values the distance to the right of the track. The upper x axis presents the same distance but normalized by the radius of maximum wind, which is 25 km in this case. Color scale represents temperature anomaly [°C].

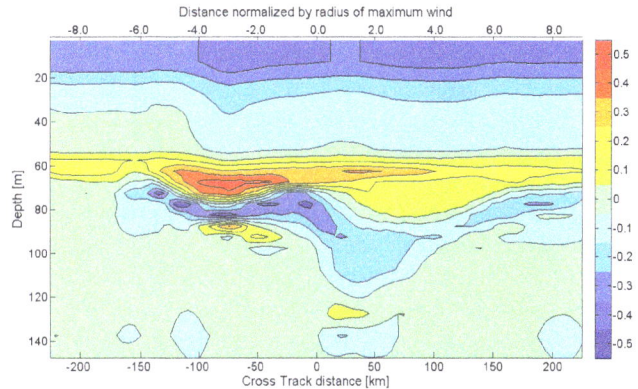

Figure 13. Vertical section of the double-TC temperature anomaly (see Eq. 1) averaged over an inertial period (40 h). The lower x axis represents distance from the Jangmi track in kilometers; the upper x axis is the same distance normalized by radius of maximum wind, which is 25 km in this case; the ordinate represents ocean depth in meters. The 0 position is at the typhoon center. Color scale represents temperature anomaly [°C].

cooling (Price, 1981). The data plotted in Fig. 6 show that the temperature at 100 m depth is about 27 °C for both of the initial profiles. However, comparing the temperature at 100 m to the SST in each case we find a 2.4 °C difference for the pre-Hagupit profile and 1.6 °C for the pre-Jangmi profile. The bigger temperature differences (less stable conditions) suggest that a higher cooling potential may exist in the Hagupit case. A shallower initial MLD causes a bigger cooling in the ocean for pre-Hagupit case in comparison to the pre-Jangmi case. In addition, presence of the barrier layer in pre-Jangmi profile is influencing the results: the pre-Jangmi salinity profile (squares at middle plot in Fig. 5) shows that, within the 80 m deep ML, there exists a shallower layer of fresher (~ 0.1 psu) water, and such a structure may reduce the amount of cooling associated with the typhoon passage (Wang et al., 2011).

Another possible explanation for weaker ocean response to stronger typhoon Jangmi is counter-coupling with inertial currents forced on the upper ocean by typhoon Hagupit passage. It is known (Price, 1981) that interaction between wind field curl and inertial motions is one of the key factors for inhomogeneous response of the upper ocean to typhoon-like forcing resulting in development of the cold wake to the right side of typhoon track (in the Northern Hemisphere). Therefore coupling with inertial oscillations that preexisted prior to the typhoon approach might either benefit or suppress mixing. We will address this point by the analysis of HHH experiments in the next section.

The overall comparisons with the Argo float indicate that the 3DPWP model does a good job in simulating the mixed-layer temperature and vertical structure changes during the individual typhoon passages. It can be used to investigate the ocean response to two sequential typhoon-like forcing.

4.2 Influence of typhoon Hagupit on typhoon Jangmi ocean response

The close proximity of the Hagupit and Jangmi tracks (140 km) and passage (6 days) over Argo float 5901579 raises a natural question: what influence did typhoon Hagupit have on the oceanic response caused by typhoon Jangmi? To investigate this relationship, we use the 3DPWP model to simulate the oceanic response caused by the typhoon Jangmi using the pre-Hagupit conditions as the typhoon's initial state (HJ) and by Hagupit and Jangmi with a 6-day gap between them (HHJ). We should stress that the effects of typhoon Hagupit are not only limited to the vertical ocean mixing, but may also include changes in the freshwater flux due to precipitation, inertial ocean motions, and other effects that may depend on a characteristic time of relaxation (Price et al., 2008). However, we concentrate here on contribution from the wind-driven forcing, and therefore shortwave flux and freshwater flux were set to 0 in this experiment. The initial oceanic state is a horizontally homogeneous ocean based on the Argo profile from 19 September.

To quantify the difference between these two experiments, let us define the double-TC anomaly difference as

$$T(t) = [T_{HJ}(t) - T_{HJ}(t = 0)] - [T_{HHJ}(t) - T_{HHJ}(t = t_1)], \quad (1)$$

where indices HJ and HHJ refer to "experiment 1" and "experiment 2" simulations respectively. Ocean state at time t_1 is disturbed by typhoon Hagupit and represents upper ocean structure prior to the arrival of typhoon Jangmi.

It can be seen (Figs. 12 and 13) that cooling caused by typhoon Jangmi would have been significantly stronger if it had not been preceded by typhoon Hagupit. The cold wake behind Jangmi in the HJ run is situated on the right side of the track and has a width of roughly 100 km. Figure 12 shows that temperature anomaly would be magnified but also on the opposite side. Strongest difference is visible 50–100 km to the left of the track of typhoon Jangmi where the cold wake of typhoon Hagupit is located. In HHJ run this part of the ocean was strongly disturbed by Hagupit; thus mixed layer was cooled and deepened. Additional cooling due to passage of Jangmi was only 0.3 °C. If there were no Hagupit, downward momentum transport and mixing enhanced by typhoon Jangmi would have accounted for twice as strong mixing in this location. The different horizontal distributions of temperature anomaly simulated with the same atmospheric forcing suggest that disturbed and undisturbed ocean initial conditions, especially the initial MLD, influenced not only the amplitude of the anomaly but also its spatial distribution. Figure 13 presents the double-TC temperature anomaly vertical cross section averaged over an inertial period. Since a typhoon passage produces inertial oscillations in the ocean, we average the simulation results over 40 h period (roughly the period of inertial oscillations at 17.5° N) to remove them. The abscissa in the plot represents the distance from the Jangmi track, and the ordinate represents the ocean depth. The ordinate position is at the typhoon center. The surface cold anomaly is seen to be approximately 0.4 °C both to the right and left side of the track. Beneath the surface, there is an alternating positive and negative double-TC anomalies extending across the typhoon track with a larger and more complex pattern on the left side of typhoon. This is due to a disturbed initial upper ocean on the left of Jangmi due to Hagupit. The subsurface net cooling down to about 50 m would have been stronger if there were no typhoon Hagupit.

A second question that can be examined is how the final ocean state after the passage of two sequential cyclones might differ from that obtained with the passage of just one typhoon. Let us introduce the following final-state temperature anomaly:

$$T(t) = [T_{HJ}(t) - T_{HJ}(t = 0)] - [T_{HHJ}(t) - T_{HHJ}(t = 0)], \quad (2)$$

where the first term defines the final ocean state after the passage of Jangmi without any influence from Hagupit. The second term represents a sequence of oceanic anomalies caused by the passage of Hagupit followed by Jangmi in the single model run. Temperature defined in Eq. (2) is presented in Fig. 14. An overall effect on the ocean of two sequential typhoons is stronger then the effect of typhoon Jangmi itself (indicated by positive values), though substantial variation in temperature anomaly is visible. On the left side of Jangmi's track the anomaly is stronger, which means that two collocated typhoons caused cooling of 0.25 °C stronger then Jangmi alone. This can be explained by the absence of Hagupit's forcing on the ocean; note that −100 km is in the peak region of the Hagupit-induced cold wake. On the right side of typhoon Jangmi's track, the joint effect of Hagupit and Jangmi is only 0.06 °C stronger then cooling from Jangmi alone – which is small. This accounts for only

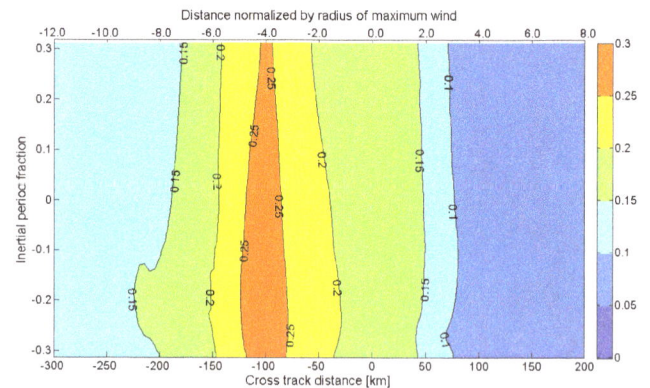

Figure 14. The HJ SST anomaly with respect to the total HHJ run anomaly (see Eq. 1). Here, distance perpendicular to the best track is plotted on the lower x axis; negative values are to the left of the track, and positive values are to the right of the track. The upper x axis presents the same distance but normalized by the radius of maximum wind, which is 25 km in this case. On the y axis there is distance along the trajectory represented as time normalized by the inertial oscillation period (40 h). The cross-track distance measures distance from the trajectory of typhoon Jangmi. The position of typhoon Hagupit is at −140 km. It can be seen that there is positive anomaly to the left and negative anomaly to the right of Jangmi's track. Color scale represents temperature anomaly [°C].

5 % of temperature anomaly in the cold wake. Note that cooling reported in this study depends on momentum transfer and mixing exclusively. Factors like heavy precipitation, which is known to form barrier layers and therefore limit magnitude of anomaly (Wang et al., 2011), were omitted.

Up to now we discussed how the presence of typhoon Hagupit influences ocean response to typhoon Jangmi. The next set of numerical experiments is designed to answer the question about the role of distance between typhoon centers on strength of oceanic cooling effects. To simplify the problem, we performed a series of simulations using two identical Hagupit-like wind fields following the same trajectory but with different time lags ranging from 55 to 195 h between them (HHH). Results (Fig. 15) are averaged over the inertial oscillation period. Panel A shows cross-track SST anomaly from different lags; an oscillation with period of ∼ 40 h is visible in area of the maximum cooling ($x = 60$ km) with magnitude of ∼ 0.2 °C. The oscillation period agrees with the inertial oscillation period at this latitude. The first SST anomaly minimum of 0.1 °C is visible at 85 h lag. As in panel A, panel B shows the vertical cross section of temperature anomaly along the cold wake ($x = 60$ km in panel A). One can notice the dependence of the temperature anomaly on lag time between sequential typhoons in the surface layer as well as below the thermocline. This is due to the coupling between wind stress curl and inertial motions in the mixed layer (Price, 1981). The results show, for Hagupit-like typhoon, the wind stress curl and inertial motion coupling effect on the upper ocean temperature cooling occurred after two inertial

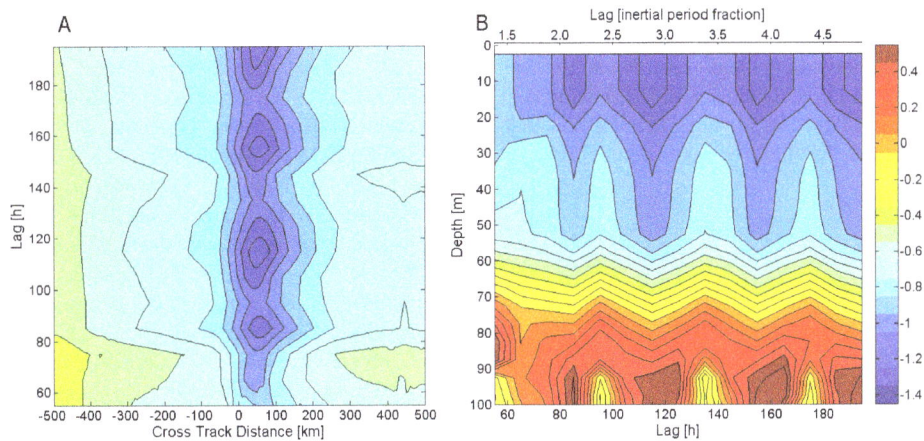

Figure 15. The HHH run results. Panel (**A**) is SST anomaly. On x axis is cross-track distance in kilometers, and on y axis is lag between Hagupit like forcing in h. Panel (**B**) is vertical cross section along cold wake ($x = 60$ km in panel **A**). On x axis is lag between Hagupit like forcing in h (lower x axis) and as fraction of inertial period (upper x axis). On y axis is depth in meters. Color scale represents temperature anomaly [°C]. Both panels use the same color scale.

oscillations (80 h). The maximum surface cold anomaly is about 1.4 °C. It is interesting to note that typhoon Jangmi occurred approximately 5 days (120 h) after typhoon Hagupit, which is within the range of maximum anomaly seen from HHH experiments. This set of numerical experiments provides some evidence of ocean response to sequential TC in terms of the inertial oscillation effect.

5 Summary and discussion

We consider the interaction of two typhoons closely collocated in time and space on the resulting observed and simulated ocean cold wake structure. We show unique observations of high-repetition Argo floats that collected in situ observations in close proximity to both typhoons. The numerical model was validated by imposing realistic atmospheric forcing using satellite-derived (CIRA) winds and observed cyclone tracks, and compared model results with observed ocean structure using the high-repetition Argo float. We show that cooling caused by the second typhoon (typhoon Jangmi) would have been about 0.5 °C (30 %) stronger if it had not been preceded by typhoon Hagupit. Additionally we also see a different spatial distribution with stronger negative SST anomalies in both cold wake region and on the left side of the track. Similar to previous studies, our results confirm the importance of initial ocean condition on the ocean response to typhoon forcing. The disturbed ocean conditions set up by a previous typhoon such as the case studied here deepen the mixed-layer depth and weaken the ocean response for the second typhoon following a similar track. This, in turn, may produce a stronger second typhoon due to less ocean cooling. We show that the amount of pre-storm upper-ocean kinetic energy in the ocean is as important as the pre-storm mixed layer depth to affect the magnitude of the ocean cooling during the storm stage. Finally, we show that the magnitude of

the cooling may also depend on the time lag between successive typhoons. Results suggest that the magnitude of the anomaly oscillates with the period of the inertial current and that the first maximum cooling occurs after two inertial oscillation periods (80 h).

We present model simulations of two typhoons' impact on the upper ocean and their possible intricate relationship driven by both undisturbed and disturbed initial oceanic thermodynamical and dynamical states. This paper indicates that studies of successive ocean–atmosphere tropical cyclone interaction will benefit from more high-repetition Argo float measurements such as these reported here and from a higher density network of such floats. More coupled model studies are also warranted in order to further understand the impact of the ocean processes discussed here on typhoon's track and intensity change.

Copyright statement

Acknowledgements. This research was sponsored by the Naval Research Laboratory under Program Element 0601153N. D. B. Baranowski was supported by the ONR grant (award number N00014-10-1-0898) and ONR Global NICOP (award number N62909-11-1-7061). The Japan Meteorological Agency best track data can be found at http://www.jma.go.jp/jma/jma-eng/jma-center/rsmc-hp-pub-eg/besttrack.html. We would like to thank Dean Roemmich and John Gilson for providing gridded Argo data used here for deriving climatology as well as answering many questions on Argo data as well as their hospitality during D. B. Baranowski's stays at Scripps Institution of Oceanography.

D. B. Baranowski would like to extend his gratitude to Hiroyuki Yamada for his help during the R/V *Mirai* cruise in summer of 2010 and hands-on experience with the high-repetition Argo floats. John Knaff and Mark deMaria provided us with CIRA wind data. We would like to thank Jim Price for providing the 3DPWD code, and James Ridout and Jerome Schmidt for proofreading this paper.

Edited by: E. J. M. Delhez

References

Babin, S. M., Carton, J. A., Dickey, T. D., and Wiggert, J. D.: Satellite evidence of hurricane-induced phytoplankton blooms in an oceanic desert, J. Geophys. Res.-Oceans, 109, C03043, doi:10.1029/2003jc001938, 2004.

Black, P. G., D'Asaro, E. A., Drennan, W. M., French, J. R., Niiler, P. P., Sanford, T. B., Terrill, E. J., Walsh, E. J., and Zhang, J. A.: Air-sea exchange in hurricanes – Synthesis of observations from the coupled boundary layer air-sea transfer experiment, Bull. Am. Meteorol. Soc., 88, 357–374, doi:10.1175/bams-88-3-357, 2007.

Brand, S.: The Effects on a Tropical Cyclone of Cooler Surface Waters Due to Upwelling and Mixing Produced by a Prior Tropical Cyclone, J. Appl. Meteorol., 10, 865–874, doi:10.1175/1520-0450(1971)010<0865:teoatc>2.0.co;2, 1971.

Cione, J. J. and Uhlhorn, E. W.: Sea surface temperature variability in hurricanes: Implications with respect to intensity change, Mon. Weather Rev., 131, 1783–1796, 2003.

D'Asaro, E. A., Sanford, T. B., Niiler, P. P., and Terrill, E. J.: Cold wake of Hurricane Frances, Geophys. Res. Lett., 34, L15609, doi:10.1029/2007gl030160, 2007.

D'Asaro, E. A., Black, P., Centurioni, L., Harr, P., Jayne, S., Lin, II, Lee, C., Morzel, J., Mrvaljevic, R., Niiler, P. P., Rainville, L., Sanford, T., and Tang, T. Y.: Typhoon-Ocean Interaction in the Western North Pacific: Part 1, Oceanography, 24, 24–31, 2011.

Elsberry, R. L. and Harr, P. A.: Tropical Cyclone Structure (TCS08) Field Experiment Science Basis, Observational Platforms, and Strategy, Asia-Pac., J. Atmos. Sci., 44, 209–231, 2008.

Emanuel, K., DesAutels, C., Holloway, C., and Korty, R.: Environmental control of tropical cyclone intensity, J. Atmos. Sci., 61, 843–858, 2004.

Falkovich, A. I., Khain, A. P., and Ginis, I.: Motion and Evolution of Binary Tropical Cyclones in a Coupled Atmosphere–Ocean Numerical Model, Mon. Weather Rev., 123, 1345–1363, doi:10.1175/1520-0493(1995)123<1345:maeobt>2.0.co;2, 1995.

Firing, E., Lien, R. C., and Muller, P.: Observations of strong inertial oscillations after the passage of tropical cyclone Ofa, J. Geophys. Res.-Oceans, 102, 3317–3322, 1997.

Fisher, E. L.: Hurricanes and the Sea-Surface Temperature Field, J. Meteorol., 15, 328–331, 1958.

Frank, W. M. and Young, G. S.: The interannual variability of tropical cyclones, Mon. Weather Rev., 135, 3587–3598, doi:10.1175/mwr3435.1, 2007.

Hong, X. D., Chang, S. W., and Raman, S.: Modification of the loop current warm core eddy by Hurricane Gilbert (1988), Nat. Hazards, 41, 501–514, doi:10.1007/s11069-006-9057-2, 2007.

Jacob, S. D. and Koblinsky, C. J.: Effects of precipitation on the upper-ocean response to a hurricane, Mon. Weather Rev., 135, 2207–2225, doi:10.1175/mwr3366.1, 2007.

Lander, M. A.: An Exploratory Analysis of the Relationship between Tropical Storm Formation in the Western North Pacific and Enso, Mon. Weather Rev., 122, 636–651, 1994.

Lin, I., Liu, W. T., Wu, C. C., Wong, G. T. F., Hu, C. M., Chen, Z. Q., Liang, W. D., Yang, Y., and Liu, K. K.: New evidence for enhanced ocean primary production triggered by tropical cyclone, Geophys. Res. Lett., 30, 1718, doi:10.1029/2003gl017141, 2003.

Mrvaljevic, R. K., Black, P. G., Centurioni, L. R., Chang, Y.-T., D'Asaro, E. A., Jayne, S. R., Lee, C. M., Lien, R.-C., Lin, I. I., Morzel, J., Niiler, P. P., Rainville, L., and Sanford, T. B.: Observations of the cold wake of Typhoon Fanapi (2010), Geophys. Res. Lett., 316–321, doi:10.1002/grl.50096, online first, 2013.

Powell, M. D., Vickery, P. J., and Reinhold, T. A.: Reduced drag coefficient for high wind speeds in tropical cyclones, Nature, 422, 279–283, doi:10.1038/Nature01481, 2003.

Price, J. F., Morzel, J., and Niiler, P. P.: Warming of SST in the cool wake of a moving hurricane, J. Geophys. Res.-Oceans, 113, C07010, doi:10.1029/2007jc004393, 2008.

Price, J. F.: Upper Ocean Response to a Hurricane, J. Phys. Oceanogr., 11, 153–175, 1981.

Price, J. F., Weller, R. A., and Pinkel, R.: Diurnal Cycling – Observations and Models of the Upper Ocean Response to Diurnal Heating, Cooling, and Wind Mixing, J. Geophys. Res.-Oceans, 91, 8411–8427, 1986.

Price, J. F., Sanford, T. B., and Forristall, G. Z.: Forced Stage Response to a Moving Hurricane, J. Phys. Oceanogr., 24, 233–260, 1994.

Price, J. F.: Metrics of hurricane-ocean interaction: vertically-integrated or vertically-averaged ocean temperature?, Ocean Sci., 5, 351–368, doi:10.5194/os-5-351-2009, 2009.

Ramage, C. S.: Typhoons of October 1970 in South China Sea – Intensification, Decay and Ocean Interaction, J. Appl. Meteorol., 13, 739–751, doi:10.1175/1520-0450, 1974.

Roemmich, D. and Gilson, J.: The 2004-2008 mean and annual cycle of temperature, salinity, and steric height in the global ocean from the Argo Program, Prog. Oceanogr., 82, 81–100, doi:10.1016/j.pocean.2009.03.004, 2009.

Sanford, T. B., Price, J. F., and Girton, J. B.: Upper-Ocean Response to Hurricane Frances (2004) Observed by Profiling EM-APEX Floats, J. Phys. Oceanogr., 41, 1041–1056, doi:10.1175/2010jpo4313.1, 2011.

Schade, L. R. and Emanuel, K. A.: The ocean's effect on the intensity of tropical cyclones: Results from a simple coupled atmosphere-ocean model, J. Atmos. Sci., 56, 642–651, doi:10.1175/1520-0469(1999)056<0642:TOSEOT>2.0.CO;2, 1999.

Uhlhorn, E. W. and Shay, L. K.: Loop Current Mixed Layer Energy Response to Hurricane Lili (2002), Part I: Observations, J. Phys. Oceanogr., 42, 400–419, doi:10.1175/jpo-d-11-096.1, 2011.

Uhlhorn, E. W. and Shay, L. K.: Loop Current Mixed Layer Energy Response to Hurricane Lili (2002), Part II: Idealized Numerical Simulations, J. Phys. Oceanogr., 43, 1173–1192, doi:10.1175/jpo-d-12-0203.1, 2013.

Wada, A. and Usui, N.: Importance of tropical cyclone heat potential for tropical cyclone intensity and intensification in the Western North Pacific, J. Oceanogr., 63, 427–447, doi:10.1007/s10872-007-0039-0, 2007.

Wang, X. D., Han, G. J., Qi, Y. Q., and Li, W.: Impact of barrier layer on typhoon-induced sea surface cooling, Dyn. Atmos. Oceans, 52, 367–385, doi:10.1016/j.dynatmoce.2011.05.002, 2011.

Wu, C. C., Lee, C. Y., and Lin, I. I.: The effect of the ocean eddy on tropical cyclone intensity, J. Atmos. Sci., 64, 3562–3578, doi:10.1175/Jas4051.1, 2007.

14

The surface thermal signature and air–sea coupling over the Agulhas rings propagating in the South Atlantic Ocean interior

J. M. A. C. Souza[1,*], B. Chapron[1], and E. Autret[1]

[1]Laboratoire d'Oceanographie Spatiale (LOS), IFREMER, Centre Brest, 29280, Plouzané, France
[*]now at: Department of Oceanography, School of Ocean and Earth Science Technology (SOEST), University of Hawaii, 1000 Pope Rd., MSB, Honolulu, 96822 HI, USA

Correspondence to: J. M. A. C. Souza (jsouza@soest.hawaii.edu)

Abstract. The surface signature of Agulhas rings propagating across the South Atlantic Ocean is observed based on three independent data sets: Advanced Microwave Scanning Radiometer for the Earth Observing System/Tropical Rainfall Measuring Mission (TRMM) Microwave Imager (TMI) (TMI/AMSR-E) satellite sea surface temperature, Argo profiling floats and a merged winds product derived from scatterometer observations and reanalysis results. A persistent pattern of cold (negative) sea surface temperature (SST) anomalies in the eddy core, with warm (positive) anomalies at the boundary, is revealed. This pattern contrasts with the classical idea of a warm core anticyclone. Taking advantage of a moving reference frame corresponding to the altimetry-detected Agulhas rings, modifications of the surface winds by the ocean-induced currents and SST gradients are evaluated using satellite SST and wind observations. As obtained, the averaged stationary thermal expression and mean eddy-induced circulation are coupled to the marine atmospheric boundary layer, leading to surface wind anomalies. Consequently, an average Ekman pumping associated with these mean surface wind variations consistently emerges. This average Ekman pumping is found to explain very well the SST anomaly signatures of the detected Agulhas rings. Particularly, this mechanism seems to be the key factor determining that these anticyclonic eddies exhibit stationary imprints of cold SST anomalies near their core centers. A residual phase with the maximum sea surface height (SSH) anomaly and wind speed anomaly is found to the right of the mean wind direction, apparently maintaining a coherent stationary thermal expression coupled to the marine atmospheric boundary layer.

1 Introduction

The Agulhas rings are important features of the South Atlantic circulation, with particular relevance for the great ocean conveyor belt. It is believed that virtually all the upper layer overturning circulation in the Atlantic originates from Indian Ocean leakage through Agulhas rings. Although previous studies have estimated the Agulhas rings general characteristics and pathways in the South Atlantic (e.g., Dencausse et al., 2010; Souza et al., 2011a), these do not usually include their thermal signature on the surface. Moreover, studies that have observed the Agulhas rings at sea (e.g., van Aken et al., 2003) are usually restricted to newly formed eddies and do not represent their modifications along their lifetimes. The eddy surface thermal signatures and their evolution over time have important consequences on the air–sea fluxes, impacting the overlying marine atmospheric boundary layer (MABL) with possible consequences on the regional atmospheric circulation.

As now available, long-term observations of concurrent satellite measurements of sea surface height (SSH), sea surface temperature (SST), and scatterometer surface vector winds enable detailed studies of the ocean–atmosphere coupling (e.g., Chelton et al., 2011; Siegel et al., 2011). High-resolution SSH fields obtained by merging measurements from several operating satellite altimeters help detect and track coherent rotating mesoscale ocean structures and better quantify their dynamics. As has been reported (e.g., Kudryavtsev et al., 2005), the near-surface wind field adjustment to SST changes across oceanic eddies occurs within an internal boundary layer developing on the downwind side of the sea front. Statistically, observed SST-induced perturbations on the surface wind stress or the scatterometer wind curl and divergence fields have been found to be linearly well correlated with small scale perturbations in the crosswind and downwind SST gradients, respectively (e.g., Chelton et al., 2001, 2007; O'Neill et al., 2005, 2010). Moreover, important consequences on cloud fraction and water content of the MABL, and a reduction in rainfall have been observed for cyclones in the Southern Ocean by Frenger et al. (2013).

Considering these ocean–atmosphere interactions, Businger and Shaw (1984, see the schematic diagram in their Fig. 10) earlier sketched several interactions of eddy SST imprints with the atmospheric boundary layer, and numerous studies have clearly evidenced air–sea couplings over oceanic mesoscale coherent structures (e.g., Chelton et al., 2001, 2007; Small et al., 2008). As summarized in Small et al. (2008), the main coupling mechanisms are related to (i) changes in the near-surface stability and surface stress, (ii) vertical transfer of momentum from higher atmospheric levels to the surface due to an increase of the turbulence in the boundary layer, (iii) secondary circulations associated with perturbations in the surface atmospheric pressure over the SST fronts and (iv) the impact of the oceanic eddy currents on the net momentum transferred between the atmosphere and the ocean.

The objective of this study is to characterize the surface thermal signature of Agulhas rings propagating across the South Atlantic Ocean and the associated wind perturbation. An air–sea coupling mechanism over the oceanic mesoscale eddies emerges from the relationship observed between the wind and SST patterns.

To analyze the air–sea coupling over Agulhas rings, the present work takes advantage of a moving reference frame (Lagrangian) corresponding to altimetry-detected mesoscale Agulhas rings propagating in the South Atlantic basin. This approach permits the evaluation of the eddy-tracked average modifications to surface winds by the sea surface temperature (SST) gradients, using satellite SST and wind observations. Agulhas rings are the largest mesoscale eddies in the world, with large radii (165 ± 69 km) and durations (497 ± 233 days) (see Souza et al., 2011a). Accordingly, the Lagrangian approach can easily be implemented to reveal the persistent structures of SSH, SST and wind anomalies using standard medium-resolution satellite products, as presented in Sects. 2 and 3. Our study follows earlier investigations on the modification of the surface winds over Gulf Stream rings (Park et al., 2006), but benefit from the higher temporal sampling of surface wind and SST information over isolated altimetry-detected rings.

Three independent data sets – TMI/AMSR-E SST satellite observations, Argo profiling floats and a merged wind product derived from scatterometer measurements and a reanalysis – are used to describe the imprint of Agulhas rings on the SST field in Sect. 4. The evidence of the mean air–sea coupling over the detected rings is presented in Sect. 5, clearly exhibiting the mean eddy/wind interactions. Maintaining the expectation that stationary thermal expression and relative air–water velocity are well expressed in the near-surface marine atmospheric boundary layer, the corresponding average Ekman pumping associated with the mean surface wind variations can then be evaluated. As revealed, the persistent eddy-induced Ekman pumping compares very well with the observed SST anomalies, as discussed in Sect. 6. Conclusions derived from the present analysis follow in Sect. 7.

2 Data

For our purposes, wind information from a blend of Quick Scatterometer (QuikSCAT) observations and reanalysis results, gridded SSH from Archiving, Validation and Interpretation of Satellite Oceanographic (AVISO) data, blended TMI/AMSR-E SST over a four-year period (between January 2005 and December 2008) and Argo float temperature profiles are used.

2.1 Blended wind fields

The blended wind product used in the present study is available for download in NetDCF format at (ftp://ftp.ifremer.fr/ifremer/cersat/products/gridded/psi-concentration/), and a complete description is provided by Bentamy et al. (2007). To enhance the spatial and temporal resolutions ($0.25°$ at synoptic times 00:00, 06:00, 12:00 and 18:00 UTC), a method has been implemented to blend the remotely sensed wind observations derived by the satellite QuikSCAT scatterometer and by three Special Sensor Microwave Imagers (SSM/I) on board Defense Meteorological Satellite Program (DMSP) satellites F13, F14 and F15 with the European Centre for Medium-Range Weather Forecasts (ECMWF) analyses. This provides a uninterrupted, high-resolution (~ 25 km) data set of wind velocities over the eddy tracks during the analyzed period.

2.2 AVISO SSH data

Four years of SSH data from the AVISO reference product were used to identify and track the Agulhas rings. It corresponds to a merged satellite product, projected on a $1/3°$ horizontal resolution Mercator grid, in time intervals of 7 days. Sea level anomalies (SLA) were obtained by subtracting the temporal mean SSH (2006–2009) from each grid point. A low-pass Hanning filter with a window of 175 days was used to remove seasonal and inter annual signals. Since the focus of the present study is on the mesoscale oceanic eddies, a Lanczos filter was applied to eliminate variability with length scales larger than 1000 km.

2.3 TMI/AMSR-E SST fields

An optimally interpolated SST (OI-SST) product, combining TRMM Microwave Imager (TMI) and Advanced Microwave Scanning Radiometer for EOS (AMSR-E) data sets, is used to provide the SST over the Agulhas rings. This product provides daily "cloud free" SST maps, with approximately 25 km spatial resolution – compatible with the wind data set. The data can be downloaded and details about the optimal interpolation procedural can be found at www.remss.com. This data set is produced by Remote Sensing Systems and is sponsored by the National Oceanographic Partnership Program (NOPP), the NASA Earth Science Physical Oceanography Program and the NASA Measures Discover Project.

2.4 Argo profiling floats

Argo is a global ocean observing system for the 21st century (www.argo.net) composed of approximately 3000 profiling floats, used to create $3° × 3° × 10$-day resolution array between $60°$ S to $60°$ N. These autonomous floats measure temperature and salinity down to 2000 m depth, every 10 days. A complete description of the Argo program and access to the data can be obtained at http://www.ARGOucsd.edu/. In the present study, we take advantage of the Argo profiles colocated with the Agulhas rings in space and time, and a derived mean vertical temperature anomaly structure, derived by Souza et al. (2011b).

3 Methods

The present work takes advantage of a Lagrangian point of view along the eddy tracks to quantify the anomalies in the surface variables (SSH, SST and wind) over the large mesoscale Agulhas rings propagating in the South Atlantic basin.

3.1 Eddy identification and tracking algorithm

An automated method based on the SLA (Chaigneau et al., 2009) was used to identify and track the Agulhas rings. The eddy identification process is performed at each time step, or SLA map, through two stages: (1) the identification of local SLA modulus maxima corresponding to the eddy centers and (2) the selection of closed SLA contours associated with each eddy center. The outermost contour embedding only one center is considered as the eddy edge. This method is similar to the one applied by Chelton et al. (2011), and showed improved results and performance in terms of processing time when compared to other classical methods (Souza et al., 2011a).

A total of 16 Agulhas rings with durations larger than six months were identified. The results from the tracking algorithm together with the general ring characteristics are discussed by Souza et al. (2011a). A good agreement with previous work was achieved, with "typical" Agulhas rings shed every 2–3 months and with a diameter of 150–200 km (van Sebille et al., 2010; van Aken et al., 2003). It is important to emphasize that due to the "noisy" SSH in the Agulhas Current retroflexion region, the automated method does not identify the structures in the exact moment of their formation. Therefore, the derived SST and wind anomaly signatures correspond to Agulhas rings propagating in the South Atlantic interior after leaving their formation region. This is important since the SST signature of newly formed Agulhas rings is in fact warm (e.g., Gladyshev et al., 2008; Arhan et al., 2011). However, when the structures propagate northwestward into the warmer South Atlantic basin their SST anomalies in relation to the surrounding waters become cold, as revealed in the next section.

3.2 Data cropping and spatial high-pass filtering

The SST anomaly (T') and SST-driven wind speed variability, or wind anomaly (V'), were isolated by removing the large spatial scale variability using a "loess" smoothing function with half-power filter cutoffs of $5°$ latitude by $5°$ longitude. The cutoff wavelengths were chosen to better isolate the influence of the mesoscale eddies from the mean field and larger-scale processes, such as propagating Rossby waves. After the eddy identification and tracking processes, the ring positions are linearly interpolated from the 7-day SLA derived axis to daily positions compatible with the SST and wind data sets. For the comparison between wind and SST, these data sets were linearly interpolated to a 25 km resolution grid in a region of 700 km (north–south and west–east) centered over the eddy core position at each time are determined from the altimetry. The properties were then projected on a horizontal grid normalized by the eddy diameter at each time step. That way, the obtained spatial patterns are isolated from the variability of the eddy diameters between distinct structures and over time.

The mean wind direction over each ring at each moment was calculated and used as a reference. For the comparison between the wind and SST, all data fields were rotated to have the x axis oriented to the mean wind at each time. Crosswind and downwind SST gradients ($(\partial T/\partial c)'$ and $(\partial T/\partial d)'$, respectively) and the wind divergence and vorticity ($(\nabla.\boldsymbol{u})'$ and $(\nabla \times \boldsymbol{u})'$, respectively) were calculated at each grid point following the method described by O'Neill et al. (2010).

4 Surface thermal signature of the Agulhas rings

The Agulhas rings are strong anticyclonic mesoscale eddies, referred to as the largest mesoscale eddies in the world by Biastoch et al. (2008). They play an important role in the transport of Indian Ocean waters to the South Atlantic. These anticyclones are usually associated with warm (positive) T' in their formation region, near the Agulhas Current retroflexion

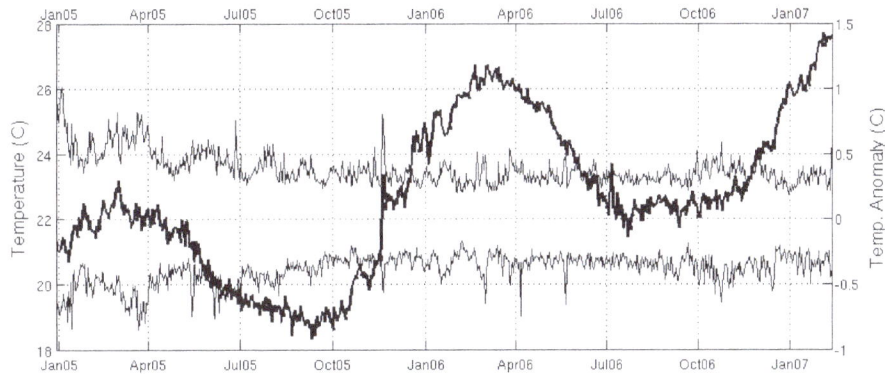

Figure 1. Time series of the maximum (warm) and minimum (cold) T' inside the eddy (thin lines) and absolute SST at the eddy center (thick line) for an Agulhas ring tracked between January 2005 and February 2007.

in the southern tip of the African continent (e.g., Arhan et al., 2011; Schmid et al., 2003; Garzoli et al., 1999). As an example, Arhan et al. (2011) and Gladyshev et al. (2008) observed anticyclonic eddies with core temperatures between 11.8 °C and 12.5 °C approximately at 43–40° S embedded in climatological SSTs lower than 10 °C.

But when these rings leave their formation region and move to lower latitude warmer waters, the positive signature on the T' becomes less clear. Moreover, although the waters present in the eddy cores are trapped and relatively laterally isolated from the surrounding ocean (Souza et al., 2011b), fluxes in the air–sea interface and the base of the mixed layer are expected to act modifying the upper layer temperature.

Taking as an example an Agulhas ring tracked between December 2004 and February 2007, it is possible to observe in Fig. 1 that while the absolute SST at the eddy center (determined from the SLA) present a seasonal cycle with a warm trend as the ring propagates northwestward, cold and warm signatures in T' are present inside the eddy and decrease together over time without a clear seasonal influence. This is similar to the behavior observed for the SLA at the eddy core by Souza et al. (2011a). While these authors obtained a decrease rate of 9.33×10^{-5} m day^{-1} for the SLA, a linear fit to the T' modulus results in an amplitude decrease rate of 3.1×10^{-4} °C day^{-1}. Observing the anomaly curves in Fig. 1, one can see that this decrease is, in fact, faster in the early days of the eddy lifetime and becomes smoother after the first year. The absolute values for T' are typically lower than 0.5 °C after the first year of the eddy formation. This makes it difficult to track the eddy signature in SST satellite images, especially when it approaches the warm waters of the Brazil Current.

The T' present similar magnitude cold and warm values inside the ring core that stay in phase during the eddy lifetime. However, in contrast to the warm core observed near the formation region by Arhan et al. (2011), the Hovmoller diagram of Fig. 2 reveals a persistent cold (negative) T' at

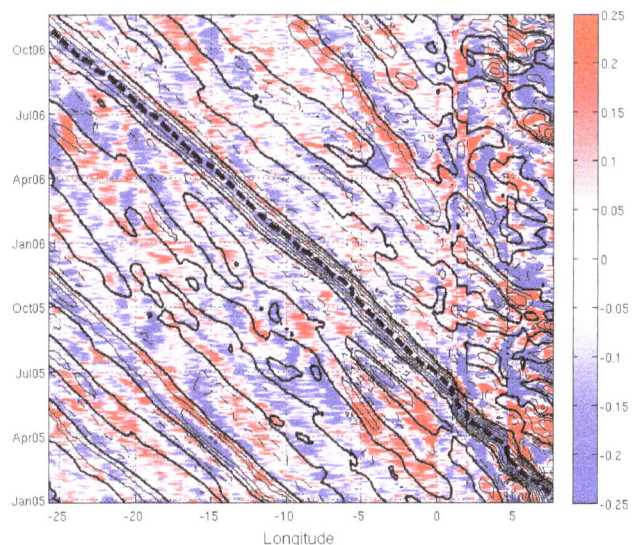

Figure 2. Hovmoller diagram of the SLA and SST anomaly (T') along the trajectory of an Agulhas ring first observed on 24 January 2005. The colors correspond to the SST anomalies, while the black contours present the SLA with 0.1 m interval. The thick black lines are the zero SLA contours, while the thick black dashed line present the trajectory of the eddy center. It is possible to observe that the eddy core is generally associated with cold (negative) SST anomalies.

the eddy center identified from the altimetry, and warm (positive) anomalies near the eddy boundaries.

To analyze these anomalies taking into account all the 16 Agulhas rings identified in the present study, let us first define the eddy "thermal center" as the position of the maximum T' modulus closest to the SLA maximum. The time series of the mean T' at the eddy thermal centers (Fig. 3) show that these negative (cold) anomalies are persistent during observed period, with a mean value of approximately −0.56 °C. The range presented by the thin gray lines in Fig. 3, with the maximum and minimum values of T' between the different rings, indicate that all observed eddies presented persistent cold cores.

Such pattern of cold T' in the eddy core is also present in the mean Agulhas ring vertical structure calculated from the Argo float temperature profiles by Souza et al. (2011b) (Fig. 4) for the same 16 rings analyzed in the present work. Focusing on the top layer of the water column, the mean Argo temperature anomaly section shows an eddy core dominated by cold (negative) T' while the warm (positive) T' is shifted to the left (west) of the center. The warm anomaly occupies the eddy core with depth due to the tilt between the temperature and velocity anomaly vertical structures. Although the mean vertical structure obtained from the Argo floats is the product of a series of simplifications used to combine profiles from different stages of the eddy lives, and it is restricted to a zonal section, it agrees with the mean spatial pattern of the eddy surface thermal signature as obtained from the TMI/AMSR-E SST product (Fig. 5). This mean surface signature also shows a cold T' core with warm anomalies in the boundary, especially on the southern portion of the eddy. Although a large standard deviation (sd) is observed for T' this mean picture is robust, especially in the eddy core as shown by the ratio between the eddy averaged mean and sd of T' (see the Supplement). A similar distribution in the blended winds shows weaker (negative) V' close to the eddy center and stronger (positive) V' at the eddy boundary. Such a relationship between V' and T' has being previously observed by several authors (a review is provided by Small et al., 2008). Thus, the obtained distribution of V' reinforces the observed spatial pattern in T'. Moreover, it suggests that a coupling mechanism between the upper ocean and the MABL is responsible for sustaining the observed structures.

5 Evidence of air–sea coupling over the Agulhas rings

As averaged, the mean spatial patterns of V' and T', Fig. 5, exhibit a correlation of 0.61 with an apparent phase shift. Simply shifting the wind field position in relation to the SST and recalculating the resulting correlations, a maximum value of 0.87 is obtained for a spatial shift on the order of one quarter of the ring radius (~ 50 km) to the right of the mean wind direction (downward). Since the Agulhas rings propagation region is dominated by the westerly winds, this is the same to say the wind is shifted equatorward (approximately 0.5°) to the T' center. A spatial shift between the wind and the SST was already reported in previous studies. Considering four distinct regions (Kuroshio Current, Gulf Stream, Brazil–Malvinas confluence zone and the Agulhas Return Current), O'Neill et al. (2010) estimated a mean shift of 1° (~ 110 km). In our case, two processes contribute to the air–sea coupling over the rings: (a) the wind accelerates (decelerates), with slight changes in wind direction, over warm (cold) SST anomalies and (b) the impact of the oceanic eddy velocities on the relative air–water velocity on diametrically

opposite sides of the eddies systematically act to modify the wind field.

Indeed, the spatial pattern for the wind perturbations in Fig. 5 suggests an influence of the oceanic eddy velocities, reducing (augmenting) the wind in the upper (lower) part of the ring. It is important here to remember that the influence of the ocean currents was removed from the scatterometer winds prior to the analysis.

Accordingly, a coincidence in the spatial patterns of $(\nabla . u)'$ and $(\partial T / \partial d)'$ is expected and observed, Fig. 6a. Taking into account the same spatial shift observed between V' and T', the correlation changes from 0.29 to 0.83. For the relationship between $(\nabla \times u)'$ and $(\partial T / \partial c)'$ (Fig. 6b), a shift of 25 km (1 grid cell) enhances the anti-correlation from -0.39 to -0.67. However, the present results are statistically robust and favor the dominant effect of the eddy-induced circulation to explain both the weaker correlation between $(\nabla \times u)'$ and $(\partial T / \partial c)'$, and the very large linear regression coefficient between $(\partial T / \partial d)'$ and shifted $(\nabla . u)'$, as discussed in the next section.

5.1 Linear relationship between the wind and the SST

Linear relationship between the wind stress curl and divergence, and the crosswind and downwind SST gradients were originally observed by Chelton et al. (2001) over energetic regions of the world ocean, where strong and persistent SST fronts are present. However, it is still not clear if the same relationships hold for the less energetic ocean interior, with the sporadic occurrence of eddies. This is the case of the Agulhas rings that propagate across the South Atlantic interior. Following this idea, linear fits between the binned properties were obtained and their respective slopes calculated. To investigate the relationship between the wind and SST, the wind properties V', $(\nabla \times u)'$ and $(\nabla . u)'$ were binned in function of T', $(\partial T / \partial c)'$ and $(\partial T / \partial d)'$, respectively. All the data including the complete time span of all 16 identified Agulhas rings were used, without spatial shift correction.

The linear relationship between T' and V' in Fig. 7 shows a similar slope (0.31) as the relationship observed for the Gulf Stream region (0.27) by O'Neill et al. (2010). The T' values are concentrated in the range ± 1 °C, corresponding to perturbations in V' of ± 0.6 m s^{-1}. These values are smaller than observed by previous authors, reflecting the fact these rings represent a weaker perturbation in the SST field when compared to western boundary current frontal systems. But the comparable values of the slopes indicate a similar coupling between the SST and the wind speed over the structures, suggesting the same mechanisms are dominant.

The linear fits between $(\nabla . u)' \times (\partial T / \partial d)'$ and $(\nabla \times u)' \times (\partial T / \partial c)'$ in Fig. 8 showed stronger slopes (1.47 and -1.26) than observed by previous authors. O'Neill et al. (2010) obtained slope values between 0.44 and 0.73 for $(\nabla . u)' \times (\partial T / \partial d)'$ and between -0.21 and -0.45 for $(\nabla \times u)' \times (\partial T / \partial c)'$. This indicates stronger coupling between the wind

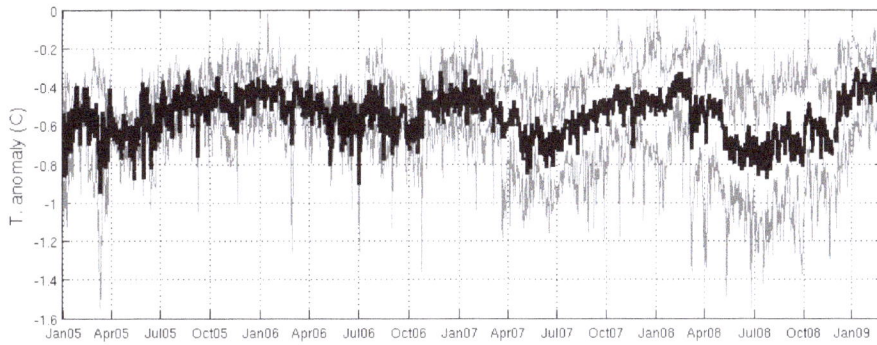

Figure 3. Time series of the mean T' in the eddy thermal centers (thick line) showing the persistence of negative (cold) values. The thin gray lines indicate the maximum and minimum values of T' at the eddy thermal centers between the different rings identified at each time, showing that all observed structures presented cold cores.

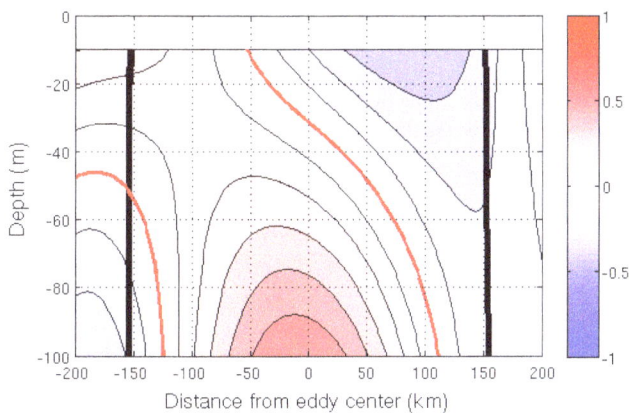

Figure 4. Top 100 m of the Agulhas rings mean T' vertical structure (c.i. 0.1 °C) obtained from the Argo float vertical profiles by Souza et al. (2011b) (c.i.: contour interval). The red contours indicate the 0 °C T', while the thick black lines delimit the water trapped region following the criteria by Flierl (1981). The cold (negative) T' dominates the eddy core near the surface, while the warm (positive) T' is shifted to the left (west).

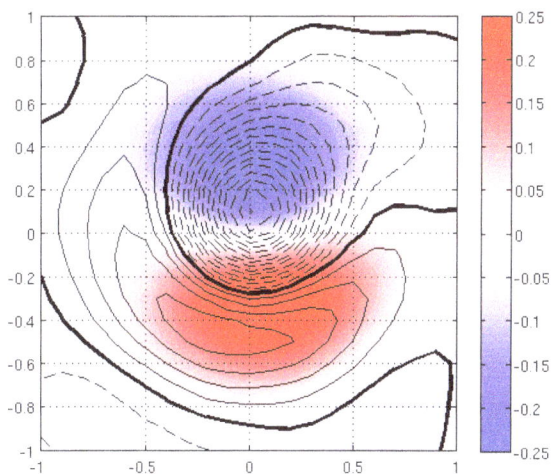

Figure 5. Map of the mean perturbation wind speed (V' color contours) and T' (black contours – c.i. 0.01 °C) obtained from the 16 Agulhas rings observed between January 2005 and December 2008. The thick black line represents the 0 °C T', the continuous lines positive values and the dashed lines negative values of T'. The figures' axes represent the distances from the eddy centers normalized by the eddy diameters.

and the SST gradients over the Agulhas rings. In particular, the case of the $(\nabla.\boldsymbol{u})' \times (\partial T/\partial d)'$ shows a high slope, which means the wind convergence presents a strong response to the SST structure. The observed differences in the linear fit slope indicate that characteristics other than the SST may be influencing the MABL winds. Aspects such as the mean atmospheric circulation and latitude impact the air–sea coupling and are analyzed in the next section.

5.2 Mean characteristics influencing the coupling

Several physical mechanisms influence the air–sea coupling over oceanic mesoscale structures. Small et al. (2008) propose as main processes the modification of the near-surface stability, the mixing of momentum, heat and humidity from higher levels into the MABL due to the increased turbulence, the formation of secondary circulation cells related to

perturbations in the pressure field and the impact of oceanic eddy velocities on the surface shear. Diverse environmental properties influence these mechanisms, altering their relative balance over the eddies.

Using the differences between the environmental conditions influencing the 16 Agulhas rings identified in the present study it is possible to investigate the importance of the environmental properties on the coupling. To achieve this, the correlations and slopes between $V' \times T'$, $(\nabla.\boldsymbol{u})' \times (\partial T/\partial d)'$ and $(\nabla \times \boldsymbol{u})' \times (\partial T/\partial c)'$ of each ring were compared to their correspondent mean wind speed (V), latitude, eddy diameter and amplitude. The results are shown in Table 1, where the correlations statistically relevant to the 95 % level are emphasized. It is observed that the latitude is

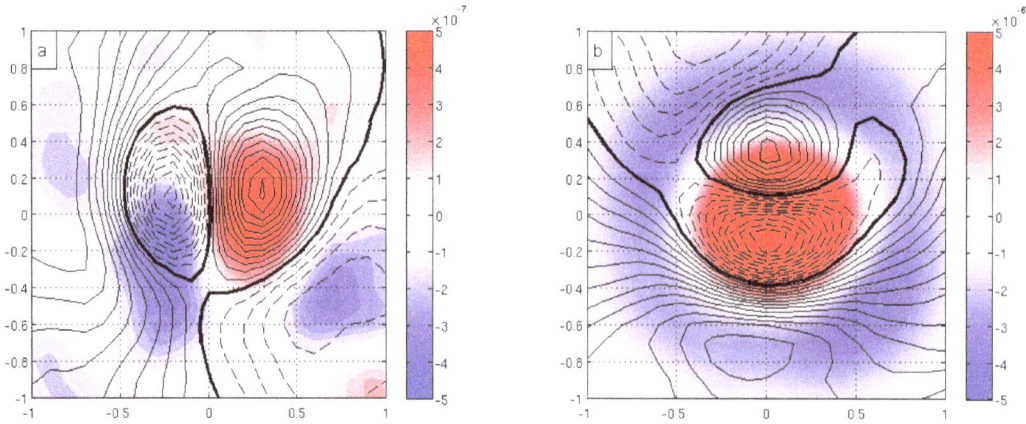

Figure 6. (a) $(\nabla.\boldsymbol{u})'$ as color contours and $(\partial T/\partial d)'$ as black contours – c.i. 1×10^{-7} °C per 100 km. **(b)** $(\nabla \times \boldsymbol{u})'$ (color contours) and $(\partial T/\partial c)'$ (black contours – c.i. 1×10^{-7} °C per 100 km). The thick black lines represent the 0 °C per 100 km contours, the continuous lines positive values and the dashed lines negative values of T' gradient. The figures' axes represent the distances from the eddy centers normalized by the eddy diameters.

Table 1. Correlations between the slopes and correlation factors that indicate the air–sea coupling over the 16 Agulhas rings observed in the present study. The emphasized values correspond to the correlations statistically relevant to the 95 % level.

	Slope			Correlation		
	$T' \times V'$	$-(\nabla \times \boldsymbol{u})' \times (\partial T/\partial c)'$	$(\nabla.\boldsymbol{u})' \times (\partial T/\partial d)'$	$T' \times V'$	$-(\nabla \times \boldsymbol{u})' \times (\partial T/\partial c)'$	$(\nabla.\boldsymbol{u})' \times (\partial T/\partial d)'$
$\bar{V}$	**60 %**	**74 %**	25 %	35 %	**45 %**	**50 %**
Mean Latitude	**−70 %**	**−56 %**	−11 %	**−52 %**	**−51 %**	**−57 %**
Mean Diameter	−19 %	−22 %	−15 %	−2 %	−4 %	−7 %
Mean Amplitude	11 %	−14 %	−3 %	11 %	−21 %	12 %

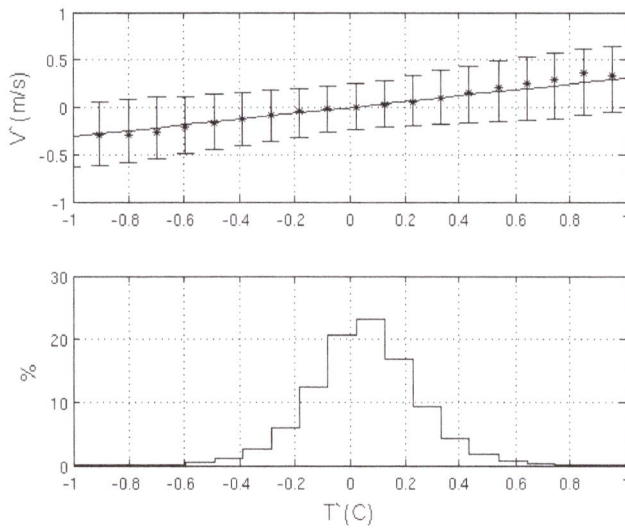

Figure 7. Binned scatter plots of the V' as function of T'. The bin averages were computed using all the data points from the 16 observed Agulhas rings. The points and error bars represent the mean within each bin and the standard deviation, and the lines are the linear fits. The histogram in the lower panel presents the relative distribution of the data between the bins.

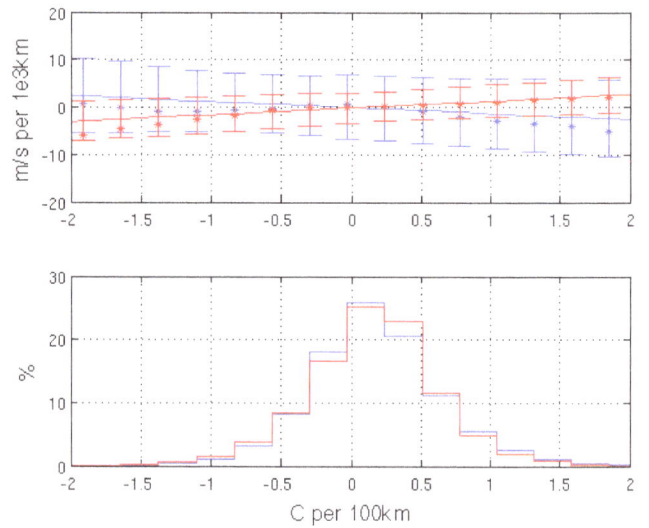

Figure 8. Binned scatter plots of the $(\nabla.\boldsymbol{u})' \times (\partial T/\partial d)'$ in red and $(\nabla \times \boldsymbol{u})' \times (\partial T/\partial c)'$ in blue. The bin averages were computed using all the data points from the 16 observed Agulhas rings. The points and error bars represent the mean within each bin and the standard deviation, and the lines are the linear fits. The histogram in the lower panel presents the relative distribution of the data between the bins.

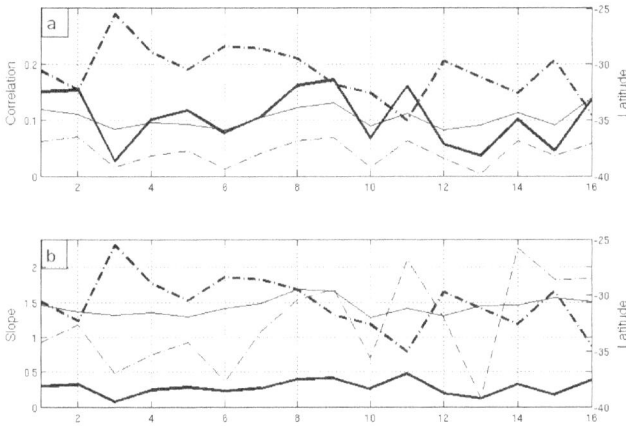

Figure 9. Correlations (**a**) and slopes (**b**) between $T' \times V'$ as thick continuous lines, $(\nabla.u)' \times (\partial T/\partial d)'$ as thin continuous lines and $-(\nabla \times u)' \times (\partial T/\partial c)'$ as thin dash-dotted lines and the mean Latitude (thick dash-dotted line) for the 16 Agulhas rings identified in the present study. A remarkable anti-correlation of the correlations and slopes with the latitude is observed.

negatively correlated with both the slopes and correlations (Fig. 9), while the mean wind intensity is positively correlated (Fig. 10). Since in this case we want to explore the influence of the mean atmospheric circulation on the coupling, the mean wind is defined as the temporal mean over the eddies. The eddy diameters and amplitudes presented low correlations with the analyzed properties that are not statistically relevant.

The influence of the latitude on the coupling can be explained through its impact on the thermal wind balance over the rings. A "temperature island" effect occurs due to the modifications in the heat flux over the SST anomalies. This heat flux generates pressure gradients in the lower MABL that create secondary cells of circulation. Following the thermal wind equation ($u'_g = \nabla P/\rho f$), the magnitude of the geostrophic wind perturbation (u'_g) due to this pressure gradient (∇P) is inversely proportional to the Coriolis factor (f). Since f is larger in higher latitudes, u'_g is smaller, that is, the pressure gradient mechanism is less effective. Chelton et al. (2004) also observed the influence of the latitude in the air–sea coupling and affirmed that the SST influence on the mean wind stress is mostly restricted to regions poleward of 40° of latitude and the tropics, though the authors associated this effect to the persistence of the SST fronts over time. In the present study, we show that such effects are observed in lower latitudes as well ($\sim 30°$ S) and transient fronts, which can also be interpreted from the results of O'Neill et al. (2010).

The influence of the wind velocity is associated with a different process. It promotes modifications in the near-surface stability or, in the case of the wind curl, generates vertical velocities in the ocean that impact the SST (Ekman pumping). While the first case is the subject of extensive research

(e.g., Kudryavtsev et al., 1996; Park and Cornillon, 2002; Liu et al., 2007; O'Neill, 2012) and will not be treated in the present work, the second process explains an important feedback mechanism from the MABL into the ocean and is discussed in the next section.

6   Induced Eddy–Ekman pumping

As statistically revealed, wind modifications, and consequently wind stress, could possibly enhance vertical velocities through an Ekman pumping mechanism in the ocean. This further impacts the resulting evolution of the SST field and again the surface stress. Such a coupled feedback has been less explored than the impacts of the SST gradients on the surface winds. Liu et al. (2007) discuss the feedback of the wind speed anomalies on the ocean in terms of modifications in the wind stress, impacting the surface currents. The authors argue that since the neutral wind vorticity anomalies are small compared to those of the ocean currents, the ocean should dominate the mesoscale coupling in the long term. The feedback on the SST is then presented as function of the modifications in the surface latent heat flux. The role of the surface heat fluxes was further emphasized by Haack et al. (2008). From another point of view, using an idealized ocean–atmosphere coupled model to study the shelf-break frontogenesis, Chen et al. (2003) showed a positive feedback mechanism that strengthens both the wind and the front. Their results indicate that although the latent heat flux appears to be important, it is not an essential coupling mechanism. Jin et al. (2009) used the empirical relationship between the wind and SST presented by Chelton et al. (2007), leading to a simple representation of the mesoscale ocean–atmosphere coupling in an idealized upwelling system where the SST-induced wind curl intensification strengthen the offshore upwelling through Ekman pumping.

Under the present analysis, a more direct estimation can be made. To quantify the mean eddy–wind interactions, an f plane approximation in the eddy-centered reference frame is considered to estimate the mean Ekman vertical velocities (w_{EK}), also taking into account the oceanic geostrophic eddy vorticity (ζ) changes obtained from the geostrophic velocities derived from the observed SLA in the rings area (e.g., Stern, 1965):

$$w_{EK} = \frac{1}{\rho_0} \left(\underbrace{\frac{\nabla \times \tau}{f + \zeta}}_{\text{term 1}} + \underbrace{\frac{\nabla \zeta}{(f + \zeta)^2} \times \tau}_{\text{term 2}} \right), \qquad (1)$$

where ρ_0 is the averaged oceanic mixed layer density, obtained considering a mean salinity of 35 and the observed SST, and τ is the wind stress simply obtained from a bulk relationship:

$$\tau_{x,y} = \rho_{AIR} C_D V |V|, \qquad (2)$$

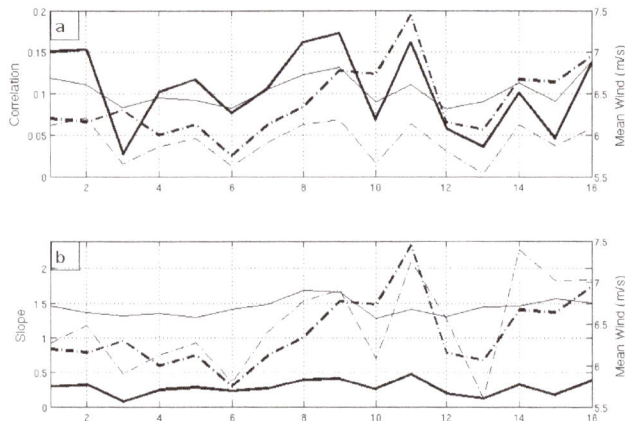

Figure 10. Correlations (**a**) and slopes (**b**) between $T' \times V'$ (thick continuous line), $(\nabla . \boldsymbol{u})' \times (\partial T/\partial d)'$ (thin continuous line) and $-(\nabla \times \boldsymbol{u})' \times (\partial T/\partial c)'$ (thin dash-dotted line) and the mean wind speed (thick dash-dotted line) for the 16 Agulhas rings identified in the present study. A remarkable concordance between the correlations and slopes with the wind speed is observed.

where ρ_{AIR} is the air density, C_D is the drag coefficient and V is the measured surface wind. C_D was calculated following Yelland and Taylor (1996). It is important to notice that for the Ekman pumping calculation the effect of the ocean currents on the observed scatterometer winds was not removed since the movement in relation to the ocean surface is the relevant quantity.

The two terms of $\boldsymbol{w}_{\text{EK}}$ in Eq. (1) take into account the particular contribution of the resulting wind stress curl, through the measured surface wind curl anomalies (term 1), and the oceanic vorticity divergence (term 2). As presented in Fig. 11, $\boldsymbol{w}_{\text{EK}}$ estimations overlay very well with the T' field.

The high correlation (0.9) between the two fields is illustrated in Fig. 12. The linear fit obtained between these two fields presents a -0.04 slope. In this figure it is possible to observe that this linear relationship is strong in the region close to the eddy core, weakening with the distance from it. This means that this process holds only for the region under the influence of the eddy.

As clearly revealed, the Ekman pumping explains a persistent feedback mechanism linking the eddy-induced mean wind and SST anomalies on the Agulhas rings. Along their propagation in the South Atlantic basin, a residual persistent upward velocity acts to support a cooler temperature flux to sustain the mean observed cool anomalies in the core of the eddies. This mechanism is mostly driven by the observed resulting spatial changes in wind stress, but taking into account that the oceanic vorticity gradient improves the overall correlation between $\boldsymbol{w}_{\text{EK}}$ and T', particularly contributing to the enhancement of the pumping asymmetry along cross-wind mean directions. As such, this observed asymmetry might lead to weaker SSH signatures, also impacting the ring propagation and attenuation (Dewar and Flierl, 1987). In fact, the

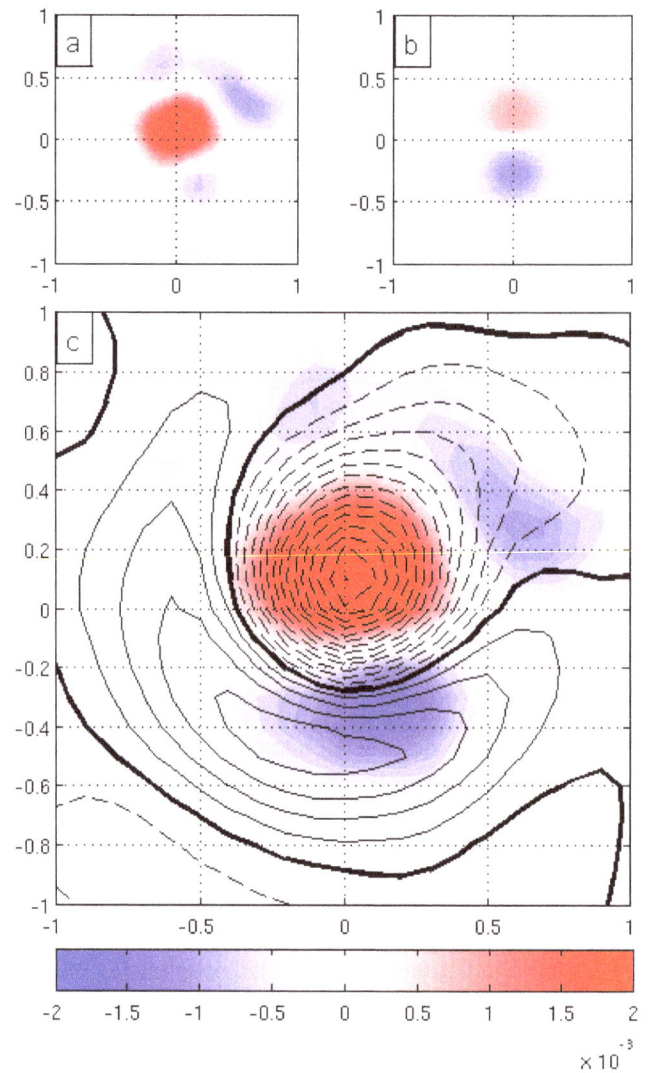

Figure 11. Results obtained for (**a**) term 1, (**b**) term 2 and (**c**) the total Ekman pumping velocity $\boldsymbol{w}_{\text{EK}}$ (m day^{-1}) for the mean estimated Agulhas ring. A remarkable concordance is observed between the mean vertical velocities (color contours) and the observed mean T' (black contours – c.i. 0.01 °C). The $\boldsymbol{w}_{\text{EK}}$ is dominated by the $(\nabla \times \boldsymbol{\tau})'$ (term 1), though the $(\nabla \times \boldsymbol{\zeta})'$ (term 2) makes an important contribution. The figures' axes represent the distances from the eddy center normalized by the eddy diameters.

relative contribution of term 2 of Eq. (1) to the total $\boldsymbol{w}_{\text{EK}}$ displayed in Fig. 13 clearly shows that this term is responsible for up to 95 % of the total vertical velocities in the lower (southern) portion of the ring. Considering the role of the mean eddy, term 2 accounts for, approximately, 30 % of the total $\boldsymbol{w}_{\text{EK}}$.

The feedback mechanisms through Ekman pumping have important consequences not only on the air–sea fluxes, but also on the primary productivity of the waters trapped inside the eddy. The present results support similar findings of the enhanced chlorophyll concentration inside an Agulhas ring

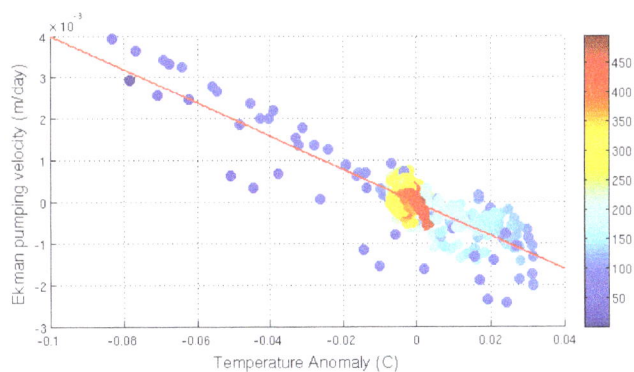

Figure 12. Relationship between the Ekman pumping vertical velocities (w_{EK}) and T'. The color code represent the distance from the eddy center (km), showing that the linear relationship between these two fields is stronger closer to the eddy core, as demonstrated by the linear regression illustrated by the red line.

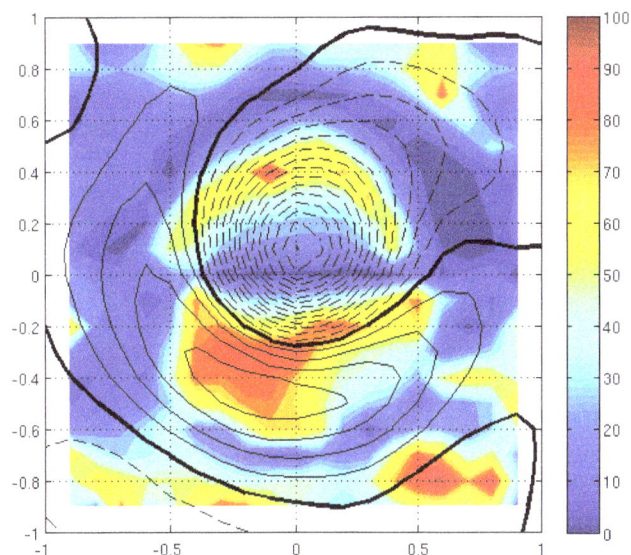

Figure 13. Relative contribution (%) of term 2 of Eq. (1) to the total Ekman pumping vertical velocity (w_{EK}) in a mean Agulhas ring and the observed mean T' (black contours – c.i. 0.01 °C). It is possible to ascribe the particular importance of this term to the asymmetry of w_{EK} and T' with higher contributions (up to 95 %) in the lower (southern) part of the eddy.

core as described by Lehahn et al. (2011), who observed a chlorophyll patch transported by an eddy along ~ 1500 km in ~ 11 months. Based on a ring first identified in April 2006, and using a particle transport model, these authors conclude that the water trapped inside the eddy core and changes in the mixed layer depth are responsible for the observed patches. The increase in the vertical mixing by particularly strong wind events is described as responsible for high frequency increments on the chlorophyll concentration. However, the present results revealed a persistent mean mechanism, sustaining positive w_{EK} at the eddy cores that are consistent with the observed high productivities. As showed in Fig. 10, the mean wind speed is in fact well correlated with both the slope and correlation of the $(\nabla \times \boldsymbol{u})' \times (\partial T / \partial c)'$, which explains the concordance between the chlorophyll concentration and the wind speed observed by Lehahn et al. (2011).

7 Conclusions

Using three independent data sets (TMI/AMSR-E SST fields, Argo profiling floats and a blended winds product), the present study reveals a particular surface thermal signature for the anticyclonic Agulhas rings as they propagate across the South Atlantic Ocean: cold T' in the eddy cores with warm anomalies at the boundaries. Consequently, in the interior of the South Atlantic basin the Agulhas rings are better described as T' dipoles, presenting strong positive and negative anomalies, than by the classical view of a warm core anticyclone that is consistent close to their formation region in the Cape Basin. These anomalies decrease in amplitude along the eddy tracks, but are still distinguishable during their whole lifetimes, estimated from the SLA.

A statistical analysis of the surface wind modifications over anticyclonic Agulhas rings is explored. The joint use of SSH, SST, and wind available products with improved spatiotemporal resolution helps to reveal persistent effects taking place over these long-lived large mesoscale rings. Impacts of their thermal and circulation surface signatures on the wind field were ascertained in the V', $(\nabla . \boldsymbol{u})'$ and $(\nabla \times \boldsymbol{u})'$ wind anomaly fields. The mean spatial patterns of V' and T' show remarkable correlation, maximum value of 0.92, with a spatial shift on the order of one quarter of the ring radius to the right of the mean wind direction. Since the Agulhas rings propagate in a region dominated by the westerly winds, this is to say that the wind anomaly is shifted poleward (approximately 0.5°) to the eddy thermal center.

A feedback mechanism explaining how the wind anomalies further influence the upper ocean thermal expression emerges. The wind stress curl $(\nabla \times \boldsymbol{\tau})'$ and ring vorticity gradient $(\nabla \times \zeta)'$ lead to an averaged Ekman pumping with a residual persistent upward velocity acting to support a cooler temperature flux at the oceanic mixed layer basis. This mechanism helps to sustain the mean observed cold anomalies and could be related to previous observations of high chlorophyll concentrations near the eddy cores. As revealed by the residual phase between the maximum SSH and cold SST anomalies, to the left of the mean wind direction, this residual Ekman pumping can affect the ring propagation and attenuation.

Acknowledgements. The authors would like to thank Patrice Klein, Clement de Boyer Montegut, Cecile Cabanes and Pierre-Yves Le Traon for the fruitful discussions we had about this work. We also acknowledge the AVISO service website maintained by CLS for making publicly available their sea surface height data. Argo data were collected and made freely available by the International Argo Program and the national programs that contribute to it (http://www.argoucsd.edu, http://argo.jcommops.org). João Marcos Souza was funded through an IFREMER postdoctoral fellowship.

Edited by: S. Josey

References

Arhan, M., Speich, S., Messager, C., Dencausse, G., Fine, R., and Boye, M.: Anticyclonic and cyclonic eddies of subtropical origin in the subantarctic zone of south of Africa, J. Geophys. Res., 116, C11004, doi:10.1029/2011JC007140, 2011.

Bentamy, A., Ayina, H.-L., Queffeulou, P., Croize-Fillon, D., and Kerbaol, V.: Improved near real time surface wind resolution over the Mediterranean Sea, Ocean Sci., 3, 259–271, doi:10.5194/os-3-259-2007, 2007.

Biastoch, A., Boning, C. W., and Lutjeharms, J. R. E.: Agulhas leakage dynamics affects decadal variability in the Atlantic overturning circulation, Nature, 456, 489–492, 2008.

Businger, J. A. and Shaw, W. J.: The response of the marine atmospheric boundary layer to mesoscale variations in sea-surface temperature, Dyn. Atmos. Oceans, 8, 267–281, 1984.

Byrne, D. A., Gordon, A. L., and Haxby, W. F.: Agulhas eddies: A synoptic view using Geosat ERM data, J. Phys. Oceanogr., 25, 902–917, 1995.

Chaigneau, A., Eldin, G., and Dewitte, B.: Eddy activity in the four major upwelling systems from satellite altimetry (1992–2007), Prog. Oceanogr., 83, 117–123, 2009.

Chelton, D. B., Esbensen, S. K., Schlax, M. G., Thum, N., Freilich, M. H., Wentz, F. J., Gentemann, C. L., McPhaden, M. J., and Schopf, P. S.: Observations of coupling between surface wind stress and sea surface temperature in the eastern tropical Pacific, J. Climate, 14, 1479–1498, 2001.

Chelton, D. B., Schlax, M. G., Freilich, M. H., and Milliff, R. F.: Satellite measurements reveal persistent small-scale features in ocean winds, Science, 303, 978–983, 2004.

Chelton, D. B., Schlax, M. G., and Samelson, R. M.: Summertime coupling between sea surface temperature and wind stress in the California Current System, J. Phys. Oceanogr., 37, 495–517, 2007.

Chelton, D. B., Schlax, M. G., and Samelson, R. M.: Global Observations of Nonlinear Mesoscale Eddies, Prog. Oceanogr., 91, 167–216, 2011.

Chen, D., Timothy, W., Tang, W., and Wang, Z.: Air-sea interaction at an oceanic front: Implications for frontogenesis and primary production. Geophys. Res. Lett., 30, 1745, doi:10.1029/2003GL017536, 2003.

Dencausse, G., Arhan, M., and Speich, S.: Routes of Agulhas rings in the southeastern Cape Basin, Deep-Sea Res I, 57, 1406–1421, 2010.

Dewar, W. K. and Flierl, G. R.: Some effects of the wind on rings, J. Phys. Oceanogr., 17, 1653–1667, 2007.

Flierl, G. R.: Particle motions in large-amplitude wave fields, Geophys. Astro. Fluid, 18, 39–74, 1981.

Frenger, I., Gruber, N., Knutti, R., and Münnich, M.: Imprint of Southern Ocean eddies on winds, clouds and rainfall, Nat. Geosci., 6, 608–612, doi:10.1038/NGEO1863, 2013.

Garzoli, S. L., Richardson, P. L., Duncombe Rae, C. M., Fratantoni, D. M., Goni, G. J., and Roubicek, A. J.: Three Agulhas rings observed during the Benguela Current Experiment, J. Geophys. Res., 104, 20971–20985, 1999.

Gladyshev, S., Arhan, M., Sokov, A., and Speich, S.: A hydrographic section from South Africa to the southern limit of the Antarctic Circumpolar Current at the Greenwich meridian, Deep-Sea Res. I, 55, 1284–1303, 2008.

Haack, T., Chelton, S. D., Pullen, J., Doyle, J., and Schlax, M.: Air-sea interaction from U. S. West Coast summertime forecasts. J. Phys. Oceanogr., 38, 2414–2437, 2008.

Jin, X., Dong, C., Kurian, J., and McWilliams, J.: SST-Wind interaction in coastal upwelling: Oceanic simulation with empirical coupling, J. Phys. Oceanogr., 39, 2957–2970, doi:10.1175/2009JPO4205.1, 2009.

Kudryavtsev, V. N., Grodsky, S. A., Dulov, V. A., and Malinovsky, V. V.: Observations of atmospheric boundary layer evolution above the Gulf Stream frontal zone, Bound-Lay. Meteorol., 79, 51–82, 1996.

Kudryavtsev, V. N., Akimov, D., Johannessen, J., and Chapron, B.: On radar imaging of current features: 1. Model and comparison with observations, J. Geophys. Res., 110, C07016, doi:10.1029/2004JC002505, 2005.

Lehahn, Y., d'Ovidio, F., Lévy, M., Amitai, Y., and Heifetz, E.: Long range transport of a quasi isolated clorophyll patch by an Agulhas ring, Geophys. Res. Lett., 38, L16610, doi:10.1029/2011GL048588, 2011.

Liu, W. T., Xie, X., and Niiler, P.: Ocean-atmosphere interaction over Ahulhas Extension meanders, J. Climate, 20, 5784–5797, 2007.

O'Neill, L. W.: Wind speed and stability effects on the coupling between surface wind stress and SST observed from buoys and satellite, J. Climate, 25, 1544–1569, doi:10.1175/JCLI-D-11-00121.1, 2012.

O'Neill, L. W., Chelton, D. B., and Esbensen, S. K.: High-Resolution Satellite Measurements of the Atmospheric Boundary Layer response to SST Variations along the Agulhas Return Current, J. Climate, 18, 2706–2723, 2005.

O'Neill, L. W., Chelton, D. B., and Esbensen, S. K.: The effects of SST-induced surface wind speed and direction gradients on midlatitude surface vorticity and divergence, J. Climate, 23, 255–281, 2010.

Park, K. and Cornillon, P. C.: Stability-induced modification of sea surface winds over Gulf Stream rings, Geophys. Res. Lett., 29, 2211, doi:10.1029/2001GL014236, 2002.

Park, K.-A., Cornillon, P., and Codiga, D. L.: Modification of surface winds near ocean fronts: Effects of Gulf Stream rings on scatterometer (QuikSCAT, NSCAT) wind observations. J. Geophys. Res., 111, C03021, doi:10.1029/2005JC003016, 2006.

Schmid, C., Boebel, O., Zenk, W., Lutjeharms, J. R. E., Garzoli, S. L., Richardson, P. L., and Barrong, C.: Early evolution of an Agulhas Ring, Deep-Sea Res. II, 50, 141–166, 2003.

Siegel, D. A., Peterson, P., McGillicuddy Jr., D. J., Maritorena, S., and Nelson, N. B.: Bio-optical footprints created by mesoscale eddies in the Sargasso Sea, Geophys. Res. Lett., 38, L13608, doi:10.1029/2011GL047660, 2011.

Small, R. J., deSzoeke, S. P., Xie, S. P., O'Neill, L., Seo, H., Song, Q., Cornillon, P., Spall, M., and Minobe, S.: Air-sea interaction over ocean fronts and eddies, Dyn. Atmos. Oceans, 45, 274–319, 2008.

Souza, J. M. A. C., de Boyer Montégut, C., and Le Traon, P. Y.: Comparison between three implementations of automatic identification algorithms for the quantification and characterization of mesoscale eddies in the South Atlantic Ocean, Ocean Sci., 7, 317–334, doi:10.5194/os-7-317-2011, 2011a.

Souza, J. M. A. C., de Boyer Montégut, C., Cabanes, C., and Klein, P.: Estimation of the Agulhas ring impacts on meridional heat fluxes and transport using ARGO floats and satellite data, Geophys. Res. Lett., 38, L21602, doi:10.1029/2011GL049359, 2011b.

Stern, M. E.: Interaction of a uniform wind stress with a geostrophic vortex, Deep-Sea Res., 12, 355–367, 1965.

van Aken, H. M., van Veldhoven, A. K., Veth, C., de Ruijter, W. P. M., van Leeuwen, P. J., Drijfhout, S. S., Whittle, C. P., and Rouault, M.: Observations of a young Agulhas Ring, Astrid, during MARE in March 2000. Deep-Sea Res. II, 50, 167–195, 2003.

van Sebille, E., van Leeuwen, P. J., Biastoch, A., and Ruijter, W. P. M.: On the fast decay of Agulhas rings. J. Geophys. Res., 115, C03010, doi:10.1029/2009JC005585, 2010.

Yelland, M. and Taylor, P.: Wind stress measurements from the open ocean, J. Phys. Oceanogr., 26, 541–558, 1996.

Permissions

All chapters in this book were first published in Ocean Science, by Copernicus Publications; hereby published with permission under the Creative Commons Attribution License or equivalent. Every chapter published in this book has been scrutinized by our experts. Their significance has been extensively debated. The topics covered herein carry significant findings which will fuel the growth of the discipline. They may even be implemented as practical applications or may be referred to as a beginning point for another development.

The contributors of this book come from diverse backgrounds, making this book a truly international effort. This book will bring forth new frontiers with its revolutionizing research information and detailed analysis of the nascent developments around the world.

We would like to thank all the contributing authors for lending their expertise to make the book truly unique. They have played a crucial role in the development of this book. Without their invaluable contributions this book wouldn't have been possible. They have made vital efforts to compile up to date information on the varied aspects of this subject to make this book a valuable addition to the collection of many professionals and students.

This book was conceptualized with the vision of imparting up-to-date information and advanced data in this field. To ensure the same, a matchless editorial board was set up. Every individual on the board went through rigorous rounds of assessment to prove their worth. After which they invested a large part of their time researching and compiling the most relevant data for our readers.

The editorial board has been involved in producing this book since its inception. They have spent rigorous hours researching and exploring the diverse topics which have resulted in the successful publishing of this book. They have passed on their knowledge of decades through this book. To expedite this challenging task, the publisher supported the team at every step. A small team of assistant editors was also appointed to further simplify the editing procedure and attain best results for the readers.

Apart from the editorial board, the designing team has also invested a significant amount of their time in understanding the subject and creating the most relevant covers. They scrutinized every image to scout for the most suitable representation of the subject and create an appropriate cover for the book.

The publishing team has been an ardent support to the editorial, designing and production team. Their endless efforts to recruit the best for this project, has resulted in the accomplishment of this book. They are a veteran in the field of academics and their pool of knowledge is as vast as their experience in printing. Their expertise and guidance has proved useful at every step. Their uncompromising quality standards have made this book an exceptional effort. Their encouragement from time to time has been an inspiration for everyone.

The publisher and the editorial board hope that this book will prove to be a valuable piece of knowledge for researchers, students, practitioners and scholars across the globe.

List of Contributors

A. Schneider
GEOMAR Helmholtz Center for Ocean Research Kiel, Kiel, Germany

T. Tanhua
GEOMAR Helmholtz Center for Ocean Research Kiel, Kiel, Germany

W. Roether
Institute of Environmental Physics, University of Bremen, Bremen, Germany

R. Steinfeldt
Institute of Environmental Physics, University of Bremen, Bremen, Germany

L. Ursella
Istituto Nazionale di Oceanografia e di Geofisica Sperimentale (OGS), B.go Grotta Gigante 42/c, 34010 Sgonico (TS), Italy

V. Kovačević
Istituto Nazionale di Oceanografia e di Geofisica Sperimentale (OGS), B.go Grotta Gigante 42/c, 34010 Sgonico (TS), Italy

M. Gačić
Istituto Nazionale di Oceanografia e di Geofisica Sperimentale (OGS), B.go Grotta Gigante 42/c, 34010 Sgonico (TS), Italy

L. Nagel
Institute of Environmental Physics, University of Heidelberg, Im Neuenheimer Feld 229, 69120 Heidelberg, Germany

K. E. Krall
Institute of Environmental Physics, University of Heidelberg, Im Neuenheimer Feld 229, 69120 Heidelberg, Germany

B. Jähne
Institute of Environmental Physics, University of Heidelberg, Im Neuenheimer Feld 229, 69120 Heidelberg, Germany
Heidelberg Collaboratory for Image Processing, University of Heidelberg, Speyerer Straße 6, 69115 Heidelberg, Germany

A. Stigebrandt
Dept. of Earth Sciences, University of Gothenburg, Gothenburg, Sweden

R. Rosenberg
Dept. of Biological and Environmental Sciences – Kristineberg, University of Gothenburg, Kristineberg Fiskebäckskil, Sweden

L. Råman Vinnå
EPFL ENAC IIE APHYS, Lausanne, Switzerland

M. Ödalen
Dept. of Meteorology, University of Stockholm, Stockholm, Sweden

E. Mesarchaki
Max-Planck-Institut für Chemie (Otto-Hahn-Institut) Hahn-Meitner-Weg 1, 55128 Mainz, Germany

C. Kräuter
Institut für Umweltphysik Universität Heidelberg, Im Neuenheimer Feld 229, 69120 Heidelberg, Germany

K. E. Krall
Institut für Umweltphysik Universität Heidelberg, Im Neuenheimer Feld 229, 69120 Heidelberg, Germany

M. Bopp
Institut für Umweltphysik Universität Heidelberg, Im Neuenheimer Feld 229, 69120 Heidelberg, Germany

F. Helleis
Max-Planck-Institut für Chemie (Otto-Hahn-Institut) Hahn-Meitner-Weg 1, 55128 Mainz, Germany

J. Williams
Max-Planck-Institut für Chemie (Otto-Hahn-Institut) Hahn-Meitner-Weg 1, 55128 Mainz, Germany

B. Jähne
Institut für Umweltphysik Universität Heidelberg, Im Neuenheimer Feld 229, 69120 Heidelberg, Germany
Heidelberg Collaboratory for Image Processing (HCI), Universität Heidelberg, Speyerer Straße 6, 69115 Heidelberg, Germany

C. G. Piecuch
Atmospheric and Environmental Research, Inc., Lexington, MA 02421, USA

R. M. Ponte
Atmospheric and Environmental Research, Inc., Lexington, MA 02421, USA

P. Mehra
CSIR-National Institute of Oceanography (NIO), Goa, India

M. Soumya
CSIR-National Institute of Oceanography (NIO), Goa, India

P. Vethamony
CSIR-National Institute of Oceanography (NIO), Goa, India

K. Vijaykumar
CSIR-National Institute of Oceanography (NIO), Goa, India

T. M. Balakrishnan Nair
Indian National Centre for Ocean Information Services (INCOIS), Hyderabad, India

Y. Agarvadekar
CSIR-National Institute of Oceanography (NIO), Goa, India

K. Jyoti
CSIR-National Institute of Oceanography (NIO), Goa, India

K. Sudheesh
CSIR-National Institute of Oceanography (NIO), Goa, India

R. Luis
CSIR-National Institute of Oceanography (NIO), Goa, India

S. Lobo
CSIR-National Institute of Oceanography (NIO), Goa, India

B. Harmalkar
CSIR-National Institute of Oceanography (NIO), Goa, India

K. E. Krall
Institute of Environmental Physics, University of Heidelberg, Im Neuenheimer Feld 229, 69120 Heidelberg, Germany

B. Jähne
Institute of Environmental Physics, University of Heidelberg, Im Neuenheimer Feld 229, 69120 Heidelberg, Germany
Heidelberg Collaboratory for Image Processing, University of Heidelberg, Speyerer Straße 6, 69115 Heidelberg, Germany

I. Fer
Geophysical Institute, University of Bergen, Bergen, Norway
Bjerknes Centre for Climate Research, Bergen, Norway

M. Müller
Norwegian Meteorological Institute, Oslo, Norway

A. K. Peterson
Geophysical Institute, University of Bergen, Bergen, Norway
Bjerknes Centre for Climate Research, Bergen, Norway

T. Stöven
Helmholtz Centre for Ocean Research Kiel, GEOMAR, Kiel, Germany

T. Tanhua
Helmholtz Centre for Ocean Research Kiel, GEOMAR, Kiel, Germany

D. J. Webb
National Oceanography Centre, Southampton SO14 3ZH, UK

K. Schneider-Zapp
Ocean Research Group, School of Marine Science and Technology, Newcastle University, Newcastle upon Tyne, UK

M. E. Salter
Department of Applied Environmental Science, Stockholm University, Stockholm, Sweden

R. C. Upstill-Goddard
Ocean Research Group, School of Marine Science and Technology, Newcastle University, Newcastle upon Tyne, UK

D. B. Baranowski
Institute of Geophysics, Faculty of Physics, University of Warsaw, Warsaw, Poland

P. J. Flatau
Scripps Institution of Oceanography, University of California San Diego, San Diego, CA, USA

S. Chen
Naval Research Laboratory, Monterey, CA, USA

P. G. Black
Science Applications International Corporation, Monterey, CA, USA

J. M. A. C. Souza
Laboratoire d'Oceanographie Spatiale (LOS), IFREMER, Centre Brest, 29280, Plouzané, France

B. Chapron
Laboratoire d'Oceanographie Spatiale (LOS), IFREMER, Centre Brest, 29280, Plouzané, France

E. Autret
Laboratoire d'Oceanographie Spatiale (LOS), IFREMER, Centre Brest, 29280, Plouzané, France

www.ingramcontent.com/pod-product-compliance
Lightning Source LLC
Chambersburg PA
CBHW080649200326
41458CB00013B/4787